Food
Antioxidants

FOOD SCIENCE AND TECHNOLOGY

A Series of Monographs, Textbooks, and Reference Books

1. Flavor Research: Principles and Techniques, *R. Teranishi, I. Hornstein, P. Issenberg, and E. L. Wick*
2. Principles of Enzymology for the Food Sciences, *John R. Whitaker*
3. Low-Temperature Preservation of Foods and Living Matter, *Owen R. Fennema, William D. Powrie, and Elmer H. Marth*
4. Principles of Food Science
   Part I: Food Chemistry, *edited by Owen R. Fennema*
   Part II: Physical Methods of Food Preservation, *Marcus Karel, Owen R. Fennema, and Daryl B. Lund*
5. Food Emulsions, *edited by Stig E. Friberg*
6. Nutritional and Safety Aspects of Food Processing, *edited by Steven R. Tannenbaum*
7. Flavor Research: Recent Advances, *edited by R. Teranishi, Robert A. Flath, and Hiroshi Sugisawa*
8. Computer-Aided Techniques in Food Technology, *edited by Israel Saguy*
9. Handbook of Tropical Foods, *edited by Harvey T. Chan*
10. Antimicrobials in Foods, *edited by Alfred Larry Branen and P. Michael Davidson*
11. Food Constituents and Food Residues: Their Chromatographic Determination, *edited by James F. Lawrence*
12. Aspartame: Physiology and Biochemistry, *edited by Lewis D. Stegink and L. J. Filer, Jr.*
13. Handbook of Vitamins: Nutritional, Biochemical, and Clinical Aspects, *edited by Lawrence J. Machlin*
14. Starch Conversion Technology, *edited by G. M. A. van Beynum and J. A. Roels*
15. Food Chemistry: Second Edition, Revised and Expanded, *edited by Owen R. Fennema*

Food
Antioxidants

Technological, Toxicological,
and Health Perspectives

edited by

D. L. Madhavi
University of Illinois
Urbana, Illinois

S. S. Deshpande
Idetek, Inc.
Sunnyvale, California

D. K. Salunkhe
Utah State University
Logan, Utah

CRC Press
Taylor & Francis Group
Boca Raton London New York

CRC Press is an imprint of the
Taylor & Francis Group, an **informa** business

CRC Press
Taylor & Francis Group
6000 Broken Sound Parkway NW, Suite 300
Boca Raton, FL 33487-2742

First issued in paperback 2019

ISBN-13: 978-0-8247-9351-7 (hbk)
ISBN-13: 978-0-367-40153-5 (pbk)

Visit the Taylor & Francis Web site at
http://www.taylorandfrancis.com

and the CRC Press Web site at
http://www.crcpress.com

Preface

Autoxidation in food and biological systems has varied implications not only for human health and nutritional status but also for the vast area of food science and technology. Antioxidants play a crucial role in preventing or delaying autoxidation and have attracted a lot of attention as food additives. Both synthetic and natural antioxidants are widely used in many food products, and numerous research articles have been published and patents granted on the application of antioxidants and the synthesis and discovery of new antioxidants. Currently, the use of some naturally occurring antioxidants in preventive and therapeutic medicine is gaining popularity. However, the use of some common antioxidants such as BHA and BHT has become a controversial issue because of adverse toxicological data.

The aim of this book is to review in depth the accumulated research covering the technological, toxicological, and nutritional aspects of food antioxidants. The first chapter is an overview of the various dietary and biological mechanisms leading to autoxidation and an introduction to the preventive mechanisms of antioxidants. The second and third chapters include general information on lipid oxidation; sources of natural and novel antioxidants; classification; chemistry of the mechanism of action, synergism, degradation in food systems; and analytical techniques for their identification. Chapter 4, on technological aspects, discusses the manufacture, chemistry, and mode of utilization of individual antioxidants. Chapter 5, on toxicological aspects, provides comprehensive information on the various toxicity studies and the most recent recommended dietary allowances of individual antioxidants. An effort is made to bring out the controversial research data. Chapter 6, on nutritional and health aspects, gives insight into the therapeutic uses of various nutritional and dietary antioxidants in a number of diseases, including cancer, heart disease, and aging. Chapter 7 summarizes and makes conclusions regarding this state-of-the-art information and identifies future research needs in this exciting field.

We feel that this book will be a useful reference for research workers and students of food technology, toxicology, nutrition, and anyone else interested in the area of food antioxidants.

The editors and authors gratefully acknowledge Ms. Cheryl Della Pietra and Mr. Rod Learmonth of Marcel Dekker, Inc., whose cooperation and assistance were invaluable in completing this project.

D. L. Madhavi
S. S. Deshpande
D. K. Salunkhe

Contents

Contributors

S. S. Deshpande, Ph.D.   Department of Research and Development, Idetek, Inc., Sunnyvale, California

U. S. Deshpande, M.S.   Cor Therapeutics, South San Francisco, California

S. J. Jadhav, Ph.D.   Alberta Agriculture, Food and Rural Development, Food Processing Development Center, Leduc, Alberta, Canada

A. D. Kulkarni, Ph.D.   Department of Sugar Chemistry and By-products, Vasantdada Sugar Institute, Pune, Maharashtra, India

P. R. Kulkarni, Ph.D.   Food and Fermentation Technology Division, Department of Chemical Technology, University of Bombay, Bombay, Maharashtra, India

D. L. Madhavi, Ph.D.   Department of Horticulture, University of Illinois, Urbana, Illinois

S. Narasimhan, Ph.D.   Department of Sensory Science, Central Food Technological Research Institute, Mysore, Karnataka, India

S. S. Nimbalkar, Ph.D.   Department of Sugar Chemistry and By-products, Vasantdada Sugar Institute, Pune, Maharashtra, India

D. Rajalakshmi, M.Sc.   Department of Sensory Science, Central Food Technological Research Institute, Mysore, Karnataka, India

D. K. Salunkhe, Ph.D.   Department of Nutrition and Food Science, Utah State University, Logan, Utah

R. S. Singhal, Ph.D.   Food and Fermentation Technology Division, Department of Chemical Technology, University of Bombay, Bombay, Maharashtra, India

1

Introduction

D. L. Madhavi
University of Illinois, Urbana, Illinois

S. S. Deshpande
Idetek, Inc., Sunnyvale, California

D. K. Salunkhe
Utah State University, Logan, Utah

Autoxidation of lipids and the generation of free radicals are natural phenomena in biological and food systems. In biological systems, various biochemical defense mechanisms involving enzymes, trace minerals, and antioxident vitamins protect the cellular components from oxidative damage. In food systems, naturally occurring antioxidants impart a certain amount of protection against oxidation. However, natural antioxidants are often lost during processing or storage, necessitating the addition of exogenous antioxidants. Antioxidants effectively retard the onset of lipid oxidation in food products. In fact, antioxidants have become an indispensable group of food additives mainly because of their unique properties of enhancing the shelf life of a host of food products without any damage to sensory or nutritional qualities. The use of antioxidants dates back to the 1940s. Gum guaiac was the first antioxidant approved for the stabilization of animal fats, especially lard. The effective use of antioxidants requires a basic understanding of the chemistry of lipid oxidation, the mechanism of action of antioxidants, and other important properties such as synergism and degradation.

In biological systems, the formation of reactive organic free radicals is mediated by a number of agents and mechanisms such as high oxygen tension, radiation, and xenobiotic metabolism. The free radicals formed are highly reactive with molecular oxygen, forming peroxy radicals and hydroperoxides and thus initiating a chain reaction. Prooxidant states cause cellular lesions in all major organs by damaging cellular components, including polyunsaturated fatty acids, phospholipids, free cholesterol, DNA, and proteins. The health implications of tissue lipid oxidation are numerous and well documented. Lipids deteriorate in food products during processing, handling, and storage. Oxidation of unsaturated lipids in the food system is catalyzed by heat, light, ionizing radiation, trace metals, and metalloproteins and also enzymatically by lipoxygenase. Lipid oxidation is the major cause of the development of off-flavor compounds and rancidity as well as a number of other reactions that reduce the shelf life and nutritive value of food products. In recent years, the possible pathological significance of dietary lipid oxidation products has attracted the attention of biochemists, food scientists, and health professionals. Though studies indicate that lipid oxidation products have cytotoxic, mutagenic, carcinogenic, atherogenic, and angiotoxic effects, the mechanism still remains unclear. Chapter 2 presents a broad outline of the mechanism of lipid oxidation, factors responsible for biological and dietary lipid oxidation, and their health implications.

Both synthetic and natural antioxidants are used in food products to retard lipid oxidation. Synthetic antioxidants are mainly phenolic, for example, butylated hydroxyanisole (BHA), butylated hydroxytoluene (BHT), tert-butyl hydroquinone (TBHQ), and the gallates. Of the natural antioxidants, two important groups, the tocopherols and ascorbic acid, are highly effective in many food products. Due to concern over the safety of synthetic compounds, extensive work is being carried out to identify novel naturally occurring compounds as replacements for potentially toxic synthetic antioxidants. Many natural products such as plant phenolic compounds, spice extracts, amino acids, and proteins have significant antioxidant activity. Food processing may also result in the formation of antioxidant compounds such as Maillard reaction products. The addition of antioxidants or their chemical transformation products also becomes incidental during storage of the food product in plastic packaging materials. Antioxidants used for polymer stabilization have a tendency to leach out to the contacting environment because of their low molecular mass and high mobility in the polymer. Studies are under way to develop polymers with chemically bound antioxidants to effectively reduce migration to the surrounding environment. Antioxidants function mainly by disrupting the free-radical chain reaction or by decomposing the lipid peroxides formed into stable end products. In general, antioxidants are effective at very low levels, 0.01% or less. At higher levels most of them behave as prooxidants because of their involvement in the initiation reactions.

Chapter 3 deals with the sources of antioxidants, mechanism of action, analytical methods for detection and determination or various antioxidants, and their antioxidative activity. Numerous chemical and instrumental methods have been developed for the determination of the activity of antioxidants in terms of oxidative stability of model or food systems. However, in recent years sensory evaluation methods have been gaining more importance as the changes in food quality due to oxidation invariably lead to changes in sensory quality. Sensory quality in general is considered the most important property in judging quality under conditions of product formulation, processing, and storage. Chapter 3 highlights the importance of sensory evaluation and describes some evaluation methods as a means of determining oxidative stability and consequently the efficacy of food antioxidants.

In recent years, highly advanced techniques of chemical synthesis have made it possible to develop new synthetic antioxidants such as Trolox-C, a derivative of α-tocopherol, and Anoxomer, a polymeric compound. The chemical synthesis of natural products like ascorbic acid and α-tocopherol on a commercial scale has been successful. The synthetic approach has been targeted at developing nontoxic, nonabsorbable compounds that are applicable to a wide range of food products. Chapter 4 compiles the commercial manufacturing methods or extraction from natural sources of various food antioxidants, their properties, and food applications.

Approval of an antioxidant for food use requires extensive toxicological studies, including studies of mutagenic, carcinogenic, and teratogenic effects. Even though a number of compounds, both synthetic and natural, have antioxidant properties, only some have been accepted as "generally recognized as safe" (GRAS) substances for food use by international bodies such as the Joint FAO/WHO Expert Committee on Food Additives (JECFA) and the European Community's Scientific Committee for Food (SCF). The toxicology of food antioxidants has become a controversial area especially after recent long-term studies indicated that BHA and BHT could produce tumors in animals at high doses. Even α-tocopherol, a natural compound, has been shown in long-term studies to cause hemorrhagic mortality at very high doses. Such studies indicate that the effects are very much dose-dependent and confirm the prooxidant properties of the antioxidants. Hence, accurate estimates of daily intake and the establishment of an acceptable daily intake level for each antioxidant become crucial. In general, the high dosages used and the duration of the experiments in toxicity studies make it difficult to assess the possible risks to humans consuming low levels over a very long period. Chapter 5 deals with toxicological studies, recommended dietary allowances, and the GRAS status of food antioxidants.

The biological effects of dietary antioxidants have generated a lot of interest in recent years. Tissue lipid oxidation is efficiently inhibited by the synergistic action of various endogenous enzymes such as superoxide dismutase and glutathione peroxidase and various dietary antioxidants such as selenium, ascorbic acid, tocopherols, β-carotene, flavonoids, and glutathione. Extensive studies are under way to determine whether these dietary antioxidants can be used in preventive as well as therapeutic medicine in many diseases including cardiovascular disease, cancer, and arthritis and even in the aging process. The 1990s could very well be the decade in which some solid evidence emerges concerning the preventive properties of dietary antioxidants in cardiovascular disease and cancer. β-Carotene has already shown some positive effects in large-scale human studies conducted by the Harvard Physicians Health Study, and the experiments will continue for several years. By the end of this decade, three other human studies will help to assess the chemoprotective properties of β-carotene and the tocopherols. A large-scale French study, SU.VI. MAX, will test the effects of a mixture of dietary antioxidants at low levels in humans. Another human study called CARET is being carried out for carotenes and retinoic acid. Antioxidant pills are already being marketed. Chapter 6 gives an insight into the biological effects of dietary antioxidants and their possible preventive and therapeutic properties.

This book offers an up-to-date broad overview of the technological, toxicological, and nutritional aspects of food antioxidants. It will be a useful reference for research workers and students of food technology, toxicology, nutrition, and medicine and for anyone else interested in the subject of food antioxidants.

2

Lipid Oxidation in Biological and Food Systems

S. J. Jadhav
Alberta Agriculture, Food Processing Development Center, Leduc,
Alberta, Canada

S. S. Nimbalkar and A. D. Kulkarni
Vasantdada Sugar Institute, Pune, Maharashtra, India

D. L. Madhavi
University of Illinois, Urbana, Illinois

2.1 INTRODUCTION

Lipids form one of the major bulk constituents in food and other biological systems. This group of organic biomolecules can be classified into three groups: simple lipids (triglycerides, steryl esters, and wax esters), compound lipids (phospholipids, glycolipids, sphingolipids, and lipoproteins), and derived lipids (fatty acids, fat-soluble vitamins and provitamins, sterols, terpenoids, and ethers). Lipids occur in animals and plants either as storage lipids, which are potential sources of energy by beta oxidation, or as membrane lipids. Storage lipids are triglycerides, whereas membrane lipids include phospholipids, sterols, sphingolipids, and glycolipids. Many foods of plant origin contain highly unsaturated lipids. Lipids of animal origin have lower levels of unsaturated lipids, but they contain certain amounts of the higher unsaturated fatty acids.

Unsaturation in fatty acids makes lipids susceptible to oxygen attack leading to complex chemical changes that eventually manifest themselves in the development of off-flavors in food. This process, known as autoxidation, is a free-radical process. The length of the induction period of autoxidation is sensitive to the presence of minor components that either extend the induction period and are known as antioxidants or shorten the induction period and are known as pro-oxidants.

In addition to its role in food deterioration, there is growing interest in the

5

problem of lipid oxidation as related to health status. Lipid oxidation is believed to play an important role in coronary heart disease (CHD), atherosclerosis, cancer, and the aging process. A complex antioxidant defense system normally protects cellular systems from the injurious effects of free radicals. The efficiency of this antioxidant defense system depends to some extent on an adequate intake of foods enriched in the desired antioxidants and micronutrient cofactors. The use of antioxidants in lipid-containing foods minimizes rancidity, retards the formation of toxic oxidation products, and allows maintenance of nutritional quality and an increase in the shelf life of a variety of lipid-containing foods.

Since the process of lipid oxidation or peroxidation has implications in chemical, biochemical, biological, and technological fields, certain aspects of the mechanism of lipid oxidation, biological and dietary lipid oxidation, and the role of antioxidants are highlighted in the present chapter.

2.2 MECHANISM OF LIPID OXIDATION

Atmospheric oxygen can react spontaneously with a number of organic compounds and cause structural degradation, which is ultimately responsible for the loss of quality of numerous chemical products of economic or industrial importance. With the discovery of this phenomenon, researchers began to investigate the chemistry of oxidative degradation and the mechanism of its inhibition or control. In food systems, the spontaneous oxidative reaction resulted in the deterioration of lipids. This direct reaction of a lipid molecule with a molecule of oxygen, termed *autoxidation*, is a free-radical chain reaction.

The basic mechanism of the lipid autoxidation reaction has been well established and has been depicted in several reviews (1–8). The mechanism of autoxidation can be distinguished in three distinct steps: initiation, propagation, and termination.

2.2.1 Initiation

The autoxidation of a fat is thought to be initiated with the formation of free radicals. When in contact with oxygen, an unsaturated lipid gives rise to free radicals (Eq. 2.1). Initiation reactions take place either by the abstraction of a hydrogen radical from an allylic methylene group of an unsaturated fatty acid or by the addition of a radical to a double bond.

$$RH \rightarrow R^{\bullet} + H^{\bullet} \tag{2.1}$$
$$ROOH \rightarrow RO^{\bullet} + HO^{\bullet} \tag{2.2}$$
$$2ROOH \rightarrow RO^{\bullet} + ROO^{\bullet} + H_2O \tag{2.3}$$

The formation of lipid radical $R^{\bullet}$ is usually mediated by trace metals, irradiation, light, or heat. Also, lipid hydroperoxide, which exists in trace quantities prior to the oxidation reaction, breaks down to yield radicals as shown by Eqs. (2.2) and

(2.3). That is, the hydroperoxides undergo homolytic cleavage to form alkoxy radicals (RO$^\bullet$) or undergo bimolecular decomposition. Lipid hydroperoxides are formed by various pathways including the reaction of singlet oxygen with unsaturated lipids or the lipoxygenase-catalyzed oxidation of polyunsaturated fatty acids.

2.2.2 Propagation

In propagation reactions, free radicals are converted into other radicals. Thus, a general feature of the reactions of free radicals is that they tend to proceed as chain reactions, that is, one radical begets another and so on. Thus, the initial formation of one radical becomes responsible for the subsequent chemical transformations of innumerable molecules because of a chain of events. In fact, propagation of free-radical oxidation processes occurs in the case of lipids by chain reactions that consume oxygen and yield new free-radical species (peroxy radicals, ROO$^\bullet$) or by the formation of peroxides (ROOH) as in Eqs. (2.4) and (2.5).

$$R^\bullet + {}^3O_2 \rightarrow ROO^\bullet \tag{2.4}$$
$$ROO^\bullet + RH \rightarrow ROOH + R^\bullet \tag{2.5}$$

The products R$^\bullet$ and ROO$^\bullet$ can further propagate free-radical reactions.

Lipid peroxy radicals (ROO$^\bullet$) initiate a chain reaction with other molecules, resulting in the formation of lipid hydroperoxides and lipid free radicals. This reaction, when repeated many times, produces an accumulation of hydroperoxides. The propagation reaction becomes a continuous process as long as unsaturated lipid or fatty acid molecules are available.

Lipid hydroperoxides may also be formed by the reaction of an unsaturated fatty acid such as linoleic acid with oxygen in the singlet excited state or enzymatically by the action of lipoxygenase. Lipid hydroperoxides, the primary products of autoxidation, are odorless and tasteless.

The chain-propagating system here involves a bimolecular reaction of a radical with a molecule. Since lipid radicals are also highly reactive, they can easily undergo propagation reactions by two mechanisms: by reaction with an oxygen molecule in the triplet ground state (Eq. 2.4) or by removal of a hydrogen atom (Eq. 2.5). Molecular oxygen is particularly susceptible to radical attack. This radical–oxygen reaction is very fast as it requires almost zero activation energy. In other words, addition of most carbon-centered radicals to an oxygen diradical is basically as fast as any reaction can be in liquid solution. The addition is diffusion- or encounter-controlled. In effect, this reaction leads to the formation of a peroxy radical (ROO$^\bullet$) whose concentration becomes greater than that of R$^\bullet$ in most food systems containing oxygen. The formation of hydroperoxide in reaction (2.5) is conducive to the radical initiation steps shown earlier.

Radical reactions are more exothermic when the new bond is not appreciably weaker than, or is even stronger than, the bond being broken. The bond formation can be explained on the basis of bond strengths or bond dissociation energies (see Table 2.1).

It is evident that the O—H bond in ROOH is weaker than aliphatic C—H bonds except for C—H bonds in the allylic position in a hydrocarbon structural unit and more particularly C—H bonds in a doubly allylic situation. Naturally occurring

Table 2.1 Bond Dissociation Energies (Bond Strengths) for Some Representative Bonds to Hydrogen

Bond[a]	Bond strength[b] (kcal/mol)
H—H	104.2
CH_3—H	104.3
R \\ CH—H / R	96.3

RCH=CH
   \\
     CH—H
   /
  R
	85

RCH=CH
   \\
     CH—H
   /
RCH=CH
(Skipped diene)
	76

⬡—H	111
CH_3S—H	91.8
HO—H	119.3
ROO—H	88

[a]R = Substituent grouping attached through a carbon atom that is connected to other atoms only by single bonds.
[b]1 kcal = 4.18 kJ.
Source: Ref. 9.

lipids that contain a linoleic or linolenic acid unit have this type of doubly allylic C—H bonds, which, being weak, are susceptible to hydrogen abstraction by a free radical. Thus, a peroxy radical (ROO·) can readily abstract hydrogen from the central CH_2 group in the skipped diene unit (—CH=CH—CH$_2$—CH=CH—) of linoleic or linolenic acid residues. The bond weakening in the allylic system as recognized by Farmer et al. (10) and Sutton (11) can be explained on the basis of a stabilizing interaction between the electrons in the double bonds and the single unpaired electron of the free radical. This mechanism is of great importance, as many of the radical reactions in food matrices belong to this category.

Chain-propagation steps can also be induced by the addition of radicals to unsaturated bonds. β-Carotene, which contains conjugated carbon–carbon double bonds, is susceptible to radical addition. The biosynthesis of prostaglandins from arachidonic acid is marked by an intramolecular addition of a peroxy radical to a double bond. Ingold (12) presents more detailed information on relationships between structure and reaction rate in bimolecular radical-propagating processes. In short, some rate constants measured for a range of free-radical steps involving atom transfer, when studied in context with the data in Table 2.2, exhibit the manner in which the rates of reaction depend on both the strength of bond being formed and the strength of the bond being broken.

2.2.3 Termination

A free radical has been defined as a molecular entity possessing an unpaired electron. Free radicals are electrically neutral, and solvation effects are generally very small. They are considered to be bonding-deficient and hence structurally unstable. Radicals therefore tend to react whenever possible to restore normal bonding. That is why a free radical is highly reactive. When there is a reduction in the amount of unsaturated lipids (or fatty acids) present, radicals bond to one another, forming a stable nonradical compound. Thus the termination reactions

$$R· + R· \rightarrow R—R \tag{2.6}$$
$$R· + ROO· \rightarrow ROOR \tag{2.7}$$
$$ROO· + ROO· \rightarrow ROOR + O_2 \tag{2.8}$$

lead to interruption of the repeating sequence of propagating steps of the chain reaction.

It is interesting to note that radicals react with other radicals as quickly as radicals react with oxygen at a diffusion-controlled rate. Radical coupling results in the release of energy equivalent to the strength of the bond being formed. Subsequently, the released energy is dissipated as heat. When the coupling of two radicals involves identical species, the reaction is recognized as dimerization. Although radical coupling is associated with a very low enthalpy of activation, the occurrence of termination reactions is controlled by radical concentration, which

Table 2.2 Second-Order Rate Constants for Representative Hydrogen Transfer Reactions

Radical[a]	Hydrogen atom donor	T (K)	k (mol^{-1} s^{-1})
HO$^{\cdot}$	PhCHCH$_3$ | H	300	2×10^9
t-BuO$^{\cdot}$	PhCHCH$_3$ | H	295	1×10^6
t-BuOO$^{\cdot}$	PhCHCH$_3$ | H	303	0.1
N—O$^{\cdot}$	PhCHCH$_3$ | H	300	$\sim 10^{-9}$
ROO$^{\cdot}$	α-Tocopherol	300	2×10^7
RCH$_2$OO$^{\cdot}$	RCH—CH=CH$_2$ | H	303	0.5
RCH$_2$OO$^{\cdot}$	CH$_2$=CH—CH—CH=CH$_2$ | H	303	7.0

[a]R = Saturated hydrocarbon grouping.
Source: Ref. 9.

is responsible for the frequency of encounters between radicals, and also by stereochemistry, which causes radicals to collide with the correct orientation. The formation of polymers in frying oils indicates the importance of termination reactions in edible oils heated to elevated temperatures. Spontaneous decomposition of hydroperoxides at 160°C (13) makes the radical concentration relatively high under this condition.

2.2.4 Extent of Oxidation: Detection and Measurement

The extent of lipid oxidation can be measured by chemical, sensory, and instrumental methods. The principles of some analytical methods are presented in Table 2.3. Fats

Table 2.3 Analytical Methods to Determine the Degree of Oxidation of Fats and Oils

Property measured	Method	Principle of reaction
Acid value	Titration by alkali	$R-CH_2-COOH + KOH \rightarrow R-CH_2-COOK + H_2O$
Peroxide value	Iodometry	$R-CH_2\!-\!CH-CH{=}CH- + 2KI \rightarrow R-CH_2-CH-CH{=}CH- + I_2 + K_2O$ (OOH on reactant; OH on product) $I_2 + 2Na_2S_2O_3 \rightarrow Na_2S_4O_6 + 2\ NaI$
Carbonyl value	2,4-DNPH method	$R-CHO + NO_2-\!\!\!\bigcirc\!\!\!-NH-NH_2 \rightarrow R-CH{=}N-NH-\!\!\!\bigcirc\!\!\!-NO_2 + H_2O$ $R-CH{=}N-NH-\!\!\!\bigcirc\!\!\!-NO_2 \rightarrow R-CH{=}N-N{=}\!\!\!\bigcirc$
	Hydroxylamine method	$R-CHO + H_2N-OH \rightarrow R-CH{=}N-OH$

Table 2.3 *(Continued)*

Property measured	Method	Principle of reaction
TBA number	Thiobarbituric acid method	
Amount of oxygen absorbed	Warburg's manometer; oxygen analyzer; measurement of weight	

2,4-DNPH = 2,4-Dinitrophenylhydrazine; TBA = thiobarbituric acid.
Source: Ref. 14.

and oils can be evaluated in terms of acid values, peroxide values, and carbonyl values. Since the human senses are more sensitive than chemical analysis in many cases, sensory tests are considered to be appropriate in the evaluation of the flavor of an oil. The accelerated oxidation method for testing the stability of oils or fats subjected to aeration at 97.8°C measures their peroxide values at specified time intervals. The stability of fat or oil is expressed as the time it takes for the peroxide value to reach 100 meq/kg oil or 20 meq/kg in the case of lard. Induction periods can be measured by an automatic instrument known as a Rancimat under conditions similar to those of the AOM (15). For reducing the incubation time to reach the rancidity point, heating at an elevated temperature has been proposed. The oven test is conducted at 63°C, and readings are expressed in terms of the number of days required for the sample to undergo a deterioration detectable by a sensory test or days required for the peroxide value to reach a given limit. The stability of fats and oils can also be determined by light beam irradiation and oxygen absorption methods. In the above stability tests, relatively short induction times are an indication of lipid oxidation.

Lipid oxidation leads to the formation of various products that can act as oxidation indicators. For practical purposes, a number of tests have been developed for assessing the state of oxidation of a food. The tests listed in Table 2.4 produce satisfactory analytical results as most of the listed products are more or less stable final reaction products generated after thermal decomposition of the lipid peroxides. Malondialdehyde, the split product of an endoperoxide of unsaturated fatty acids, is allowed to react with thiobarbituric acid (TBA), and the reaction product is measured photometrically at 532 nm. At least three isolated double bonds are required to be present in a fatty acid to produce malondialdehyde. The TBA assay is thought to be not very specific if very little free malondialdehyde is formed during the lipid peroxidation process (28–31). Also, many more compounds than malondialdehyde react with TBA. Lipid peroxidation results in diene conjugation, which can be measured by UV absorbance corresponding to conjugated double

Table 2.4 Analytical Tracers Commonly Used for Evaluating the State of Oxidation of Foods

Name	Tracer	Ref.
TBA	Malonaldehyde	16,17
Anisidine value	2-Alkenal	18
2,4-DNPH	Aldehyde (carbonyls)	19
UV index (conjugation index)	Conjugated diones and triones	20–22
Headspace gas analysis	Hydrocarbons; aldehydes, ketones	4, 23–26

Source: Ref. 27.

bonds. Another approach is to study the second derivative spectra to diene conjugates as in the case of autoxidized polyunsaturated fatty acids (see Corongiu and Milia (32)). Biological materials are difficult to assay by the diene conjugation method (33). Isolation and identification of peroxidation products, particularly lipid hydroperoxides that have been reduced to the hydroxy compounds, are still used. However, assays based on measuring hydroperoxides are not very reliable because of the unstable nature of the hydroperoxides. Fluorescent products derived from the reaction between lipid aldehydes and amino acids of proteins (34,35) have also been measured as an indication of lipid peroxidation. However, this method lacks specificity.

The formation of volatile hydrocarbons such as pentane and ethane is dependent on the breakdown of lipid hydroperoxides by metal ions or heme in the presence of oxygen (36–38). A gas chromatogram of a headspace gas analysis of precooked wheat flakes stored at 20°C for 3 months indicated pentane as the predominant hydrocarbon derived from the decomposition of lipid hydroperoxides (Fig. 2.1) (39). The minor hydrocarbon components are ethane, propane, butane, and hexane. Ethane is produced from linolenic acid; pentane is derived from arachidonic acid as well as from linoleic acid.

Electron spin resonance (ESR) is a powerful technique for the detection and study of free radicals. This technique is based on the principle that when placed in a strong magnetic field a free radical absorbs microwave radiation of particular frequencies. Although ESR is highly sensitive (threshold detection limit 10^{-6} M or

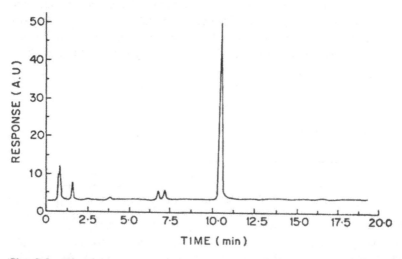

Fig. 2.1 Gas chromatogram of a headspace gas analysis of precooked wheat flakes stored at 20°C for 3 months. Ethane ($t = 1.5$ min); propane ($t = 3.8$ min, trace); butane ($t = 7.2$ min, none); pentane ($t = 10.5$ min); hexane ($t = 13.5$ min, trace). (From Ref. 27.)

less), it may encounter difficulties in the detection of very low instantaneous concentrations of reactive radicals existing in a food matrix. Under such conditions, the spin trapping technique is recommended (40,41). When added to the system under investigation, small quantities of such nonradical traps form new radicals (i.e., spin adducts) that are readily detectable by ESR methods. Since radiation generates radicals, various analytical procedures are important for evaluating the radiation dose intended for food preservation. The nitrones PBN (α-phenyl-N-tert-butylnitrone) and DMPO (5,5-dimethylpyrrolidone-N-oxide) and the nitroso compound MNP (2-methyl-2-nitrosopropane) are typical spin traps, and the nitrones have proved to be effective with oxygen-centered radicals. Although spin trapping is useful in identification of the trapped radical, it may lead to incorrect identifications in some cases.

2.3 BIOLOGICAL LIPID OXIDATION

In cellular systems, lipid peroxidation can occur mainly in biomembranes, where the content of unsaturated fatty acids is relatively high. Lipid peroxidation is a very complicated chemical and biochemical reaction process involving free radicals, oxygen, metal ions, and, in biological systems, a number of other factors.

Lipid peroxidation is mainly studied in isolated microsomes from the liver. Initiation and propagation of lipid peroxidation are catalyzed by iron and microsomal NADPH–cytochrome P-450 reductase (42). This enzyme is responsible for the formation of a superoxide anion, a species formed by the addition of an extra electron onto the diatomic oxygen molecule, which catalyzes the reduction of iron ions. It is likely that cytochrome P-450 is also involved in this reduction reaction in whole microsomes (43). However, according to Davis et al. (44) and Kappus and Kostrucha (38), this is not the case in reconstituted systems and the reductase alone is capable of catalyzing the peroxidation of the lipids in liposomes. There is no clear-cut mechanism of lipid peroxidation in the microsomal enzyme system. Also, the initiation species of lipid peroxidation are ambiguous as in autoxidation. Researchers have yet to ascertain if microsomal lipid peroxidation is induced by hydroxyl radicals derived from the Haber–Weiss reaction or iron–oxygen complexes (45–47).

This difference of opinion appears to be due to the site-specific formation of a reactive species in the membrane or a membrane-like structure (48). However, in microsomes the peroxidation products are the same as in a lipid autoxidation process. Although xanthine oxidase, which forms superoxide anions, catalyzes lipid peroxidation, the reactive species initiating lipid peroxidation are unknown (49). Mitochondrial lipid peroxidation, which seems to involve NADH oxidase, can also be induced by iron ions. However, the mitochondrial process has not been investigated as systematically as the corresponding microsomal process (50). Lipid peroxidation in vivo depends largely on iron ions, the source of which is not very clear.

2.3.1   Factors Inducing Prooxidant States

The major activated or reactive species of oxygen are the superoxide anion radical (O_2^{-}) and its conjugate acid the hydroperoxy radical (HO_2^{-}), hydrogen peroxide (H_2O_2), singlet oxygen $(^1O_2)$, and the hydroxyl radical $(HO^{\cdot})$. The electronic structure of these activated forms of oxygen facilitate their reactivity with biological compounds. The strongest electrophiles (electron seekers), $HO^{\cdot}$ and $^1O_2$, are the most reactive forms of active oxygen, followed in reactivity by superoxide anion radical and then hydrogen peroxide. Some of the active forms can be interconverted, and these processes can be facilitated by the presence of specific catalysts.

Activated forms of oxygen can also be formed by gamma irradiation. The radiation causes a split in one of the covalent bonds in water to create a hydroxyl radical and a hydrogen radical. The former, being the most reactive radical, can attack and damage living systems as it can pluck an electron from virtually any organic macromolecule situated in its vicinity (51). The reactions of $HO^{\cdot}$ leave behind a legacy in the cell in the form of free-radical chain reactions. As a consequence of $HO^{\cdot}$ attack, free-radical chain reactions occur within DNA, and the deoxyribose, purines, and pyrimidines undergo chemical alterations (52). This can cause mutations and disruption of the DNA strand (53). Imperfect repair of DNA damage can result in oncogene activation and carcinogenesis. Much circumstantial evidence suggests that free-radical biochemistry must be considered to have major significance in carcinogenesis.

In cellular prooxidant states, the intracellular contents of the reactive species of oxygen increase due to their overproduction or lack of destruction. The typical reactions responsible for the prooxidant states are the initiation of autoxidation chain processes by hydroxyl and hydroperoxy radicals and of branching reactions by alkoxy radicals, the addition of hydroxyl radicals and singlet oxygen to double bonds, hydrogen abstraction from allylic carbon atoms by hydroxy radicals, and oxidation of sulfhydryl, thioether, and amino functions. The biological effects are mutations, sister chromatid exchanges, chromosomal aberrations, cytotoxicity, carcinogenesis, and cellular degeneration related to aging.

Prooxidant states may vary depending on the inducing agent and the target cell. This is presumably because of the variation in the quantity, type, and intra- and extracellular distribution of the reactive oxygen species. The different classes of agents and mechanisms that can cause prooxidant states have been well documented by Cerutti (54). These aspects as applied to biological systems are discussed here.

Hyperbaric Oxygen Tension

The formation of reactive oxygen species in small amounts may result from increases in oxygen tension. Thus, oxygen tension above 40% can interfere with macromolecular synthesis and cell division and is cytotoxic, mutagenic, and clastogenic.

Radiation

Aerobic ionizing radiation causes cellular macromolecular (including DNA) damage. Also, it is mutagenic, clastogenic, and carcinogenic. Radiation-induced carcinogenesis is considered to be due to peroxidation of membrane lipids. It seems that the mechanism of action depends on wavelength. The near-UV radiation is effective in the generation of active oxygen and causes ubiquitous oxidative damage to macromolecules. This type of wavelength range is mutagenic, clastogenic, and carcinogenic for cultured cells. That solar radiation is the causative agent is apparent from the epidemiology of nonmelanoma skin cancer. Thus near-UV radiation may have a promotional role in tumorigenesis.

Xenobiotic Metabolism and Fenton-Type Reagents

The metabolism of certain xenobiotics leads to the formation of active oxygen. The products of these reactions, active oxygen and xenobiotic radicals, can cause damage to macromolecules. Hence the biological effects of xenobiotic metabolism are similar to those of ionizing radiation. Certain potent carcinogens and carcinostatic drugs are included in this group. Xenobiotic radicals with quinoid structures can participate in redox cycles in which semiquinone intermediates are oxidized to quinones with simultaneous reduction of oxygen to superoxide anion radicals. Some examples are daunorubicin, streptonigrin, adriamycin, mitomycin C, and certain polycyclic aromatic hydrocarbons, the last three with a property of forming covalent DNA adducts. A prooxidant state by way of quinone metabolism can affect the concentration of NADPH and the release of mitochondrial calcium ions. The disturbance of Ca ion homeostasis can lead to changes in cytoskeletal function and structure. The reductive metabolism of 4-nitroquinoline-N-oxide (4-NQO) produces radical intermediates that can transfer an electron to oxygen. The mechanism of action of the carcinogen is attributed to the radicals involved. The antitumor antibiotic bleomycin acts by a mechanism that involves Fenton-type reactions. It binds iron ions, and the complex will degrade DNA in the presence of oxygen and a reducing agent such as ascorbic acid (55,56). Other agents that are involved in the reduction of oxygen to superoxide anion radicals and are responsible for lipid peroxidation and hemolysis are azo anions and hydrazyl and bipyridylium radicals. Examples of these are sulfonazo III, phenylhydrazine, and paraquat. The trichloromethyl radical derived from carbon tetrachloride can initiate lipid peroxidation and is a potent hepatotoxin and hepatopromoter and a complete hepatocarcinogen.

Electron Transport Chain

The cytochrome electron transport chain can release large amounts of active oxygen when influenced by such agents as inhibitors, uncouplers, inducers, and pseudosubstrates. This leads to prooxidant states that can cause damage to lipids and cytochromes. Rotenone, the respiration inhibitor, is hepatocarcinogenic, whereas phenobarbital, the P-450 inducer, is a hepatopromoter.

Peroxisome Proliferators

Peroxisomes are organelles that contain hydrogen peroxide–generating oxidases, catalase, and enzymes associated with the beta oxidation of fatty acids. Xenobiotics such as clofibrate, nafenopin, and di(2-ethylhexyl) phthalate activate the biosynthesis of peroxisomes, causing overproduction of hydrogen peroxide. The resulting prooxidation states lead to the formation of lipofuscin (age pigment) in the liver subsequent to prolonged exposure to peroxisome proliferators. This condition is considered a measure of lipid peroxidation.

As some hydrogen peroxide escapes the catalase in the peroxisome (57–60), it contributes to the supply of oxygen radicals, which also results from other metabolic sources (60–64). Oxygen radicals derived from hydrogen peroxide in the presence of iron-containing compounds in the cell (61) can damage DNA and can initiate the rancidity chain reaction leading to the production of mutagens, promoters, and carcinogens such as fatty acid peroxides, cholesterol hydroperoxide, endoperoxides, cholesterol and fatty acid epoxides, enals and other aldehydes, and alkoxy and hydroperoxy radicals. Certain trans fatty acids and C-22:1 fatty acids seem to cause peroxisomal proliferation as they are poorly oxidized in mitochondria and are preferentially oxidized in the peroxisomes, though they appear to be selective for heart or liver (65–68). It is possible that abnormal fatty acids could cause perturbations in the mitochondrial or peroxisomal membranes, increasing the flux of superoxide and hydrogen peroxide.

Peroxisome proliferators are hepatocarcinogenic and hepatopromoters (57,69) and are likely to cause chromosomal damage by indirect action. The prooxidant state induced by di(2-ethylhexyl) phthalate may be caused by stimulation of peroxisome biosynthesis or by inhibition of the electron transport chain by its metabolite monoethylhexyl phthalate (70).

Inhibition of Antioxidant Defense Systems

In cells, the removal of reactive oxygen species such as superoxide anion radical and hydrogen peroxide is carried out by enzyme systems as part of the antioxidant defense. The removal of $O_2^{\cdot-}$ is catalyzed by superoxide dismutase (SOD).

$$2\,O_2^{\cdot-} + 2\,H^+ \;\rightarrow\; H_2O_2 + O_2 \tag{2.9}$$

Glutathione peroxidase removes H_2O_2 by using it to convert reduced glutathione (GSH) into oxidized glutathione (GSSG).

$$H_2O_2 + 2GSH \rightarrow 2H_2O + GSSG \tag{2.10}$$

Catalase, which is largely sequestered in peroxisomes, may have some significance in the general intracellular scavenging of hydrogen peroxide by disproportionation.

$$2H_2O_2 \rightarrow 2H_2O + O_2 \tag{2.11}$$

Thus, SOD enzymes work in conjunction with glutathione peroxidase and catalase (60) and maintain the steady-state concentrations of active oxygen at acceptable levels under physiological conditions. Inhibition of these enzymes can induce a prooxidant state. The inhibitors of SOD and catalase are diethyldithiocarbamate and azide, hydroxylamine, and aminotriazole, respectively (71). These drugs appear to act via the reduction of the antioxidant defense (72).

The promoter phorbol-12-myristate-13-acetate (PMA) induces a reduction in the amount of SOD and catalase but not glutathione peroxidase in mouse epidermal cells in vivo (73). Low molecular weight nonprotein sulfhydryls such as glutathione, cysteine, and cysteinylglycine have a role to play in the cellular defense against active oxygen. Moreover, vitamins C and E, β-carotene, and urates are important physiological molecules that contribute to the natural antioxidant defense (74).

Membrane Active Agents

Membrane active agents of xenobiotic and endogenous origin share the property of membrane activity. They can affect various membranes and interact with receptors or disturb membrane conformation in a less specific manner. Membrane lipids may become susceptible to autoxidation if chaotropic agents perturb membrane conformation. Stimulation of membrane phospholipases by certain agents can result in the formation of free arachidonic acid and its metabolites. In fact, biosynthesis of the biologically important prostaglandins and hydroxyarachidonic acid proceeds through unstable intermediates that release active oxygen. Membrane active agents can also elicit an oxidative burst by the activation of a NADPH-dependent oxidase in specialized phagocytic leukocytes. Some examples of membrane active agents that are of particular importance in carcinogenesis are peptide hormones; growth factors; lectins; tumor promotors PMA, telecocidin, and mezerein; aflatoxin B_1; certain bacteria and viruses; and asbestos and silica particulates.

2.3.2 Health Implications of Biological Lipid Oxidation

In cellular systems, lipid peroxidation is of great importance, especially in biomembranes where most of the oxygen-activating enzymes are present. As a result, these enzymes become responsible for the induction of oxidation stress, which leads to the formation of reactive oxygen species such as superoxide anion, hydroxyl radical, hydrogen peroxide, and singlet oxygen. Although the cells are protected from these reactive oxygen species by a number of cellular defense mechanisms, lipid peroxidation does occur in biomembranes under certain conditions that overcome the cellular defense system. Lipid peroxidation in membranes destroys the membrane structure and causes loss in the function of the cell organelles (42). Also, receptors present in the membrane are released or inactivated. These changes occur both intracellularly and in plasma membranes. The secondary effects of lipid peroxidation are the initiation of new free-radical reactions derived from the lipid peroxides diffused within the cell or transported into the bloodstream. The hydroxyalkenals, the products of lipid peroxidation, are able to bind covalently to nucleic acids, thereby inducing changes in the DNA (75). In higher organisms, the hydroxyalkenals have chemotactic properties that induce inflammatory reactions (76). The process of lipid peroxidation activates phospholipases and removes the peroxidized lipids from the membranes (42,77). Other biological effects of phospholipases are well recognized. Furthermore, activation of lysosomal proteases, nucleosidases, and other lipases causes degradation of biological molecules and subsequently cell death. Thus, lipid peroxidation activates a process of cell death by degradation of cellular components and inactivates the cellular defense systems. Obviously, the biological effects of lipid peroxidation can appear in the system at sites far from the site of initiation.

Induction of lipid peroxidation has been linked to a number of diseases (51,78, 79). A list of diseases related to the process of lipid peroxidation is given in Table 2.5. Although these diseases are linked to some kind of oxidative stress, lipid peroxidation cannot be considered the only cause of this stress. Reactive oxygen species could be responsible for biological toxicity, and lipid peroxidation may occur as a consequence of these changes.

Oxygen radicals and the lipid peroxidation theory of aging (62,63,91–97) suggest that the major cause of oxidative stress is damage to DNA (98) and other macromolecules. Thus, progressive defects in the protective mechanism against free-radical-initiated damage to macromolecules finally lead to tissue damage. A causative role for oxygen radicals in the aging process is apparent from a large body of experimental data, and there exists a positive correlation between the tissue concentration of specific antioxidants and lifespan. The relevant antioxidants are carotenoids, vitamin E, urate, and enzymes such as superoxide dismutase. Cancer and other degenerative diseases such as heart disease appear to be caused by the above destructive process. Lipofuscin (age pigment) accumulates in mammalian species as they age and has been linked to lipid peroxidation (62,63,93,94,99). The fluorescent products in age pigments are formed by malondialdehyde (a mutagen and carcinogen and a major end product of rancidity) cross-linking proteins and lipids (93).

DNA-damaging agents and promoters are considered to play an important role

Table 2.5 Lipid Peroxidation–Induced Diseases and Effects

Disease	Remarks
1. Hemochromatosis	Organ damage due to Fe overload leading to increased lipid peroxidation.
2. Keshan disease	Selenium deficiency causes decrease in glutathione peroxidase activity leading to increased lipid peroxiation.
3. Rheumatoid arthritis	Due to Fe-induced lipid peroxidation.
4. Atherosclerosis	Lipid peroxides and the reaction products of lipid peroxidation such as hydroxyalkenals alter low-density lipoproteins (LDLs), which are important in the development of atherosclerotic lesions (80–83).
5. Ischaemia	Occurs during reperfusion injury of heart and brain; also results in lipid peroxidation, probably by transformation of xanthine dehydrogenase to xanthine oxidase and by the production of reactive oxygen species (84,85).
6. Aging	May be due to lipid peroxidation, but has been confirmed in erythrocytes (86).
7. Carcinogenesis	Wide speculation about the involvement of lipid peroxidation in carcinogenesis; this is due to genotoxic effects of lipid peroxides (87–90).

in carcinogenesis. It is suggested that certain promoters of carcinogenesis act by generating oxygen radicals, which results in lipid peroxidation (99–106). Lipid peroxidation involves protein cross-linking (91,107) and affects cell organization including membrane and surface structure and the mitotic apparatus. Promoters are able to produce oxygen radicals. Fats and hydrogen peroxides are among the most important promoters.

2.4 DIETARY LIPID OXIDATION

Lipids occur in almost all foodstuffs, and most of them (more than 90%) are in the form of triglycerides, which are esters of fatty acids and glycerol. These natural fatty acids contain straight-chain even-numbered aliphatic carboxylic acids, which may be saturated or unsaturated with up to six double bonds. The latter are normally arranged along the chain, separated from each other by methylene groups and with cis conformations. It has been well established that the carbon chain length and the degree of unsaturation of the fatty acids are most critical in determining the oxidative stability of the lipids. In addition to triglycerides, foods also contain other types of lipids such as phospholipids, glycolipids, sphingolipids, lipoproteins, sterols, and hydrocarbons.

Lipids can undergo deterioration in various ways during processing, handling, and storage of food products. Two of the typical reactions of lipids are hydrolysis and oxidation. Hydrolytic rancidity of lipids is caused by hydrolysis of the ester linkage resulting in the formation of free fatty acids and glycerol. This reaction is catalyzed by high temperatures, acids, lipolytic enzymes, and high moisture content in vegetable oils. However, hydrolytic cleavage is not of great influence in the development of off-flavors due to modern refining techniques. Lipids are suscep-tible to oxidation that occurs at the unsaturated bonds in the fatty acid part of the glycerides. Lipid oxidation reactions are much more complicated than the simple addition of oxygen exemplified by the oxidation of metals. However, it should be recognized that oxidation is generally treated as the most frequently occurring form of lipid deterioration, which leads to the development of rancidity, off-flavor compounds, polymerization, reversion, and a number of other reactions causing reduction in the shelf life and nutritive value of the food product. Antioxidants can effectively retard the process of oxidation, but they cannot reverse it. Also, they are not effective in suppressing hydrolytic rancidity. The overall mechanism of lipid oxidation is presented in Fig. 2.2 (1).

In food systems, oxidation reactions can be broadly divided into two categories. In the first category, oxidation of highly unsaturated fats, particularly polyunsatu-rated types, results in polymeric end products. The second category deals with the oxidation of moderately unsaturated fats, which results in rancidity, reversion, and other types of off-flavors and odors (108).

A contemporary definition of oxidation is that it is the process by which oxygen is added or hydrogen or electrons are withdrawn. The component that is reduced and gains electrons is the oxidant. In foods, oxygen is the most common oxidant, although other endogenous and added chemicals can also serve as oxidants. As indicated earlier, the oxidation of unsaturated lipids in foods is often caused by free-radical reactions. These chain reactions are influenced by several factors such as heat, light, ionization reactions, trace metals (especially copper and iron), and metalloproteins such as heme and also enzymatically by lipoxygenase.

2.4.1   Factors Inducing Prooxidant States

Heat Treatment

As is true of most chemical reactions, the rate of oxidation increases with temper-ature and thus with heat treatment. It should also be recognized that the rate of oxidation reactions is greatly influenced by the type, relative concentrations, and relative reactivities of prooxidants and antioxidants in the food. Thermally induced oxidation reactions can occur in both saturated and unsaturated lipids at tempera-

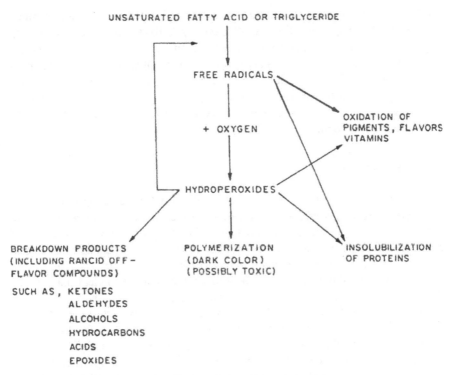

Fig. 2.2 Overall mechanism of lipid oxidation. (From Ref. 1.)

tures encountered during processes such as deep fat frying (109). Oxidation generally proceeds via the initial formation of hydroperoxides. The high temperatures can cause many isomerization and scission reactions to take place, producing a myriad of secondary or breakdown products such as epoxides, dihydroperoxides, cyclized fatty acids, dimers, and, with scission reactions, aldehydes and ketones. Saturated fatty acids are relatively stable at the temperatures used in conventional canning operations, but unsaturated fats deteriorate, under the conditions of oxygen and heat, to form a large number of volatile compounds, which give rise to both desirable and undesirable flavors (110). Drying (dehydration) brings food component molecules into closer proximity, thereby increasing the likelihood that they will interact (111). The removal of water from a food material increases its physical accessibility to atmospheric oxygen through microcapillaries that open up through to the center of the material. Also, a food's chemical susceptibility to reaction with atmospheric oxygen increases as a result of the removal of the sheath of water molecules protecting reactive sites on the food components. Lipids, particularly those that contain unsaturated fatty acids, are readily attacked by oxygen-containing free radicals, which gain access to them through the open structure of many dried foods. Autoxidation of dried foods is accelerated by several factors, including exposure to light and elevated temperatures. However, the effect of pH is complex, as different prooxidation factors may operate in different pH ranges. Autoxidation occurs most rapidly at very low water activity (a_w) levels, with higher levels producing a diminution in rates until the a_w 0.4–0.5 range is reached. Further increases in a_w tend to decrease oxidation rates, probably as a result of the dilution of oxidative components (112).

Photosensitization

It appears unlikely that the direct reaction of a lipid molecule (R—H) with a molecule of oxygen would lead to the formation of hydroperoxide with ease. In fact, the lipid–oxygen reaction conflicts with the basic concept of conservation of spin angular momentum because the lipid molecule occurs in a singlet electronic state whereas the oxygen molecule is in a triplet ground state. As a result, the initiation of autoxidation by way of the direct reaction of the singlet lipid with the triplet oxygen is highly improbable. Since the energy of activation of the reaction is very high (146–272 kJ/mol) (113), the expected reactivity of the reactants or the thermodynamic equilibrium would not be achieved. Although the hydroperoxide is produced during the propagation reaction (Eq 2.5), the role of an initiator such as catalytic lipoxygenase enzyme or photosensitizer is very important from the view-point of the original lipid hydroperoxide formation. The well-known photo-sensitizers are chlorophyll, hematoporphyrins, or flavins (including riboflavin), which are capable of converting triplet oxygen to singlet oxygen (114). The mechanism of this conversion is initiated by the transfer of the photosensitizer to its electronically excited state due to the absorption of light in the visible or near-UV region. Subsequently, the photosensitizer is able to transfer its excess energy to an oxygen molecule, giving rise to singlet oxygen. Thus, the singlet oxygen can react with a lipid molecule to yield a hydroperoxide (115–117). The sequence of events is indicated by the reactions

$$S \xrightarrow{hv} {}^1S \rightarrow {}^3S* \tag{2.12}$$

$$^3S* + {}^3O_2 \rightarrow {}^1O_2 + {}^1S* \tag{2.13}$$

$$^1O_2^* + RH \rightarrow ROOH \tag{2.14}$$

where ^1S is the singlet-state sensitizer, $^1S*$ is the excited singlet-state sensitizer, $^3S*$ is the excited triplet-state sensitizer, $^3O_2$ is normal triplet oxygen, $^1O_2^*$ is excited singlet-state oxygen, and hv is ultraviolet light energy in photons.

It is observed that singlet oxygen reacts about 1000–10,000 times as fast as normal oxygen with methyl linoleate. An increase in temperature (118–120) or UV light with the proper sensitizer could be responsible for direct attack of oxygen via the formation of singlet-state oxygen. Chahine and deMan (121) observed very rapid oxidation of corn oil in UV light. The energy of activation was just about 17 kJ/mol, a substantial drop from the 146–272 kJ/mol expected for the interaction between R—H and oxygen. These calculations favor the formation of singlet oxygen, perhaps due to photosensitization with the trace of chlorophyll in the corn oil or the trace metals.

Lipoxygenase

Hydroperoxide can also be formed via an alternate route in which the reaction of polyunsaturated lipid with oxygen is catalyzed by the enzyme lipoxygenase (122). Lipoxygenase specifically oxidizes polyethenoid acids containing methylene-interrupted double bonds that are in the cis geometrical configuration such as those in linoleic, linolenic, and arachidonic acids but not in oleic acid. Free-radical intermediates occur during lipoxygenase catalysis, and these can lead to cooxidation of easily oxidized compounds such as carotenoids and polyphenols. Lipoxygenases are present in spices, wheat flour, and vegetables and will catalyze the oxidation of unsaturated fats in dried and drying produce, increasing the rate at which peroxides and volatile breakdown products are generated.

Metals: Iron and Copper

Transition metal ions are remarkably good promoters of free-radical reactions (123) because of single electron transfer during their change in oxidation states. In a thorough review of the kinetics of metal-catalyzed lipid oxidation (124), transition metal ions having variable oxidation numbers (iron as Fe^{2+} or Fe^{3+}; copper as Cu^+ or Cu^{2+}; Mn, Ni, Co, etc.) were implicated as enhancing the rate of oxidation. This action is due to the reduction of the initiation step activation energy down to 63–104 kJ/mol. Also, a direct reaction between a metal catalyst and a lipid molecule is envisaged in the chain initiation step (Eq. 2.15). This process has been shown to be exothermic for methyl linoleate (125).

$$M^{n+1} + R\text{—}H \rightarrow M^{n+} + H^+ + R^{\cdot} \tag{2.15}$$

However, the decomposition of hydroperoxide seems to be the main initiation process. A simple homolytic cleavage of the weak O—O bond (activation energy for cleavage, 184 kJ/mol) (126,127) in hydroperoxide can lead to the initial formation of radicals (Eq. 2.2) that is responsible for a chain reaction. Spontaneous homolytic cleavage is likely to be insignificant in foods except when they are heated during cooking. The bimolecular process (Eq. 2.3) is more important than homolytic cleavage for generating free radicals (128,129). In the presence of transition metal ions, the catalytic decomposition of hydroperoxides by the metal appears to be the major source of free radicals. Transition metal ions such as iron and copper can accelerate peroxidation by decomposing lipid hydroperoxide in both their lower (Eq. 2.16) and higher (Eq. 2.17) oxidation states (130–134). The alkoxyl and peroxyl radicals that are produced during these reactions can abstract hydrogen and perpetuate the chain reaction (Eq. 2.18) of lipid peroxidation (133, 134).

$$ROOH + Fe^{2+} (Cu^+) \xrightarrow{Fast} RO^\cdot + Fe^{3+} (Cu^{2+}) + OH^- \qquad (2.16)$$
$$\text{(Alkoxyl radical)}$$

$$ROOH + Fe^{3+} (Cu^{2+}) \xrightarrow{Slow} ROO^\cdot + Fe^{2+} (Cu^+) + H^+ \qquad (2.17)$$
$$\text{(Peroxyl radical)}$$

$$RO^\cdot + RH \rightarrow R^\cdot + ROH \qquad (2.18a)$$
$$ROO^\cdot + RH \rightarrow R^\cdot + ROOH \qquad (2.18b)$$

These reactions may proceed in a cycle, so that even trace quantities of metal ions may be adequate for effectively generating free radicals. Reducing agents like ascorbic acid or superoxide anion radical ($O_2^{\cdot-}$) can accelerate these metal ion–dependent reactions, as Fe^{2+} and Cu^+ ions appear to react with hydroperoxide faster than Fe^{3+} and Cu^{2+}, respectively. Waters (135) concluded that the reactions catalyzed by metal ions in their higher oxidation states (Eq. 2.17) are unlikely to be significant in autoxidation in aqueous solution, although the metal ions may oxidize the lipid (R—H) or a reaction product (ROH or $R^\cdot$). He also felt that in aqueous systems the peroxide may displace the water from the coordination shell of the metal and enter the ligand shell as concluded by Ochiai (136) on the basis of infrared, electron spin resonance, and visible spectroscopic data. Since the peroxide in the shell is in close coordination with the metal, cleavage reactions become favorable to form free radicals. Several iron ion–oxygen complexes have also been claimed to stimulate lipid peroxidation (46), although their role is not yet clearly understood (137).

Labuza (1) reviewed the published work on coordination spheres, solvent systems, and catalysis as applied to the study of lipid oxidation of food, and drew the following conclusions:

1. Most foods oxidize through a metal-catalyzed reaction because metals are present in the necessary trace amounts.
2. Most likely, a metal hydroperoxide complex is formed, which then decomposes to form free radicals that enter into the coordination sphere of the metal.
3. The oxidation state of the metal is considered to be important in reduction of the hydroperoxide decomposition activation energy.
4. Competition for the metal by other coordination species may alter the oxidation rate by changing the electronic structure of the outer shell of the complex.

5. Certain chelating complexes may reduce the catalytic role of a metal by steric hindrance even though some coordination positions are not filled.
6. The solvent system has a very significant effect on metal catalysis. Water may prevent metal from complexing with hydroperoxides; the solvent may form an inert insoluble compound with the metal or increase the mobility of the metal catalyst.

Metalloproteins

The oxidation of lipids involving the uptake of oxygen can occur in the presence of catalysts such as hemoproteins as initiated by heat or light. According to Possami et al. (138), the hemoglobin–oxygen reaction produces singlet oxygen, which is responsible for the lipid oxidation. The heme-catalyzed oxidation of lipids showed maximum rate kinetics (139–141). However, at very high catalyst concentrations, an inhibition of the rate is likely to occur (142) due to the formation of a stable metal–hydroperoxide complex (143). Chalk and Smith (144) also supported the metal–hydroperoxide complex theory with a series of chelated metal complexes in the autoxidation of several olefins. Betts and Uri (145) proposed the formation of an inactive chelate of the type $(MX_2)_2 \cdot 2\ RO_2$ in this reaction in which the $ROO^{\bullet}$ free radicals act as bridges. The state of the trace metal catalysis has a profound effect on the rate and course of the lipid oxidation reaction. This was evident when Watts (146,147) found that heating meat to 110°C destroys the heme structure, causing a faster development of rancidity. It should be remembered that the state of the solvent system around the lipid and metal is also important for creating an environment for antioxidant or prooxidant effect of certain food components.

In beef, the lean tissues contain about 2–4% of total fat as phospholipids, which, being unsaturated, become responsible for rancidity development mainly due to their close proximity to the heme catalysts of the mitochondria. The meat pigments, which can act as potent catalysts, become intimately mixed with the lipids in ground meat and cause the onset of rancidity at a faster rate than in whole cuts of raw meat (148,149). Fresh beef also develops rancidity when cooked and stored.

Based on published work, Labuza (1) made certain observations and concluded that heating may remove the steric factor caused by the protein part of heme structures (catalase and peroxidase), thereby increasing the oxidation rate of linoleic acid. Unfolding of the protein may allow easy exposure of the iron to the peroxide. Enzymatic removal of the protein portion also enhances the catalytic activity of cytochrome c (a heme) by severalfold. The oxidative reactivity order cytochrome c > hemin > hemogloblin > catalase corresponds to the order of decreasing molecular size of the protein. Thus, it is increasingly clear that heating may be important in releasing either the pigment or the lipid from a protected compartment, causing more intimate contact. The form of the oxidizable substrate is also important. It can therefore be expected that the lipid somehow forms a complex with the heme pigment before it breaks down as proposed by Tappel (150,151).

Labuza (1) also emphasized the electronic configuration involved in the interaction of the heme pigment in oxidation. Heme contains porphyrin rings around iron as a square planar chelate. The structure attains stability as iron forms complexes with two more ligands in coordination positions 5 and 6. The iron is free within this structure to be oxidized to the ferric form. The electronic configuration of the outer shells of Fe^{2+} is

The following configuration shows iron in bound form with the nitrogen N atoms of heme (i.e., two covalent and two coordinate bonds).

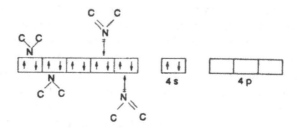

In hemoglobin or myoglobin, the 5- and 6-positions are occupied by N of histidine and water. Thus, the shells tend to be closer to the noble gas structure.

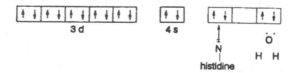

In an oxidation reaction, a hydroperoxide can replace a water molecule and the two oxygens in the peroxide can occupy the empty 4p positions. This could lead to a stable complex. However, the two oxygen atoms, being highly electronegative and side by side, tend to undergo homolytic fission due to energy transfer from the iron porphyrin ring.

Radiation

Radiation generates radicals, including hydroxyl radicals. The variety of analytical procedures established for evaluating the radiation dose for food products largely depend on reactions that integrate the flux of radicals created during irradiation.

The measurement of one or more of the products resulting from the radiation-generated hydroxyl radical on DNA can serve as an index of oxidative damage to DNA in the whole organism or in isolated cells and also that of the prooxidant state. Examples are the measurements of thymine and thymidine glycol excretion in human urine (152) and 8-hydroxyguanine in DNA isolated from mice (153). The free-radical-induced modified bases in DNA are detectable using a combination of gas chromatography, mass spectrometry, and selected ion monitoring (52,154–156).

2.4.2 Major Reactions of Oxidized Lipids in Food Systems

Apart from browning reactions and microbial spoilage, the oxidative breakdown of lipids has a profound effect on the shelf life of many products. Lipid oxidation is a complicated chemical and biochemical reaction process that leads to the formation of a multitude of products. In the following paragraphs, only pertinent reactions are highlighted.

Protein Damage

Proteins, peptides, and amino acids are susceptible to oxidative changes caused by several agents such as light, gamma irradiation, peroxidizing lipids, metal ions, the products of enzymatic and nonenzymatic browning reactions, and certain food additives. Mane et al. (157) observed the degradation of amino acids caused by hydrogen peroxide.

Parkin and Damodaran (109) and Belitz (158) present overviews on the oxidative changes in proteins and amino acids that are summarized here. Peroxide treatment of proteins leads to the formation of methionine sulfoxide, which is further oxidized to methionine sulfones (Eq. 2.19). Peroxides or other forms of activated oxygen can oxidize cysteine residues to sulfenic (Cy—SOH), sulfinic (Cy—SO$_2$H), and sulfonic (Cy—SO$_3$H) acid derivatives. On the other hand, cystine residues in proteins yield mono-, di-, tri-, and tetrasulfoxides when subjected to oxidation.

$$-CH_2-CH_2-S-CH_3 \rightleftharpoons -CH_2-CH_2-\overset{\displaystyle O}{\overset{\displaystyle \|}{S}}-CH_3 \rightarrow -CH_2-CH_2-\overset{\displaystyle O}{\overset{\displaystyle \|}{S}}-CH_3 \qquad (2.19)$$

The oxidation of histidine, cysteine, methionine, tryptophan, and tyrosine by superoxide ion, hydrogen peroxide, and singlet oxygen can occur when food containing photosensitizers such as riboflavin and chlorophyll is exposed to light. Gamma irradiation of foods produces hydrogen peroxide as a result of the radiolysis of water in the presence of oxygen, which in turn causes oxidative changes in proteins.

In fact, free amino acids and amino acid residues undergo substantial oxidation in the presence of peroxidizing lipids. Methionine, cysteine, histidine, and lysine appear to be the most susceptible amino acids or amino acid residues. The extent of loss of amino acids in a protein in the presence of oxidized lipids is presented in Table 2.6. Degradation products obtained in model systems of pure amino acids and oxidized lipids are listed in Table 2.7. There are two probable mechanisms to be considered in the oxidation of proteins by peroxidizing lipids. One involves alkoxy (RO˙) and peroxy (ROO˙) radicals, and the other involves reactions with malondialdehyde and other carbonyl compounds. In the first mechanism, reaction of the lipid free radicals with proteins (P) yields protein free radicals (P˙), and this is followed by polymerization of protein molecules (Eqs. 2.20–2.24). Also, the lipid peroxides formed during the reaction can oxidize methionine, cysteine, histidine, and tryptophan residues. As malondialdehyde is highly reactive, it reacts with amino groups of lysyl residues, giving rise to intermolecular cross-links.

Table 2.6 Amino Acid Losses Occurring in Protein Reaction with Peroxidized Lipids

Reaction system		Reaction conditions		
Protein	Lipid	Time	Temp. (C)	Amino acid [loss,%]
Cytochrome c	Linolenic acid	5 h	37	His [59], Ser [55], Pro [53], Val [49], Arg [42], Met [38], Cys [35][a]
Trypsin	Linoleic acid	40 min	37	Met [83], His [12][a]
Lysozyme	Linoleic acid	8 days	37	Trp [56], His [42], Lys [17], Met [14], Arg [9]
Casein	Linoleic acid ethyl ester	4 days	60	Lys [50], Met [47], Ile [30], Phe [30], Arg [29], Asp [29], Gly [29], His [28], Thr [27], Ala [27], Tyr [27][a,b]
Ovalbumin	Linoleic acid ethyl ester	24 h	55	Met [17], Ser [10], Lys [9], Ala [8], Leu [8][a,b]

[a]Tryptophan analysis was not performed.
[b]Cysteine analysis was omitted.
Source: Ref. 158.

Table 2.7 Amino Acid Products Formed in Reaction with Peroxidized Lipid

Reaction system		Compounds formed from amino acids
Amino acid	Lipid	
Histidine	Methyl linoleate	Imidazolelactic acid, imidazoleacetic acid
Cystine	Ethyl arachidonate	Cystine, hydrogen sulfide, cysteic acid, alanine, cystine disulfoxide
Methionine	Methyl linoleate	Methionine sulfoxide
Lysine	Methyl linoleate	Diaminopentane, aspartic acid, glycine, alanine, α-aminoadipic acid, pipecolinic acid, 1,10-diamino-1,10-dicarboxydecane

Source: Ref. 158.

$$ROO^{\bullet} + P \rightarrow ROOP \qquad (2.20a)$$
$$ROOP + O_2 \rightarrow {}^{\bullet}OOROOP \qquad (2.20b)$$
$${}^{\bullet}OOROOP + P \rightarrow POOROOP \qquad (2.20c)$$
or
$$RO^{\bullet} + P \rightarrow ROH + P^{\bullet} \qquad (2.21)$$
$$ROO^{\bullet} + P \rightarrow ROOH + P^{\bullet} \qquad (2.22)$$
$$P^{\bullet} + P \rightarrow P\!-\!P^{\bullet} \qquad (2.23)$$
$$P\!-\!P^{\bullet} + P \rightarrow P\!-\!P\!-\!P^{\bullet} \qquad (2.24)$$

When proteins react with lipid hydroperoxides and their degradation products, some of their properties are changed. This is evidenced by changes in food texture, a decrease in protein solubility (formation of a protein network or cross-links), a change in color (browning), and changes in nutritive value (loss of essential amino acids).

Off-Flavor

It is important to realize that the lipid hydroperoxides formed during the autoxidation of polyunsaturated lipids are flavorless. Nevertheless, thermal degradation of hydroperoxides generates off-flavors, mainly due to the formation of carbonyl compounds as oxidation-fragmentation products. Labuza (1) has described an overall picture of rancidity. According to him, rancidity is the development of an off-flavor that makes the food unacceptable on the consumer market level. It is therefore obvious that various types of off-flavors are produced by different mechanisms. In the present context, rancidity means those objectionable flavors that are developed as a result of the reaction between oxygen and an unsaturated fatty acid, which may or may not be involved in the molecular structure of triglyceride or phospholipid.

Oxidation of linoleic acid can be initiated by two basic mechanisms, abstraction (autoxidation) and "ene" addition. Abstraction leads to the formation of 9- and 13-OOH isomers, (hydroperoxides), whereas "ene" addition caused by singlet oxygen produces 9-, 10-, 12-, and 13-OOH isomers. Thus, the type of oxidative mechanism and the degree of unsaturation determine the type of hydroperoxide isomer. These different hydroperoxides are quite unstable and can break down and initiate a chain reaction in an explosive manner. In fact, this breakdown mechanism is solely responsible for the rancid off-flavors and off-odors in foods caused by the simple aldehydes at very low concentrations. The fragmentation of linolenate residues via peroxidation is shown in Fig. 2.3, which indicates the formation of cis-3-hexenal (9). The formation of C-13 hydroperoxide produces the conjugated 9,11-diene system. In alkoxyl radical I, bond a to the vinylic carbon is much stronger than bond b to the saturated allylic carbon. Hence the fragmentation of bond b will be favored. This possibility does not exist in an adduct like II.

Grosch (159) observed a number of products from reactions of the different hydroperoxide radicals. A hydrocarbon (i.e., pentane) is formed from 13-hydroperoxylinoleic acid by thermal scission of the 13–14 bond while hexanal is produced by thermal scission of the 12–13 bond. Pentane is flavorless and odorless in the concentration range that is normally obtained by the autoxidation of food lipids. However, its formation is parallel to the formation of the objectionable hexanal. Moreover, pentane formation is of interest only for analytical reasons as its determination is more convenient than that of hexanal.

Linolenic acid hydroperoxyl radical or hydroperoxide, being unstable, can be subject to secondary reactions giving rise to malondialdehyde (MDA) or aldehydes as shown in Fig. 2.4 (109). MDA is often used by food scientists as an indicator of the degree of lipid oxidation in foods. The low molecular weight compounds formed have very low odor threshold values. If they are objectionable in odor, only a few parts per million (ppm) or parts per billion (ppb) are required to impart an unacceptable odor to the food. Hexanal has been implicated as a major breakdown off-flavor compound in many foods including potatoes (160,161). It has been detected in oil at 150 ppb and in milk at 50 ppb (162). This means that less than 0.00002% of the fat has to be oxidized to form an objectionable off-odor. In oil, the threshold value of cis-3-hexenal is 100 ppb (162). Some specific breakdown products found in rancid food and derived from the respective hydroperoxides are listed in Table 2.8.

Linoleate is the major unsaturated fatty acid in most foods and oxidizes 10–15 times faster than oleate. The degree of oxidation of linoleate is measured in terms of hexanal formed (160). The monocarbonyl compounds derived from autoxidation of unsaturated fatty acids readily condense with protein free amino groups, forming

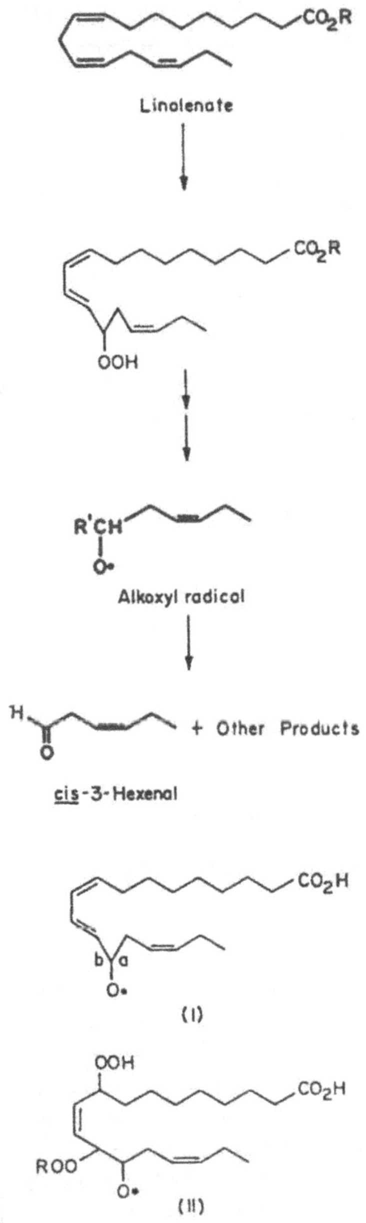

Fig. 2.3   Schematic of the oxidative degradation of linolenate to *cis*-3-hexenal. (From Ref. 9.)

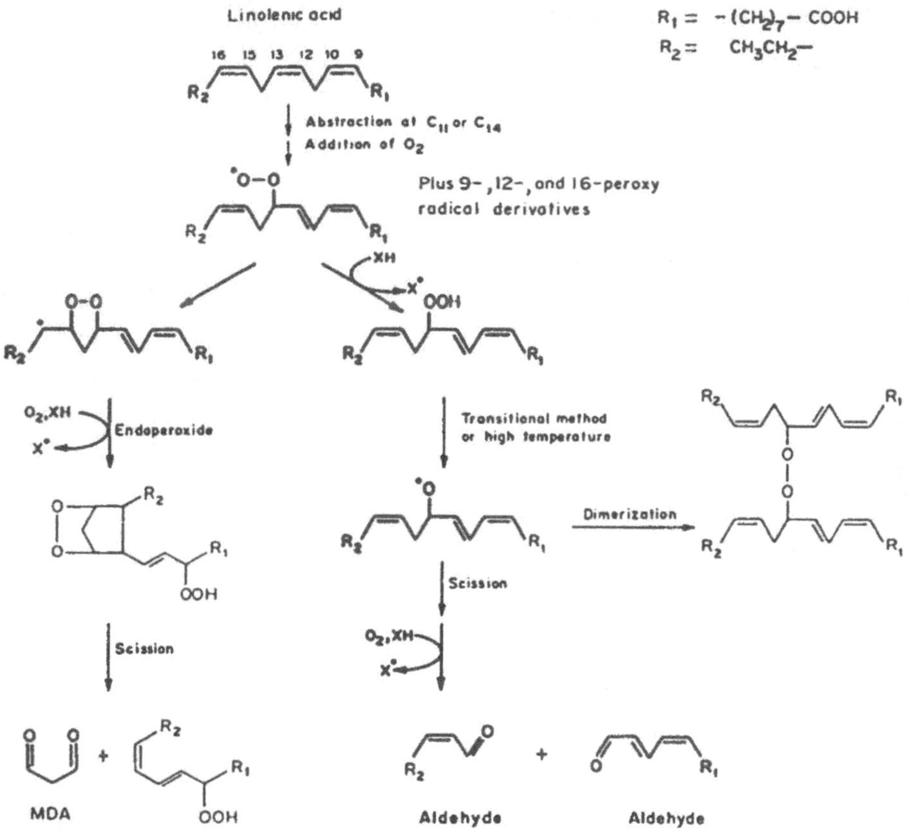

Fig. 2.4 Initiation and secondary reactions for the oxidation of linolenic acid. (From Ref. 109.)

Schiff bases that can yield brown polymers by repeated aldol condensations (158). The polymers are often nitrogen-free as the amino compound can be easily eliminated by hydrolysis. When hydrolysis occurs in the early stages of aldol condensations (i.e., after the first or second condensation) and the released aldehyde, a powerful off-flavor compound, does not reenter the reaction, the condensation process leads to a change not only in coloration but also in aroma.

As indicated earlier, vegetable oils and fats deteriorate during the frying process at high temperature and in the presence of oxygen. Thermal and oxidation reactions occur simultaneously, producing both volatile and nonvolatile decomposition products. The nature and quantity of these products are largely influenced by the frying conditions and the type of food being fried. However, accumulation of these degradation products causes off-flavors in the food and subsequently results in

Table 2.8 Typical Breakdown Products

Fatty acid	Hydroperoxide	Aldehyde formed
Oleate	C_8	Undec-2-enal
	C_9	2-Decenal
	C_{10}	Nonanal
	C_{11}	Octanal
Linoleate	C_9	2,4-Decadienal; 2-octenal
	C_{13}	Hexanal
Linolenate	C_9	2,4,7-Decatrienal
	C_{12}	2,4-Heptadienal
	C_{13}	3-Hexenal
	C_{16}	Propanal

Source: Ref. 1.

smoking, foaming, and discoloration of the frying medium. The probable chemical changes taking place in frying oils are shown in Fig. 2.5 (163).

Loss of Vitamins

Vitamin E (α-tocopherol) is a primary antioxidant that is consumed during the induction period of fat. Ascorbic acid, vitamin A, and β-carotene also behave as antioxidants and are prone to oxidation. The environmental factors such as temperature, oxygen, metal ions, light, and radiation, which are conducive to lipid oxidation, and also cause the destruction of certain vitamins. Ascorbic acid can be degraded by active oxygen and by reactions initiated by transition metals. As an antioxidant, ascorbic acid removes oxygen in systems where oxygen is present in limited amounts and gets oxidized to dehydroascorbic acid. Ascorbic acid quenches various activated oxygen species and also reduces free radicals and primary antioxidant radicals. Besides oxidation by molecular oxygen in the presence of metal ions, the antioxidant behavior also enhances the loss of ascorbic acid.

Although relatively immune to losses due to vaporization, tocopherol can undergo rapid losses due to thermal oxidation when used in a medium offering a large surface area in contact with air. The availability of the fat-soluble vitamins A, D, and E as well as that of vitamin C and folate can be reduced during lipid oxidation. The most likely reason for loss of fat-soluble vitamins in dehydration and dried food storage is the interaction between them and the free radicals generated during the oxidation of lipids. β-Carotene, like tocopherol, has been identified as a sacrificial inhibitor of the propagation stage of lipid oxidation that becomes used up during dry food storage. Vitamin A and C are sensitive to the presence of oxygen, whereas riboflavin is sensitive to light. Thus, exclusion of

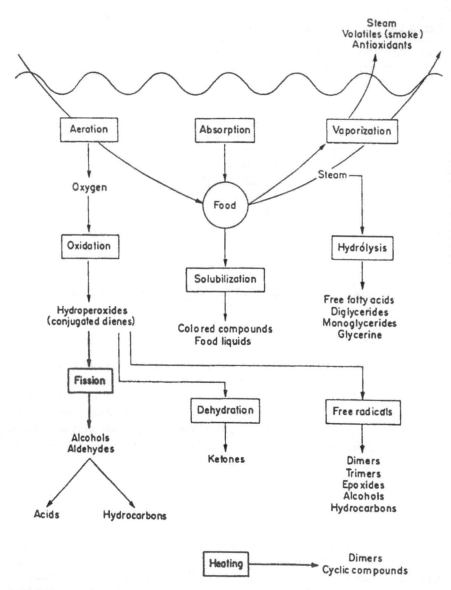

Fig. 2.5   Diagram of oil changes during frying. (From Ref. 163.)

oxygen and light can reduce not only the risk of lipid oxidation but also that of vitamin losses.

Oxidation of Pigments

Lipid oxidation produces several products, some of which react with other components of the foodstuff such as pigments and amino acids, resulting in further discoloration or undesirable flavors. Thus, the risk of oxidation damage is not limited to high-fat foods; proteins and pigments are also subject to this process. In plant foods, oxidative browning of pigments is particularly undesirable. It is now well recognized that autoxidation is accompanied by reactions with proteins, the formation of off-flavors or flavorless products, and the degradation of pigments such as β-carotene, which is a quencher of singlet oxygen. Carotenoids are generally found complexed with either proteins or fatty acids, which protect them from oxidation. The breakdown of these complexes during processing leads to degradation of carotenoids, resulting in blanching or discoloration of the product.

2.4.3 Health Implications of Dietary Lipid Oxidation Products

Addis and Warner (164) classified the possible health aspects of dietary lipid oxidation products into two areas of research, one dealing with the possible toxic effects of thermally altered lipids and the other related to the possible role of dietary cholesterol oxides and fatty acid oxidation products in coronary heart disease (CHD). According to several researchers, the deleterious effects attributed to heated fat are unrealistic. Also, this abused or heated fat requires reexamination using more precise and realistic heat treatments, relevant levels of intake, and appropriate pathological endpoints (164) so that any doubts about arterial injury, atherosclerosis, or CHD may be dealt with.

The relative atherogenicity of lipids and lipid oxidation products (i.e., cholesterol vs. cholesterol oxides and fatty acids vs. fatty acid hydroperoxides and/or secondary oxidation products) is another controversial issue. Questions have been raised about the extent of dietary influence on serum cholesterol and the degree to which serum cholesterol is related to CHD. The lipid hypothesis, which states that hypercholesterolemia is the primary cause of CHD, is promoted by the National Heart, Lung and Blood Institute (NHLBI) and the American Heart Association (AHA). However, in recent years, the lipid oxidation products–CHD connection has been gaining considerable support because of accumulated evidence by way of a large number of research publications. Thus, the research findings support the idea that oxidized lipids are more deleterious to arterial health than native lipids themselves. Hence cholesterol is considered to be harmless unless it is converted to one or more of a number of autoxidation products. Excellent reviews have appeared in the past that tackle the health aspects of lipid oxidation products in food (165–180).

Coronary Heart Disease

It should be recognized that the concept that atherosclerosis is a straightforward accumulation of cholesterol in the arteries is not true. Several researchers, therefore, have rejected the "lipid hypothesis" in favor of the competing "response-to-injury" hypothesis. It is realized that the latter hypohesis generally displays excellent agreement with the concept that dietary lipid oxides are a key risk factor in CHD (178). However, the lipid hypothesis supporting the diet–CHD relationship appears to have created a favorable impact on the food and animal industries interested in developing low-calorie, low-fat, low-cholesterol, high-fiber and fat-substituted foods useful to the consumer. It is obvious that lowering dietary lipid content can reduce the exposure of humans to lipid oxidation products. Addis and Park (178) divided CHD into three arbitrary phases—arterial injury, atherosclerosis, and myocardial infarction—and noted that adverse effects of oxidation products have been reported for all three phases.

Lipid Peroxides

Absorption studies have established that lipid oxidation products and the secondary oxidation products are absorbed from dietary sources (181,182). Kanazawa et al. (183) demonstrated in rats the absorption and uptake of linoleic acid hydroperoxides and secondary oxidation products by the liver. Evidence of liver hypertrophy, an increase in hepatic peroxide levels, and increased serum transaminase activities were also reported. Even in the absence of absorption, dietary oxidation products may pose a risk to the intestinal mucosa (164). Kanazawa et al. (184) observed that oral administration of peroxidation products to rats induced acute toxic reactions resulting in hemorrhage and diarrhea. Peters and Boyd (185) observed that rats fed a mixture of rancid cottonseed and cod-liver oil developed listlessness, anorexia, proteinuria, oligodipsia, diarrhea, diuresis, and loss in body weight.

Lipid hydroperoxides are also known to have an effect on a number of enzyme systems. Linoleic acid hydroperoxides have been reported to inhibit the in vitro activity of ribonuclease, pepsin, trypsin, and pancreatic lipases (186). Kanazawa and Ashida (187) reported that the hepatic dysfunction caused by lipid peroxides may be due to decreased activity of hepatic enzymes such as glucose-6-phosphate dehydrogenase, glucokinase, mitochondrial NAD-dependent aldehyde dehydrogenase, glyceraldehyde phosphate dehydrogenase, and levels of coenzyme A (CoA). They also concluded that the enzymes and CoA were targets of direct attack by the peroxidation products. Lipid peroxides have also been implicated in the etiology of atherosclerosis. Lipid hydroperoxides alter the low-density lipoproteins, which play a major role in the development of atherosclerotic lesions (80,188–190). Sasaguri et al. (191) reported cellular damage and vacuolization of cultured endothelial cells on incubation with linoleic acid hydroperoxide. Linoleic acid hydroperoxides inhibit prostacyclin production in cultured endothelial cells, which may lead to the development of coronary artery disease, especially myocardial infarction, in vivo (192). Fujimoto et al. (193) demonstrated a strong reaction between DNA and several primary peroxides, indicating the possible role of the peroxides in carcinogenesis. It is suggested that dietary lipid peroxides may function as promoters of carcinogenesis (172). Studies of Cutler and Schneider (194) have shown that oxidized linoleic acid induced cervical sarcoma and malignant and benign mammary tumors in rats fed a diet high in fat. Experiments also suggest that lipid peroxides induce injection-site sarcomas in mice (195,196). Oxidized linoleic acid increased teratogenicity in the urogenital system in rats. In mice, oxidized linoleic acid applied directly to the ovaries induced fetal malformations in first-generation litters and a significant increase in embryonic resorption in second-generation litters (197). Though studies indicate the possible role of lipid peroxides in carcinogenesis, the mechanism still remains unclear.

Cholesterol Oxidation Products

Cholesterol oxidation products have received a lot of attention because of their involvement in the development of coronary artery disease, and numerous studies have been published. Nearly 30 oxidation products have been detected in commercial cholesterol (198). Typical oxidation products include cholestanetriol, enantiomeric 5,6-epoxides, 7-ketocholesterol, isomeric 7-hydroxycholesterol, and 25-hydroxycholesterol. Cholesterol oxidation products are readily absorbed and incorporated into high-density, low-density, and very low density lipoproteins (164,199). Cholesterol oxides are reported to inhibit the biosynthesis of cholesterol in cultured aortic smooth muscle cells of rabbits (200), which may result in a decrease in membrane cholesterol levels and lead to impaired functionality and cellular injury. Cholesterol oxides also inhibit the activity of 5'-nucleotidases and Na^+- and K^+-ATPases (201). Cholestanetriol and 25-hydroxycholesterol have been reported to be the most atherogenic of the cholesterol oxidation products (202). Oxysterols are also reported to affect ion transport across membranes, particularly in calcium channels (203,204). Oxysterols affect membrane permeability to divalent cations and glucose (205,206). The toxic effects of oxysterols have been extensively reviewed by Smith and Johnson (179). Oxysterols induce reduction in growth, loss of body weight, diminished appetite, necrosis, and acute inflammation in experimental animals. Oxysterols inhibit de novo synthesis of sterols by inhibiting 3-hydroxy-3-methylglutaryl CoA reductase activity. Several other enzymes are also affected by oxysterols, including cholesterol 5,6-epoxide hydrolase, cholesterol 7α-hydroxylase, and acylcholesterol acyl CoA O-transferase. The angiotoxic effects of cholesterol oxidation products have been demonstrated both in vivo (207) and in vitro (208). Specific oxysterols and epoxides are reported to be mutagenic (179,209) and to react with DNA (210). Cholesterol oxidation products may also have carcinogenic potential (179,211). The studies indicate that the oxidation products of cholesterol and not pure cholesterol itself are responsible for the observed cytotoxic, atherogenic, angiotoxic, and enzyme effects.

Pure cholesterol is not atherogenic, even in a sensitive animal such as the rabbit (202). Dried egg yolk frequently contains high levels of both cholesterol and its oxides (178), and the biological effects of such impure sources of cholesterol are difficult to interpret. The cytotoxic and atherogenic effects of oxysterols are significant in the initiation of atherosclerosis, which supports the role of oxysterols in CHD and the response-to-injury hypothesis. Cholesterol oxidation products are known to cause endothelial injury (172,178,212). The observations made by Peng et al. (212) are consistent with the response-to-injury hypothesis of atherosclerosis. An interesting study by Henning and Boissonneault (213) led them to conclude that dietary cholesterol oxides may be an important risk factor for CHD. Matthias et al. (214) observed triol-induced aortic smooth muscle cell toxicity and damage to the endothelial cells of rats, a species relatively resistant to atherosclerosis. Although most of the research on oxysterols has focused on arterial injury, an important study indicated that dietary oxysterols are responsible for a greater elevation of plasma cholesterol than purified cholesterol in the rabbit (215). Morin and Peng (216) noted a net accumulation of cholesterol ester in cultured rabbit aortic smooth muscle cells due to 25-hydroxycholesterol, which is considered an important intermediate in the formation of plaque.

A complete review by Smith and Johnson (179) deals with the recent literature of oxysterols and in vivo and in vitro cytotoxicity and atherogenicity. In short, it can be stated that oxysterols are far more atherogenic than the native sterol counterpart and that oxysterol-induced atherogenicity does not depend on serum cholesterol level. In other words, cholesterol in the diet of humans can neither be atherogenic nor hypercholesterolemic. Hence prevention of lipid oxidation rather than cholesterol removal is important (178) when dealing with future research in this area. Addis and Warner (164) postulated and hypothesized mechanisms whereby dietary lipid oxidation products could accelerate atherosclerotic arterial changes and contribute to CHD. The mechanism involves the modification of low-density lipoprotein (LDL) to modified low-density lipoprotein (mLDL).

Malonaldehyde

Malonaldehyde, a major secondary product of lipid oxidation, has been reported to be toxic to living cells (169,217), an initiator, a complete carcinogen, and a mutagen (169,218–220). Malonaldehyde can cross-link with lipids and proteins (93,221), inactivate ribonuclease (222), and bind covalently to nucleic acids

(193,223,224). It has been observed that consumption of unpreserved cod-liver oil increases the urinary malonaldehyde concentration in humans, indicating the absorption of malonaldehyde (225). Incubation of skin fibroblast cells with malonaldehyde resulted in cytoplasmic vacuolization, karyorrhexis, micronucleation, and a reduction in the protein-synthesizing capacity of the cells (226). Malonaldehyde also induced chromosomal aberrations in cultured mammalian cells (227). Mukai and Goldstein (220) reported that malonaldehyde is mutagenic in the Ames *Salmonella* testing system. It is postulated that the fluorescent products in the age pigment lipofuscin are formed by the interaction of malonaldehyde with lipids and proteins (93). Free malonaldehyde in tissues can be quantitated by an improved HPLC method (228) and this technique is found to be more accurate, sensitive, and specific than the thiobarbituric acid (TBA) method.

2.5 ROLE OF ANTIOXIDANTS IN INHIBITING LIPID PEROXIDATION

2.5.1 Antioxidants and Their Action

Antioxidants are chemical compounds that are capable of donating hydrogen radicals. As a consequence, they reduce the primary radicals to nonradical chemical species and are thus converted to oxidized antioxidant radicals. The molecular structure of an antioxidant is so suitable that it not only donates a hydrogen atom but also forms a radical with low reactivity and no further possibility of reaction with lipids. Antioxidants are of two types: primary or chain-breaking and secondary or preventive. Gordon (8) described the mechanism of antioxidant action in vitro. Primary antioxidants, when present in trace amounts, can react wih peroxyl radicals before they react with further unsaturated lipid molecules and convert them to more stable products. Secondary antioxidants are compounds that retard the rate of chain initiation by various mechanisms other than the pathway followed by the primary antioxidants. The secondary antioxidants reduce the rate of autoxidation of lipids by such processes as binding metal ions, scavenging oxygen, decomposing hydroperoxides to nonradical products, absorbing UV radiation, and deactivating singlet oxygen. They usually require the presence of another minor component for effective action. Typical examples are sequestering agents, metal ions, reducing agents (ascorbic acid), and tocopherols or other phenolics.

Phenolics

It is important to note that the best-known substances that possess antioxidant properties are phenolics, which fall into the category of primary antioxidants. The presence of electron-donating groups at the ortho and para positions in phenol enhances its antioxidant activity by an inductive effect. Although the chemical antioxidant activity of various antioxidants depends on a number of factors, the stability or reactivity of the primary antioxidant radical formed after hydrogen abstraction is more important than any other factors. Thus, the delocalization (stabilization) of the solitary electron over the aromatic ring of the phenolic α-tocopheroxyl radical is shown by way of various mesomeric structures (229). The important antioxidant function of hydrogen donation can be regulated by varying the nature and number of substituent groups attached directly to the aromatic ring (BHT) or attached somewhat away on the structure of the compound (carnosic acid). The presence of bulk groups in the vicinity of the hydroxyl group can influence the antioxidant activity due to steric hindrance. The effect of antioxidant concentration on autoxidation rate is governed by the structure of the antioxidant, oxidation conditions, and the nature of the sample being affected. At high concentrations, phenolic antioxidants suffer from loss of activity and become prooxidants due to their participation in the initiation process. Phenolic antioxidants are effective in oils that have undergone relatively less deterioration but are ineffective when added to severely deteriorated oils (230,231). The antioxidant effects of chlorophyll pigment in the dark occur by a mechanism that is similar to that of phenolics (232).

Radical Trapping Agents

β-Carotene and related carotenoids are effective quenchers of singlet oxygen (233–235) and can act as antioxidants by preventing the formation of hydroperoxides (8). The antioxidant action is limited to low oxygen partial pressures, less than 150 torr, and at higher oxygen pressures β-carotene may become prooxidant. Burton and Ingold (236) proposed that β-carotene and related carotenoids can act as effective antioxidants at low oxygen pressures where the singlet oxygen is not formed and attributed this activity to their rapid reaction with the chain-carrying peroxyl radical leading to a resonance-stabilized carbon-centered radical. β-Carotene at less than a parts-per-million level reduces the rise in peroxide value of soybean oil induced by photooxidizing conditions (237). Lipid oxidation initiated by xanthine oxidase can be inhibited by β-carotene because of the latter's ability to quench singlet oxygen (238).

Chelating Agents

Chelating agents bind metal ions and greatly increase the energy of activation of the initiation reactions. A chelator that forms sigma bonds with a metal is considered an effective secondary antioxidant because of the stabilized oxidation form of the metal ion. Chelators that improve the shelf life of lipid-containing foods are EDTA, citric acid, and phosphoric acid derivatives. An EDTA—metal ion complex, which has six coordinate bonds and strainless five-membered rings, forms thermodynamically stable structures. Although citric acid is a weaker chelator than EDTA, it is effective in retarding the oxidative degradation of lipids in foods and is commonly added to vegetable oils after it has been deodorized. The effect of esterification on the antioxidant activity indicates that the carboxyl groups are involved in chelation of metal ions (239). As for oligophosphates, their metal-chelating activity increases with the number of phosphate residues in the range of 2–6 (240).

Oxygen Scavengers

Ascorbic acid, ascorbyl palmitate, and erythorbic acid or its sodium salt are capable of stabilizing fat-containing foods because of their function as oxygen scavengers. Ascorbic acid reacts directly with oxygen to form dehydroascorbic acid and thus depletes or eliminates the supply of oxygen available to effect autoxidation. In this regard, ascorbic acid (vitamin C) is widely added to fruit drinks and increasingly to beer or to canned and bottled products with a headspace of air. Yourga et al. (241) observed that erythorbic acid in model solutions is oxidized more rapidly than ascorbic acid. However, erythorbic acid lacks the vitamin C activity needed for biological systems and does not occur in nature. Ascorbyl palmitate is often more effective against oxidative degradation in fatty foods because of its increased solubility in the lipid phase. Cort (242) has shown that ascorbyl palmitate at a level of 0.01% is more effective than BHA or BHT in controlling the development of rancidity in vegetable oils. Peroxide values (listed in Table 2.9) indicate that ascorbyl palmitate can effectively protect vegetable oils from autoxidation because of the natural occurrence of tocopherols in the oils. As lard contains no tocopherols, a high concentration of ascorbyl palmitate is required to achieve low peroxide values. From the biochemical and physiological points of view, ascorbyl palmitate can be easily hydrolyzed in the gastrointestinal tract and assimilated as ascorbic acid and palmitic acid, which are components of the human diet.

Enzymes

Superoxide anion radical ($O_2^{\cdot-}$) is a reactive oxygen species formed by the transfer of an extra electron to the oxygen molecule. It is one of the various radicals that can contribute to lipid oxidation in biological systems. It may be produced by the

Table 2.9 Antioxidant Effects of Ascorbyl Palmitate at 80°C Determined by Peroxide Values

Concentration of ascorbyl palmitate (ppm)	Peroxide values after x days			
	Peanut oil 2 days	Soy oil 2 days	Palm oil 3 days	Lard 5 days
0	17.1	100	29.5	>400
100	7.6	43	24	>400
200	6.5	12	14	86
500	3.7	2	10	1.2

Source: Ref. 27.

enzyme xanthine oxidase and hydrogen peroxide (238). Part of the superoxide anion radical formation in vivo may be accidental, but it is mostly functional. The formation of superoxide anion radical in vivo is substantiated by the discovery of superoxide dismutase (SOD), an enzyme that removes the superoxide anion radical by catalyzing a dismutation reaction (244). It appears that SOD is essential to normal aerobic life. The enzymes catalase and glutathione peroxidase contribute to the removal of hydrogen peroxide from animal cells. The removal of hydrogen peroxide as well as superoxide anion radical is considered an important antioxidant defense in cells. The presence of xanthine oxidase and SOD affords stability to milk (245).

Methyl Silicone and Sterols

The use of methyl silicone (polydimethylsiloxane) in edible oils has certain advantages. It retards the oxidative degradation of oils when they are heated on a hot plate but not when they are heated in an oven (246). Moreover, in conjunction with citric acid, it is superior to TBHQ for controlling room odor of heated soybean oil (247). Martin (248) reported that 0.03 ppm of methyl silicone was sufficient to inhibit oxidative changes in frying oils. According to Freeman et al. (249), a monolayer of methyl silicone on the oil surface is adequate for antioxidant activity. These authors further suggested probable reasons for the protective action of methyl silicone: (a) It forms a physical barrier against entry of oxygen into the oil; (b) it serves the role of inert surface; (c) it oxidizes antioxidants and inhibits free-radical propagation, or (d) it inhibits convection currents in the surface layer.

Although most sterols do not act as antioxidants, Δ^5-avenasterol, fucosterol, and citrostadienol were found to be effective antioxidants in oils heated at 180°C (250,251). The proposed mode of action of sterol antioxidants indicated that the allylic methyl group in the side chain donates a hydrogen atom and produces a free radical that is stabilized by isomerization to a tertiary allylic free radical (252). It is interesting to note that Δ^5-avenasterol and methyl silicone exhibit similar effects at 100°C or 180°C when oil is heated on a hot plate or in an oven. Also, this sterol tends to increase in concentration in a layer at the surface but shows no effect at room temperature. It therefore appears that avenasterol behaves like a chemical antioxidant that gets oxidized and inhibits the propagation of free-radical chains, with its effectiveness arising from its concentration in the surface where oxidation takes place.

Phospholipids

Although the secondary antioxidant effects of phospholipids have been occasionally attributed to their metal chelating property (253,254), this mechanism does not justify the increased activity of dipalmitoyl phosphatidylethanolamine in the presence of α-tocopherol with phospholipid concentrations up to 6%, which is more than what is required for chelating metal ions (255). Similarly, not all phospholipids follow one specific mechanism of antioxidant action. In one instance, phosphatidylethanolamine is effective as a synergist in combination with polyhydroxyflavones (256) whereas in another instance only phosphatidylethanolamine, phosphatidyl inositol, phosphatidylserine, and phosphatidic acid showed effectiveness in protecting α-tocopherol during the autoxidation of methyl linoleate (257). However, phosphatidylcholine was ineffective in both cases. The other possible mode of action of phospholipids is by way of releasing protons and decomposing hydroperoxides without the formation of free radicals (258,259) or via the regeneration of primary antioxidants (253,260).

Maillard Reaction Products

In foodstuffs, the Maillard reaction normally occurs between reducing sugars and various amino acids, peptides, or proteins and produces innumerable products with different structures depending upon the reactants, processing conditions, and the degree of browning. Several such products exhibit antioxidant properties (261, 262), including intermediate reductones (261) and high molecular weight melanoidins (263) with a variety of mechanisms. Reductone may act by breaking the radical chain via donation of a hydrogen atom (264) while Maillard reaction products can act as metal ion chelators (265,266).

Biological Antioxidant Systems

Antioxidant synergism between α-tocopherol and vitamin C has been well established. Both compounds have been implicated in vivo and in vitro with redox reactions related to autoxidation. Cell membranes and plasma lipoproteins contain α-tocopherol, which acts as a chain-breaking antioxidant. It is a highly lipophilic molecule that is located in the interior of biological membranes. Vitamin C is water-soluble. The chain-breaking reaction converts α-tocopherol into its radical. This radical, being unreactive, is not able to react with the adjacent fatty acid side chains. However, the tocopherol radical tends to migrate to the membrane surface and is converted back to α-tocopherol by reaction with vitamin C (ascorbic acid). Thus, α-tocopherol and vitamin C begin to counteract the consequences of lipid peroxidation in lipoproteins and in membranes upon initiation of this process (267,268). Löliger (27) stated that several researchers had developed evidence for the lipid autoxidation–dependent radical exchange reactions between α-tocopherol (vitamin E) and vitamin C, both in lipid–water systems. These radical exchange reactions between lipid radical and vitamins E and C have been shown by experiments in vivo and in model systems using micellar or liposomal pseudophases in which vitamin E and oxidizable substrate (linoleate) are present in the pseudophase and the vitamin C is present in the aqueous phase (Fig. 2.6).

Small molecules such as uric acid, carnosine, lipoic acid, and bilirubin have

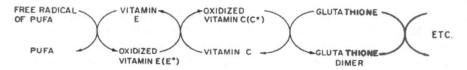

Fig. 2.6 Lipid radical detoxification cascade as postulated by Tappel (269). (From Ref. 27.)

been suspected of functioning as antioxidants in vivo, but their physiological significance needs to be assessed carefully (270). In spite of antioxidant defense, cell systems are capable of repairing DNA damage caused by radicals (53), degrading proteins attacked by radicals (271), and metabolizing lipid hydroperoxides (51). As indicated earlier, the removal of hydrogen peroxide as well as superoxide anion radical is an important antioxidant defense in cells.

Synthetic Antioxidants

A large number of synthetic antioxidants are available for the stabilization of nonfood materials such as plastics, rubber, and polymers. However, the use of synthetic antioxidants in foods requires that different aspects be taken into consideration, including technological necessity (272), the toxicology of lipid oxidation products (172), and the toxicology of antioxidants (273). Since food additives are subjected to the most stringent toxicological testing procedures, only a few synthetic antioxidants have been used in foods for a long time. In fact, antioxidants are extensively tested for the absence of carcinogenicity and other toxic effects in themselves, in their oxidized forms, and in their reaction products with food constituents; for their effective low concentration; and for the absence of the ability to impart an unpleasant flavor to the food in which they are used. Among the various synthetic antioxidants studied with a view to preventing the oxidation of fats, some are used as food additives; see in Tables 2.10 and 2.11.

Most of the synthetic antioxidants are of the phenolic type. The differences in their antioxidant activities are related to their chemical structures, which also influence their physical properties such as volatility, solubility, and thermal stability. The commercially available and currently used synthetic antioxidants are BHA, BHT, TBHQ, and propyl gallate (PG). In addition to the structure–function criteria, there are several factors such as the nature of the lipid requiring antioxidant protection, the physical state of the food, storage conditions, and water activity that govern the efficiency of the various antioxidants. TBHQ and gallate esters are useful in situations where steam-volatile antioxidants such as BHA and BHT are readily lost during spray, vacuum, or drum drying. In general, BHT, BHA, and TBHQ are used at levels of 100–200 ppm, and gallates are used at levels up to 200–250 ppm. The effectiveness of BHA and propyl gallate in the stabilization of

Table 2.10 Permitted Antioxidants in Various Countries[a]

Country	Tocopherol	Guaiagum	PG	BG	OG	DG	NDGA	BHA	BHT	THBP	HMBP	TBHQ
Australia			+		+	+		+				
Austria	+	+	+	+	+	+	+	+	+			
Belgium	+		+		+	+		+	+			+
Brazil		+	+		+	+		+	+			
Canada	+		+									
Czechoslovakia			+									
Denmark	+	+	+		+	+	+	+	+			
Finland	+		+		+	+	+	+	+			
France			+		+	+		+	+			
Greece			+		+	+						
Haiti			+		+	+	+	+	+			
Hong Kong		+	+		+	+		+	+			
India	+		+		+	+	+	+	+			+
Italy	+		+		+	+		+	+			
Jamaica	+		+				+	+	+			
Japan			+					+	+			
Korea			+		+	+	+	+	+			
Malaysia		+	+					+				
Mexico	+		+		+	+		+	+			
Morocco			+		+	+		+	+			
Holland	+		+		+	+		+	+			
New Zealand		+	+					+				
Nicaragua	+		+		+	+	+	+	+			
Norway	+	+	+		+	+	+	+				
Pakistan	+		+		+	+	+	+				
Peru			+					+	+			+

Table 2.10 (*Continued*)

Country	Tocopherol	Guaiagum	PG	BG	OG	DG	NDGA	BHA	BHT	THBP	HMBP	TBHQ
Poland	+		+									
Rumania		+			+							
South Africa	+	+	+		+	+		+				
Sri Lanka			+		+	+		+	+			
Spain			+		+	+		+	+			
Sweden	+	+	+		+	+		+	+			
Switzerland	+		+		+	+		+				
Turkey	+		+		+	+		+	+			
Taiwan			+		+	+		+	+			
UK		+	+		+	+		+	+			
USA	+		+			+		+	+	+	+	+
USSR					+	+		+	+			
Yugoslavia			+		+		+	+				

*Guaiagum, guaiacol gum; PG, propyl gallate; BG, butyl gallate; OG, octyl gallate; DG, dodecyl gallate; NDGA, nordihydroguaiaretic acid; BHA, butylated hydroxyanisole; BHT, butylated hydroxytoluene; THBP, trihydroxybutylphenone; TBHQ, *t*-butyl hydroquinone; HMBP, hydroxymethylbutylphenol.
Source: Ref. 274.

Table 2.11 Japanese Restrictions on the Use of Antioxidants

Antioxidant	Limitation or restriction	Maximum permitted level (mg/kg)
Butylated hydroxyanisole (BHA)[a]	Butter	200
	Fats and oils	200
	Frozen fish, shellfish, and whale meat (for dipping solution)	1000
	Mashed potato (dried)	200
	Salted fish and shellfish	200
	Dried fish and shellfish	200
Butylated hydroxytoluene (BHT)[b]	Butter	200
	Chewing gum	750
	Fats and oils	200
	Frozen fish, shellfish, and whale meat (for dipping solution)	1000
	Mashed potato (dried)	200
	Salted fish and shellfish	200
	Dried fish and shellfish	200
Isopropyl citrate (as monopropyl citrate)	Fats and oils	100
EDTA CaNa$_2$, EDTA Na$_2$ (as EDTA CaNa$_2$)[c]	Canned or bottled soft drinks	35
	Canned or bottled food (except for soft drinks)	250
Erythorbic acid	Only for antioxidant use	
Sodium erythorbate	Only for antioxidant use	
Nordihydroguaiaretic acid	Butter	100
	Fats and oils	100
Propyl gallate	Butter	100
	Fats and oils	100
Resin guaiac	Butter	1000
	Fats and oils	1000
$(\pm)$-α-Tocopherol	Only for antioxidant use	

[a]If used in combination with BHT, total amount of both antioxidants.
[b]If used in combination with BHA, total amount of both antioxidants, except for chewing gum.
[c]To be chelated with calcium ions before completion of the final food.
Source: Ref. 274.

animal fats and vegetable oils against oxidative degradation, according to the Rancimat oxidation test, is shown in Table 2.12. The selected results indicate that chicken fat and safflower oil show a difference in stability due to their differing fatty acid composition and the presence of different levels of oxidation promoters and inhibitors. In addition to this limited number of antioxidants, numerous ready-to-use mixtures are available for the stabilization of lipids in various food applications. These mixtures essentially contain a solvent (i.e., vegetable oil, liquid monoglyceride, or propylene glycol), citric acid, and one or more of the above synthetic antioxidants.

Natural Antioxidants

A wide range of natural antioxidants have been shown to occur in plants and animals. Although these compounds can be synthesized, they are treated as if they belong to natural sources. In recent years, there has been a greater awareness of the promotion and use of natural antioxidants over the typical synthetic antioxidants such as BHA or BHT. In the recent past, natural antioxidants attracted the attention of many food manufacturers as a result of necessity created by a situation where it was extremely difficult to use synthetic antioxidants and also by the obligation to produce healthy foods. Moreover, new toxicological data on some of the synthetic antioxidants cautioned against their use. However, it is difficult to formulate a safe procedure for the use of either synthetic or natural antioxidants in food systems. In spite of the well-established synthetic antioxi-

Table 2.12 Efficiency of Synthetic Antioxidants for the Stabilization of Animal Fats and Vegetable Oils Against Oxidative Degradation by the Rancimat (15) Accelerated Test at 100°C and 20 mL/min Flow[a]

Lipid	Antioxidant	Concn (ppm)	Induction time (h)
Rendered lard	BHA	100	36
	Propyl gallate	100	20
	Blank	—	5
Refined chicken fat	BHA	100	13.5
	Propyl gallate	100	14
	Blank	—	4
Safflower oil	BHA	100	10
	Propyl gallate	100	15
	Blank	—	8

[a]Determined graphically.
Source: Ref. 27.

dants, there exist a wide range of natural antioxidants some of which are currently used effectively for the protection of foods against oxidative degradation. It should be very well recognized that a natural chain-breaking antioxidant has essentially the same mechanism of action as a synthetic chain-breaking antioxidant. Both of them act as radical scavengers and donate hydrogen atoms to reduce by a one-electron transfer reaction the primary radicals formed during the autoxidation reactions. Table 2.13 lists some potential sources of natural antioxidants grouped by their origin.

Numerous chemicals that are found in animal or plant tissues and that are also available as synthetic molecules are used in several food applications. Among these chemicals, vitamin E, vitamin C, β-carotene, and uric acid are very interesting synthetic products of natural origin capable of participating in the in vivo radical defense mechanism (51,269,275–281). The vitamin E–vitamin C mixes became the basis for several approaches to the stabilization of oils and other foods because of the radical exchange reactions (Fig. 2.6) between lipid radicals, vitamin E, and vitamin C. It is important to note that, like vitamin E, vitamin C, is not soluble in the lipid phase, which is most susceptible to oxidation. Hence esters of L-ascorbic acid (i.e., ascorbyl palmitate), which are lipid-soluble, were developed (243,282, 283). A mixture consisting of lecithin, vitamin E, and vitamin C is able to stabilize marine oil to such an extent that induction times as indicated by oxygen absorption at 80°C are extended about 25-fold for 500 ppm vitamin E and 1000 ppm vitamin C (284).

Besides biologically active compounds, pure chemical substances of plant origin treated as part of the daily food intake can be added as stabilizing agents against oxidative degradation. Amino acids, which are responsible for producing Maillard reaction products, belong to this group. However, it is important that the Maillard browning reactions be carried out under controlled conditions of heat treatment in order to avoid undue loss of the protein quality of the food. Nordihydroguaiaretic acid with excellent antioxidant properties is available from either synthesis or plant sources, but shows no practical applications in commercially available foods yet. Vanillin, the most widely used food-grade flavoring agent, has been recently recognized for its antioxidant potential (39,285,286). Its antioxidant properties can be justified on the basis of comparison of its chemical structure with those of known phenolic antioxidants.

Plant extracts contain a variety of natural products, but certain active principles are responsible for their antioxidant activity. Oats are an easily available source of antioxidants and appear to be promising in low concentrations for effective retardation of the rate of oxidation. Tea antioxidants could prove to be useful as natural antioxidants if large-scale applications become successful.

Herbs and spices occupy a special position in foods as traditional food ingredients and hence are appropriately used directly for their antioxidant characteristics. Among herbs, rosemary (*Rosmarinus officinalis* L.) is the only herb that is

Table 2.13 Potential Sources of Natural Antioxidants

Plant extracts

Cocoa shells	Leaf lipids	Oats
Tea	Olive	Garlic
Red onion skin	Wheat gliadin	Apple cuticle
Korum rind	Licorice	Nutmeg
Mustard leaf seed	Chia seed	Peanut seed coat
Rice hull	Birch bark	Carob pod
Myristica fragrans	*Silbum marianum*	
	Seed oil	

Spices and herbs

Rosemary	Pepper	Clove
Oregano	Spices, general	

Chemicals in animal or plant tissues

Vitamin E	Vitamin C	β-Carotene
Uric acid	Phenolic acids	L-Ascorbic acid ester
Phytic acid	Amino acids	Sesame seed oil
Flavones	Saponin	Soya saponins
Vanillin	NDGA (nordihydroguaiaretic acid)	*Osbeckia chinensis* tannins

Fermentation products

Penicillium commune	Tempeh oil	Microorganisms
Penicillium herquei		

Food components or products of heat-processed foods

Peptide & sugar	Phospholipids	Amino acids
Rosmarinic acid and phospholipids	Ascorbyl palmitate, lecithin, tocopherols	Maillard reaction products
Soy protein hydrolysates		

Source: Compiled from Ref. 27.

Table 2.14   Antioxidant Activity of Rosmaridiphenol in Steamed Prime Lard[a]

	Peroxide value after x days			
	7 days	14 days	21 days	28 days
Rosmaridiphenol	1.6	2.3	3.1	4.1
BHT	1.3	1.9	2.7	3.4
BHA	2.7	6.5	12.1	17.0
Control	4.7	10.0	30.0	12.0

[a]Peroxide value in lard at 60°C.
Source: Ref. 27.

exploited commercially as an antioxidant. Rosemary antioxidants and vanillin, unlike vitamin E and vitamin C, are not involved in any way in the antioxidant defense mechanism. Since 1952, several scientific reports have appeared at regular intervals that recognize the antioxidant properties of rosemary, thus attracting a lot of attention for various reasons (287–297). Both crude and refined extracts of *R. officinalis* L. have exhibited excellent antioxidant properties. The crude extract, however, imparts an objectionable color, odor, and taste to the food system under stabilization. Hence, it is desirable to purify the crude extract by a suitable chromatographic technique or molecular distillation. It is observed that approximately 90% of the antioxidant activity of rosemary can be attributed to carnosol, a C_{20} isoprenoid with a phenolic structure. The other effective antioxidant components include carnosic acid, rosmanol, rosmaridiphenol, and rosmariquinone. The antioxidant activities of the rosemary constituents can be evaluated by comparing their efficiencies with those of BHA and BHT (Tables 2.14 and 2.15).

Table 2.15   Optical Density of Emulsions Containing 200 ppm of the Indicated Rosemary Constituent for 100 ppm BHT

Antioxidant	Optical density
Carnosol	0.08
Rosmanol	0.07
Rosmaridiphenol	0.14
BHT	0.09
Control	0.93

Source: Ref. 27.

REFERENCES

1. T. P. Labuza, *CRC Crit. Rev. Food Technol.*, 2:355 (1971).
2. W. Grosch, *Lebensm. Ger. Chem.*, 38:81 (1984).
3. E. N. Frankel, *J. Amer. Oil Chem. Soc.*, 61:1908 (1984).
4. E. N. Frankel, in *Flavor Chemistry of Fats and Oils* (D. B. Min and T. H. Smouse, Eds.), American Oil Chemists' Society, Champlain, IL, 1985, p. 1.
5. J. Pokorny, in *Autoxidation of Unsaturated Lipids* (H. W. S. Chan, Ed.), Academic Press, London, 1987, p. 141.
6. C. E. Eriksson, in *Autoxidation of Unsaturated Lipids* (H. W. S. Chan, Ed.), Academic Press, London, 1987, p. 207.
7. H. Kappus, in *Free Radicals and Food Additives* (O. I. Aruoma and B. Halliwell, Eds.), Taylor and Francis, London, 1991, p. 59.
8. M. H. Gordon, in *Food Antioxidants* (B. J. F. Hudson, Ed.), Elsevier, New York, 1990, p. 1.
9. H. Kaur and M. J. Perkins, in *Free Radicals and Food Additives* (O. I. Aruoma and B. Halliwell, Eds.), Taylor and Francis, London, 1991, p. 17.
10. E. H. Farmer, H. P. Koch, and D. A. Sutton, *J. Chem. Soc.*, 1943:541 (1943).
11. D. H. Sutton, *J. Chem. Soc.*, 1944:242 (1944).
12. K. U. Ingold, in *Free Radicals*, Vol. 1 (J. K. Kochi, Ed.), Interscience, New York, 1973, p. 37.
13. H. W. S. Chan, F. A. A. Prescott, and P. A. T. Swoboda, *J. Amer. Oil Chem. Soc.*, 53:572 (1976).
14. H. Narita, in *Encyclopaedia of Food Science, Food Technology, and Nutrition* (R. Macrae, R. K. Robinson, and M. J. Sadler, Eds.), Academic Press, London, 1993, p. 226.
15. M. W. Laeubli, P. A. Bruttel, and E. Schalch, *Fette Wiss. Technol.*, 90:56 (1988).
16. R. O. Sinnhuber and T. C. Yu, *Food Res.*, 23:626 (1958).
17. J. D. Manwaring and A. S. Csallany, *Lipids*, 23:651 (1988).
18. IUPAC, *Standard Methods of Analysis of Oils and Fats and Their Derivation*, Pergamon Press, Oxford, 1979.
19. A. S. Henick, M. F. Benca, and J. H. Mitchell, *J. Amer. Oil Chem. Soc.*, 31: 488 (1954).
20. BSM 684, British Standard Method 684, Milton Keynes, British Standard Institution, 1978.
21. L. J. Parr and P. A. T. Swoboda, *J. Food Tech.*, 11:1 (1976).
22. E. H. Farmer and D. A. Sutton, *J. Chem. Soc.*, 1943:122 (1943).
23. J. Löliger, Food safety: new methods for research and control, *Proc. Int. Symp.*, *Milan*, 1988, pp. 22–23.
24. J. M. Snyder, E. N. Frankel, and E. Selke, *J. Amer. Oil Chem. Soc.*, 62:1675 (1985).
25. A. J. St. Angelo, H. P. Dupuy, and G. J. F. Flick, *J. Food Quality*, 10:393 (1988).
26. G. Hall and H. Lingnert, in *Developments in Food Science*, Vol. 12, *The Shelf Life of Foods and Beverages* (G. Charalambous, Ed.), Elsevier, Amsterdam, 1986, p. 735.
27. J. Löliger, in *Free Radicals and Food Additives* (O. I. Aruoma and B. Halliwell, Eds.), Taylor and Francis, London, 1991, p. 121.
28. D. R. Janero and B. Burghardt, *Lipids*, 23:452 (1988).
29. K. Kikugawa, T. Kato, and A. Iwata, *Anal. Biochem.*, 174:512 (1988).
30. J. A. Knight, R. K. Pieper, and L. McClellan, *Clin. Chem.*, 34:2433 (1988).

31. H. Kosugi, T. Kato, and K. Kikigawa, *Lipids*, *23*:1024 (1988).
32. F. P. Corongiu and A. Milia, *Chem. Biol. Interact.*, *44*:289 (1983).
33. S. Thompson and M. T. Smith, *Chem. Biol. Interact.*, *55*: 357 (1985).
34. K. Fukuzawa, K. Kishikawa, A. Tokumura, H. Tsukatani, and M. Shibuya, *Lipids*, *20*:854 (1985).
35. H. Shimasaki, N. Hirai, and N. Veta, *J. Biochem.*, *104*:761 (1988).
36. J. Kostrucha and H. Kappus, *Biochim. Biophys. Acta*, *879*:120 (1986).
37. R. Reiter and R. F. Burk, *Biochem. Pharmacol.*, *36*:925 (1987).
38. H. Kappus and J. Kostrucha, in *Eicosanoids, Lipid Peroxidation and Cancer* (S. K. Nigam and D. C. H. McBrien, Eds.), Springer-Verlag, Berlin, 1988, p. 227.
39. J. Burri, M Graf, P. Lambelet, and J. Löliger, *J. Sci. Food Agric.*, *110*:153 (1989).
40. M. J. Perkins, *Adv. Phys. Org. Chem.*, *17*:1 (1980).
41. G. R. Buettner, *Free Radical Biol. Med.*, *3*:259 (1987).
42. H. Kappus, in *Oxidative Stress* (H. Sies, Ed.), Academic Press, London, 1985, p. 273.
43. L. A. Morehouse and S. D. Aust, *Free Radical Biol. Med.*, *4*:269 (1988).
44. H. W. Davis, T. Suzuki, and J. B. Schenkman, *Arch. Biochem. Biophys.*, *252*:218 (1987).
45. J. G. Goddard and G. D. Sweeney, *Arch. Biochem. Biophys.*, *259*:372 (1987).
46. G. Minotti and S. D. Aust, *Chem. Phys. Lipids*, *44*:191 (1987).
47. J. Klimek, *Biochim. Biophys. Acta*, *958*:31 (1988).
48. G. F. Vile and C. C. Winterbourn, *FEBS Lett.*, *215*:151 (1987).
49. V. M. Samokyszyn, C. E. Thomas, D. W. Reif, and S. D. Aust, *Drug Metab. Rev.*, *19*:283 (1989).
50. A. Bindoli, *Free Radical Biol. Med.*, *5*:247 (1988).
51. B. Halliwell and J. M. C. Gutteridge, *Free Radicals in Biology and Medicine*, 2nd ed., Clarendon Press, Oxford, 1989.
52. O. I. Aruoma, B. Halliwell, and M. Dizdaroglu, *J. Biol. Chem.*, *264*:13024 (1989).
53. L. H. Breimer, *Brit. J. Cancer*, *57*:6 (1988).
54. P. A. Cerutti, *Science*, *227*:375 (1985).
55. J. M. C. Gutteridge, D. A. Rowley, and B. Halliwell, *Biochem. J.*, *199*:263 (1981).
56. J. M. C. Gutteridge and B. Halliwell, *Life Chem. Rep.*, *4*:113 (1987).
57. J. K. Reddy, J. R. Warren, M. K. Reddy, and N. D. Lalwani, *Ann. N.Y. Acad. Sci.*, *386*:81 (1982).
58. H. Horie, H. Ishii, and T. Suga, *J. Biochem.(Tokyo)*, *90*:1691 (1981).
59. D. P. Jones, L. Eklow, H. Thor, and S. Orrenius, *Arch. Biochem. Biophys.*, *210*:505 (1981).
60. B. Chance, H. Sies, and A. Boveris, *Physiol. Rev.*, *59*:527 (1979).
61. W. A. Pryor, *Free Radicals in Biology*, Vols. 1–5, Academic Press, New York, 1976–1982.
62. D. Harman, *Proc. Natl. Acad. Sci. USA*, *78*:7128 (1981).
63. D. Harman, in *Free Radicals in Biology*, Vol. 5 (W. A. Pryor, Ed.), Academic Press, New York, 1982, p. 255.
64. E. Emerit, M. Keck, A. Levy, J. Feinggold, and A. M. Michelson, *Mutat. Res.*, *103*:105 (1982).
65. C. E. Neat, M. S. Thomassen, and H. Osmundsen, *Biochem. J.*, *196*:149 (1981).
66. J. Bremer and K. R. Norum, *J. Lipid Res.*, *23*:243 (1982).

67. M. S. Thomassen, E. N. Christiansen, and K. R. Norum, *Biochem. J.*, *206*:195 (1982).
68. H. Osmundsen, *Int. J. Biochem.*, *14*:905 (1982).
69. J. Reddy, D. Azarnoff, and C. Hignite, *Nature* (*Lond.*), *283*:397 (1980).
70. J. Ward, J. Rice, D. Creasia, P. Lynch, and C. Riggs, *Carcinogenesis*, *4*:1021 (1983).
71. B. Goldstein, M. Rozen, J. Quintawala, and M. Amoruso, *Biochem. Pharmacol.*, *28*:27 (1979).
72. A. Rannug and U. Rannug, *Chem. Biol. Interact.*, *49*:329 (1984).
73. V. Solanki, R. Rana, and T. Sluga, *Carcinogenesis*, *2*:1141 (1982).
74. B. N. Ames, *Science*, *221*:1256 (1983).
75. C. K. Winter, H. J. Segall, and W. F. Haddon, *Cancer Res.*, *46*:5682 (1986).
76. H. Esterbauer, H. Zollner, and R. J. Schaur, *Biochemistry*, *1*:311 (1988).
77. F. J. G. M. Van Kuijk, A. Sevanian, G. J. Handelman, and E. A. Dratz, *Trends Biochem. Sci.*, *12*:31 (1987).
78. B. Halliwell, and J. M. C. Gutteridge, *Mol. Aspect Med.*, *8*:89 (1985).
79. B. Halliwell, in *Free Radicals and Food Additives* (O. I. Aruoma and B. Halliwell, Eds.), Taylor and Francis, London, 1991, p. 37.
80. G. Jürgens, J. Lang, and H. Esterbauer, *Biochim. Biophys. Acta*, *875*:103 (1986).
81. H. Esterbauer, G. Jürgens, O. Quehenberger, and E. Koller, *J. Lipid Res.* *28*:495 (1987).
82. H. Esterbauer, O. Quehenberger, and G. Jürgens, in *Eicosanoid, Lipid Peroxidation and Cancer* (S. K. Nigam, D. C. H. McBrien, and T. Stater, Eds.), Springer-Verlag, Berlin, 1988, p. 203.
83. B. Henning and C. K. Chow, *Free Radical Biol. Med.*, *4*:99. (1988).
84. S. Yoshida, R. Busto, B. D. Waston, M. Santiso, and M. D. Ginsberg, *J. Neurochem.*, *44*:1593 (1985).
85. T. Nishida, H. Shibata, M. Koseki, K. Nakao, Y. Kawashima, Y. Yoshida, and K. Tagawa, *Biochim. Biophys. Acta*, *890*:82 (1987).
86. S. K. Jain, *Biochim. Biophys. Acta*, *937*:205 (1988).
87. S. Akasaka, *Biochim. Biophys. Acta*, *867*:201 (1986).
88. G. Brambilla, A. Martelli, E. Cajelli, R. Canonero, and U. M. Marinari, in *Eicosanoids, Lipid Peroxidation and Cancer* (S. K. Nigam, D. C. H. McBrien, and T. F. Slater, Eds.), Springer-Verlag, Berlin, 1988, p. 243.
89. C. G. Fraga and A. L. Tappel, *Biochem. J.*, *252*:893 (1988).
90. C. E. Vaca, J. Wilhelm, and M. Harma-Ringdahl, *Mutat. Res.*, *195*:137 (1988).
91. P. Hochstein and S. K. Jain, *Fed. Proc. Fed. Amer. Soc. Exp. Biol.*, *40*:183 (1981).
92. J. R. Totter, *Proc. Natl. Acad. Sci. USA*, *77*:1763 (1980).
93. A. L. Tappel, in *Free Radicals in Biology*, Vol. 4 (W. A. Pryor, Ed.), Academic Press, New York, 1980, p. 1.
94. R. S. Sohal, *Age Pigments*, Elsevier/North-Holland, Amsterdam, 1981, p. 303.
95. J. E. Fleming, J. Miquel, S. F. Cottrell, L. S. Yengoyan, and A. C. Economos, *Gerontology*, *28*:44 (1982).
96. J. M. Tolmasoff, T. Ono, and R. G. Cutler, *Proc. Natl. Acad. Sci. USA*, *77*:2777 (1980).
97. J. L. Sullivan, *Gerontology*, *28*:242 (1982).
98. H. L. Gensler and H. Bernstein, *Quart. Rev. Biol.*, *6*:279 (1981).
99. H. B. Demopoulos, D. D. Pietronigro, E. S. Flamm, and M. L. Seligman, *J. Environ. Pathol. Toxicol.*, *3*:273 (1980).

100. J. L. Marx, *Science*, *219*:158 (1983).
101. B. D. Goldstein, G. Witz, M. Amoruso, D. S. Stone, and W. Troll, *Cancer Lett.*, *11*:257 (1981).
102. W. Troll, in *Environmental Mutagens and Carcinogens* (T. Sugimura, S. Kondo, and H. Takebe, Eds.), Univ. Tokyo Press, Tokyo and Liss, New York, 1982, p. 217.
103. B. N. Ames, M. C. Hollstein, and R. Cathcart, in *Lipid Peroxide in Biology and Medicine* (K. Yagi, Ed.), Academic Press, New York, 1982, p. 339.
104. I. Emerit and P. A. Cerutti, *Proc. Natl. Acad. Sci. USA*, *79*:7509 (1982).
105. I. Emerit and P. A. Cerutti, *Nature (Lond.)*, *293*:144 (1981).
106. I. Emerit, A. Levy, and P. Cerutti, *Mutat. Res.*, *110*:327 (1983).
107. J. Funes and M. Karel, *Lipids*, *16*:347 (1981).
108. B. N. Stuckey, in *CRC Handbook of Food Additives*, Vol. 1 (T. E. Furia, Ed.), CRC Press, Boca Raton, FL, 1972, p. 185.
109. K. L. Parkin and S. Damodaran, in *Encyclopaedia of Food Science, Food Technology, and Nutrition* (R. Macrae, R. K. Robinson, and M. J. Sadler, Eds.), Academic Press, New York, 1993, p. 3375.
110. R. J. Pitcher, *Encyclopaedia of Food Science, Food Technology, and Nutrition* (R. Macrae, R. K. Robinson, and M. J. Sadler, Eds.), Academic Press, New York, 1993, p. 649.
111. W. F. A. Horner, in *Encyclopaedia of Food Science, Food Technology, and Nutrition* (R. Macrae, R. K. Robinson, and M. J. Sadler, Eds.), Academic Press, New York, 1993, p. 1485.
112. J. A. Troller, in *Encyclopaedia of Food Science, Food Technology, and Nutrition* (R. Macrae, R. K. Robinson, and M. J. Sadler, Eds.), Academic Press, New York, 1993, p. 4846.
113. O. S. Privett and M. L. Blank, *J. Amer. Oil Chem. Soc.*, *39*:465 (1962).
114. M. B. Korycka-Dahl and T. Richardson, *CRC Crit. Rev. Food Sci. Nutr.*, *10*:209 (1978).
115. H. R. Rawls and P. J. van Santen, *J. Amer. Oil Chem. Soc.*, *47*:121 (1970).
116. H. R. Rawls and P. J. van Santen, *Ann. N.Y. Acad Sci.*, *171*:135 (1971).
117. C. S. Foote, *Pure Appl. Chem.*, 27:635 (1971).
118. J. L. Bolland and P. TenHave, *Trans. Faraday Soc.*, *43*:201 (1947).
119. G. A. Russel, *J. Amer. Chem. Soc.*, *78*:1041 (1956).
120. J. R. Shelton and D. Vincent, *J. Amer. Chem. Soc.*, *85*:2433 (1963).
121. M. H. Chahine and J. M. deMan, *Can. Inst. Food Tech. J.*, *4*:24 (1971).
122. N. A. M. Eskin, S. Grossman, and A. Pinsky, *CRC Crit. Rev. Food Sci. Nutr.*, *9*:1 (1977).
123. H. A. O. Hill, *Phil. Trans. Roy. Soc. Lond.*, Ser. B, *294*:119 (1981).
124. K. U. Ingold, in *Lipids and Their Oxidation* (H. W. Schultz, Ed.), Avi, Westport, CT, 1962, p. 93.
125. F. W. Heaton and N. Uri, *J. Lipid Res.*, *2*:152 (1961).
126. S. W. Benson, *J. Chem. Phys.*, *40*:1007 (1964).
127. R. Hiatt and K. C. Irwin, *J. Org. Chem.*, *33*:1436 (1968).
128. L. Bateman and H. Hughes, *J. Chem. Soc.*, *1952*:4594 (1952).
129. C. Walling and L. Heaton, *J. Amer. Chem. Soc.*, *87*:48 (1965).
130. R. Hiatt, K. C. Irwin, and C. W. Gould, *J. Org. Chem.*, *33*:1430 (1968).

131. R. Hiatt, T. Mill, and F. R. Mayo, *J. Org. Chem.*, *33*:1416 (1968).
132. C. E. H. Bawn, *Disc. Faraday Soc.*, *14*:181 (1953).
133. B. Halliwell and J. M. C. Gutteridge, *Biochem. J.*, *219*:1 (1984).
134. B. Halliwell and J. M. C. Gutteridge, *Lancet*, *1*:1396 (1984).
135. W. A. Waters, *J. Amer. Oil Chem. Soc.*, *48*:427 (1971).
136. E. Ochiai, *Tetrahedron*, *20*:1819, (1964).
137. O. I. Aruoma, B. Halliwell, M. J. Laughton, J. G. Quinlan, and J. M. C. Gutteridge, *Biochem. J.*, *258*:617 (1989).
138. L. D. Possami, R. Banerjee, C. Balny, and P. Douzou, *Nature*, *226*:861 (1970).
139. A. L. Tappel, *Arch. Biochem. Biophys.*, *44*:378 (1953).
140. A. L. Tappel, *J. Biol. Chem.*, *217*:721 (1955).
141. A. L. Tappel, *J. Amer. Oil Chem. Soc.*, *32*:252 (1955).
142. W. D. Brown, L. Harris, and H. Olcott, *Arch. Biochem. Biophys.*, *101*:14 (1963).
143. S. E. Lewis and E. D. Willis, *Biochim. Biophys. Acta*, *70*:336 (1963).
144. A. J. Chalk and J. F. Smith, *Trans. Faraday Soc.*, *53*:1235 (1957).
145. A. T. Betts and N. Uri, *Makro Chem.*, *95*:22 (1966).
146. B. M. Watts, *Adv. Food Res.*, *5*:1 (1954).
147. B. M. Watts, Campbell Flavor Chem. Symp., Campbell Soup Co., USA, 1965.
148. B. E. Green, *J. Food Sci.*, *34*:110 (1969).
149. B. Watts, in *Lipids and Their Oxidation* (H. W. Scultz, Ed.), AVI, Westport, CT, 1962, p. 202.
150. A. L. Tappel, *Food Res.*, *21*:195 (1956).
151. A. L. Tappel, in *Lipids and their Oxidation* (H. W. Schultz, Ed.), AVI, Westport, CT, 1962, p. 122.
152. R. Adelman, R. L. Saul, and B. N. Ames, *Proc. Natl. Acad. Sci. USA*, *85*:2706 (1988).
153. H. Kasai, P. F. Crain, Y. Kuchino, S. Nishimura, A. Ootsuyama, and H. Tannoka, *Carcinogenesis*, *7*:1849 (1946).
154. M. Dizdaroglu, *BioTechniques*, *4*:536 (1986).
155. M. Dizdaroglu, in *Handbook of Free Radicals and Antioxidants in Biomedicine* (J. Miguel, A. T. Quintanilha, and H. Weber, Eds.), CRC Press, Boca Raton, FL, 1989, p. 153.
156. A. F. Fuciarelli, B. J. Wegher, E. Gajewski, M. Dizdaroglu, and W. F. Blakley, *Rad. Res.*, *119*: 219 (1989).
157. J. D. Mane, S. P. Phadnis, and S. J. Jadhav, *Int. Sugar J.*, *94*(1128):322 (1992).
158. H. D. Belitz, in *Encyclopaedia of Food Science, Food Technology, and Nutrition* (R. Macrae, R. K. Robinson, and M. J. Sadler, Eds.), Academic Press, New York, 1993, p. 3815.
159. W. Grosch, in *Autoxidation of Unsaturated Lipids* (H. W. S. Chan, Ed.), Academic Press, London, 1987, p. 95.
160. R. Buttery, G. C. Hendel, and M. Boggs, *J. Agric. Food Chem.*, *9*:245 (1961).
161. M. Karel and T. P. Labuza, *Mechanisms of Deterioration and Formulation of Space Diets*, AF Contract 49:609, 1967.
162. J. E. Kinsella, *Soc. Chem. Ind.*, *11*:36 (Jan. 1969).
163. J. Podmore, in *Encyclopaedia of Food Science, Food Technology, and Nutrition* (R. Macrae, R. K. Robinson, and M. J. Sadler, Eds.), Academic Press, New York, 1993, p. 4699.

164. P. B. Addis and G. J. Warner, in *Free Radicals and Food Additives* (O. I. Aruoma and B. Halliwell, Eds.), Taylor and Francis, London, England, 1991, p. 77.
165. M. G. Simic and M. Karel, *Autoxidation in Food and Biological Systems*, Plenum Press, New York, 1980.
166. L. L. Smith, *Cholesterol Autoxidation*, Plenum Press, New York, 1981.
167. G. A. Dhopeshwarkar, *Prog. Lipid Res.*, *19*:197 (1981).
168. R. Ross, *Atherosclerosis*, *1*:293 (1981).
169. P. B. Addis, A. S. Csallany, and S. E. Kindom, in *Xenobiotics in Foods and Feeds* (J. W. Finley and D. E. Schwass, Eds.), ACS Symp. Ser, 234, Am. Chem. Soc., Washington, DC, 1983, pp. 85–98.
170. J. C. Alexander, in *Xenobiotics in Foods and Feeds* (J. W. Finley and D. E. Schwass, Eds.), ACS Symp. Ser, 234, Am. Chem. Soc., Washington, DC, 1983, pp. 129–148.
171. E. T. Finocchiaro and T. Richardson, *J. Food Prot. 46*:917 (1983).
172. P. B. Addis, *Food Chem. Toxicol.*, 24:1021 (1986).
173. R. Ross, *N. Engl. J. Med.*, *314*:488 (1986).
174. L. L. Smith, *Chem. Phys. Lipids*, *44*:87 (1987).
175. R. J. Morin, T. Zemplenyi, and S. K. Peng, *Pharmacol. Ther.*, *32*:237 (1987).
176. K. Yagi, *Chem. Phys. Lipids*, *43*:337 (1987).
177. K. Yagi, in *The Role of Oxygen in Chemistry and Biochemistry* (W. Ando and Y. Moro-oka, Eds.), Elsevier, Amsterdam, 1988, p. 383.
178. P. B. Addis and S. W. Park, in *Food Toxicology: A Perspective on the Relative Risks* (S. L. Taylor and R. A. Scanlan, Eds.), Marcel Dekker, New York, 1989, p. 297.
179. L. L. Smith and B. H. Johnson, *Free Radical Biol. Med.*, 7:285 (1989).
180. D. Steinberg, S. Parthasarathy, T. E. Carew, J. C. Khoo, and J. L. Witztum, *N. Engl. J. Med.*, *320*:915 (1989).
181. J. Glavind, F. Christensen, and C. Sylven, *Acta Chem. Scand.*, *25*:3220 (1971).
182. M. Naruszewicz, E. Wonzy, E. Mirkiewicz, G. Nowioka, and W. B. Szostak, *Atherosclerosis*, *66*:45 (1987).
183. K. Kanazawa, E. Kanazawa, and M. Natake, *Lipids*, *20*:412 (1985).
184. K. Kanazawa, H. Ashida, S. Minamoto, G. Danno, and M. Natake, *J. Nutr. Sci. Vit.*, *34*:363 (1988).
185. J. M. Peters and E. M. Boyd, *Food Cosmet. Toxicol.*, 7:197 (1969).
186. S. Matsuhita, *J. Agric. Food Chem.*, 23:150 (1975).
187. K. Kanazawa and H. Ashida, *Arch. Biochem. Biophys.*, *288*:71 (1991).
188. K. Yagi, H. Ohkawa, N. Ohishi, M. Yamashita, and T. Nakashima, *J. Appl. Biochem*, *3*:58 (1981).
189. N. Yagi, T. Inagaki, Y. Sasaguri, R. Nakano, and T. Narasimha, *J. Clin. Biochem. Nutr.*, *3*:87 (1987).
190. I. Nishigaki, M. Hagihara, M. Maseki, Y. Tomoda, K. Nagayama, T. Nakashima, and K. Yagi, *Biochem. Int.*, *8*:501 (1984).
191. Y. Sasaguri, T. Nakashima, M. Morimatsu, and K. Yagi, *J. Appl. Biochem.*, *6*:144 (1984).
192. Y. Sasaguri, M. Morimatsu, T. Nakashima, O. Tokunaga, and K. Yagi, *Biochem. Int.*, *11*:517 (1985).
193. K. Fujimoto, W. E. Neff, and E. N. Frankel, *Biochem. Biophys. Acta*, *795*:100 (1984).
194. M. G. Cutler and R. Schneider, *Food Chem. Toxicol.*, *11*:443 (1973).

195. B. L. Van Duuren, N. Nelson, L. Orris, E. D. Palmes, and F. L. Schmitt, *J. Natl. Cancer Inst.*, *31*:41 (1963).
196. B. L. Van Duuren L. Langseth, L. Orris, M. Baden, and M. Kuschner, *J. Natl. Cancer Inst.*, *39*:1213 (1967).
197. M. G. Cutler and R. Schneider, *Food Chem. Toxicol.*, *11*:935 (1973).
198. L. L. Smith, W. S. Mathews, V. C. Price, R. C. Bachmann, and B. Reynolds, *J. Chromatogr, 27*:187 (1967).
199. J. Bascoul, N. Domerue, M. Olle, and A. Crastes De Paulet, *Lipids*, *21*:383 (1986).
200. S. K. Peng, R. J. Morin, P. Tham, and C. B. Taylor, *Artery*, *13*:144 (1985).
201. S. K. Peng and R. J. Morin, *Artery, 14*:85 (1987).
202. C. B. Taylor, S. K. Peng, N. T. Werthessen, P. Tham, and K. T. Lee, *Amer. J. Clin. Nutr.*, *32*:40 (1979).
203. S. K. Peng, R. J. Morin, and S. Sentovich, *J. Amer. Oil Chem. Soc.*, *62*:634 (1985) (Abstr.).
204. L. Neyes, M. Stimpel, R. Locher, W. Vetter, and R. Streuli, *J. Amer. Oil Chem. Soc.*, *62*:634 (1985) (Abstr.).
205. R. P. Holmes and N. L. Yoss, *Biochim. Biophys. Acta, 770*:285 (1984).
206. J. J. H. Theunissen, R. L. Jackson, H. J. M. Kempen, and R. A. Daniel, *Biochim. Biophys. Acta, 860*:66 (1986).
207. H. Imai, N. J. Werthessen, V. Subramanyam, P. W. Lequesne, A. H. Soloway, and M. Kanasawa, *Science, 207*:651 (1980).
208. S. K. Peng, C. B. Taylor, P. Tham, N. J. Werthessen, and B. Mikkelson, *Arch. Pathol. Lab. Med.*, *102*:57 (1978).
209. A. Sevanian and A. R. Peterson, *Food Chem. Toxicol.*, *24*: 1103 (1986).
210. G. M. Blackburn, A. Rashid, and M. H. Thompson, *J. Chem. Soc. Chem. Commun.*, *9*:421 (1979).
211. K. Suzuki, W. R. Bruce, J. Baptisa, R. Furrer, D. J. Vaughan, and J. J. Krepinsky, *Cancer Lett.*, *33*:307 (1986).
212. S. K. Peng, C. B. Taylor, J. C. Hill, and R. J. Morin, *Atherosclerosis, 54*:121 (1985).
213. G. Henning and G. A. Boissonneault, *Atherosclerosis, 68*:255 (1987).
214. D. Matthias, C. H. Becker, W. Gödicke, R. Schmidt, and K. Ponsold, *Atherosclerosis, 63*:115 (1987).
215. V. A. Kosyth, V. Z. Lankin, E. A. Podrez, D. K. Novikov, S. A. Volgushev, A. V. Victorov, V. S. Repin, and V. N. Smirnov, *Lipids, 24*:109 (1989).
216. R. J. Morin and S. K. Peng, *Lipids, 24*:217 (1989).
217. A. M. Pearson, J. I. Gray, A. M. Wolzak, and N. A. Horensten, *Food Technol.*, *37*:121 (1983).
218. A. K. Basu and L. J. Marnett, *Carcinogenesis, 4*:331 (1983).
219. R. J. Shamberger, T. L. Andrione, and C. E. Willis, *J. Natl. Cancer Inst.*, *53*:1771 (1974).
220. F. H. Mukai and B. D. Goldstein, *Science, 191*:868 (1976).
221. T. W. Kwon and W. D. Brown, *Fed. Proc. Fed. Amer. Soc. Exp. Biol.*, *24*:592 (1965) (Abstr.).
222. D. B. Menzel, *Lipids, 2*:83 (1967).
223. C. E. Vaca, J. Wilhelm, and M. Harms-Ringdahl, *Mutat. Res.*, *195*:137 (1988).

224. A. K. Basu, S. M. O'Hara, P. Valladier, K. Stone, O. Mols, and L. J. Marnett, *Chem. Res. Toxicol.*, *1*:53 (1988).
225. L. A. Piche, H. H. Draper, and P. D. Cole, *Lipids*, *23*:370 (1988).
226. R. P. Bird and H. H. Draper, *J. Toxicol. Environ. Health*, *6*:811 (1980).
227. R. P. Bird and H. H. Draper, and P. K. Basrur, *Mutat. Res.*, *101*:237 (1982).
228. A. S. Csallany, M. D. Guan, J. D. Manwaring, and P. B. Addis, *Anal. Biochem.*, *142*:277 (1984).
229. G. W. Burton and K. U. Ingold, *Acc. Chem. Res.*, *19*:194 (1986),.
230. G. Scott, *Atmospheric Oxidation and Antioxidants*, Elsevier, New York, 1985.
231. A. F. Mabrouk and L. R. Dugan, *J. Amer. Oil Chem. Soc.*, *38*:692 (1961).
232. Y. Endo, R. Usuki, and T. Kareda, *J. Amer. Oil Chem. Soc.*, *62*:1387 (1985).
233. C. S. Foote and R. W. Denny, *J. Amer. Oil Chem. Soc.*, *90*:6233 (1968).
234. C. S. Foote, in *Free Radicals in Biology*, Vol. 2 (W. A. Pryor, Ed.), Academic Press, New York, 1976, p. 85.
235. J. Terao, *Lipids*, *24*:659 (1989).
236. G. W. Burton and K. U. Ingold, *Science*, *224*:569 (1984).
237. A. H. Clements, R. H. Van den Engh, D. T. Frost, and K. Hoogenhout, *J. Amer. Oil Chem. Soc.*, *50*:325 (1973).
238. E. W. Kellogg III and I. Fridovich, *J. Biol. Chem.*, *250*:8812 (1975).
239. K. Täufel and F. Linow, *Fette Seifen Anstrichm.*, *65*:795 (1963).
240. B. M. Watts, *J. Amer. Oil Chem. Soc.*, *27*:48 (1950).
241. F. J. Yourga, W. B. Esselen, and C. R. Fellers, *Food Res.*, *9*:188 (1944).
242. W. M. Cort, *J. Amer. Oil Chem. Soc.*, *51*:321 (1974).
243. H. Kläui and G. Pongracz, in *Vitamin C, Ascorbic Acid* (J. N. Counsell and D. H. Horning, Eds.), Applied Science, London, 1981, p. 139.
244. I. Fridovich, *Adv. Enzymol.*, *41*:35 (1974).
245. M. B. Korycka-Dahl and T. Richardson, *J. Dairy Sci.*, *63*:1181 (1980).
246. S. P. Rock, L. Fisher, and H. Roth, *J. Amer. Oil Chem. Soc.*, *44*:102A (1967).
247. K. Warner, T. L. Mounts, and W. F. Kwolek, *J. Amer. Oil Chem. Soc.*, *62*:1483 (1985).
248. J. B. Martin, U.S. Patent 2,634,213 (1953).
249. I. P. Freeman, F. B. Padley, and W. L. Sheppard, *J. Amer. Oil Chem. Soc.*, *50*:101 (1973).
250. R. J. Sims, J. A. Fiorti, and M. J. Kanuk, *J. Amer. Oil Chem. Soc.*, *49*:298 (1972).
251. D. Boskou and I. D. Morton, *J. Sci. Food Agric.*, *27*:928 (1976).
252. M. H. Gordon and P. Magos, *Food Chem.*, *10*:141 (1983).
253. P. Brandt, E. Hollstein, and C. Franzke, *Lebensm. Ind.*, *20*:31 (1973).
254. F. Linow and G. Mieth, *Nahrung*, *20*:19 (1976).
255. B. J. F. Hudson and M. Ghavami, *Lebensm. Wiss. Technol.*, *17*:191 (1984).
256. B. J. F. Hudson and J. I. Lewis, *Food Chem.*, *10*:111 (1983).
257. Y. Ishikawa, K. Sugiyama, and K. Nakabayashi, *J. Amer. Oil Chem. Soc.*, *61*:950 (1984).
258. P. T. Tai, J. Pokorny, and G. Janicek, *Z. Lebensm. Unters. Forsch.*, *156*(5):257 (1974).
259. J. Pokorny, H. Poskocilova, and J. Davidek, *Nahrung*, *25*:K29 (1981).
260. S. Z. Dziedzic, J. L. Robinson, and B. J. F. Hudson, *J. Agric. Food Chem.*, *34*:1027 (1986).
261. K. Eichner, *Prog. Food Nutr. Sci.*, *5*:441 (1981).

262. H. Lingnert and C. E. Eriksson, *Prog. Food Nutr. Sci.*, *5*:453 (1981).
263. N. Yamaguchi, K. Koyama, and M. Fujimaki, *Prog. Food Nutr. Sci.*, *5*:429 (1981).
264. W. O. Lundberg, *Autoxidation and Antoxidants*, Vol. 1, Interscience, New York, 1961.
265. G. Kajimoto and H. Yoshida, *Yukaguku*, *24*:297 (1975).
266. T. Gomyo and M. Horikoshi, *Agric. Biol. Chem.*, *40*:33 (1976).
267. H. Esterbauer, G. Striegl, H. Puhl, and M. Rotheneder, *Free Radical Res. Commun.*, *6*:67 (1989).
268. H. Esterbauer, G. Jürgens, H. Puhl, and O. Quehenberger, in *Medical, Biochemical and Chemical Aspects of Free Radicals* (O. Hayaishi, E. Niki, M. Kondo, and T. Yoshikawa, Eds.), Elsevier, Amsterdam, 1989, p. 1203.
269. A. L. Tappel, *Geriatrics*, *23*(10):97 (1968).
270. B. Halliwell, *Free Radical Res. Commun.*, *9*:1 (1990).
271. O. Marcillat, Y. Zang, S. W. Lin, and K. J. A. Davies, *Biochem. J.*, *254*:677 (1988).
272. J. W. Finley and P. Given, *Food Chem. Toxicol.*, *24*(10/11):991 (1986).
273. R. Haigh, *Food Chem. Toxicol.*, *24*(10/11):1031 (1986).
274. H. Narita, in *Encyclopaedia of Food Science, Food Technology, and Nutrition* (R. Macrae, R. K. Robinson, and M. J. Sadler, Eds.), Academic Press, New York, 1993, p. 216.
275. L. A. Wittig, in *Free Radicals in Biology* (W. A. Pryor, Ed.), Academic Press, London, 1980, p. 295.
276. J. R. Walton and L. Packer, in *Vitamin E: Comprehensive Treatise* (L. J. Machlin, Ed.), Marcel Dekker, New York, 1980, p. 495.
277. D. L. Gilbert, *Oxygen and Living Processes: An Interdisciplinary Approach*, Springer, New York, 1981.
278. F. L. Boschke, *Radicals Biochemistry*, Springer-Verlag, Berlin, 1983.
279. R. L. Wilson, *Biology of Vitamin E*, Ciba Foundation Symposium, Pitman, London, 1983, p. 19.
280. B. N. Ames, R. Cathcart, E. Schwiers, and P. Hochstein, *Proc. Natl. Acad. Sci. USA*, *78*:6858 (1981).
281. B. N. Ames, R. Cathcart, E. Schwiers, and P. Hochstein, *Carcinogens Mutat. Environ*, *2*:69 (1983).
282. G. Pongracz, *Int. J. Vitam. Nutr. Res.*, *43*:517 (1973).
283. M. Takahashi, E. Niki, A. Kawakami, A. Kumasaka, Y. Yamamoto, A. Kamiya, and K. Tanaka, *Bull. Chem. Soc. Jpn.*, *59*:3179 (1986).
284. J. Löliger, in *Rancidiy in Foods*, 2nd ed. (J. C. Allen and R. J. Hamilton, Eds.), Applied Science, London, 1989, p. 105.
285. J. Burri, M. Graf, P. Lambelet, and J. Löliger, Eur. Patent 340500 (1988).
286. O. I. Aruoma, P. J. Evans, H. Kaur, L. Sutcliffe, and B. Halliwell, *Free Radical Res. Commun.*, *10*:143 (1990).
287. J. R. Chipault, G. R. Mizuno, J. M. Kawkins, and W. O. Lundberg, *Food Res.*, *17*:46 (1952).
288. R. Viani, Offendegungsschrift 2445345, Oct. 16, 1973.
289. J. W. Wu, M. H. Lee, C. T. Ho, and S. S. Chang, *J. Amer. Oil Chem. Soc.*, *59*:339 (1979).
290. S. S. Chang, B. Ostric-Matijaseric, O. A. L. Hsieh, and C. Huang, *J. Food Sci.*, *42*:1102 (1977).

291. S. S. Chang, C. T. Ho, and C. M. Houlihan, U.S. Patent Appl. 639,899 (1987).
292. V. E. Hartmann, P. Racine, J. Garnero, and Y. Tollard d'Audiffret, *Perfums Cosmetiq. Aromes*, *36*:33 (1980).
293. U. Bracco, J. Löliger, and J. L. Viret, *J. Amer. Oil Chem. Soc.*, *58*:686 (1981).
294. R. Inatani, N. Nakatami, H. Fuwa, and H. Sato, *Agric. Biol. Chem.*, *46*:1661 (1982).
295. R. Inatani, N. Nakatami, H. Fuwa, *Agric. Biol. Chem.*, *47*:521 (1983).
296. C. M. Houlihan, C. T. Ho, and S. S. Chang, *J. Amer. Oil Chem. Soc.*, *61*:1036 (1984).
297. C. M. Houlihan, C. T. Ho, and S. S. Chang, *J. Amer. Oil Chem. Soc.*, *62*:96 (1985).

3

Food Antioxidants: Sources and Methods of Evaluation

D. Rajalakshmi and S. Narasimhan

Central Food Technological Research Institute, Mysore, Karnataka, India

3.1 INTRODUCTION

Food products undergo a chain of change in the natural matrix due to ripening, harvesting, primary processing, and storage. These changes are caused by several factors including browning reactions, microbial spoilage, and autoxidation of lipids. Of the various factors, autoxidation of lipids contributes significantly to the deterioration and reduction in the shelf life of many products. Lipid oxidation is a free-radical chain reaction that causes a total change in the sensory properties and nutritive value of food products. Changes in color, texture, odor, and flavor; loss of vitamins; and damage to proteins are some of the effects of lipid oxidation. The onset of lipid oxidation can be delayed by the addition of antioxidants.

Antioxidants are a group of chemicals effective in extending the shelf life of a wide variety of food products. The use of antioxidants dates back to the 1940s. Natural substances like gum guaiac were used initially to preserve fats and oils. They were soon replaced by many synthetic compounds that had better antioxidant activity and were more easily available. The use of antioxidants was also extended to a wide variety of food products including high-fat foods, cereals, and even products containing very low levels of lipids. Most of the raw materials used for food manufacturing contain natural antioxidants. However, during processing or storage, the natural antioxidants are depleted, necessitating the addition of antioxidant chemicals. In recent years, naturally occurring antioxidants have come to be preferred by both consumers and food manufacturers mainly because of concern raised over the safety of synthetic antioxidants. In general, antioxidants function by reducing the rate of initiation reactions in the free-radical chain reactions and are functional at very low concentrations, 0.01% or less. Antioxidants cannot, however, revert the oxidative process or prevent hydrolytic rancidity.

The use of antioxidants in food products is governed by regulatory laws of the country or by international standards. Even though many natural and synthetic compounds have antioxidant properties, only a few of them have been accepted as "generally recognized as safe" (GRAS) substances for use in food products by international bodies such as the Joint FAO/WHO Expert Committee on Food Additives (JECFA) and the European Community's Scientific Committee for Food (SCF). Table 3.1 presents the antioxidants permitted for use in food products.

Antioxidants should satisfy several requirements before being accepted for incorporation into food products (1,2). The antioxidant should be soluble in fats; it should not impart foreign color, odor, or flavor to the fat even on long storage; it should be effective for at least 1 year at a temperature of 25–30°C; it should be stable to heat processing and protect the finished product (carry-through effect); it should be easy to incorporate; and it should be effective at low concentrations. The safety of the antioxidant must be established. According to Lehman et al. (1), an antioxidant is considered safe if it fulfills two conditions: Its LD_{50} must not be less than 1000 mg/kg body weight, and the antioxidant should not have any significant effect on the growth of the experimental animal in long-term studies at a level 100 times the level proposed for human consumption. Approval of an antioxidant for food use also requires extensive toxicological studies of its possible mutagenic, teratogenic, and carcinogenic effects.

Because of the widespread use of antioxidants and the stringent regulations regarding the level of incorporation, a number of chemical and instrumental

Table 3.1 Antioxidants Permitted in Foods

Ascorbic acid, sodium, calcium salts	Glycine
Ascorbyl palmitate and stearate	Gum guaiac
Anoxomer	Lecithin
Butylated hydroxyanisole	Ionox-100
Butylated hydroxytoluene	Polyphosphates
tert-Butyl hydroquinone[a]	Propyl, octyl, and dodecyl gallates
Citric acid, stearyl, and isopropyl esters	Tartaric acid
Erythorbic acid and sodium salt	Thiodipropionic acid, dilauryl and
Ethoxyquin	distearyl esters
Ethylenediaminetetraacetic acid and	Tocopherols
calcium disodium salt	Trihydroxy butyrophenone

[a]Not permitted for food use in European Economic Community countries.

methods have been developed to estimate antioxidant levels in food products. Various methods have been developed to determine the oxidative stability of food lipids as a means of evaluating the effectiveness of both synthetic and natural antioxidants. In recent years, sensory evaluation methods have been gaining in importance as they are directly related to consumer acceptance of the food product. This chapter presents an overview of the sources of antioxidants, their mechanism of action, and instrumental and chemical methods of evaluation and highlights the importance of sensory evaluation methods for the determination of antioxidant effectiveness in terms of oxidative stability of food lipids.

3.2 CLASSIFICATION OF FOOD ANTIOXIDANTS

Based on their function, food antioxidants are classified as primary or chain-breaking antioxidants, synergists, and secondary antioxidants (Fig. 3.1). The mechanism of antioxidant activity has been reviewed by a number of investigators (3–5).

3.2.1 Primary Antioxidants

Primary antioxidants terminate the free-radical chain reaction by donating hydrogen or electrons to free radicals and converting them to more stable products. They may also function by addition in reactions with the lipid radicals, forming lipid–antioxidant complexes. Both hindered phenolic (e.g., BHA, BHT, TBHQ, and tocopherols) and polyhydroxyphenolic (e.g., gallates) antioxidants belong to this group. Many of the naturally occurring phenolic compounds like flavonoids, eugenol, vanillin, and rosemary antioxidant also have chain-breaking properties. Primary antioxidants are effective at very low concentrations, and at higher levels they may become prooxidants.

Primary antioxidants may either delay or inhibit the initiation step by reacting with a fat free radical or inhibit the propagation step by reacting with the peroxy or alkoxy radicals.

$$AH + L^\bullet \rightarrow A^\bullet + LH \qquad (3.1)$$
$$AH + LOO^\bullet \rightarrow A^\bullet + LOOH \qquad (3.2)$$
$$AH + LO^\bullet \rightarrow A^\bullet + LOH \qquad (3.3)$$

The antioxidant free radical may further interfere with the chain-propagation reactions by forming peroxy antioxidant compounds.

$$A^\bullet + LOO^\bullet \rightarrow LOOA \qquad (3.4)$$
$$A^\bullet + LO^\bullet \rightarrow LOA \qquad (3.5)$$

Bickel et al. (6) isolated some alkyl peroxy antioxidant compounds (LOOA) with the addition of the peroxide at the 4- or 6-position on the aromatic ring during

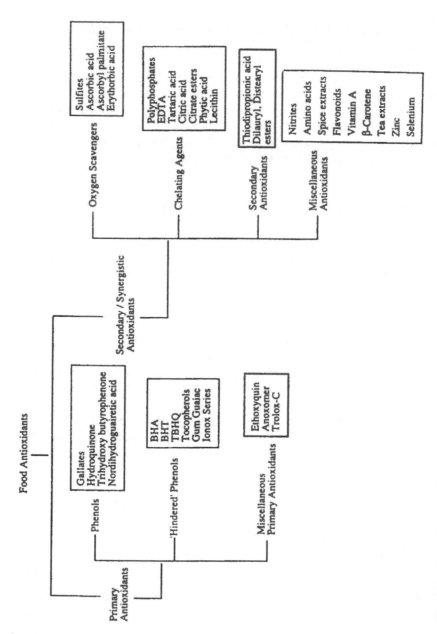

Fig. 3.1   Classification of food antioxidants. (From Ref. 4.)

the oxidation of BHT. The compounds were stable in the dark but were decomposed by either light or heat.

Hindered phenolic antioxidants have substituted alkyl or electron-releasing groups in the ortho or para positions in the aromatic ring, which decreases the reactivity of the —OH group. The electron-donating groups increase the electron density on the —OH group by an inductive effect and increase the reactivity with the lipid radicals. Substitution with butyl or ethyl groups at the para position enhances the activity compared to that of methyl groups. However, substitution with electron-attracting groups (e.g., nitro) or by longer chain or branched alkyl groups at the para position reduces the activity due to steric hindrance. Reaction of a hindered phenolic antioxidant with a free radical results in the formation of a free phenoxy radical. The phenoxy radicals are stabilized by delocalization of the unpaired electron in the aromatic ring and are further stabilized by bulky groups at the ortho position. Substitution also reduces the number of propagation reactions that may occur involving antioxidant free radicals:

$$AOO^{\cdot} + LH^{\cdot} \rightarrow AOOH + L^{\cdot} \tag{3.6}$$
$$A^{\cdot} + LH^{\cdot} \rightarrow AH + L^{\cdot} \tag{3.7}$$
$$A^{\cdot} + O_2 \rightarrow AOO^{\cdot} \tag{3.8}$$

Much less is known about the chemistry or mechanism of action of polyhydroxy phenolic antioxidants. There is general agreement that the reaction of such an antioxidant with a peroxy radical is likely to give an orthoquinone (4).

3.2.2. Synergistic Antioxidants

Synergistic antioxidants can be broadly classified as oxygen scavengers and chelators. Synergists function by various mechanisms. They may act as hydrogen donors to the phenoxy radical, thereby regenerating the primary antioxidant. Hence phenolic antioxidants can be used at lower levels if a synergist is added simultaneously to the food product. Synergists also provide an acidic medium that improves the stability of primary antioxidants.

Oxygen scavengers such as ascorbic acid, ascorbyl palmitate, sulfites, and erythorbates react with free oxygen and remove it in a closed system. Ascorbic acid and ascorbyl palmitate also act as synergists with primary antioxidants, especially with tocopherols. Chelators like ethylenediaminetetraacetic acid, citric acid, and phosphates are not antioxidants, but they are highly effective as synergists with both primary antioxidants and oxygen scavengers. An unshared pair of electrons in their molecular structure promotes the chelating action. They form stable complexes with prooxidant metals such as iron and copper, which promote initiation reactions and raise the energy of activation of the initiation reactions considerably.

3.2.3   Secondary Antioxidants

Secondary or preventive antioxidants such as thiodipropionic acid and dilauryl thiodipropionate function by decomposing the lipid peroxides into stable end products. Dilauryl thiodipropionate is also effective in inactivating peracids in model systems that contain performic acid and nonanal, ethyl oleate, or cholesterol. In this system dilauryl thiodipropionate is preferentially oxidized to the sulfoxide and prevents the formation of nonanoic acid, 9-epoxyethyl oleate, and 5,6-epoxy-cholesterol, respectively (7).

3.2.4   Miscellaneous Antioxidants

Compounds listed under miscellaneous antioxidants such as flavonoids and related compounds and amino acids function as both primary antioxidants and synergists. Nitrites and nitrates, which are used mainly in meat curing, probably function as antioxidants by converting heme proteins to inactive nitric oxide forms and by chelating the metal ions, especially nonheme iron, copper, and cobalt, that are present in meat. β-Carotene and related carotenoids are effective quenchers of singlet oxygen and also prevent the formation of hydroperoxides. Zinc strongly inhibits lipid peroxidation at the membrane level, possibly by altering or preventing iron binding. Selenium is necessary for the synthesis and activity of glutathione peroxidase, a primary cellular antioxidant enzyme. The enzymes glucose oxidase and catalase function by removing dissolved or headspace oxygen and preventing the accumulation of hydrogen peroxide, respectively.

3.3   SOURCES OF FOOD ANTIOXIDANTS

3.3.1   Synthetic Antioxidants

Synthetic antioxidants are mainly phenolic and include butylated hydroxyanisole (BHA), butylated hydroxytoluene (BHT), *tert*-butyl hydroquinone (TBHQ), and propyl, octyl, and dodecyl gallates (Fig. 3.2). Polymeric antioxidants such as Anoxomer, Ionox-330, and Ionox-100, a derivative of BHT, have also been introduced, but they are not being used commercially. In general, the use of primary antioxidants is limited to 100–200 ppm of BHA, BHT, or TBHQ or 200–500 ppm of the gallates for the stabilization of fats and oils. Commercially, a number of ready-to-use formulations containing a food grade solvent (propylene glycol or glycerol monooleate), a synergist like citric acid, and one or more phenolic antioxidants are available (Table 3.2).

3.3.2   Antioxidants from Food Packaging Materials

Plastics are used extensively in food packaging. Plastic films, containers, and coating materials include high-density and low-density polyethylenes, polyvinyl

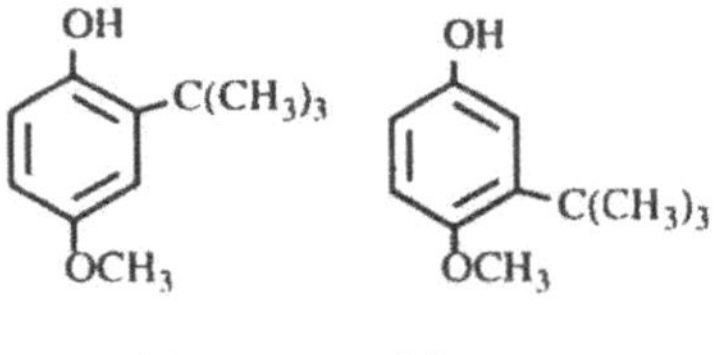

OH
$C(CH_3)_3$

OH

$C(CH_3)_3$
OCH_3

OCH_3

$CO(CH_2)_3 CH_3$
OH

HO
OH

(a)   (b)

Butylated hydroxyanisole. (a) 3-BHA. (b) 2-BHA.   Trihydroxy butyrophenone.

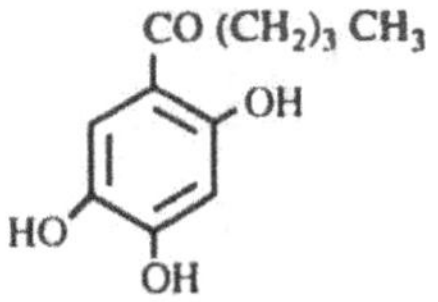

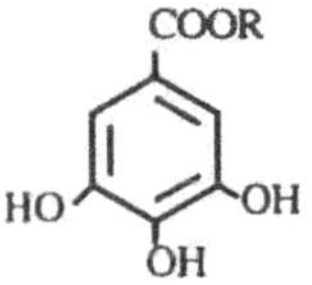

COOR

HO   OH
OH

OH
$(CH_3)_3C$ $C(CH_3)_3$

CH_3

R : C_3H_7 Propyl gallate   Butylated hydroxytoluene.
       C_8H_{17} Octyl gallate
       $C_{12}H_{25}$ Dodecyl gallate
       Gallates.

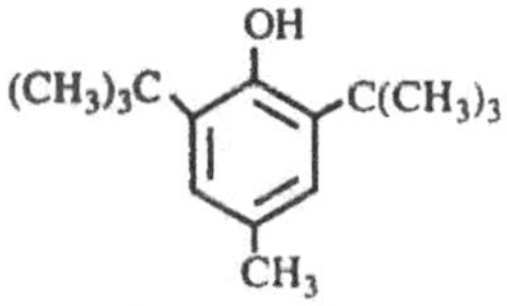

OH
$C(CH_3)_3$

OH

OH
$(CH_3)_3C$ $C(CH_3)_3$

CH_2OH

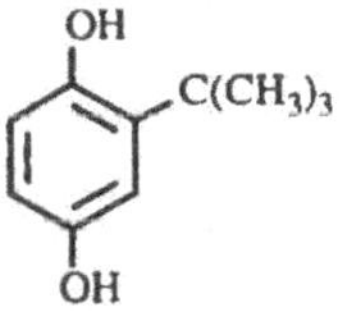

Tertiary butyl hydroquinone. 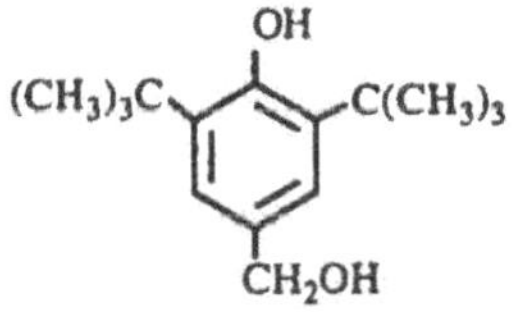   Ionox-100

Fig. 3.2   Some synthetic food antioxidants.

chloride, polypropylene, and copolymers of ethylene and vinyl acetate. Some of the additives added to plastics are antioxidants, heat stabilizers, plasticizers, and UV absorbers. The additives are added during processing and fabrication or after molding. Synthetic chain-breaking and preventive antioxidants are added to prevent oxidative degradation of polymers during fabrication and long-term storage. Chain-breaking antioxidants include hindered phenols, aromatic amines, quinones, and nitro compounds. Preventive antioxidants include phosphite esters, sulfides, and dithiophosphates. Some of these antioxidants or their degradation products may migrate to the food product as unintentional additives.

Addition of antioxidants via migration from the packaging materials can also be intentional. Because of their steam volatility, BHA and BHT are important additives incorporated into packaging materials from which they migrate into the packaged food. In this application, they are added to the packaging material either in the paraffin wax used for making waxed inner liners or applied to the packaging

Table 3.2 Some Commercial Antioxidant Preparations

	BHA (%)	BHT (%)	Propyl gallate (%)	TBHQ (%)	Propylene glycol (%)	Citric acid (%)	Vegetable oil (%)	Glyceryl monooleate (%)	Sorbitan monooleate (%)	Water (%)	Ethyl alcohol (%)	Citrate monoglyceride (%)
Eastman Tenox BHT		X										
Eastman Tenox BHA	X											
Eastman Tenox 2	20		6		70	4						
Eastman Tenox 4	20	20					60					
Eastman Tenox 6	10	10	6		12	6	28	20	8			
Eastman Tenox 7	28		12		34	6		20				
Eastman Tenox 20				20	70	10						
Eastman Tenox 22	20			6	70	4						
Eastman Tenox 26	10	10		6	12	6	28	28				
Eastman Tenox R	20				60	20						
Eastman Tenox S-1			20		70	10						
UOP-BHA	X											
UOP-Sustane		X										
UOP-Sustane 3F	66.7		20			13.3						
UOP-Sustane 6	18	22					60					
UOP-Sustane E	10						40		2.5	47.5		
UOP-Sustane W	10	10	6		8	6	28	30				
UOP-Sustane P	20	20									60	
Shell Ionol		X			X							
Griffths G-16	6	13.5	5.5			2						X

Source: Ref. 8.

board as an emulsion. This is especially useful in low-fat foods such as breakfast cereals or cake mixes, in which it is difficult to achieve proper contact with the fatty phase. Since BHA and BHT are volatile even at ambient temperatures, they gradually diffuse from the wax into the product (9,10).

3.3.3 Natural Antioxidants

In recent years, consumers and food manufacturers have been opting for products with "all natural" labels. The volume of such products increased 175% from 1989 to 1990, and the number of products claiming to be without additives or preservatives rose by 99% during the same period (11). Consequently, a lot of emphasis was given to the identification and incorporation of novel, natural antioxidants in food products. The area of natural antioxidants developed enormously in the past decade mainly because of the increasing limitations on the use of synthetic antioxidants and enhanced public awareness of health issues. In general, natural antioxidants are preferred by consumers because they are considered safe. Pokorny (12) lists some of the advantages and disadvantages of natural antioxidants compared to synthetic antioxidants (Table 3.3).

Many common food ingredients contain antioxidant compounds (Table 3.4). However, such ingredients can be used only in products when they are compatible with the texture, color, and flavor of the end product. Hence, identification and further purification of the antioxidant compounds become essential for the effective use of natural antioxidants on a commercial basis. Table 3.5 lists some of the naturally occurring antioxidant compounds. Many reviews have appeared on the sources and activities of specific compounds (12–16).

Some of the natural antioxidants listed in Table 3.5 such as the tocopherols and vitamin C are used widely in food products. They have also been synthesized on a commercial scale. Synthetic α-tocopherol and its acetate are designated dl-α-to-

Table 3.3 Advantages and Disadvantages of Natural Antioxidants in Comparison with Synthetic Antioxidants

Advantages	Disadvantages
Readily accepted by the consumer, as considered to be safe and not a "chemical."	Usually more expensive if purified and less efficient if not purified.
No safety tests required by legislation if a component of a food that is "generally recognized as safe" (GRAS).	Properties of different preparations vary if not purified.
	Safety often not known.
	May impart color, aftertaste, or off-flavor to the product.

Source: Ref. 12.

Table 3.4 Natural Antioxidants in Some Food Ingredients

Source	Antioxidant
Oils and oilseeds	Tocopherols and tocotrienols; sesamol and related substances; olive oil resins; phospholipids
Oat and rice brans	Various lignin-derived compounds
Fruits and vegetables	Ascorbic acid; hydroxycarboxylic acids; flavonoids; carotenoids
Spices, herbs, tea, cocoa	Phenolic compounds
Proteins and protein hydrolysates	Amino acids; dihydropyridines; Maillard reaction products

Source: Ref. 12.

copherol and *dl*-α-tocopheryl acetate but are actually mixtures of four racemates. Trolox-C, a synthetic derivative of α-tocopherol and *d*-α-tocopheryl polyethylene glycol 1000 succinate, a synthetic water-soluble form of α-tocopherol, have been introduced. Very recently, Koga and Terao (17) reported the synthesis of a novel phosphatidyl derivative of α-tocopherol that is more effective than α-tocopherol in lard and in a methyl linoleate–methyl laurate model system. Citric acid, tartaric acid, and lecithin are other common compounds present in food ingredients that are used as synergists in food products. Phytic acid, a major component of all seeds, is a potent chelating agent. It is used widely as a food additive in Japan. Figure 3.3 presents some of the naturally occurring antioxidants used in food products.

Natural antioxidants obtained from sources other than food ingredients have

Table 3.5 Some Naturally Occurring Antioxidants

Amino acids	β-Carotene
Carnosine	Carnosol
Citric acid	Curcumin
Eugenol	Flavonoids
Lecithin	Lignans
Nordihydroguaiaretic acid	Phenolic acids
Phytic acid	Protein hydrolysates
Proteins	Rosmarinic acid
Saponins	Sterols
Tartaric acid	Uric acid
Turmerin	Vitamin C
Vitamin E	Vanillin

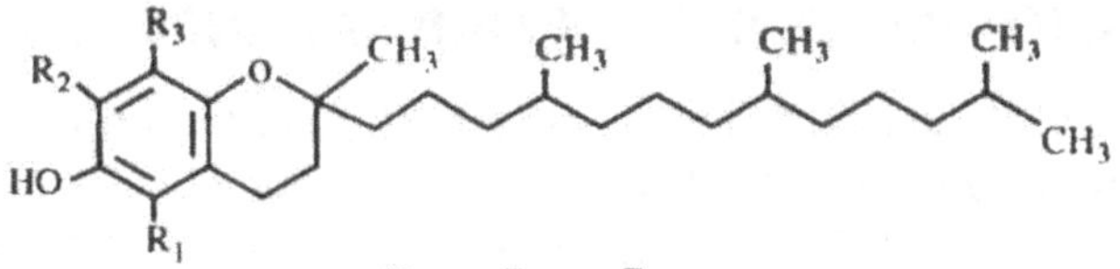

	R_1	R_2	R_3	
	CH_3	CH_3	CH_3	α-tocopherol
	CH_3	H	CH_3	β-tocopherol
	H	CH_3	CH_3	γ-tocopherol
	H	H	CH_3	δ-tocopherol

Tocopherols.

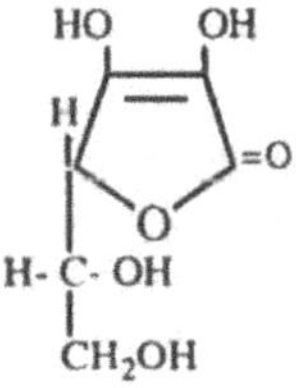

Ascorbic acid.

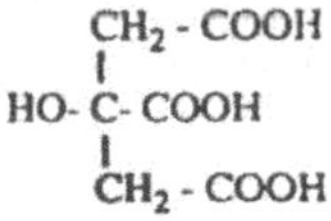

Citric acid.

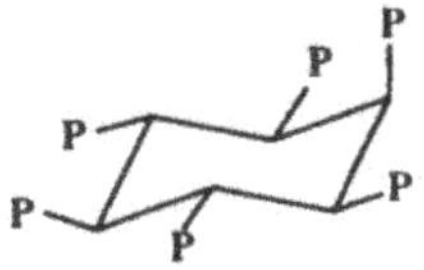

$P : H_2PO_4$

Phytic acid

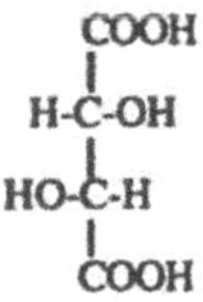

Tartaric acid.

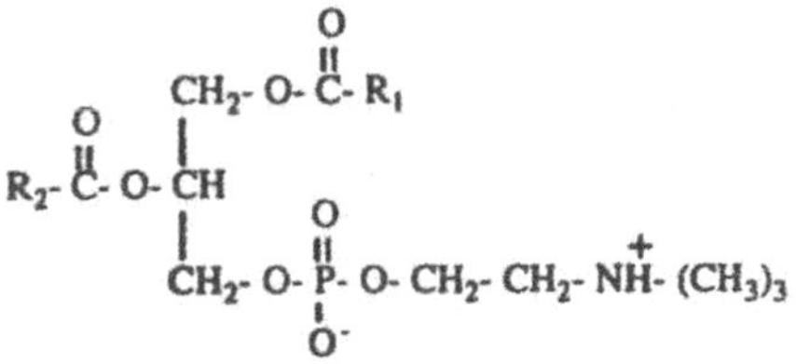

Phosphatidyl choline.

Fig. 3.3   Some naturally occurring food antioxidants.

also been used in food products. Gum guaiac is a naturally occurring phenolic antioxidant from the wood of *Guajacum officinale* L. or G. *sanctum* L. It was widely used in the 1940s as an antioxidant for oils and fats, especially for the stabilization of lard. Nordihydroguaiaretic acid is phenolic natural antioxidant obtained from the creosote bush (*Larrea divaricata*) that was widely used in food products in the 1950s and 1960s.

3.4 POTENTIAL SOURCES OF NATURAL ANTIOXIDANTS

In recent years, numerous reports have been published on the identification of novel, naturally occurring antioxidants from plants, animals, microbial sources, and processed food products. An exhaustive listing of the various sources has been compiled by Loliger (13). Table 3.6 presents a list of plant antioxidant sources. Recent reports in this area discuss young green barley leaves (18); leaves of *Polygonum hydropiper,* a medicinal herb (19); pea bean (20); leaves of *Vernonia amygdalina* (21); and wild rice (22).

In addition to identification of the sources, numerous reports have appeared on further identification and isolation of the active compounds from various sources. Most natural antioxidants are phenolic compounds that, with the exception of the tocopherols, contain ortho-substituted active groups, whereas the synthetic antioxidants, with the exception of the gallates, are para-substituted (12). Some of the major active compounds reported so far are flavonoids and related compounds in plant extracts; phenolics in spices and herbs; and proteins, protein hydrolysates, peptides, amino acids, and Maillard reaction products. This chapter presents a brief summary of the important sources and compounds and some recent studies wherever possible.

3.4.1 Flavonoids and Related Compounds

Flavonoids are one of the most widely occurring groups of secondary metabolites in plants. They are found in almost all parts of the plant. The chemical structures are based on a C_6–C_3–C_6 skeleton. Various subgroups are classified on the basis

Table 3.6 Some Potential Sources of Natural Antioxidants from Plant Extracts

Apple cuticle	Oats
Birch bark	Olives
Carob pod	Oregano
Chia seed	Peanut seed coat
Cloves	Pepper
Cocoa shells	Red onion skin
Garlic	Rice hull
Korum rind	Rosemary
Leaf lipids	Sesame seed oil
Licorice	*Silybium marianum* seed oil
Mustard leaf seed	Tea
Myristica fragrans	Wheat gliadin
Nutmeg	

Source: Compiled from Loliger (13).

of the substitution patterns of ring C and the position of ring B. The major subgroups are flavonols, flavones, isoflavones, catechins, proanthocyanidins, and anthocyanins (Fig. 3.4). Chalcones, flavanones, leucoanthocyanins, and dihydroflavonols are the common precursors for the different subgroups in the biosynthetic pathway. Cinnamic and phenolic acids are closely related to flavonoids, and some of them are precursors for the flavonoid biosynthetic pathway. Most of these compounds have shown marked antioxidant activity in model systems. The antioxidant properties reported for numerous extracts from leaves, seed hulls, seeds, fruits, and stems are mainly due to flavonoids and cinnamic acids. Pratt and Hudson (15) conducted a general survey of the mechanisms and compounds involved and the structure–activity relationships of these compounds. Flavonoids function as primary antioxidants, chelators, and superoxide anion scavengers. The presence of hydroxyl groups at the 3′-, 4′-, and 5′-positions in the B ring enhances the antioxidant activity compared to that of a single hydroxyl group. Also, the presence of a 3-hydroxyl group and the 2–3 double bond in the C ring seems to have an effect on the antioxidant properties.

3,4-Dihydroxychalcones such as butein and okanin are particularly effective antioxidants in the range of concentrations 0.025–0.1% in lard. They are more effective than the corresponding flavonones. Butein at 0.02% is twice as active as quercetin and α-tocopherol and about six times as active as BHT in lard. It has been suggested that the effectiveness of 3,4-dihydroxychalcones is dependent on the formation of extended forms of resonance-stabilized free radicals (14). Dihydroquercetin has the same antioxidant activity as quercetin. It is effective in lard, cottonseed oil, and butter oil (24). Dihydroquercetin has been identified as one of the antioxidant compounds in peanuts (25).

The antioxidant activity of the flavones luteolin and isovitexin, a C-glycosyl flavone, have been reported. Luteolin is particularly active in lard and stripped corn oil (15,23). It has also been identified as the active compound in peanut seed hulls and in the leaf extracts of *Vernonia amygdalina* (21,26). Igile et al. (21) showed that luteolin is a significantly more potent antioxidant than BHT at the same concentration (15 mg/mL) in the coupled oxidation of β-carotene and linoleic acid. Luteolin was reported to be as active as BHA and stronger than α-tocopherol (26). Isovitexin was identified as the active component in the methanolic extracts of long-life rice hulls (*Oryza sativa* var. Katakutara) (27). Isovitexin from young green barley leaves has also been reported. Its antioxidant activity was almost equivalent to that of α-tocopherol in an ethyl linoleate model system (18).

The flavonols quercetin, myricetin, robinetin, and gossypetin also have potent antioxidant properties. Quercetagetin, gossypetin, 3,5,8,3′,4′-pentahydroxyflavone, and 3,7,8,2′,5′-pentahydroxyflavone are the most potent antioxidants reported in nonaqueous systems (15). Quercetin has been effective in inhibiting copper-catalyzed oxidation of lard and in dry milk products and potato flakes

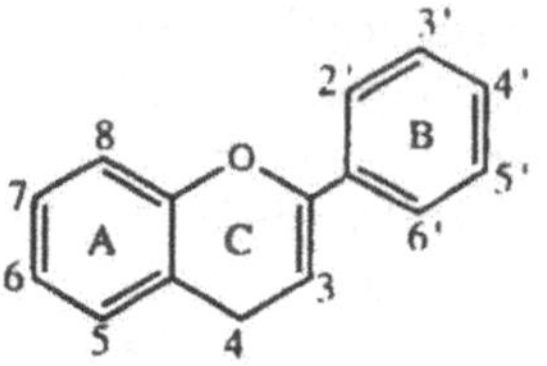

C$_6$-C$_3$-C$_6$ configuration of flavonoids

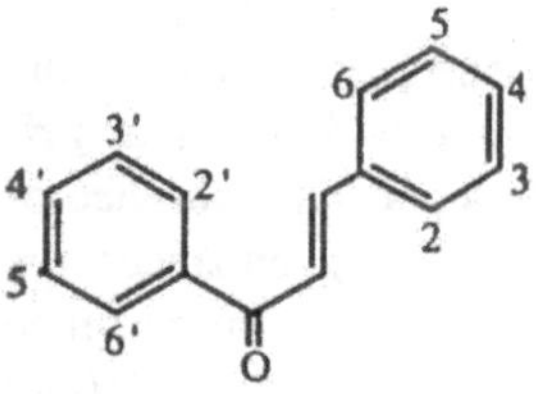

Chalcones
Butein 2'=4'=3=4=OH
Okanin 2'=3'=4'=3=4=OH

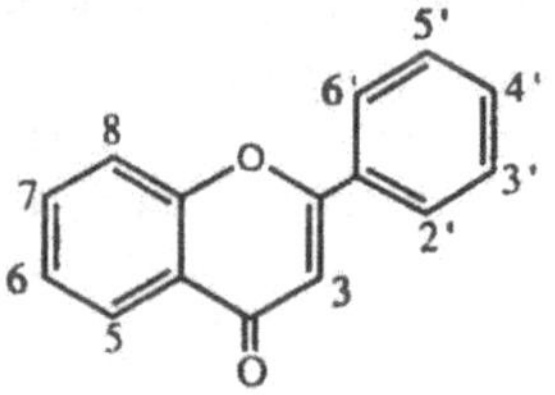

Flavones
Luteolin 5=7=3'=4'=OH
Isovitexin 4'=5=7=OH, 6=Glucose

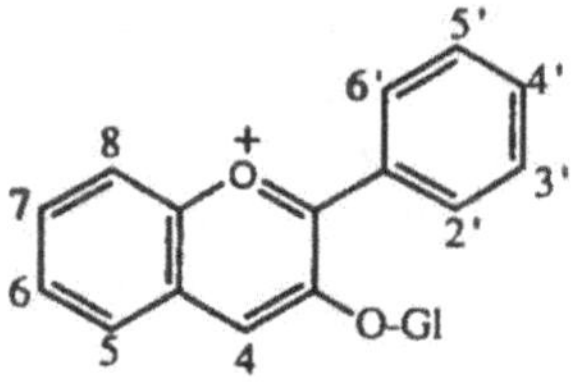

Anthocyanins

Cyanidin-3-glucoside 5'=4'=5=7=OH
Malvidin-3-glucoside 5=7=4'=OH, 3'=5'=OCH$_3$

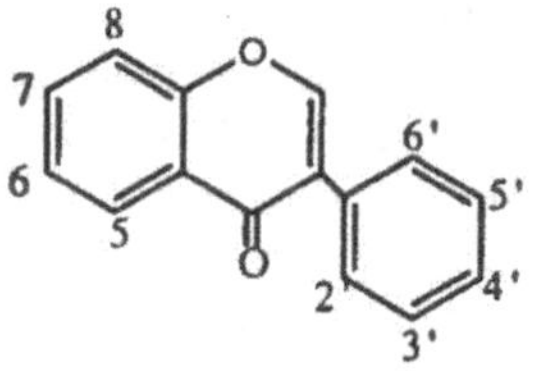

Isoflavones

Daidzein 7=4'=OH
Genistein 5=7=4'=OH

Fig. 3.4 Some antioxidant flavonoids and related compounds.

(28–30). Quercetin, 7,4'-dimethylquercetin, 3'-methylquercetin, and isoquercitrin (quercetin glucoside) were identified as the active compounds in the leaf extracts of *Polygonum hydropiper*, a medicinal herb (19). The acetone extracts of red onion skin have antioxidant properties that can be attributed to the presence of quercetin (2.5–6.5% as aglycone) (31,32).

Isoflavones from soybeans consist primarily of 7-*O*-monoglucosides of genistein, daidzein, and glycitein. The aglycones have comparable antioxidant properties. Studies by Dziedzic and Hudson (23) showed a pronounced synergism

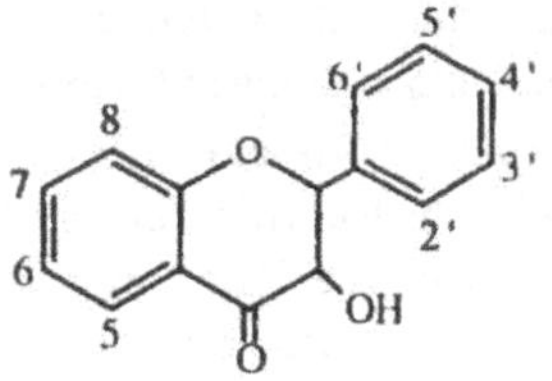

Dihydroflavonols
Dihydroquercetin 3=5=7=3'=4'=OH

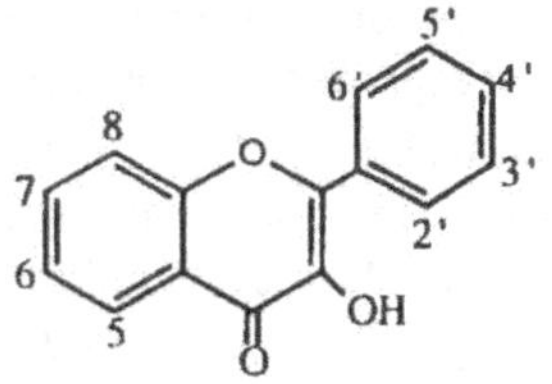

Flavonols
Quercetin 3=5=7=3'=4'=OH
Myricetin 3=5=7=3'=4'=5'=OH
Gossypetin 3=5=7=8=4'=5'=OH

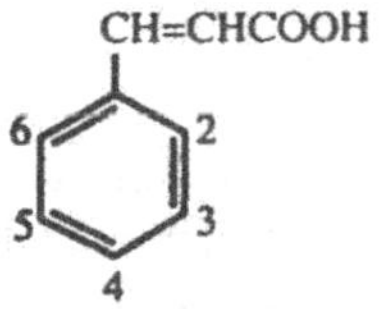

Cinnamic acids
Ferulic acid 4=OH, 3=OCH₃
Caffeic acid 3=4=OH

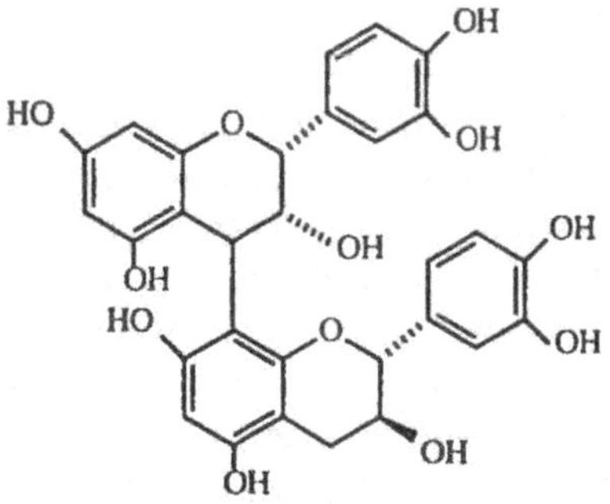

Procyanidin B-1

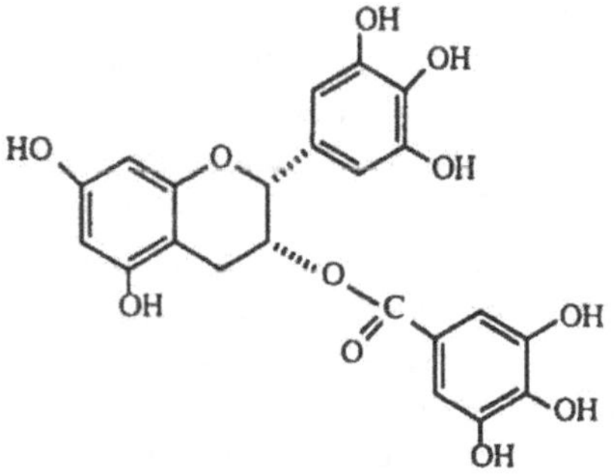

(-)-Epigallocatechin gallate

Fig. 3.4 (*Continued*).

between genistein and phosphatidylethanolamine in lard. Both the 4'- and 5-hydroxy groups are needed for significant activity. In general, isoflavones show a relatively low order of antioxidant activity compared to the flavonols, flavones, flavonones, and chalcones (15).

Of the various leaf extracts reported, green tea extracts have the potential for large-scale application as natural antioxidants. The ethanol and acetone extracts of green tea are highly effective in soybean, corn, palm, and peanut oils and lard (33). The major active compounds in green tea are catechins with the following order of activity: epigallocatechin gallate > epigallocatechin > epicatechin gallate > epicatechin. They were superior to BHA and α-tocopherol in lard and salad oil. Epigallocatechin gallate also showed synergism with ascorbic acid, α-tocopherol, citric acid, and tartaric acid (34).

The antioxidant activity of water-soluble tannins from *Osbeckia chinensis* has been reported (35). Two proanthocyanidin dimers B-1 and B-3 having a higher antioxidant activity than α-tocopherol in a linoleic acid–β-carotene–water system were isolated from red beans (36). Esters of gallic acid, the main constituent of tannins, are being used as food antioxidants. Proanthocyanidins from fruits such as grapes, black currants, and bilberries function as superoxide anion scavengers in model systems (37). Proanthocyanidins are effective in salad oil, frying oils, and lard (38).

Anthocyanins, the major coloring pigments in higher plants, also have antioxidant activity. The antioxidant properties of the major pigment in grapes, malvidin-3,5-diglucoside, have been reported (39). Igarashi et al. (40) also reported the antioxidant properties of nasunin, the acylated anthocyanin of egg-plant. The higher antioxidant activity of nasunin compared to the corresponding delphinidin aglycone was attributed to the presence of *p*-coumaric acid moiety in nasunin. Recently, Tsuda et al. (20) reported the antioxidant activity of anthocyanins cyanidin-, pelargonidin-, and delphinidin-3-glucosides from the seed coat of pea bean.

Phenolic acids with antioxidant activity occur widely in oilseeds and leaf extracts. The various active components identified in mustard seeds and rapeseeds include cinnamic, ferulic, caffeic, prototocatechuic, sinapic, salicylic, and vanillic acids (41). Phenolic acids with antioxidant properties have also been reported in soybeans, soybean flour, protein concentrate, protein hydrolysate, cottonseed, cottonseed flour, peanuts, and peanut flour. The antioxidant activity is determined by the number of hydroxyl groups in the molecule (15). The antioxidant properties of germinated oat (*Avena sativa*) grains or their aqueous extracts have been attributed to the presence of dihydrocaffeic acid and phospholipids (42). Recently, Ohta et al. (43) identified diferulic acid and ferulic acid sugar esters as active components in corn bran hemicellulose fragments. Ohta et al. (44) also identified ferulic acid glycosides as active compounds in sake.

3.4.2 Phenolics from Spices and Herbs

Antioxidants from spices and herbs have the potential for large-scale applications. Spices have been used not only for their flavoring properties but also for their food-preserving ability for hundreds of years. Early scientific studies on their antioxidant activity were done using whole spices or their extracts (45–49). These studies indicated that a wide range of extracts have antioxidant properties. Later studies have concentrated on isolation and identification of active components from various herbs and spices such as rosemary, sage, thyme, mace, oregano, turmeric, ginger, pepper, cloves, and capsicum. Most of these compounds have been reported to be as effective as BHA, BHT, or α-tocopherol. Some of the major drawbacks with spice extracts are their color, flavor, and taste. Several studies have been carried out to identify and isolate odorless, tasteless compounds. Figure 3.5 presents the structures of some of the active compounds in spices and herbs.

Four odorless and tasteless diterpenelactones—rosmanol, carnosol, epirosmanol, and isorosmanol—have been isolated from rosemary (*Rosmarinus officinalis* L.) and sage (*Salvia officinalis* L.). Other active molecules so far identified in rosemary are rosmarinic acid, rosmariquinone, and rosmaridiphenol. Carnosol and rosmarinic acid are the most active components in rosemary (50–53). Tanshen (*Salvia miltiorrhiza* Bunge) is a wild herb used in Chinese medicine. Recent studies have identified a number of quinones, including dehydrorosmariquinone, miltirone, and rosmariquinone, that have antioxidant properties (54,55). In oregano, the active components are a phenolic derivative of caffeic acid, rosmarinic acid, and its derivative (56,57). Thymol and its biphenol derivatives have been reported to be active in thyme (58,59).

In capsicum, in addition to capsaicin and dihydrocapsaicin, a new tasteless compound with antioxidant properties has been isolated (60). Five antioxidative phenolic amides have been isolated from black pepper. Of these, ferulic acid amide of tyramine and a piperine-related compound with an open methylenedioxy ring showed stronger activity than α-tocopherol (61). Gallic acid and eugenol have been identified as the major active compounds in cloves (62). Recent studies on ginger extracts revealed that in addition to the pungent principles shogaol, zingerone, and gingerol, several gingerol-related compounds and diarylheptanoids have higher activity than α-tocopherol (63). In addition to curcumin, several tasteless antioxidant compounds have recently been isolated from turmeric. Tetrahydrocurcumin, which is closely related to curcumin structurally, is colorless, heat-resistant, and antioxidative (14). Papua mace contains 2-allylphenols and a number of lignans that were found to possess strong antioxidant properties (64). Vanillin, used extensively as a food flavoring agent, has potent antioxidant properties even in dry mixes like rice flakes (65).

Though many antioxidant compounds have been identified in herbs and spices, only with rosemary and sage large-scale processing methods been developed to produce a purified product that is bland, odorless, and tasteless (66,67). Rosemary antioxidant is available commercially as a fine powder that is soluble in fats and oils and insoluble in water. It is recommended that 200–1000 mg be used per kilogram of the food product to be stabilized.

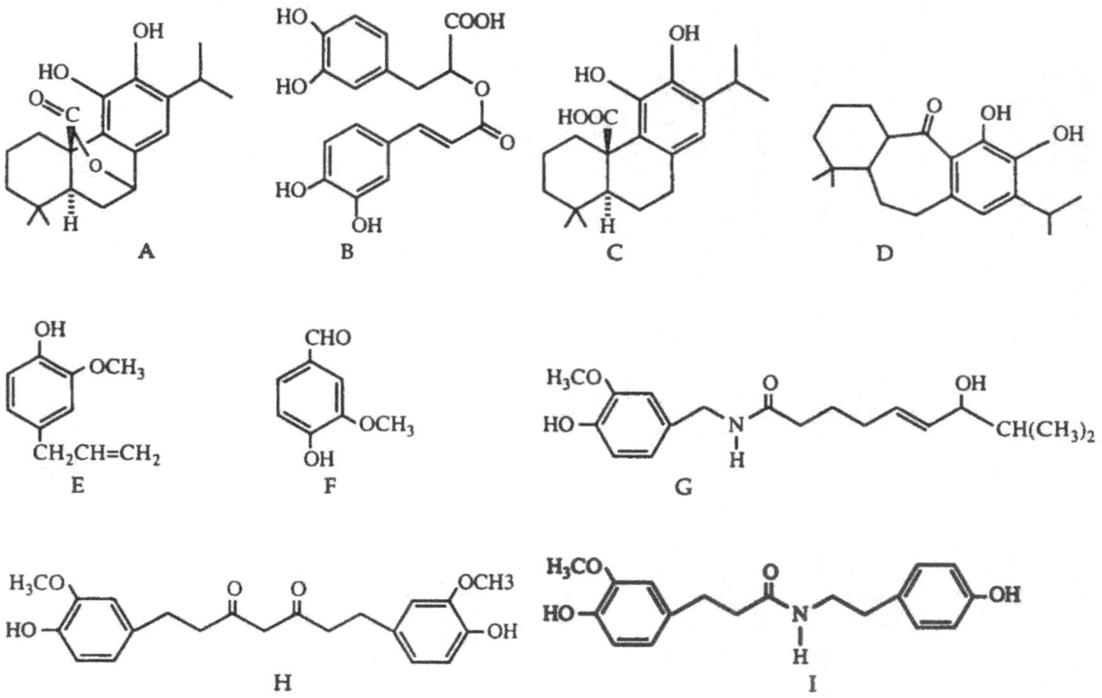

Fig. 3.5 Some antioxidant compounds from spices. (A) Carnosol from rosemary; (B) rosmarinic acid from rosemary; (C) carnosic acid from rosemary; (D) rosmaridiphenol from rosemary; (E) eugenol from cloves; (F) vanillin from vanilla pods; (G) capsaicin from capsicum; (H) tetrahydrocurcumin from turmeric; (I) ferulic acid amide from black pepper.

3.4.3 Proteins, Peptides, Amino Acids, and Maillard Reaction Products

Proteins, protein hydrolysates, peptides, amino acids, and Maillard reaction products are known to have significant antioxidant properties. In general, they function as synergists or primary antioxidants. Gluten, egg albumen, and casein were effective in safflower and sardine oil model systems (68). A combination of soy protein isolate and gliadin was effective in safflower oil (69). Soy proteins were more effective than pea fiber in preserving the color of minced beef (70). Soy protein hydrolysate, autolyzed yeast proteins, and hydrolyzed vegetable proteins function as synergists with primary antioxidants. When used alone, 10% autolyzed yeast protein was equivalent to 0.02% BHA in corn oil. But when combined with only 0.005% BHA, the total effect was as great as that of 50% protein isolate or twice that of 0.02% BHA (71,72).

The antioxidant activity of peptides has been reported. Turmerin, a water-soluble peptide containing 40 amino acid residues obtained from turmeric (*Curcuma longa*), was found to be effective in model systems. The peptide inhibited lipid oxidation by 70–88%, as compared to 70–85% for curcumin and 80% for BHA, in L-α-phosphatidylcholine vesicles, liposomes, and erythrocyte ghosts (73). Carnosine, a β-alanine-histidine dipeptide found in skeletal muscles, was found to be very effective in inhibiting oxidative rancidity in muscle foods. In frozen salted ground pork, carnosine was more effective than BHT, α-tocopherol, and sodium tripolyphosphate. Carnosine may function as chelator or as a primary antioxidant (74,75).

Most of the amino acids have antioxidant properties depending on the pH of the medium and their concentration. Studies by Marcuse (76) indicated that several amino acids were effective even in the absence of primary antioxidants in a linoleic acid model system. Glycine, methionine, histidine, tryptophan, proline, and lysine have been found to be effective in fats and oils. Glycine has been listed as a GRAS substance for addition to fats and oils at concentrations up to 0.01%. Amino acids also function synergistically with other antioxidants. Numerous reports have appeared on the synergistic effects of amino acids in model systems of fats, oils, dairy products, and meat products. Troloxyl-amino acids formed by a covalent attachment of Trolox-C with various amino acids were more effective than Trolox-C in a linoleate emulsion system (77). Popovici et al. (78) reported that a combination of histidine and ascorbic acid is highly effective in corn oil. Combinations of cystine with BHT or tocopherols were highly effective in milk fat (79,80).

Maillard reaction products formed during heat processing or storage of food products are known to have significant antioxidant properties. They seem to function as metal chelators (81). Some of the recent studies in this area have been on the identification and evaluation of the antioxidant heterocyclic volatile compounds produced in Maillard reactions. Macku and Shibamoto (82) identified 1-methylpyrrole as an antioxidant in the headspace of a heated corn oil/glycine model system. Eiserich et al. (83) showed that thiazoles, oxazoles, and furanones formed in an L-cysteine/D-glucose model system had antioxidant properties. Similarly, Eiserich and Shibamoto (84) in a recent report showed that alkylthiophenes, 2-thiophenethiol, 2-methyl-3-furanthiol, and furfuryl mercaptan have antioxidant properties. The degree of unsaturation in the heterocyclic ring, as well as the substituent type, had variable effects on the antioxidative capacity of these compounds.

3.4.4 Other Natural Sources

Lignans

The antioxidative activity of various lignanphenols from sesame seed has been reported. Raw and roasted sesame seed oils are characterized by a higher oxidative stability than other vegetable oils. Fukuda et al. (85) compared the rate of autoxidation of raw and roasted sesame oils with that of various vegetable oils at 60°C. Both raw and roasted sesame oils remained stable even after 50 days of incubation, whereas corn oil, safflower oil, and common salad oil (a mixture of rapeseed and soybean oils) oxidized after 10–20 days of incubation. Some of the antioxidant compounds in sesame seed and oil have been found to be sesamol, sesamolinol, sesaminol, pinoresinol, and sesamol dimer (Fig. 3.6) (86,87). Sesamol (3,4-methylenediphenoxyphenol) is formed by the hydrolysis of sesamolin, a lignan characteristic of sesame during bleaching, hydrogenation, or other refining conditions. Sesamol is also formed during frying, which contributes to the stability of the fried product (88,89). Sesamol is readily converted to sesamol dimer and then to sesamol dimer quinone, which have been shown to exhibit antioxidative activity in lard, soybean oil, and methyl oleate (90,91).

Carotenoids

Carotenoids such as β-carotene, canthaxanthin, astaxanthin, phytoene, and lutein are present in a wide variety of food ingredients. Carotenoids are mainly used as natural colorants and also have antioxidant properties. They are effective in quenching singlet oxygen, thereby preventing the formation of hydroperoxides in the presence of singlet oxygen. It has been proposed that the carotenoids may act as primary antioxidants at low oxygen pressures under conditions where singlet oxygen is not formed. The ability to quench singlet oxygen is related to the number of double bonds in the molecule. Carotenoids with 9,10, or 11 conjugated double bonds are reported to be better quenchers than those with eight or fewer conjugated double bonds (3,92–95). Clements et al. (96) reported that β-carotene is effective in stabilizing soybean oil under photooxidizing conditions.

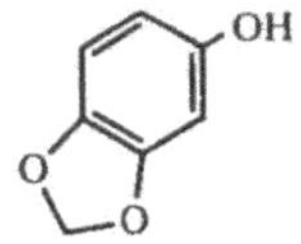

Sesamol

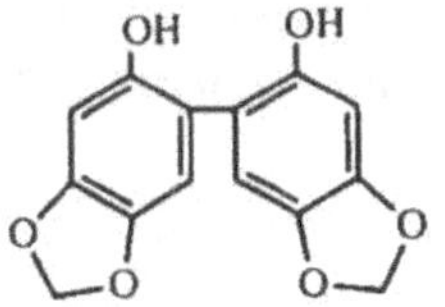

Sesamol dimer

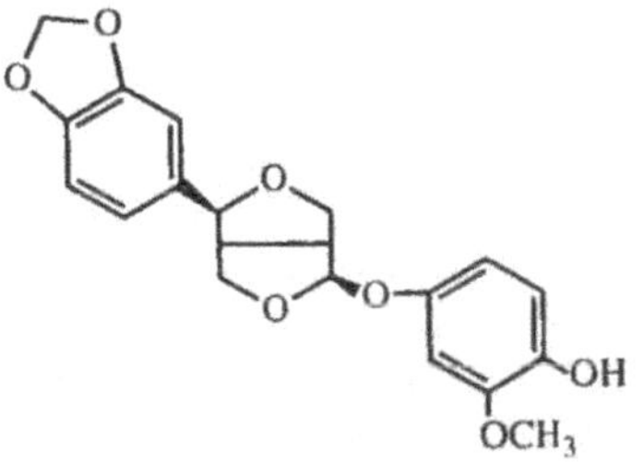

Sesamolinol

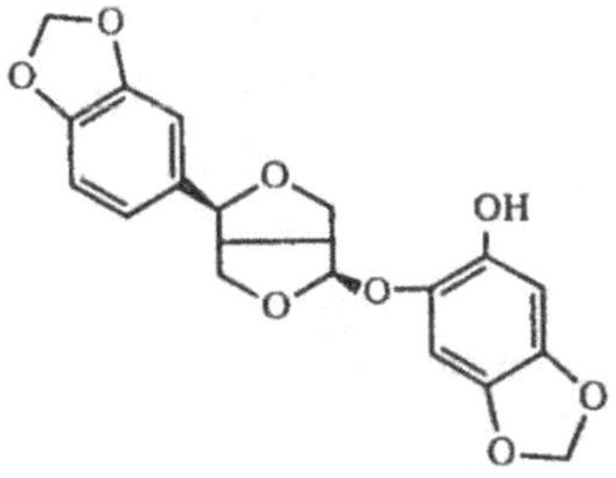

Sesaminol

Fig. 3.6   Some antioxidant compounds from sesame.

Glucose Oxidase, Catalase, and Superoxide Dismutase

Glucose oxidase is effective at removing dissolved or headspace oxygen in soft drinks, mayonnaise, and salad dressings. The main commercial source of the enzyme is a selected strain of *Aspergillus niger*. The enzyme catalyzes a reaction between oxygen and glucose, yielding D-gluconic acid and hydrogen peroxide. Commercial glucose oxidase systems contain catalase, which hydrolyzes hydrogen peroxide to water and oxygen, thus preventing its accumulation in the food (10).

The enzyme superoxide dismutase (SOD) removes superoxide radicals and plays a significant role in inhibiting biological lipid oxidation. SOD may also contribute to the stability of food products. Hill et al. (97) showed that a combination of superoxide dismutase and catalase reduced the development of oxidized flavor in heat-treated high linoleic acid milk in the absence of added copper. Lingnert (98) reported the antioxidant properties of bovine SOD and yeast (*Saccharomyces cerevisiae*) SOD in the emulsified linoleic acid system. The antioxidant effects at 1–25 units/mL ranged from 0.45 to 0.96 for bovine SOD and from

–0.04 to 0.22 for yeast SOD. Enzymatic activity remained after heat treatment (100°C) for 30 min and decreased with decreasing pH.

Porphyrin-Related Compounds

Porphyrin-related compounds such as chlorophyll, pheophytin, and bilirubin have antioxidant properties (99–101). Chlorophyll is known to be a prooxidant in light. However, Endo et al. (101) reported that both chlorophyll and pheophytin retarded the oxidative deterioration of triglycerides in rapeseed and soybean oils in the dark. Among the four chlorophyll derivatives (chlorophyll a and b and pheophytin a and b), chlorophyll a showed the strongest activity. The porphyrin structure seemed to be important for antioxidant activity.

Sterols

Sterols such as fucosterol and Δ^5-avenasterol have been shown to have antioxidant activity in oils at 180°C (102,103). Takagi and Iida (104) reported that the antioxidant activity of canary seed extract (*Phalaris canariensis*) is due to various sterols such as gramisterol, cycloartenol, sitosterol, campesterol, and triterpene alcohol esters of caffeic acid. It has been suggested that sterols function by forming a monolayer at the surface of oils and inhibit oxidation by acting as hydrogen donors (105).

Miscellaneous Sources

Other minor sources include smoke components from vegetable extracts and phenols from the neutral fraction of wood smoke (106,107). Antioxidant compounds have been isolated from microbial and algal sources. Flavoglucin, a phenolic compound from the fungus *Eurothricum chevalieri*, was found to be very effective in vegetable oils and lard (108). Antioxidant compounds, including curvulic acid and protocatechuic acid, have also been isolated from fungal sources (109).

3.5 DEGRADATION OF PRIMARY ANTIOXIDANTS

Primary antioxidants undergo degradation during the autoxidation of fats and oils. The study of the chemistry of degradation helps in understanding the mechanism of action of primary antioxidants, properties of the degradation products, and the role of synergists in regenerating the primary antioxidants. The subject has been reviewed by Kikugawa et al. (110). Degradation studies have been carried out under conditions of autoxidation of fats and oils such as thermal oxidation, oxidation by the active oxygen method, and UV/visible light. Formation of antixiodant dimers via phenoxy radical seems to be the common feature with the primary antioxidants BHA, BHT, TBHQ, and tocopherols.

The degradation of BHT in soybean oil irradiated by UV light was followed by Harano et al. (111). On irradiation, the concentration of BHT decreased, and it was completely lost at the end of the induction period of the oil. The degradation products have been identified as BHT aldehyde (3,5-di-*tert*-butyl-4-hydroxybenz-aldehyde), BHT alcohol (3,5-di-*tert*-butyl-4-hydroxybenzyl alcohol), 2,6-di-*tert*-butylbenzoquinone (BQ), and dimers 3,5,3',5'-tetra-*tert*-butyl-4,4'-dihydroxy-1,2-diphenylethylene (BE) and 3,5,3',5'-tetra-*tert*-butylstilbenequinone (SQ). Of the degradation products, BHT-alcohol, BQ, and BE still retained 56%, 23%, and 64% of the activity of BHT, respectively (Table 3.7). The degradation of 3-BHA resulted in the formation of two dimers (Fig. 3.7) that retained the antioxidant activity at lower levels than the parent compound (Table 3.8) (112,113). TBHQ, on irradiation in benzene, yielded five oxidation products identified as 2-*tert*-butyl-*p*-benzoqui-none, 2,2-dimethyl-5-hydroxy-2,3-dihydrobenzo[*p*]furan, 2-[2-(3'-*tert*-butyl-4'-hydroxy-phenoxy)-2-methyl-1-propyl]hydroquinone, 2-(2-hydroxy-2-methyl-1-propyl)hydroquinone, and 2-*tert*-butyl-4-ethoxyphenol. The oxidation products retained the antioxidant properties, and some of them had a higher level activity than TBHQ depending on the substrate (114,115). The degradation of tocopherols during chemical oxidation and in edible oils has been followed by a number of investigators and reviewed by Kikugawa et al. (110). The degradation of α-, β-, γ-, and δ-tocopherols resulted in the formation of dimers and trimers. The γ-tocopherol was found to be more susceptible to oxidation than α-tocopherol (116). The degradation of propyl gallate (PG) resulted in the formation of ellagic acid (Fig. 3.8), which still retained the antioxidant activity (117).

Table 3.7 Antioxidant Activity of Oxidation Products of BHT in Soybean Oil

Compound (0.1%)	Antioxidant activity[a] $(t - t_0)/(t_{BHT} - t_0)$
BHT	1.0
BHT Aldehyde	0
BHT Alcohol	0.56
BQ	0.23
BE	0.64
SQ	0

[a] t_0 = Induction period without antioxidant; t = induction period with the new antioxidant; t_{BHT} = induction period with the parent antioxidant.
Source: Ref. 111.

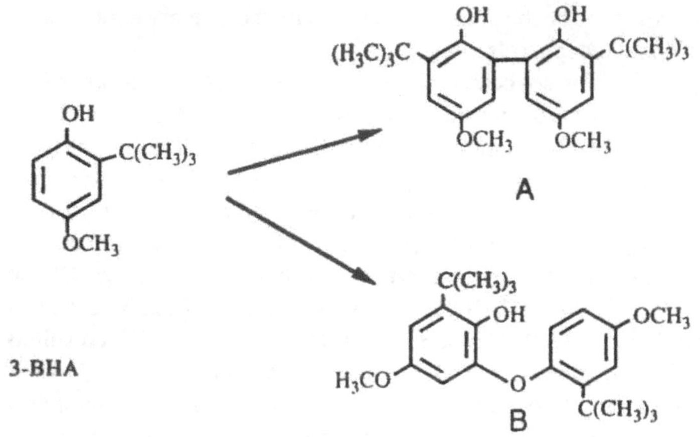

Fig. 3.7   Formation of dimers from the oxidation of 3-BHA.

The degradation of antioxidant mixtures like BHA+BHT or BHA+PG resulted in the formation of heterodimers. Irradiation of a mixture of 3-BHA and BHT resulted in the formation of a new dimer identified as 3,3′,5′-tri-*tert*-butyl-5-methoxy-2,4′-dihydroxydiphenylmethane, a biphenyl-type dimer, in addition to the oxidation products of 3-BHA and BHT. The antioxidant property of the dimer was comparable to that of BHT (Table 3.9) (118). Irradiation of a mixture of 3-BHA

Table 3.8   Antioxidant Activities of the Oxidation Products of 3-BHA

Compound	Concentration (0.02% oil)	Antioxidant activity[a] $(t - t_0)/(t_{3\text{-BHA}} - t_0)$
3-BHA	Lard	1.0
A[b]	Lard	0.33
B[b]	Lard	0.70
3-BHA	Beef tallow	1.0
A	Beef tallow	0.18
B	Beef tallow	0.52
3-BHA	Methyl oleate	1.00
A	Methyl oleate	0.24
B	Methyl oleate	0.71

[a]t_0 = Induction period without antioxidant; t = induction period with new antioxidant; $t_{3\text{-BHA}}$ = induction period with parent antioxidant.
[b]Refer to Fig. 3.7.
Source: Ref. 113.

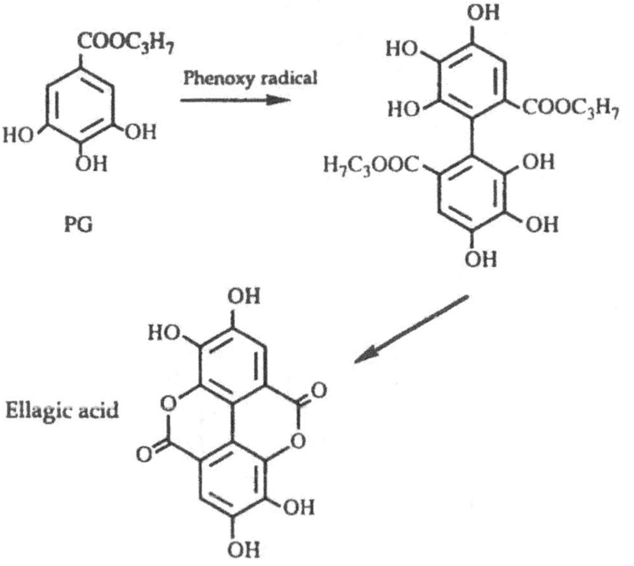

Fig. 3.8   Formation of ellagic acid from propyl gallate.

and propyl gallate resulted in the formation of two new heterodimers (Fig. 3.9). The dimers retained the antioxidant properties and were found to have a higher activity than 3-BHA in soybean oil (Table 3.10) (117).

3.6   SYNERGISM

Synergism has been observed between primary antioxidants and between primary antioxidants and nonphenolic compounds like ascorbic acid and lecithin. The

Table 3.9   Antioxidant Activity of BHT, 3-BHA, Their Mixture, and Heterodimer

Compound (1 mM)	Antioxidant activity $(t - t_0)/(t_{BHT} - t_0)$		
	Methyl oleate	Lard	Soybean oil
BHT	1.0	1.0	1.0
3-BHA	1.42	1.53	0.02
BHT + BHA	1.66	1.67	0.55
Heterodimer	1.03	1.66	1.47

Source: Ref. 118.

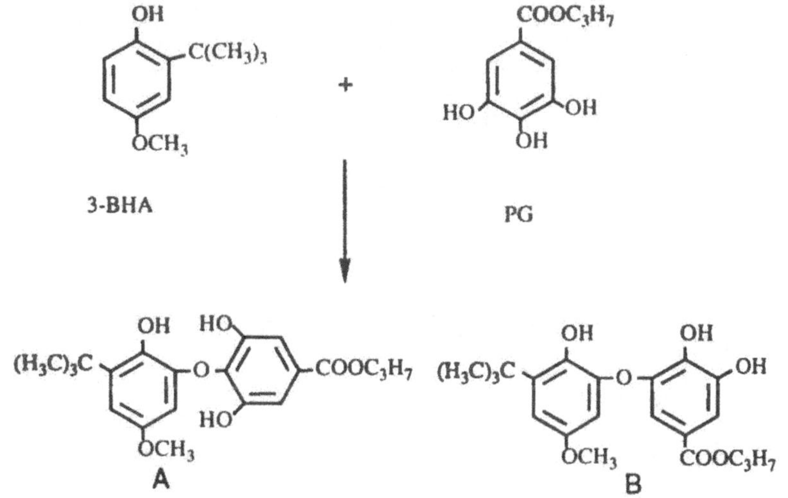

Fig. 3.9 Heterodimers from the oxidation of a mixture of 3-BHA and PG.

mechanism involved in synergism between different compounds has been reviewed by many investigators (110,119,120). Synergists function by different mechanisms. Synergists in general extend the life of a primary antioxidant, and the antioxidant effect of the mixture is greater than when the compounds are used alone. They may act as hydrogen donors to the primary antioxidant radicals, thus regenerating the primary antioxidant, or inactivate metal ions, thus neutralizing their prooxidant effects. The primary antioxidant is not destroyed so rapidly by free radicals generated by peroxide decomposition and thus remains effective for a longer period (120).

Synergism between 3-BHA and BHT has been studied in model systems

Table 3.10 Antioxidant Activity of 3-BHA, Propyl Gallate, and Heterodimers

	Antioxidant activity $(t - t_0)/(t_{3\text{-BHA}} - t_0)$		
Compound (0.01%)	Methyl oleate	Lard	Soybean oil
3-BHA	1.0	1.0	1.0
Propyl gallate	0.52	1.50	12.25
Heterodimer A[a]	0.53	0.88	4.0
Heterodimer B[a]	0.47	0.91	1.50

[a]Refer to Fig. 3.9.
Source: Ref. 117.

containing peroxy radicals prepared by the cobalt-catalyzed cleavage of *tert*-butyl hydroperoxides. It was observed that 3-BHA donated a hydrogen atom to the peroxy radical to form a phenoxy radical. A transfer of hydrogen from BHT regenerated 3-BHA from the phenoxy radical. In the process BHT was oxidized to quinone methide (Fig. 3.10) (121,122). A combination of 3-BHA and BHT showed a higher antioxidant activity than either of them used singly in soybean oil, lard, and methyl oleate (Table 3.9). Synergism between tocopherols and ascorbic acid has also been studied extensively. It was observed that ascorbic acid functions by donating hydrogen to the α-tocopheryl radical, regenerating α-tocopherol. Niki et al. (123) demonstrated the synergism between ascorbic acid and α-tocopherol in a methyl linoleate model system. When used alone, both disappeared linearly with time. When they were used in combination, ascorbic acid disappeared first, followed by α-tocopherol.

Phospholipids like phosphatidylethanolamine (PE), phosphatidylcholine, and phosphatidylserine are effective synergists. Of the three, phosphatidylethanol-amine greatly enhances the activity of a wide range of primary antioxidants including tocopherols, BHA, BHT, TBHQ, and PG at temperatures above 80°C (124). Husain et al. (125) demonstrated that at higher temperatures both phosphatidylcholine and phosphatidylethanolamine formed brown products that were effective in inhibiting the formation of linoleate hydroperoxides. Phosphatidylcholine was found to be highly effective in ternary mixtures with vitamin E and ascorbic acid (13). Similar effects have been observed in mixtures comprising

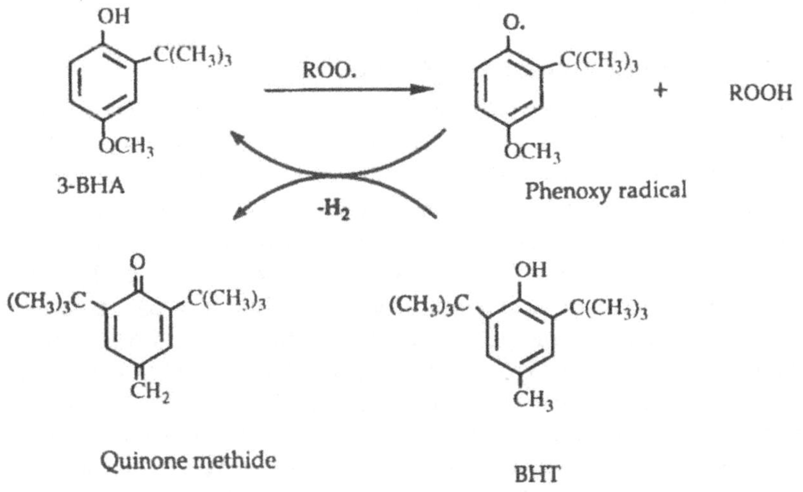

Fig. 3.10   Synergism between 3-BHA and BHT.

ascorbyl palmitate, α-tocopherol, and phosphatidylcholine (126). However, the mechanism of action of phosphatidylcholine is not clear. It is postulated that in ternary mixtures the regeneration of vitamin E by ascorbic acid or ascorbyl palmitate is mediated by phosphatidylcholine (13,127).

The synergistic action of other metal chelators like citric acid, phosphoric acid, EDTA, and tartaric acid could be due mainly to the inactivation of metal ions that exert a prooxidant effect. Hydrolyzed vegetable proteins and autolyzed yeast proteins function as potent synergists with primary antioxidants such as BHA and BHT. However, the chemical basis of the activity has not been established (128). The amino acids glycine and methionine also function as synergists. Various amino acids enhance the activity of Trolox-C in vegetable oils (77).

3.7 DETERMINATION OF ANTIOXIDANTS IN FOOD SYSTEMS

The methods of detecting and quantitatively estimating synthetic and natural antioxidants have been reviewed extensively by Endean (129), Robards and Dilli (130), and Kochhar and Rossell (131). These methods range from qualitative detection by color reactions to semiquantitative and quantitative methods such as spectrophotmetry; voltammetry; polarography; and chromatographic methods like paper, thin-layer, and column chromatography and the more advanced gas-liquid chromatography (GLC) and high-performance liquid chromatography (HPLC). The various methods are discussed in detail in Refs. 129–131. We present a brief summary of the analytical techniques developed for some of the commonly used antioxidants. Table 3.11 presents some the AOAC methods for the determination of various primary and secondary antioxidants.

3.7.1 Determination of Primary Antioxidants

Butylated hydroxyanisole, butylated hydroxytoluene, and the gallates are extracted from fats and oils with either aqueous alcohol or acetonitrile. Aqueous alcoholic extraction of antioxidants from fat has the disadvantage that losses by volatilization of BHA and BHT are likely to occur if a concentration step is incorporated. Hence, extraction with acetonitrile or dry alcohol or a combination of the two is employed. Steam distillation has also been used for the isolation of BHA and BHT. Propyl gallate can be extracted with water from a solution of a fat–antioxidant mixture in petroleum ether, hexane, or heptane. Ethoxyquin is extracted with 1 N HCl. In the case of high-fat foods, fat and antioxidant are generally removed from the dried product by Soxhlet extraction with petroleum ether or chloroform/methanol (1:1). The antioxidants are then quantitatively isolated from the fat as before. BHA and BHT are extracted from low-fat foods by direct extraction with ethyl ether or carbon disulfide. Gallates can be directly extracted from low-fat foods with alcohol or acetonitrile. A clean-up step is generally incorporated to remove the interfering

Table 3.11 AOAC Methods for the Determination of Some Primary and Secondary Antioxidants

Antioxidant	Product	Method	Reference, section
BHA, BHT, TBHQ, THBP, PG, Ionox-100	Oils, fats	HPLC, UV detection at 280 nm	133, 20.009–20.013
BHA, BHT	Breakfast cereals	GLC	133, 20.014–20.020
α-Tocopherol	Foods and feeds	Saponification, TLC, colorimetry, polarimetry	133, 43.129–43.151
Ethoxyquin	Chicken tissues, eggs	Photofluorimetry	133, 41.024
Ascorbic acid	Foods, vitamin preparations	Titrimetric, fluorimetric	133, 43.064–43.081
Sulfites	Meat	Spectrophotometry	133, 20.043, 20.129
Tartaric acid	Cheese, fruit products	Titrimetry	152, 16.234, 22.063
Citric acid	Cheese, fruit products, milk	Pentabromoacetone method, chromatographic	152, 16.237, 16.024, 22.066, 22.074
Phytic acid	Foods	Chromatographic	157, 986.11
Lecithin	Fat	Ashing/phosphorus estimation	133, 13.045

Source: Compiled from AOAC (133, 152, 157).

substances before analysis. Column chromatography using polyamide, silica gel, or Florisil is generally employed, depending on the antioxidant in the extract (129).

Qualitative and Quantitative Analytical Methods

Several rapid screening tests for qualitative analysis have been found useful for routine analysis. The presence of BHA and gallates in lard can be determined by the decolorization of 0.2% $FeCl_3$ and 0.1 M NH_4SCN. BHT can be detected in fats, oils, and bacon by the use of $K_3Fe(CN)_6/Fe_2(SO_4)_3$ (129).

Spectroscopic Methods

Various colorimetric and UV/visible spectrophotometric methods have been developed for the semiquantitative or quantitative estimation of various antioxidants. However, many of these tests are not specific and require selective extraction and calibration procedures. The colorimetric method of Emmerie and Engel (132) for the estimation of total tocopherol content is based on the reduction of $FeCl_3/2,2'$-bipyridyl to form a red color. This method is nonspecific and has also been used for the determination of BHT, PG, and BHA (129). The AOAC method for the determination of α-tocopherol involves solvent extraction, saponification, and TLC followed by colorimetry using bathophenanthroline/$FeCl_3$ reagent (133). Bukovits and Lazerovich (134) developed a second-order derivative method for determining the concentration of individual saturated tocopherols using multicomponent analysis and standard mixtures as calibrants.

Several colorimetric methods have been developed for the determination of BHA. The use of diazobenzenesulfonic acid involves the development of the characteristic red-purple color by the diazotization of BHA with Ehrlich reagent in the presence of 1 N sodium hydroxide. BHA can also be estimated by reaction with 2,6-dichloroquinone chlorimide (Gibbs' reagent), which gives a stable blue indophenol with maximum absorption at 620 nm. BHA can be determined by measuring the fluorescence in 50% ethanol-ligroin. BHT is determined by the reaction with dianisidine and nitrous acid, which gives a pink compound with an absorption maximum at 520 nm. The gallates are estimated by the reaction with ferrous tartrate, resulting in a purple complex measured at 540 nm. The AOAC method for the estimation of ethoxyquin in chicken eggs and tissue is based on fluorescence after extraction in isooctane in the presence of anhydrous Na_2SO_4 and anhydrous Na_2CO_3 (129,133).

Chromatographic Techniques

Chromatographic techniques such as paper, thin-layer, and gel permeation chromatography have been developed by various workers since the early 1950s for the detection or semiquantification of antioxidants. Quantitative or semiquantitative results have been obtained by the use of densitometry or by elution of spots followed by spectrophotometry or colorimetry. The methods employing activated silica gel layers seem to be the best for separating BHA, BHT, α-tocopherol, and the gallates. The spray reagents employed for the detection of various antioxidants are listed by Endean (129) and Robards and Dilli (130).

Several GLC methods have also been developed specifically for estimation in cereals, fats and oils, emulsions, vitamin preparations, and potato powders. Gas-liquid chromatographic methods have been developed mainly for the quantification of tocopherols, BHA, and BHT. Silyl derivatives are used for estimations. A direct injection method has also been developed for BHA and BHT. SE-30 or Carbowax-20M columns and flame ionization detectors are most frequently used. To achieve satisfactory results with GLC, it is necessary to obtain a pure quantitative extract or to select a column and operating conditions that will separate the unknown antioxidants from each other and from interfering substances. The various GLC methods are listed by Kochhar and Rossell (131) and Endean (129).

In general, the colorimetric, spectrophotometric, and chromatographic methods discussed so far are time-consuming, and many are subject to interference from other compounds in the extract. Newer approaches have arisen with the development of rapid HPLC methods that are more quantitative and use simplified sample preparation procedures. Both normal and reverse-phase HPLC methods have been developed for most of the common primary antioxidants (Table 3.12). Numerous HPLC methods have been developed for tocopherols, involving extraction of tocopherols from foods with solvents, dilution of the oils and fats with solvent without saponification, and saponification of the sample, followed by extraction and dilution with the solvent. The various methods are listed by Kochhar and Rossell (131).

For phenolic antioxidants in oils, the AOAC has recommended a reverse-phase system with a Lichrosorb RP-18 column coupled to UV detection at 280 nm. The mobile phase consists of (A) 5% acetic acid and (B) 100% acetonitrile, and elution is with a linear gradient of 30% B in A to 100% B over 10 min with a 4-min hold at a flow rate of 2 mL/min. This method is applicable to PG, BHA, BHT, TBHQ, and Ionox-100 (133).

Deschnytere et al. (135) developed a method for the analysis of tocopherols without previous extraction that can also be used for the separation of triglycerides. The simultaneous reverse-phase HPLC method developed by Andrikopoulos et al. (136) is a very useful technique. It includes separation of triglycerides and phenolic antioxidants by a modification of the AOAC method. The study reported separation of triglycerides together with nine synthetic phenolic antioxidants as well as the tocopherols and α-tocopheryl acetate. A reverse-phase C_{18} column and gradient elution with water (pH 3.0)/acetonitrile/methanol/isopropanol was used for oil dissolved in isopropanol/hexane. The best separation of antioxidants was with a 5:7 ratio of methanol and acetonitrile. Methanol was used as a polar modifier, and

Table 3.12 Some HPLC Methods Used for the Determination of Primary Antioxidants

Antioxidant	Product	Stationary phase	Mobile phase	Detection
BHA, BHT, PG, THBP, Ionox-100, OG, DG	Oil, lard, shortenings	10 μm Lichrosorb RP-18	Gradient elution: (a) 5% acetic acid in water; (b) 5% acetic acid in acetonitrile	UV, 280 nm
BHA, BHT, PG, TBHQ	Oils, fats	μ-Bondapak C$_{18}$	Gradient mixture: 1% acetic acid in water, 1% acetic acid in methanol	UV, 280 nm
BHT, BHA, PG, DG, OG	Oils, fats	10 μm Bondapak C$_{18}$	Gradient elution aqueous acetic acid and methanol	UV, 280 nm
BHA, PG, TBHQ	Oils, foods	10 μm Bondapak C$_{18}$	Methanol: 0.1M ammonium acetate (1:1)	Amperometric
BHA, PG, OG, DG, TBHQ	Oils, fats	5 μm Lichrosorb DIOL	Hexane: 1,4-dioxane: acetonitrile (62:28:10)	UV, 280 nm
BHT, BHA, TBHQ	Oils	Porasil or Rad-Pak Cyano	Hexane: dichloromethane: acetonitrile (85:9.5:5.5)	UV, 280 nm
α-, β-, γ-, δ-Tocopherols, α-, β-, γ-, δ-tocotrienols	Oils diluted with solvent	Partisil-5	Dry heptane/damp heptane/isopropanol (49.55:49.55:0.9)	Fluorescence ex. 290 nm em. 330 nm
α-, β-, γ-, δ-Tocopherols	Oils diluted with mobile phase	μ-Porasil	Hexane/isopropanol (98.5:1.5)	UV, 295 nm

Source: Compiled from Kochhar and Rossell (131), which contains the original references.

the acidic conditions prevented tailing of the antioxidants. The less polar triglycerides were eluted with isopropanol. To date, this procedure is considered worthwhile as it has the precision and capability to separate the various antioxidants and triglycerides simultaneously. Denis Page and Charbonneau (137) reported simultaneous HPLC determination of BHT, BHA, PG, TBHQ, THBP, and Ionox-100 using the AOAC method from a number of dry foods such as breakfast cereals, potato flakes, rice, and cake mixes. All dry samples were rehydrated before extraction to maximize recoveries of antioxidants.

Vinas et al. (138) developed an HPLC method with UV detection at 270 nm for the determination of ethoxyquin in paprika that eliminates the necessity of removing other colored substances. Samples were extracted with ethyl acetate and analyzed by reverse-phase HPLC using gradient elution with acetonitrile and water. The sensitivity is increased with fluorimetric detection with excitation at 311 nm and emission at 444 nm.

3.7.2 Determination of Synergistic Antioxidants

Oxygen Scavengers

In the Monier–Williams method, sulfites are separated from the food matrix by acid distillation. The sulfur dioxide is then oxidized by hydrogen peroxide to sulfuric acid, which is estimated by titration with alkali (139). An ion-exclusion chromatographic method with electrochemical detection has been reported for total sulfites that has broad applicability to many food products (140,141).

Titrimetric methods have been described for the estimation of ascorbic acid, ascorbyl palmitate, sodium ascorbate, and sodium erythorbate (142). The AOAC method for the determination of ascorbic acid involves titration with 2,6-dichlorophenol-indophenol (133). In addition, direct UV spectrophotometry, differential pulse polarography, and HPLC methods using anion-exchange or reverse-phase columns with UV detection have been developed for the estimation of ascorbic acid (131). Ascorbic acid can be determined even in the presence of other additives and colorants in soft drinks by a method using second- and third-order derivative spectrophotometric data (143). Another recent method is for the simultaneous determination of ascorbic acid and isoascorbic acid in food products by paired-ion reverse-phase HPLC (144). Food samples were extracted with metaphosphoric acid, and the compounds were separated using a mobile phase containing metaphosphoric acid, sodium acetate, and tetrabutylammonium hydrogen sulfate, pH 5.4. Detection was by amperometry using a glassy carbon electrode and a Ag/AgCl reference electrode.

Estimation of ascorbyl palmitate in fats and oils by a potentiometric titration method that does not require extraction has been reported. An HPLC method has been developed for the estimation of ascorbyl palmitate in flour, yogurt, and vegetable oils. Ascorbyl palmitate was extracted with methanol, and the analysis was carried out using a 5-μm Chromegabond diamine column and UV detection at 255 nm. The mobile phase consisted of methanol/0.02 M monobasic phosphate buffer (70:30 v/v) at pH 3.5 (131).

Chelating Agents

The ISO method for the detection of polyphosphates in meat and meat products involves extracting the samples with trichloroacetic acid, clearing the serum obtained with an ethanol–diethyl ether mixture, separating the phosphates by TLC, and using ammonium molybdate/stannous chloride reagent for detection (145). For the quantitative analysis of polyphosphates in various soft drinks, atomic absorption spectrometry and anion-exchange column chromatography have been reported (146). Matsunaga et al. (147) reported an HPLC method for the determination of polyphosphates in various food products such as processed cheese, kamaboko, chikuwa, ham, and sausage. The phosphates were extracted with cold 4% trichloroacetic acid. Linear polyphosphates with the range of polymerization from 2 to 12 were separated on an anion-exchange column, Protein Pak G-DEAE with 0.01 M HNO_3 containing a linear gradient of $NaNO_3$ as mobile phase. Polyphosphates were detected as the decrease of absorbance at 500 nm of a ferric chloride–sulfosalicylate complex by means of an on-line postcolumn reaction with 0.5 mM $FeCl_3$ solution containing 2.5 mM sulfosalicylic acid at room temperature.

Colorimetric and HPLC methods have been reported for the determination of EDTA. The colorimetric method is based on the color reaction of free Cu^{2+} ions with bathocuprone; EDTA-Cu chelate does not react under any conditions. Food samples were homogenized with 0.1 M NaOH and subjected to equilibrium dialysis against 0.02 M NaOH, and EDTA and its metal chelates were converted to EDTA-Cu by the addition of $CuSO_4$. From the difference between absorbance at 477 nm of free Cu^{2+} and total Cu^{2+} EDTA was calculated. The method is applicable to mayonnaise, salad dressing, and canned foods (148). A reverse-phase HPLC method has been reported for EDTA that can be applied to sauce, salad dressing, mayonnaise, margarine, canned lima beans, canned kidney beans, and sandwich spread (149). EDTA was extracted with water, purified on a DEAE-Sephadex column, and converted to EDTA-Fe(III) by adding ferric chloride. The chelate was separated on a Lichrosorb RP-18 column. The solvent was a mixture of 5% ethanol and 95% water containing 0.01 M tetrabutylammonium hydroxide, pH 4.5, and a UV detector at 255 nm was used to identify the peaks. Reverse-phase ion-pair HPLC methods have also been reported for the determination of EDTA as Cu^{2+}–EDTA complex in canned mushrooms and canned crustaceans (150,151).

The AOAC method for tartaric acid involves its conversion to cream of tartar and titrimetry with 0.1 N alkali (152). Colorimetric, titrimetric, and GLC methods for the determination of citric acid have been reported (131). Ashoor and Knox

(153) reported a simple, accurate HPLC method using an Aminex HPX-87H column and a mobile phase of 0.009 N sulfuric acid for the determination of citric acid in a wide variety of foods, including processed cheese, cottage cheese, powdered drinks, frozen orange juice concentrate, strawberries, and tomatoes. The samples were extracted with water, and EDTA was added as a chelating agent prior to injection. Fujita et al. (154) reported a highly sensitive and simple spectrophotometric method for the determination of citric acid in foods. The method is based on the formation of a ternary complex between citric acid, O-hydroxyhydroquinonephthalein and Fe(III) with minimum interference from other acids in the sample. Tsuji et al. (155) reported a simple and rapid GLC method for the determination of isopropyl citrates in butter, oils, and milk powder. Ushiyama et al. (156) reported a quantitative method for isopropyl citrate in cottonseed oil and butter. Isopropyl citrate was hydrolyzed to citric acid, and HPLC was used to quantify the citric acid.

The AOAC method for phytic acid involves extraction with dilute HCl. The extract is mixed with EDTA/NaOH solution and placed on an anion-exchange column (AG_1-X_4, chloride form). Phytate is eluted with 0.7 M NaCl solution and wet-digested with a mixture of concentrated HNO_3 and H_2SO_4 to release phosphorus, which is measured colorimetrically (157). Bos et al. (158) reported an improved HPLC determination of phytic acid. Phytic acid was separated with an anion-exchange column and detected at 300 nm after reaction with a ferric salt in an in-line postcolumn derivatization. The sample extracts were treated with EDTA to eliminate the interference from metal ions. The standard method for lecithin involves extraction with solvents and determination of phosphorus after ashing. Several TLC and HPLC methods have also been developed for the quantitation of phospholipids (131).

3.8 DETERMINATION OF ANTIOXIDANT ACTIVITY

The effectiveness of synthetic and natural antioxidants is measured by monitoring the oxidative stability of the food lipids. After the sample is oxidized under standard conditions, the extent of oxidation is measured by chemical, instrumental, or sensory methods. Most of these studies are aimed at measuring the extension of the induction period by the addition of the antioxidant. The extension in induction period is also expressed as an antioxidant index or protection factor. The induction period is the time required for the sample to start oxidizing rapidly and coincides with the onset of off-flavor development in the lipids.

3.8.1 Accelerated Stability Methods

Accelerated stability methods generally involve storage studies run under normal or accelerated conditions. Some of the conventional stability tests and their limitations have been reviewed by Ragnarsson and Labuza (159), Rossell (160), Sims and Fioriti (9), and recently by Frankel (161). Table 3.13 lists the conventional stability tests in increasing order of severity of the oxidation conditions used and their limitations.

According to Frankel (161), the limitations of high-temperature stability tests include the following. The rates of oxidation become dependent on oxygen concentration because the solubility of oxygen decreases at elevated temperatures; oxidation occurs rapidly and results in drastic changes in oxygen availability; the induction period occurs at an oxidation level that is too high and beyond the point at which rancid flavors are detected; side reactions such as polymerization and cyclization become important and may not be relevant to normal storage temperatures; analysis of oxidation under these conditions is of questionable value; volatile antioxidants such as BHA and BHT are subject to significant losses at elevated temperature; phenolic antioxidants in natural extracts decompose at elevated temperatures. Some of the discrepancies can be alleviated by testing at different temperatures and calibrating the tests for each formulation.

3.8.2 Measurement of Oxidative Stability

After the sample is oxidized by any of the above methods, the formation of hydroperoxides or their decomposition products is measured by a chemical, instrumental, or sensory method (Table 3.14). A number of excellent reviews have appeared describing the various methods and related literature (160–163). We present a brief summary of the main principles involved in various methods.

Table 3.13 Standard Accelerated Stability Tests

Test	Conditions	Characteristics
Ambient storage	Room temperature, atmospheric pressure	Too slow
Light	Room temperature, atmospheric pressure	Different mechanism
Metal catalysts	Room temperature, atmospheric pressure	More decomposition
Weight-gain method	30–80°C, atmospheric pressure	Endpoint questionable
Schaal oven	60–70°C, atmospheric pressure	Fewest problems
Oxygen uptake	80–100°C, atmospheric pressure	Different mechanism
Oxygen bomb (ASTM)[a]	99°C, 65–110 psi O_2	Different mechanism
Active oxygen (AOM)	98°C, air bubbling	Different mechanism
Rancimat	100–140°C	Endpoint questionable

[a]ASTM = American Society for Testing and Materials.
Source: Ref. 161.

Table 3.14 Analytical Methods to Measure Oxidative Stability of
Fats and Oils

Chemical methods	Chromatographic methods
Peroxide value	HPLC
Anisidine value	Gas chromatography
Thiobarbituric acid test	Headspace gas analysis
Kreis test	GC-MS
Carbonyl value	Measurement of oxygen absorption
Acid value	Dissolved oxygen meter
Spectrophotometric methods	Warburg's manometer
UV absorption	Weighting method
Carotene bleaching test	Sensory methods
IR spectrometry	Flavor and odor evaluations
ESR spectrometry	
Chemiluminescence	

Chemical Methods

Various chemical methods have been developed to measure the peroxides, hydro-peroxides, free fatty acids, and decomposition products, especially aldehydes, which contribute to off-flavors. Peroxide value (PV) is the most common measure of oxidative rancidity. The hydroperoxide or peroxide formed is estimated iodomet-rically. The sample is reacted with a saturated aqueous solution of potassium iodide, and the iodine liberated by the peroxides is titrated with a standard solution of sodium thiosulfate (164). The peroxide value is expressed in units of milliequivalents of oxygen per kilogram of fat. An electrochemical method has also been developed to measure the iodine. Peroxide value can be determined colorimetrically based on the oxidation of ferrous to ferric ion, with the latter determined as ferric thiocyanate (165). The acid value measures the free fatty acids formed titrimetrically with standard sodium hydroxide to a neutral endpoint. The conductivity of the low molecular weight free fatty acids can be measured by the Rancimat apparatus.

The Kreis test is a colorimetric method based on the reaction of phloroglucinol with epoxyaldehydes and malonaldehyde (166). The red color formed is measured spectrophotometrically. The thiobarbituric acid (TBA) test, which is again widely used, is based on the color reaction of TBA with malonaldehyde. The results are expressed as TBA number, which is calculated as milligrams malonaldehyde per kilogram sample. The anisidine value is based on the reaction of aldehydes, especially 2-alkenals, with p-anisidine, and the reaction products are measured spectrophotometrically at 350 nm (167). The carbonyl value is a measure of the carbonyl compounds formed during oxidation. It is based on the conversion of volatile carbonyl products to 2,4-dinitrophenylhydrazone derivatives (168,169).

Oxygen Absorption Methods

Oxygen absorption methods are based on the fact that the oxidation of fats and oils is accompanied by an uptake of atmospheric oxygen and an increase in the weight of the samples. The oxygen uptake is measured with a Warburg manometer. The apparatus is a closed system in which the sample is held under oxygen and the oxygen uptake is measured manometrically. In the weight-gain method, oxygen absorption is measured by recording the increase in the weight of the oxidizing fat under controlled conditions.

Spectrophotometric Methods

The UV absorption method essentially measures the hydroperoxides and the conjugated dienes and trienes, which absorb in the UV region from 232 to 278 nm. Since several compounds absorb in this range, this method may not be suitable for complex food systems. Infrared spectroscopy measures unusual functional groups and compounds with trans double bonds (160). Ultraviolet and infrared spectroscopy are useful in conjunction with HPLC. The carotene-bleaching method is based on the coupled oxidation of β-carotene and linoleic acid in a model system (170).

Electron spin resonance (ESR) spectroscopy detects free redicals. It is based on the absorption of microwave energy emitted by the promotion of an electron to a higher energy level when the sample is placed in a variable magnetic field. The ESR technique requires radical concentrations higher than 10^{-8} M. Since the steady-state concentrations of free radicals generally remain below 10^{-7} M in solutions, spin trapping methods are commonly followed to reduce the rate of disappearance of the radicals. The spin trapping method uses exogenously added compounds to trap the highly reactive radicals to form more stable detectable radical adducts. Nitroso or nitrone compounds are commonly used as spin traps (163). The ESR technique has been applied in biological and model systems and could become a useful technique in food systems. The spin trapping method has been used successfully by Zhao et al. (171) to study the antioxidant activity of green tea extracts, ascorbic acid, rosemary antioxidant, and other natural antioxidants.

The chemiluminescence method is based on the reaction of luminol, luciferin, and other luminescent molecules with peroxy radicals or singlet oxygen generated by the oxidation of hydroperoxides. The method is used mainly in the ultra-microanalysis of biological oxidations in conjunction with advanced detectors and HPLC (172,173).

Chromatographic Methods

Gas chromatographic methods are widely used for the measurement of volatile compounds either by headspace analysis or by direct injection of the oil. The headspace methods can be applied to both oils and fatty foods. Hexanal and pentane are commonly measured by gas chromatography, as are other volatiles such as 2-heptanal, 2-decanal, 2,4-decadienal, and 2,4-heptadienal. The combination of gas chromatography and mass spectroscopy (GC/MS) offers a highly sensitive analytical technique for the identification and quantification of off-flavor compounds. High correlation coefficients have been obtained between individual or total volatiles and flavor score (174). Section 3.9 includes some of the studies on the GC analysis and correlations with sensory quality in food products. High-performance liquid chromatography (HPLC) is another highly useful technique to measure peroxides, hydroperoxides, and secondary oxidation products.

Sensory Evaluation Methods

Sensory evaluation methods are based on flavor and odor evaluations. These evaluation methods will be discussed in detail in the forthcoming section.

The advantages and limitations of many of these methods are reviewed elsewhere (160,161,163,175). The limitations and advantages compiled by Frankel (161) include the following.

The determination of peroxide value provides an empirical measure that is less sensitive and precise than sensory and headspace methods.

Oxygen absorption methods have limited sensitivity and require high levels of oxidation as the endpoint for induction periods.

The TBA test is sensitive and precise but is not specific because other components such as browning reaction products interfere with the reaction.

The Rancimat requires a higher level of oxidation, and the endpoint oxidation indices are unreliable because high temperatures are required to obtain detectable organic acids.

The carotene-bleaching method is simple and sensitive but is subject to interference from oxidizing and reducing agents present in crude extracts.

The analysis of volatiles by gas chromatography is closely related to flavor evaluation and also provides data on the origin of flavor and odor volatiles and their precursors.

Sensory methods provide the most useful information related to consumer acceptance of the product. Although these methods are very sensitive, they are highly dependent on the quality of the training the taste panel has received.

Since numerous methods have been developed, with varying testing conditions and endpoints, it has become increasingly difficult to compare and interpret the antioxidant activity of different compounds or extracts based on published literature. This is especially true in determining the activities of natural antioxidant extracts wherein the activity depends to a large extent on the mode of extraction, the presence of inhibitors, and the nature and concentrations of active components in the crude extracts. Frankel (161) lists some of the published literature on the methods used by investigators in the evaluation of various natural antioxidants (Table 3.15).

3.9 MEASUREMENTS OF OXIDATIVE STABILITY IN TERMS OF SENSORY QUALITY

Changes in food quality due to oxidation range from a change in color to changes in appearance, odor, texture, aroma, and taste. In principle, everyone dealing with foods agrees that sensory quality is the ultimate factor in judging quality in any given situation—product formulation, processing, and storage. Surveillance of quality in these three stages is important for successful marketing of the product. Frankel (161), in his review on the search of better methods to evaluate natural oxidants and oxidative stability in food lipids, ranked the sensory method as the first among the various analytical methods for predicting stability, shelf life, and consumer acceptability with special reference to lipid oxidative stability. He also states that if the panel is well trained and the methods are appropriate, the most useful information will result from sensory methods. The methods have been ranked according to their usefulness in predicting: sensory > headspace volatiles > oxygen absorbance > peroxide value > thiobarbituric acid–reactive substances (TBARS) > carotene bleaching by oxidation with linoleic acid > conductivity caused by short-chain acids produced at elevated temperatures.

An attempt has been made here to review the literature that mentions at least one sensory parameter in the determination of changes in food quality due to oxidation. There have been more reports on animal-based foods than on other foods. This is understandable as animal-based foods are more expensive, deteriorate much faster, and hence cannot be consumed unless they are well preserved. The studies of the various workers in the field are grouped by product in the following pages. Natural antioxidants, crude extracts of natural materials, concentrated extracts, and synthetic or chemical antioxidants either individually or in combination have been used extensively in these studies.

Pork

Liu et al. (176) studied the efficacy of rosemary oleoresin and sodium salts in restructured pork steaks using TBA analysis, sensory evaluation, and hexanal content (frozen storage only) to monitor lipid oxidation in pork steaks. Sodium tripolyphosphate (STPP) significantly reduced lipid oxidation in cooked steaks during refrigerated storage (~4°C) for 8 days and in raw steaks stored at −30°C for 8 months. Lipid stability was not enhanced by treatment with oleoresin rosemary (OR) and STPP compared to STPP alone. Water-soluble OR/STPP did not result

Table 3.15 Evaluation Methods Used for Various Natural Antioxidants

Natural antioxidant	Substrate	Stability test	Temperature/conditions	Method (endpoint)
Spices	Corn oil emulsion, lard, pie crust	Warburg	40°C	O$_2$ uptake
		AOM	98°C	PV
Hydroxyflavones	Methyl linoleate emulsion	Co catalyst	Room temp.	O$_2$ uptake
		Fe catalyst	Room temp.	PV
Flavonoids	Lard	Oven	60°C	IP (PV 25)
Rosemary	Chicken fat	Oven	60°C	PV
	Soybean oil	—	Room temp.	Sensory
Polyphenols (leaves)	Lard, soybean oil	Astell	100°C	IP, PV
Rosemary, sage, cocoa shells	Chicken fat, potato flakes	Oven	90°C	IP
		—	UV	Headspace, residual O$_2$, sensory, carotenoids
Rosemary	Lard	Oven	60°C	PV
Polyhydroxyflavonoids	Lard	Rancimat	140°C	Conductivity
Hydroxyisoflavones	Lard	Rancimat	100°C	Conductivity
Rosemary	Linoleic acid	—	Room temp.	TBARS
Flavonoids (peanuts)	Carotene-linoleic acid	Bleaching	TLC	Visible (470 nm)
Rosemary quinone	Lard	Oven	60°C	PV
Carnosol, rosmanol	Lard	AOM	98°C	PV
Catechins (tea)	Lard	AOM	98°C	PV
Vanillin	Cereal flakes	—	Room	Headspace, sensory, residual O$_2$
Rosemary, carnosic acid, and synergists	Lard, peanut oil	Rancimat	100°C	Conductivity
	Lard	Rancimat	140°C	Conductivity
Flavones, flavanones, flavonols	Stripped corn oil, lard	Rancimat	100°C	Conductivity (PV 50)

Table 3.15 (*Continued*)

Natural antioxidant	Substrate	Stability test	Temperature/conditions	Method (endpoint)
Rosemary	Potato flakes, wheat flakes, pork patties	Room	20°C	Headspace (pentane)
Oat extracts	Soybean oil	Oven	32°C, 60°C, 180°C	PV (IP) Conjugated dienes
Isoflavones (soybean)	Chicken fat, chicken olein, carotene, linoleic acid	Rancimat	100°C	Conductivity
		Oven	37°C	HPLC (234 nm)
		Bleaching	UV	Visible (450 nm)
Polyphenols (green tea)	Chicken fat	Rancimat	100°C	IP rancidity
Rosemary, carnosic acid, ursolic acid	Lard, linoleic acid	Rancimat	110°C	IP
		Lipoxygenase	22°C	IC$_{50}$

*IP = induction period; IC$_{50}$ = concentration necessary for 50% inhibition; AOM = active oxygen method; PV = peroxide value; TBARS = thiobarbituric acid–reactive substances.
Source: From Frankel (161) which contains the original references.

in significantly ($P > 0.05$) greater lipid stability than oil-soluble OR/STPP treatments. Hexanal content significantly increased after 8 months of frozen storage.

Boneless pork loin roasts were either hot-processed (loins removed from carcasses 1 h postmortem) or cold-processed (loins removed after chilling the carcasses at 3°C for 24 h) and then injected with a stock solution containing 10% salt (9:1 NaCl/KCl) and 3% phosphate (1:1 hexametaphosphate/pyrophosphate) containing an antioxidant (0.5% ascorbic acid, 0.02% BHA, or 0.02% TBHQ) or no antioxidant (control) (177). The roasts were injected at 1-cm intervals to 110% of the fresh weight and allowed to equilibrate for 1 h. Chops removed from the roasts were either wrapped, stored at −18°C for 45–63 days, thawed, cooked, and then examined for palatability attributes or overwrapped with polyvinyl chloride (PVC) film, held in a retail case at 3–4°C under twin 40-W fluorescent bulbs, and examined daily for visual attributes. Results indicated that hot-processing produced chops that were more tender and had a darker red lean color than cold-processing. None of the antioxidants affected lean color. Ascorbic acid and BHA generally had no effect on discoloration or on palatability. TBHQ increased discoloration and decreased overall desirability but slightly increased the flavor desirability.

Effects of dietary tocopherol and oxidized oil on oxidative stability of membranal lipids in pig muscles and on oxidative stability of pork products during refrigerated and frozen storage were evaluated by Buckley et al. (178). Membrane-bound α-tocopherol stabilized membranal lipids and reduced the extent of lipid oxidation occurring in pork patties and pork chops during storage. Oxidized dietary oil had an adverse effect on the stability of both membranal lipids and pork products. Salt added to pork patties accelerated oxidation when the patties were stored under fluorescent light and in the dark.

Marriot and Graham (179) reported the use of BHA, ascorbic acid, sodium tripolyphosphate, KCl, and lecithin, in various combinations at various levels, for protecting restructured pork chops during frozen storage. Quality attributes evaluated included flavor, cohesiveness, appearance, juiciness, texture, color, and oxidative rancidity (TBA value). Best results were obtained by formulations including BHA, polyphosphate, and ascorbic acid. Storage time seemed to have more effect on visual quality than additives used. Lecithin did not appear to function as an antioxidant according to these studies.

Olson and Rust (180) dry-cured hams with a curing mixture of salt, sugar, and NaNO₃, a salt low in heavy metal ion content, and a proprietary salt mixture containing BHA, BHT, and citric acid as antioxidants. After aging, hams were chemically analyzed, and a taste panel evaluated them in terms of flavor, saltiness, tenderness, and overall satisfaction. The proprietary mixture showed a significant reduction of rancidity as measured by thiobarbituric acid values for fat and a greater flavor preference. However, no correlation between flavor preference and rancidity could be established because of the design of this experiment.

The efficacy of three antioxidants and a reductant for preventing deterioration contributing to the retail acceptability of bacon slices (muscle color, surface discoloration, retail appearance of lean and fat surface areas) during frozen storage and simulated retail display was examined by Jeremiah (181). The antioxidants BHA, BHT, and propyl gallate and the reductant ascorbic acid were incorporated into a dry sugar bacon cure alone and in combination. Composite results indicated that incorporation of the evaluated formulations into dry sugar bacon cures did not appear to be practical for either extending the frozen storability or retail display life of frozen and thawed bacon from an appearance aspect. However, incorporation of BHA and BHT in combination extended the retail display life of fresh bacon slices by approximately 3.5 days, based upon regression analysis.

Lard

Titarenko et al. (182) showed that addition of retinol (50–100 mg/kg) increased the biological value of pork lard and acted as an antioxidant during 6 months of storage at fluctuating temperature (16–27°C) and relative humidity (48–72%). Addition of retinol at 100 mg/kg was most effective; this resulted in acid and peroxide values remaining at acceptable levels for 4 months (2 months for control) and relatively little loss in sensory quality throughout storage.

Antioxidative and synergistic activities with antioxidants of sucrose ester of fatty acids (SEF) were investigated with respect to autoxidation of lard by Oishi et al. (183). The degree of oxidation of lard was measured by peroxide, acid, iodine, and dinitrophenylhydrazine values. To examine the stabilities of BHA, BHT, and tocopherols (Toc) added to lard, these antioxidants were determined by HPLC. Under autoxidative conditions, slight antioxidative and synergistic activities of SEF with BHA or Toc were observed. Addition of SEF increased the stabilities of BHA or Toc in lard, whereas the stability of BHT was reported similar to that without SEF.

Noodles

The effect of antioxidants on storage of instant ramen, a unique dried noodle often produced by deep-frying with lard to give the characteristic flavor, was studied by Kuwahara et al. (184). Natural vitamin E at various concentrations was tested as an antioxidant and compared to the synthetic antioxidants BHA and BHT. Natural vitamin E at a 0.01–0.05% level gave satisfactory results for preventing oxidation of lard. The lard containing 0.03% natural vitamin E and the instant ramen fried with the lard gave better results in the AOM and storage tests than the lard containing synthetic antioxidant and the instant ramen fried with it.

Chicken and Turkey

Phytic acid was shown by Empson et al. (185) to form an iron chelate that inhibits iron-catalyzed hydroxyl radical formation and lipid peroxidation. To further characterize its antioxidant properties in model food systems, the effects of phytic acid on ascorbic acid degradation in aqueous solution and on stability of oil-in-water emulsions were investigated. In both systems, 1 mM phytic acid provided significant protection against oxidative damage and increased emulsion shelf life fourfold. Further testing of the antioxidant efficacy of phytic acid in the whole food system and its effect on the development of warmed-over flavor was carried out. Phytic acid substantially inhibited O_2 uptake, malondialdehyde formation, and the development of warmed-over flavor in refrigerated chicken.

Turkey breast or thigh muscle was mixed with 2% pure salt, rock salt, or pure salt plus 50 ppm of Cu, Fe, or Mg either in combination or singly. The efficacy of two antioxidants (Tenox 6 and an antioxidant-coated salt) was tested by Salih et al. (186). Lipid oxidation was monitored during refrigerated and frozen storage of raw and cooked turkey by the TBA test. TBA results indicated that the most significant prooxidant effect was caused by salt plus Cu^{2+} plus Fe^{3+}, followed by salt plus Fe^{3+} or Cu^{2+} alone. Tenox 6 was an effective antioxidant in the presence of copper and iron ions. Thigh meat was found to be more susceptible to oxidation than breast meat. Cooking had a significant prooxidant effect, as measured by TBA.

Gluten, casein, gelatin, gliadin, and egg albumen were investigated for their ability to prevent oxidation in safflower and sardine oils in model systems held at 37°C and 30–50% relative humidity 5 weeks or less (68). The degree of oxidation was assessed as peroxide values or TBA values, and changes in triglycerides during the experimental period were examined by TLC and GC. Results indicate that gliadin and egg albumen are the most effective antioxidants of the proteins examined.

The effects of adding antioxidants (BHA, BHT, santoquin) or refined soybean oil to mechanically deboned poultry meat (hen, broiler, and duck meat obtained by a Beehive plant) were studied during frozen storage (23 weeks at -18°C). Results as reported by Pikul et al. (187) revealed that antioxidants significantly inhibited oxidative changes (peroxide number, acid number, malonaldehyde content) but produced a "warmed-over" off-flavor in the cooked meat. Addition of 9% refined soybean oil effectively inhibited fat oxidation during frozen storage without adverse effects on sensory quality.

The effects of the addition of the nonabsorbable antioxidant Poly AO-79 to ground turkey meat were studied using three carriers, corn oil, 95% ethanol, or polysorbate 20/H_2O (188). Addition of 200 ppm (fat basis) Poly AO-79 in polysorbate 20/H_2O had the same antioxidant effect as 200 ppm TBHQ in corn oil. Use of Poly AO-79 in polysorbate 20/H_2O did not cause off-flavor, off-odor, or changes in microbial quality of ground turkey meat. The antioxidant ability of Poly AO-79 was increased by using polysorbate 20 as carrier without H_2O.

Turkey meat retorted in sealed cans at 121°C for 50 min was evaluated for susceptibility to warmed-over flavor by means of organoleptic evaluation and the TBA test (189). The retorted turkey was found to have significant resistance to the development of warmed-over flavor, which was not found in less severely cooked turkey. This resistance was shown to be due to the presence of water-soluble, low molecular weight antioxidative substances. These substances are probably formed as the result of browning interactions that occur in the meat during retorting. Indications are that these compounds inhibit the development of warmed-over flavor by acting as primary antioxidants, interrupting the free-radical mechanism characteristic of lipid oxidation.

In feeding trials with 6-week-old chickens, Atkinson et al. (190) compared the effects on carcass flavor and fat composition of diets containing 20% untreated or ethoxyquin-treated fishmeal with those of a diet containing groundnut oil cake instead of fishmeal. The chickens were killed at 9 weeks of age, and the carcasses were stored at -20°C for up to 1 month. Chickens fed on the groundnut oil-cake ration or on the untreated fishmeal had a neutral flavor, while those given the ethoxyquin-treated meals had a tainted flavor that was slightly improved by the addition of α-tocopherol, squalene, or $Na_2S_2O_5$ to the diet. The fatty acid compositions of the fishmeals in the diets and those of the chicken carcasses were compared, and correlations between the content of long-chain unsaturated fatty acids and the flavor score are discussed briefly by the authors (19). The addition of 5 or 10% poultry by-product meal to a basal diet containing 5% fishmeal had no effect on the flavor of the chickens.

Turkey thighs (with skin), were ground and treated with two levels each of three antioxidants—(i) BHA (0.005 and 0.01%), (ii) 40% BHA, 52% propylene glycol (PG), and 8% citric acid (0.02 and 0.03%); and (iii) 20% BHA, 70% PG, 4% citric acid and 6% propyl gallate (0.03 and 0.06%)— and held at 3°C for less than 10 days. Flavor scores, bacterial counts, and thiobarbituric acid (TBA) values were studied (191). Flavor, scored as different from a reference sample, was moderately different at zero days and increased only slightly with 10 days storage. Total and coliform counts were initially lowest in control samples, and coliform counts in controls were less than those treated with antioxidant throughout storage. Bacterial numbers increased during storage, but counts of less than 10^8 organisms/g had little influence on panel flavor evaluations. TBA values were effectively controlled by all antioxidant treatments but also had little relationship to panel flavor scores. Significant differences between levels of treatment, among treatments, or among storage periods were not consistent.

The U.S. Army Natick Research and Development Command reported the results of an investigation of methods for introducing antioxidants into foods in which they assessed the sensory quality of two classes of foods, one with high water activity (A_w) and the other with low A_w, over a period of storage of 6 months (192). Although the study has taken into cognizance the various instrumental and sensory parameters and their methods of analysis, the methodology adopted to identify and quantify the various off-flavors sensorically appears to lack rigor. The scale used is a seven-point hedonic scale that shrinks to five points as has been proved by many workers. Specificity to off-flavors is lacking, and hence the accuracy suffers. In this extensive study, the conclusions drawn regarding frankfurters refer to the retardation of the oxidative deterioration due to use of antioxidants. No marked effect of the antioxidants was seen in the case of chicken legs and fish sticks as the products were on par in quality with the controls up to the end of 6 months. The effect of antioxidants at the levels tried (150–200 ppm) in the case of freeze-dried products was not apparent as the treated samples were not found to have a longer shelf life than the controls.

Lamb

St. Angelo et al. (193) reported the effect of injecting an infusion of 0.3 M calcium chloride plus 1% sodium ascorbate or 0.25% maltol into freshly slaughtered lambs. The results indicated an acceleration of postmortem tenderization through the activation of calcium-dependent proteases, inhibition of lipid oxidation, and retardation of warmed-over flavor in ground cooked patties made from the meat of the lambs. After storage for 2 days at 4°C, patties of meat from lamb carcasses treated with antioxidants retained more desirable flavor characteristics (meaty and musty/herby) of lamb and had less off-flavor (painty and cardboardy) intensities than controls (untreated carcass meat).

Beef

Giaccone et al. (194) tested the antioxidative activity of mixtures of additives using Milan salami. Four additive blends were used—(i) 1.5% NaCl, 1.5% KCl, 250 ppm $NaNO_3$, and 0.7% sucrose; (ii) 3% NaCl, 0.1% ascorbyl palmitate, 250 ppm KNO_3, and 0.7% sucrose; (iii) 2.1% NaCl, 0.7% KCl, 0.2% phosphates, 0.02% δ-tocopherol, 150 ppm $NaNO_3$, and 0.7% sucrose; or (iv) 2.2% NaCl, 0.8% KCl, 0.1% calcium ascorbate, 0.2% δ-tocopherol, 150 ppm $NaNO_2$, and 0.7% sucrose—along with control batches containing conventional additives (3% NaCl, 0.3% ascorbic acid, 250 ppm $NaNO_3$, and 0.5% sucrose). The salami samples were ripened for up to 90 days at 13–15°C and 85% RH. At intervals, microbiological quality, composition, TBA value, and sensory quality were assessed. All four experimental formulations had an antioxidant effect, (iii) and (iv) having the greatest effect on TBA values. These two formulations also permitted the preparation of high-quality Milan salami within a ripening time of only 45 days. None of the additive blends studied impaired any aspect of quality of the salami.

The gelation and rheological properties of minced beef frozen and stored at –18°C for 6 months without additives or with 1.5% NaCl, 0.5% tripolyphosphate, and 0.5% sodium acid pyrophosphate or 200 ppm of a BHA–BHT mixture were studied by Barbut and Mittal (195). These characteristics were also determined for nonfrozen samples. The relationships between shear rate and shear stress for the different treatments were nonlinear and resembled Bingham pseudoplastic behavior. Continuous evaluation of the modulus of rigidity (G) during cooking ($0.5°C/$ min) revealed higher G values for the nonfrozen phosphate treatments. Salt addition resulted in significantly lower G values of the stored meat compared to the control. Antioxidant addition retarded some of these effects. Among the phosphates, tripolyphosphate was the best at maintaining the same G values as control samples. Water-holding capacity (after salt addition) was increased after phosphate addition in the nonfrozen meat and did not change after storage. The control, NaCl, and antioxidant treatments gave an increase in water-holding capacity during storage.

Mikkelson et al. (196) studied the effects of different polyphosphates on the color stability of beef patties during freezer storage and interactions between added NaCl, polyphosphates, and product pH. Sodium diphosphate ($Na_4P_2O_7$), triphosphate ($Na_5P_3O_{10}$), and trimetaphosphate ($Na_3P_3O_9$) were added (at 0.2 or 0.5%) to ground beef (~13.5% fat), which was formed into patties, frozen, and stored in a retail-type freezer cabinet. Polyphosphates increased initial product redness, due to "blooming," but reduced color stability with respect to browning during freezer storage. Hunter a values of the patty surface were linearly correlated with the metmyoglobin fraction of total myoglobin in extracts from the surface layer and were used to follow the oxidation of oxymyoglobin during storage. The effects of polyphosphates on the color stability of oxymyoglobin in patties differed from those observed in aqueous solution. Lipid oxidation, measured as the production of TBA-reactive substances, was significantly lower in patties with added $Na_4P_2O_7$ and $Na_5P_3O_{10}$ than in those with added $Na_3P_3O_9$ or control patties without added condensed phosphates. Added NaCl and exposure to light during storage both had a pro-oxidant effect. Addition of $Na_4P_2O_7$ and $Na_5P_3O_{10}$ countered the pro-oxidant effect of NaCl. Added phosphates and NaCl acted synergistically to increase the color loss of patties stored in the light. Patties containing added NaCl and phosphates should be protected from light. Coupling between oxidation of pigments and lipids in stored patties is also discussed.

Ground beef patties treated with metal chelators, free-radical scavengers, rosemary, and sodium alginate were examined by chemical analysis [for 2-thiobarbituric acid–reactive substances (TBARS)], instrumental analysis (GC), and sensory means for determination of warmed-over flavor (WOF) by St. Angelo et al. (197). A highly trained analytical sensory panel evaluated the patties for desirable descriptors, such as "cooked beef brothy" (CBB), and for WOF descriptors "painty" and "cardboardy." Results showed that many of the compounds retarded lipid oxidation when judged by chemical means but not all affected development of WOF when judged by sensory means since CBB values decreased and WOF descriptors increased. The free-radical scavengers appeared overall to be the most effective inhibitors of warmed-over flavor.

Effects of antioxidants (0.01 or 0.02% of total fat), salt level (0.375 or 0.750%), and salt type (NaCl, KCl, or a 65% NaCl + 35% KCl combination) in a $2 \times 2 \times 3$ factorial arrangement on the quality of restructured beef steaks were determined by Wheeler et al. (198). Meat blends with 0.02% antioxidant had lower TBA values than those with 0.01% after 85–155 days of frozen storage. Steaks with no salt (pooled across antioxidant levels) had lower TBA values than steaks with any type of salt after 85 days of storage or either level of salt after 155 days of storage. Steaks with either level of added salt resulted in higher ratings for juiciness, saltiness, and overall palatability than steaks with no added salt. Juiciness, flavor desirability, saltiness, and overall palatability ratings generally were higher for restructured steaks made with NaCl or NaCl + KCl than for those made with KCl. KCl at 35% of total salt could serve as an NaCl substitute in restructured beef steaks.

Arganosa (199) examined the effects of cooking temperature (cooking in a water bath at 70, 85, and 100°C), refrigerated storage (0 and 3 days at 4°C), and the presence of two antioxidants, namely sodium tripolyphosphate (STPP) and soy protein isolate (SPI), on the flavor and texture of restructured beef roasts. Oxidation, as measured by TBA, increased with cooking temperature and storage time. Both STPP and SPI inhibited oxidation, STPP being more effective at higher cooking temperatures and SPI at lower cooking temperatures. Lower cooking temperature increased cook yields, moisture content, expressible moisture, and water binding. Dehydration occurred during storage. Storage reduced roast juiciness and tenderness. Both STPP and SPI increased cook yields. The flavor and textural stability can be improved by incorporating antioxidants, cooking at lower temperature, and minimizing refrigerated storage.

Two natural antioxidant formulations (containing mixed tocopherols, ascorbyl palmitate, and citric acid) and TBHQ were evaluated in restructured beef steaks formulated to contain 18% fat and 0.75% NaCl by Crackel et al. (200). Lipid oxidation was monitored over 12 months of frozen storage. TBHQ significantly reduced TBA numbers in raw and freshly cooked samples and in cooked samples held 4 h at 4°C. Natural antioxidants provided significant protection in freshly cooked meat and were as effective as TBHQ in retarding lipid oxidation. Sensory evaluation and GC quantitation of hexanal were also used to monitor oxidation, but correlation among the three methods was not significant.

Mann (201) reported on the effectiveness of a phosphate compound, either alone or in combination with other antioxidants, in preventing warmed-over flavor in precooked recombined beef roasts of two chuck muscles over extended refrigerated storage. Results indicated that phosphate alone was effective in preventing warmed-over flavor in vacuum-packaged precooked roast beef for 23 days under refrigerated storage and that the addition of Maillard reaction products may contribute to flavor maintenance over extended storage periods.

Freeze-dried, washed ground meat was mixed with lecithin and antioxidants dissolved in water, heated, and then stored under refrigeration to study phospholipid oxidation as related to warmed-over flavor (202). Type I (free-radical stoppers, primarily phenolic-type compounds that can donate an electron to a radical) and II (free-radical production preventers, mostly chelating agents, which can tie up transition metals) antioxidants decreased the amount of thiobarbituric acid-reactive substances, while most type III antioxidants (environmental factors such as redox compounds and A_W regulators) gave an increase, depending on their concentration. Similar effects on phospholipid oxidation were detected with capillary GC for representatives of each type of antioxidant—BHA, EDTA, sodium bisulfite, cysteine, and A_W regulators.

According to Dessouki et al. (203), soaking beef cuts in lime or green pepper extracts for 24 h at 4°C had no antioxidative effect on the lipids during subsequent frozen storage at –10°C in polyethylene bags. Black pepper, chlortetracycline (CTC), and propolis (a resinous substance found in beehives) proved to be efficient inhibitors of oxidation, with relative efficiencies of 15.0, 2.5, and 1.22–1.36, respectively. Glazing the meat with 0.15% propolis solution was less effective than soaking in 0.1% solution. Neither CTC nor propolis affected meat flavor.

The effects of added antioxidants on restructured combination (50: 50) beef/ pork steaks were studied by Chastain et al. (204). Steaks were formulated to contain 20% fat and 0.75% salt. Antioxidants used at a 0.02% level (based on fat content of meat) were BHA, TBHQ, and BHA + TBHQ. Cooked steaks were evaluated for sensory properties and overall acceptability, initially and after 4, 8, 12, 16, and 20 weeks of freezer storage. Steaks were also evaluated after the various storage times for 2-thiobarbituric acid (TBA) values, subjective color by panel evaluation, objective color with the Hunter color difference meter, tensile strength, shear value, and cooking loss. Flavor and overall acceptability were significantly better in treated samples than in control samples. BHA was more effective in protecting color, and TBHQ was more effective in protecting flavor ($P < 0.05$). All treated samples showed lower TBA values than control samples ($P < 0.05$).

The effect of the antioxidant *tert*-butyl hydroquinone (TBHQ) was evaluated on flavor change in cooked ground beef patties by sensory panels, and oxidative deterioration was measured by TBA determination of malonaldehyde production (205). Cooked patties, with TBHQ (0.01% of the fat by weight) or without, were stored for 2, 7, and 21 days at –20.5°C and for 2, 5, and 7 days at 2.8°C. Paired comparison panels evaluated warmed-over flavor among the treatments. Flavor deterioration was detected in untreated cooked ground beef patties stored at 2.8°C (refrigerated) for 2, 5, and 7 days and at –20.5°C (frozen) for 21 days. Refrigerated samples had more warmed-over flavor (stored, stale, rancid) than did frozen patties ($P < 0.01$). TBHQ in the amounts used decreased the formation of warmed-over flavor, for all temperature time storage periods tested. TBHQ gave less protection under refrigeration than at freezer temperatures. Malonaldehyde levels were lower in the TBHQ patties than in cooked, untreated samples, indicating the retardation of oxidative rancidity.

Enhanced antioxidant action was observed by Haymon et al. (206) for the combination of sodium tripolyphosphate (STPP) and lemon juice concentrate (LJC) in frozen meat products. The STPP/LJC antioxidant was evaluated in beef products (patties, steak, and meatloaf) and soy-extended products that were cooked and then frozen. Storage trials of the precooked frozen meat products were conducted with unrestricted O_2 at $-18°C$. All the products were reheated from the frozen state in microwave ovens. The reheated cooked products made with STPP/ LJC antioxidant were statistically preferred to other antioxidant treatments and controls as judged by expert taste panels. The antioxidant activity was monitored by expert flavor panels and thiobarbituric acid number. The combination sodium tripolyphosphate and lemon juice concentrate has been made into a dry free-flowing food ingredient for use in the meat industry.

Deteriorative changes in meat may occur from heme and lipid oxidations, producing alterations in color, flavor, and odor (207). Samples of ground beef with either low levels (about 3%) or high levels (about 10%) of polyunsaturation in added fat were examined for storage-produced changes. High polyunsaturation levels increased meat deterioration. The antioxidant effectiveness of five additives (0.005% level) derived from natural sources (α-tocopherol, ascorbic acid, 1-ascorbyl stearate, citric acid, and ascorbic acid with sodium bicarbonate) was examined during 10 days of storage. Samples were judged to be commercially unacceptable after 1–4 days storage, but monitoring was continued to determine differences in the additives' antioxidant action. Ascorbic acid exerted a definite prooxidant action. The other additives showed only a slight effect in decreasing the rate of lipid and heme oxidations compared to untreated samples. A hypothesis of coupled heme–lipid oxidation has been presented.

Fish

The effects of curing period and use of an antioxidant mixture on the quality of lakerda (salted fish) were studied by Toemek et al. (208). Presalted fillets of Atlantic bonito and large bonito were dry salted by (i) dipping in 2% ascorbic acid solution and then salting or by (ii) treating with ascorbic acid solution and a salt solution containing an antioxidant mixture of 6% propyl gallate, 14% BHT, 3% BHA, 36% vegetable oil, and 36% monoglyceride citrate. All samples were cured in brine and analyzed for changes in physicochemical and sensory properties after 1, 2, 3, 6, and 9 months. Moisture content and salt content decreased with curing period in all samples, from 56.3 to 44.4–46.9% and from 32.9–37.6 to 16.0–16.4%, respectively, after 9 months. The use of the antioxidant mixture did not have a significant effect on moisture and salt content, except for salt content in the first month of curing. Initial TBA value of 0.3 increased after 9 months to 4.6 and 2.4 in treatments (i) and (ii), respectively. Use of the antioxidant mixture was reported to have a beneficial effect on organoleptic properties of lakerda, as indicated by better sensory scores for taste, odor, and overall quality. However, consumption was not possible after 6 months of storage due to deterioration of organoleptic properties in all samples.

The antioxidant effects of natural tocopherol mixture, calcium phytate, and inositol on marinated sardine (*Sardinops melanostictus*) during cold storage were compared with the effects of BHA by Asahara (209). Salted fillets of sardine were dipped into pickling solutions with and without added antioxidants and stored at 3°C for 200 days. During storage, peroxide value (PV), carbonyl value (COV), thiobarbituric acid (TBA) value, and acid value (AV) were determined on lipids of samples. PV, COV, and TBA value of lipids in control sample without added antioxidant increased markedly during the first 5–10 days storage and then more slowly for up to 200 days storage. Rancid flavor was detected in the sample after 60 days of storage. In samples treated with either 0.10% inositol or 0.05% calcium phytate, lipid oxidation proceeded at a slightly lower rate than in controls. Tocopherol mixture (0.05%) was fairly effective in protection of lipid oxidation, especially in combination with inositol and/or calcium phytate, though somewhat inferior to 0.02% BHA. However, the effect of 0.05% tocopherol mixture combined with 0.10% inositol on the stability of linoleic acid at 40°C was similar to that of 0.05% tocopherol mixture alone; inositol did not show any synergistic effect on protection of linoleic acid oxidation.

The effects of antioxidants on the quality of fermented sardine were investigated. Both BHA and Tenox-II at 0.02% were effective in preventing lipid oxidation during the 1-month fermentation period in 20% salt concentration (210). Treatment with both antioxidants prevented lipid oxidation for 2 months, whereas the control showed marked deterioration during 1 month of fermentation. Most of the nucleotides in sardine was decomposed from adenosine triphosphate to inosine and hypoxanthine during fermentation for 1 month. According to the omission test, the main constituents of the characteristic taste of fermented sardine appeared to be free amino acids and 5'-IMP.

The problems of oxidative rancidity in fish products are reviewed by Licciardello et al. (211). The efficacy of several antioxidants in controlling oxidative flavor changes of frozen minced Atlantic whiting was studied, including 550 ppm disodium EDTA; 0.84% FP 88E (Freez-Gard); 0.15% sodium erythorbate; 15 or 75 ppm Tenox A (BHA + citric acid in propylene glycol); 30 or 150 ppm Tenox 20 (TBHQ + citric acid in propylene glycol); and 30 or 150 ppm Tenox S-1 (propyl gallate + citric acid in propylene glycol). Another sample was prepared with 0.5% commercial crab seasoning and 3.2% dry, textured vegetable protein. Antioxidant-treated minced fish was shaped into sticks (fish fingers), battered, breaded, and blanched for 30 s in vegetable oil at 204.5°C, and the frozen packaged products were stored for 20 weeks at –6°C. Samples were examined at 4-week intervals for changes in peroxide value, TBA number, and sensory quality. Taste panelists judged the most effective treatments to be 150 ppm Tenox S-1, crab seasoning, 0.15% sodium erythorbate, and 150 ppm Tenox 20. Flavor scores for the first 8 weeks of storage were correlated with TBA numbers ($r = 0.70$) and peroxide values ($r = 0.60$). Beyond 8 weeks, flavor scores changed at varying rates, showing a slow decline, remaining stationary, or even reversing the downward trend (untreated controls only).

Lake whitefish (*Coregonus clupeaformis*) was canned with one of two antioxidant mixtures, AX1 (20% BHA, 20% BHT) or AX2 (10% BHA, 10% BHT, 6% citric acid, and 6% propyl gallate), and compared to untreated control samples (212). The flavor and quality of the fish were examined at intervals during 4 days of open refrigerated storage using sensory, chemical, and physical methods. Both TBA values and the results of paired flavor comparisons by an eight-member trained sensory panel demonstrated that the greatest degree of oxidative deterioration was always associated with the untreated fish, while AX1-treated fish exhibited the least amount of off-flavor over the storage period. The relative degree of off-flavor as evaluated by the panelists tended to increase on storage; TBA values showed a linear relationship with time. Tristimulus readings using the Hunter colorimeter showed a significant difference between the control sample and the two antioxidant-treated samples with respect to the pink color of freshly opened canned whitefish.

Oxidative deterioration of crude food fats and oils may have harmful effects in refined and processed products, particularly on flavor. Crude whale oil was stored at 30°C for 146 days either alone or with antioxidant additions of (i) 0.02% BHA, (ii) 0.02% BHT, (iii) 0.02% propyl gallate, (iv) 0.02% TBHQ, (v) TBHQ + 0.01% citric acid, or (vi) 0.01% citric acid. Parallel storage tests were made with the crude oils subjected to alkali refining and steam deodorization after antioxidant treatment. Stability was determined from weight gains during storage in air, peroxide values, and anisidine values for secondary oxidation products. The most effective antioxidant in crude oil was (iv), followed in decreasing order of oxidation retardation during storage by (i), (ii), and (iii). Stability times for deodorized oils ranged from 13.5 to 18 days at 30°C using crude oils treated with (i), (ii), and (vi) but were 49.5 and 56 days, respectively, after treatment with (iv) and (v). The improved oxidative stability obtained with (iv) in crude oil and its continued effectiveness after deodorization are attributed to carry-through of the antioxidant (213). Due to the wide variations of commercial refining and deodorization practices, use of (iv) should be declared and subject to a maximum permitted level of 0.02% in the final product.

Farragut (214) reported the effects of some antioxidants and EDTA and those of vacuum packaging on the development of oxidative rancidity in frozen Spanish mackerel (*Scomberomorus maculatus*). Samples were dipped or injected with solutions containing BHA + BHT (Tenox 4); BHA, BHT, propyl gallate, citric acid, and propylene glycol (Tenox 6); EDTA; Tenox 4 + EDTA; and Tenox 6 + EDTA. The treated fillet samples were vacuum packaged in Cryovac bags and frozen immediately or glazed and frozen. All samples were stored at −10°F. Samples dipped or injected showed the same trends in peroxide and free fatty acid content throughout the experiment. Organoleptic evaluations of these samples were almost identical. Vacuum-packaged samples were rated organoleptically superior to glazed samples. Samples treated with Tenox 4, Tenox 6, Tenox 4 + EDTA, and Tenox 6 + EDTA developed rancid odors and flavors after 6 months of frozen storage. The development of rancidity in Spanish mackerel during frozen storage was retarded by the application of EDTA. The only criticism received by these samples from the taste panel was on their appearance; a yellowish discoloration developed on and adjacent to the belly flap after the sixth month of storage. The concentration of EDTA may be a factor in the development of this discoloration.

Dried herrings quickly lose their specific flavor properties during storage. These changes are caused mainly by oxidation and hydrolytic processes of the fatty and nitrogen fractions. The use of antioxidants in the form of smoking fluids or ascorbic and citric acids (or a mixture of the two) retards this process of oxidation, prolongs the period of storage, and retains the appropriate organoleptic properties of the fish. Lapshin and Teplitsyna (215) described the changes in nitrogen fractions, especially those of free amino acids, during the process of storage.

Dairy Products

Quantitative distribution of various milk components (including SNF, lactose, protein, ash, phospholipids, ascorbic acid, tocopherol, and nine minerals) in different fractions of fresh, stored, Cu-induced oxidized, and autoxidized milks was studied by Patel (216). He concluded that autoxidation of the lipid and development of oxidized flavor in milk is a complex phenomenon involving many factors. Autoxidation was studied (217) in milk from cows fed a formaldehyde-treated casein/safflower oil supplement. A rapid disappearance of dissolved oxygen (DO), measured with an oxygen electrode, from milk samples stored at 0°C in tubes without headspace coincided with the development of oxidized flavors detected organoleptically. A correlation coefficient of 0.9 ($P < 0.001$) was obtained between the amount of DO disappearing and the taste panel scores for oxidized flavors. BHA, sesamol, nordihydroguaiaretic acid, ethoxyquin, or BHA + propyl gallate or tocopherols, when added in emulsified form to the milk at the rate of 10–15 mg per liter of milk, checked the development of oxidized flavors and the rapid disappearance of DO. Other antioxidants tested either were ineffective or imparted off-flavors to milk. Samples of mare's milk neither developed oxidized flavors nor showed rapid disappearance of DO over a test period of 8 days despite containing 20% linoleic acid in the fat. The oxygen electrode provides a convenient and sensitive method for studying autoxidation in milk.

Fresh milk samples with and without added antioxidants were held for 7 days at 104°F and tested by a 10-member panel twice daily for oxidized flavor development (218). Of various antioxidants tried, propyl gallate, butylated hydroxytoluene, and thiodipropionic acid at 0.005–0.02% levels tended to be the most effective. Low levels of thiodipropionic acid seemed to be particularly effective; however, its mode of action and effectiveness in baked products using milk fat is unknown. Although it prevented oxidized flavor in fluid milk, nordihydroguaiaretic acid was found to be ineffective and tended to speed up the oxidation in milk fat.

Effects of different processing methods and antioxidants on the oxidative stability and carbon dioxide formation in low-fat dry milk stored at 65°C for 8 days were determined by measuring volatile compound formation, oxygen disappearance, and carbon dioxide formation in the gastight sample bottle by GC. Foam-spray-dried or steam-jacketed centrifugal wheel atomization dried milks had significantly less ($P < 0.05$) volatile compound formation, oxygen depletion, and carbon dioxide formation than conventional spray-dried milks. Dry milks containing 100 ppm propyl gallate were reported by Min et al. (219) to show significantly less ($P < 0.05$) oxygen disappearance, volatile compound formation, and carbon dioxide formation than dry milks containing 100 ppm BHA or TBHQ during storage as reported.

Hall and Lingnert (220) traced the changes in odor and flavor during storage of dry whole milk by a profiling technique. The milk was dried with added synthetic antioxidants (BHA, BHT) as well as with antioxidative Maillard reaction mixtures from histidine and glucose. Samples were stored in either air or nitrogen. Positive flavors and odors disappeared most rapidly and off-flavors and off-odors increased most rapidly in unprotected samples. Maillard reaction products were as effective as BHA or BHT in retarding the development of oxidative off-flavors and off-odors. In general, flavor analysis was found to be more sensitive than odor analysis in detecting changes taking place during storage. There was, however, a high correlation between odor and flavor evaluation.

Dried whole milk retained an acceptable taste for 3–4 months when stored with kaempferol and for 12 months when stored with quercetin at 20°C, whereas dihydroquercetin was a less effective antioxidant (221). Dried milk stored without antioxidant acquired a tallowy flavor after 8 months. Of the nine milk phospholipid fractions, three disappeared after more than 6 months of storage with or without antioxidant, and two were found only in samples with antioxidant.

Abbot (222) reported the effect of ascorbyl palmitate and citric acid on the keeping quality of spray-dried whole milks stored at 37°C with or without antioxidants for up to approximately 8 months. Ascorbyl palmitate (AP) alone at the 0.01% level slowed down the uptake of oxygen and the rise in peroxide value of the fat in the dried milk, but the taste became unacceptable after 70 days, compared with 44 days for the dried milk without antioxidant and 80 days for dried milk with 0.01% dodecyl gallate. When a combination of 0.01% AP and 0.01% citric acid was used with dried milk, the product was unacceptable in taste after 235 days compared with approximately 100 days for the dried milk alone or with 0.01% citric acid alone. Dried milk containing 0.05% AP had an unacceptable foreign flavor, and the synergistic effect of citric acid decreased at higher concentrations. The effectiveness of ascorbyl palmitate, a nonphenolic antioxidant, in delaying fat oxidation is limited, but together with an equal amount of citric acid, some protection is afforded for a limited period of time, such as in the rewetting process for converting milk powders to aggregated granules.

Kuzio (223) divided each batch of butter oil into five portions, retained one as control, and subjected the other four to different treatments: the addition of antioxidant and three varying degrees of hydrogenation with the I value reduced by 2, 4, and 6 units. These butter oils were used to make butter cream candies. The storage stability at 32°C of butter candies containing three food grade antioxidants—Tenox PG (propyl gallate), Tenox R (BHA 20%, citric acid 20%, propylene glycol 60%), and Tenox 4 (BHA 20%, BHT 20%, maize oil 60%)—was compared. A rapid decrease in flavor scores correlating with peroxide values was shown for the control and for Tenox 4 and Tenox R samples in the early stages of storage. Between 14 and 21 days, scores leveled off, followed by a steady decrease. Tenox PG was the best of the three antioxidants for butter cream candies. Significant differences were noted in the storage stability of the candies according to the treatments given the butter oils. Differences were small until the seventh day of storage. The control showed a much higher TBA value and higher unacceptability than the others.

The effect of a natural antioxidant, ghee residue, on the shelf life of flavored butter oil was studied by Wadhwa et al. (224). Ghee residues were prepared by clarifying butter at 120°C. Flavored butter oil was made by heating butter oil with skim milk dahi (20% w/w) at 120°C for 3 min. Incorporation of 15–20% ghee residue in the flavored butter oil improved the product shelf life.

Buffalo milk ghee, prepared from desi butter (i), ripened cream butter (ii), or sweet cream butter (iii), had mean flavor scores (max. 100) of 99.55, 98.16, and 97.16, respectively, initially and 87.16, 87, 50, and 88.0 after storage for 150 days at room temperature (18–36°C) (225). Addition of 0.02% BHA, 0.02% BHT, or BHA + BHT (0.01% each) had no effect on the initial flavor score but increased the scores at 150 days to 90.55, 91.0, and 90.44 in (i), 90.22, 90.61, and 90.16 in (ii), and 90.83, 91.77, and 91.22 in (iii). Free fatty acid (FFA) content (as oleic acid) was in the order (i) > (ii) > (iii), the mean levels being 0.354, 0.290, and 0.236%, respectively ($P < 0.05$). Mean peroxide values also differed significantly ($P < 0.05$), being 0.498, 0.417, and 0.246 meq/kg fat, respectively, in the three treatments. The addition of antioxidant retarded the development of both FFA and peroxides during storage. All three antioxidant treatments had a similar effect upon FFA development, but peroxide development was retarded to a greater extent by BHT or BHA + BHT than by BHA. It is concluded that BHT was the most useful of the antioxidants tested for improving shelf life of buffalo ghee.

Changes in FFA content and peroxide value (PV) of ghee samples containing curry, betel, and drumstick leaves at 1.0 and 3.0% were studied at monthly intervals during storage for 12 months at ambient temperature. Effects were compared with legally permitted levels of BHA, BHT, and their combination. Only curry leaves could protect ghee from hydrolytic rancidity; betel and drumstick leaves and BHA, BHT, or their combination gave higher FFA values than the control, the value being maximum for drumstick leaves. None of the natural antioxidants studied could protect ghee from oxidative deterioration but gave higher PV than the control, while chemical antioxidants were effective in this respect. Lower flavor and color scores were obtained in ghee samples with curry, and lower color scores with betel leaves (226).

Mattsson and Johansson (227) found that the addition of 5, 10, or 25 ppm ascorbyl palmitate to ripened creamery butter (pH 4.7, NaCl content 1.3%) improved the keeping quality at $-5°C$ from 6 weeks to 12, 14, and 22 weeks, respectively, but at concentrations of 25 ppm it imparted a chemical off-flavor. Butter containing 10 ppm ascorbyl palmitate could be stored for 1 year at $-25°C$ and retain its good quality with no oxidation defects.

Khoa (evaporated milk), containing approximately 70% total solids, was packed in sterile polyethylene bags and stored at 30 or $5°C$. Additions of aqueous solutions of (i) 0.05–0.25% ascorbic acid, (ii) 0.01–0.07% propyl gallate, (iii) 0.05–0.20% lecithin, (iv) 0.005–0.020% nisaplin (containing 2.5% nisin), or (v) 0.20–0.40% potassium sorbate were made, and their effects on the flavor, standard plate count, yeast and mold count, free fatty acid content, iodine value, formol titration value, and titratable acidity of the stored samples were studied by Jha et al. (228). Addition of (i)–(iv) did not improve the shelf life of khoa at either temperature and had no significant effect on total bacterial counts or yeast and mold counts compared with controls. Addition of (v) resulted in an increase in the shelf life from 2 days at $30°C$ and 20 days at $5°C$ to 10–11 days at $30°C$ and less than 40 days at $5°C$ and significant reductions in total bacterial counts (38 × 10^3/g after 4 days at $30°C$ vs. 49 × 10^6/g in controls; 10 × 10^3/g after 30 days at $5°C$ vs. 125 × 10^6/g in controls) and in yeast and mold counts (250 and 210 g^{-1} vs. 86 × 10^3 and 52 × 10^4 g^{-1} in controls under the same storage conditions as above). Addition of (v) also reduced the rates of acidity development, fatty acid liberation, and proteolysis in the khoa. Flavor deterioration in khoa appeared to be closely related to the yeast and mold content.

The antioxidant activity of α-, γ-, and δ-tocopherols in membrane lipids and the synergistic antioxidant activity of the membrane lipid obtained by churning milk with the tocopherols were studied by Kanno et al. (229). Membrane lipid fractions (buttermilk, butter serum, and total membrane lipid) without added tocopherol were rapidly autoxidized, with simultaneous development of browning and a fishy off-flavor. Peroxide value decreased after reaching a maximum. Addition of 0.01% α-, γ-, or δ-tocopherols decreased peroxide formation, browning, and off-flavor. Addition of membrane lipid (optimum concentration 10%) prolonged the induction period of the oxidation of milk fat obtained by churning. The antioxidant activity of tocopherols added to churned milk fat was enhanced by the addition of 1% membrane lipid.

Fruits

The effects of postharvest dips of 1% lecithin + 2.5% $CaCl_2$ (lec-Ca) on Granny Smith apples harvested on four dates in April and May and removed from 0°C storage in October were examined by Watkins et al. (230). Compared with dipping in water, lec-Ca dips increased CO_2 and decreased O_2 levels in the fruit, reduced the production of α-farnescene, but had little effect on flesh firmness, soluble solids, the incidence of superficial scald, or the oxidation products of α-farnescene (conjugated trienes). Addition of antioxidants (ascorbic, tannic, citric, or tartaric acid or ascorbyl palmitate) to lec-Ca dips did not reduce superficial scald of apples (stored for 30 weeks at 0°C prior to 1 week storage at 20°C) to levels comparable to those found with the commercially used antioxidant diphenylamine. The incidence of internal breakdown was markedly reduced by the use of lecithin regardless of the presence of calcium or any of the antioxidants.

Citral deteriorated by air bubbling or shaking at 38°C at pH 2.0 for up to 30 days. Acidic degradation products were isolated and characterized as their ethyl esters. In addition to ethyl geranate and 3,7-dimethyl-7-ethoxy-2-(E)-octanoate, which have been previously identified, four octenoic acid derivatives were identified: 3,7-dimethyl-7-ethoxy-2(Z)-, 7-chloro-3,7-dimethyl-2(E)-, 3,7-dimethyl-7-hy-droxy-2(E)-, and 3,7-dimethyl-7-hydroxy-2(Z)-octenoic acid (231).

Bielig and Feier (232) treated model orange juice systems and orange juice with BHA, BHT, PG, α-tocopherol, or quercetin (QC) and studied the effects on the production of volatile aroma compounds by GLC. The effects on the production of C_6 alcohols, butyraldehyde, and C_8 and C_{10} aldehydes are tabulated for the oxidation of model solutions containing palmitic, oleic, linoleic, or linolenic acid for each antioxidant. The effects of each antioxidant were studied on the production of (i) hexanal; (ii) C_8, (iii) C_{10}, and (iv) C_{12} aldehydes; and (v) C_6 alcohols by bottled orange juices stored at 18°C for 7 months. Pronounced changes in volatile aroma compounds were observed during the first 2 months of storage, and the results were not comparable with the effects observed in model solutions. BHA and PG seemed to be the most effective for inhibiting the formation of off-flavor compounds; BHA influenced (i), (ii), and (v) contents, and PG influenced (i), (iii), (iv), and (v) contents. α-Tocopherol had no observable effect despite significant results in model solutions. BHT had no significant effects, and quercetin inhibited the formation of (i) and (v). Controls stored at 4 or 18°C showed great differences in contents of some aroma compounds.

Nuts

Chopped pecan kernels were treated with various antioxidants, sealed in cans (with or without headspace modification by vacuum or nitrogen flushing), and stored at 85°F for 15 weeks. The effects of treatment on sensory properties, color, and volatiles were determined (233). The use of vacuum or a nitrogen flush was the most effective method for preserving pecan kernel quality as determined by sensory and instrumental methods. Alpha, gamma, and mixed tocopherols at 0.05% (and all these except α-tocopherol at 0.02%) had significant effects on color and reduced flavor loss, and γ-tocopherol and mixed tocopherols reduced the amount of head-space pentane produced. Tocopherol treatment was not as effective as vacuum or nitrogen flush for preserving pecan kernels. BHA, BHT, and TBHQ were ineffective.

Cavaletto and Yamamoto (234) investigated the chemical and physical changes in roasting oil from a commercial processing plant and the relationship of these changes to the quality and shelf life of roasted macadamia kernels. Oil was sampled initially and after 2, 4, and 13 weeks of roasting. Generally, the FFA content, color (absorbance at 550 nm), refractive index, and viscosity of the roasting oil increased with use. Changes in iodine number and fatty acid composition indicated that there was considerable oil exchange between roasting oil and macadamia kernels. Antioxidant loss in the roasting oil was rapid. No flavor differences were observed in kernels immediately after roasting in the various oil samples, and shelf life was not appreciably affected by continuous use of the oil for as long as 13 weeks. A second study investigated the effects of vacuum packing (0, 15, and 24 weeks in vacuum) and direct antioxidant application (approximately equal to 76 ppm BHA and BHT) on the stability of dry-roasted macadamia kernels. The stability of antioxidant-treated kernels was greater than that of untreated kernels regardless of vacuum level. Vacuum packing had no effect on antioxidant-treated kernels but showed some benefit for untreated kernels.

Oils and Fats

Tests by Svidovskii et al. (235) on margarine containing 0, 0.01, or 0.1% betaine as an antioxidant and stored at 4–6°C and 80–85% RH for 3 months showed that the addition of 0.1% betaine retarded acidity development and had a positive effect on retention of margarine organoleptic quality. Shelf life was prolonged from 1 month to 2 months.

Synergistic antioxidant effects of butyl caffeate (BC) and mixed tocopherol concentrate (m-Toc) on lard and palm oil were investigated by the oven test at 60°C and with the active oxygen method (AOM) by Aoyama et al. (236). The antioxidant effect of BC synthesized from caffeic acid and n-butyl alcohol was also studied and its purity determined as 94.5% by HPLC. Butyl caffeate showed a strong antioxidant effect in lard and synergistically enhanced the effect of m-Toc in the oven and AOM tests. These effects were more pronounced than those for caffeic acid reported earlier by the same author and coworkers. Butyl caffeate also showed a definite antioxidant effect in palm oil; however, no synergistic effect with m-Toc could be detected. Addition of BC thus retarded oxidative deterioration of tocopherol in lard and palm oil.

Linalyl acetate and undecylenic acid were studied by Yan and White (237) to determine their abilities to reduce oxidative changes in soybean oil held at frying temperature. All compounds to be tested were added to soybean oil and heated to 180°C for 56–70 h. Fatty acid changes and conjugated diene formation were monitored. Acetylation of linalool to linalyl acetate (LA) caused the formation of many by-products, which were separated by TLC into three bands. The materials isolated from the bands were tested and found to be equally effective antioxidants. Purchased LA had an antioxidant effect similar to that of the bands. The LA materials from the bands were further purified and identified by GC-MS. All the effective compounds were similar in structure to LA. Undecylenic acid provided some protective effect but less than that of LA, which had less antioxidant effect than δ-7-avenasterol and polydimethylsiloxane.

Storage stability of canola oil treated with different levels of TBHQ in the presence and absence of citric acid was evaluated and compared with the stability of a similar oil sample treated with BHA/BHT and citric acid (238). Canola oil without antioxidant served as a control sample. Samples were stored in amber and clear glass bottles in the presence of light for up to 16 weeks. Off-flavor development during storage was studied by GLC. Results obtained indicated that TBHQ, at levels as low as 100 ppm, had a substantial effect in retarding oxidative changes in canola oils. Addition of BHA/BHT was shown to be of no benefit in improving the storage stability of canola oil. Oxidative changes in stabilized and unstabilized oils stored in amber glass bottles were much lower than in comparable oils stored in clear glass bottles. Light was therefore considered a very important factor in the production of off-flavor in canola oils.

Flavor deterioration of salad dressings was investigated to determine the effect of hydrogenation of the oil, additives, and storage conditions by Warner et al. (239). Flavor quality tests were developed and were reported to be correlated with GC analyses of volatile compounds in oils separated from the dressings. Hydrogenation of soybean oil with Cu and Ni catalysts effectively increased the storage stability of salad dressings at 21°C but not at 32°C. The use of BHA as an antioxidant in the oil or EDTA as a metal inactivator in the starch base as well as nitrogen packaging were found to be effective in prolonging the storage stability of salad dressings made with unhydrogenated soybean oil. Therefore, these additives or nitrogen packaging are expected to provide economic substitutes for hydrogenation of soybean oil used in salad dressings.

Min and Wen (240) studied the qualitative and quantitative effects of the common antioxidants BHA, BHT, PG, and TBHQ on the rate of dissolved free O_2 disappearance in soybean oil during storage. The order of effectiveness was BHA < BHT < PG < TBHQ. Statistical analyses of the results showed that the effects of BHA and BHT were not significantly different from each other at the 5% level, but BHA and BHT were different from PG and TBHQ, and PG was different from TBHQ. The antioxidant effectiveness of 0, 50, 100, 150, and 200 ppm BHA, BHT,

PG, or TBHQ differed significantly at the 5% level. The higher the amount of antioxidants added, the slower the rate of disappearance of dissolved O_2 in the oil.

Flavor and oxidative stabilities in soybean oil were studied by Mounts et al. (241) using organoleptic evaluation and chemical analysis. The studies were carried out with unhydrogenated and hydrogenated soybean oils. Analyses of these oils showed iodine values of 138, 109, and 113 and percent linolenate of 8.3, 3.3, and 0.4, respectively. Each oil was deodorized by the addition of either citric acid alone or citric acid plus BHA and BHT. Addition of antioxidants did not improve the flavor stabilities of the oils in accelerated storage tests but did improve the flavor stabilities of hydrogenated oils in light exposure tests. All three oils that received the same additive treatment had equivalent flavor stability in both accelerated storage and light exposure tests. However, both hydrogenation and antioxidant treatment improved oxidative stability as measured by AOM. There was good correlation between flavor score and the logarithm of the peroxide value as determined at the time of tasting.

Slowikowska et al. (242) determined the activity of the antioxidants propyl gallate (PG) and dodecyl gallate (DG) on rapeseed oil during storage. Samples of rapeseed oil were stored with up to 0.02% of a mixture of DG and PG (1:1) without antioxidants. The stability of the oils was evaluated by AOM, and for flavor during storage for 24 weeks. It was found that the antioxidants improved oil stability as measured by the AOM. The Lea number (in the AOM test) and flavor, however, did not vary significantly during storage for oils with and without antioxidants.

Snacks

Ohyabu et al. (243) measured the degree of rancidity in aged rice crackers soaked in antioxidant-containing oils. Rice crackers were fried in soybean oil, corn oil, and lard at 240°C; the oil was extracted from the fried crackers by hexane; and the defatted crackers were soaked in oils to which 0.01% citric acid, 0.005% BHA, and 0.005% BHT were added and aged in the dark at 37 and 60°C and in light at 25°C. The development of rancidity was measured by sensory evaluation and chemical tests (peroxide value, carbonyl value, volatile carbonyl value). Results revealed that antioxidants delayed the development of rancidity but did not change the relations between chemical characteristics and flavor score.

The possibility of using TBHQ, α-tocopherol, Prolong-P (a mixture of rosemary, thyme, and marjoram) or ascorbyl palmitate as alternative antioxidants to BHA in potato flake production was investigated in detail by Baardseth (244). Potato flakes incorporating the additives were air-packed in alu-laminate (polyester, aluminium, polyethylene) and stored at 15°C for 4 months, followed by 20 months at 25°C. Reference BHA samples were nitrogen-packed in a triple laminate and stored at −80°C for up to 24 months. Flakes were analyzed for moisture, A_w, and metal content (Cu and Fe), and sensory evaluations were carried out by a panel of 12 assessors. Color measurements were performed after 12 and 24 months of storage by CIE and L*a*b* values. Tabulated results show that addition of α-tocopherol resulted in flakes that were described as rancid and haylike and hence unacceptable. Flakes with TBHQ developed a rancid flavor after 4 months, which did not then increase further until 16 months. Prolong-P was a better antioxidant than TBHQ, but a steady increase in rancidity did occur during storage. Ascorbyl palmitate gave a good antioxidative effect for up to 16 months and also protected against bleaching of the flakes (due to loss of carotenoids) to a greater extent than other antioxidants. Analysis of variance showed significant differences between antioxidants for all sensory parameters tested except texture. Therefore, none of the four tested antioxidants was able to replace BHA in potato flakes stored for up to 24 months.

Fresh green peas (cv. Frisky) were ground with dry ice, and the powder produced was stirred with 10 mM sodium phosphate buffer (pH 7.0). The crude extract obtained was filtered and centrifuged, and the resulting supernatant was again centrifuged, dialyzed against 1 mM phosphate buffer, and diluted to give a protein concentration in the crude extracts of approximately 1.5–3.0 mg/mL. The extracts were treated with antioxidants—BHA, BHT, and α-tocopherol—at various concentrations. The extracts, both with and without antioxidants, were then heated in a water bath at 90°C for 1, 2, 5, 10, and 20 min. Unheated extract containing no antioxidant served as control. The extracts were then examined for residual peroxidase activity and for changes in peroxidase isoenzyme patterns. The conclusions of Lee and Klein (245) were that a combination of antioxidant and heat treatment is more effective in inactivating peroxidase than heat treatment alone and that in the presence of an antioxidant heat treatment time can be reduced, thus permitting better retention of heat-sensitive nutrients and sensory quality in foods.

A modified atmosphere packaging system for chilled fresh potato strips, which rapidly produced O_2 levels of less than 3%, was identified by O'Beirne and Ballantyne (246). This system enclosed the strips within 25-mm low-density polyethylene (LDPE) film, heat-sealed to a fiber tray lined with Surlyn-PVDC-Surlyn, and the package was flushed with an initial atmosphere of 5% O_2/10% CO_2. An equilibrium-modified atmosphere of 3–4% CO_2/1–2% O_2 was established after 3 days of storage at 5°C. This modified atmosphere package, combined with dipping the potato strips in a 10% ascorbic acid solution, inhibited enzymatic discoloration for 1 week at 5°C. Vacuum packaging within a Surlyn/PVDC-coated polyester film, with or without dipping in ascorbic acid solution, inhibited discoloration of potato strips stored at 5°C for at least 2 weeks.

Potato crisps were made from Katahdin potatoes fried in a mixture of cottonseed, corn, and soy oils at 190°C for 115 s and then (i) given no salt, (ii) salted at 1% w/w with (iii) 30 mg ascorbic acid/100 g salt, (iv) 60 mg ascorbic acid/100 g salt, and (v) 0.4% TBHQ, 2.0% citric acid, 1.5% tricalcium phosphate, and 96.1% salt. (i)–(v) were packaged in transparent or opaque bags and stored at 10, 20, or 30°C in the dark or exposed to light for 8 weeks prior to sensory analysis. A 10-member panel scored the product for flavor. Results were tabulated and show that (iii) improved the flavor of the crisps, whereas (iv) caused a bitter flavor. Antioxidants were more effective when product was stored in the dark at low temperature in opaque bags. No-salt samples were generally ranked low and developed off-flavors quickly (247).

Oxidative stability of deep-fried instant noodles (ramyon) was studied by Park et al. (248). The noodles were prepared in rapeseed oil fortified with 0.02% antioxidants (natural tocopherols, BHA, TBHQ, or ascorbyl palmitate + citric acid) or blended at 7:3, 5:5, and 3:7 (w/w) with palm oil. Ramyon samples were stored at $35 \pm 0.5°C$ for 90 days, during which oil was extracted from the products and analyzed for indices such as peroxide value, acid value, iodine value, anisidine value, dielectric constant, and fatty acid composition. The ramyon samples were also evaluated for rancid flavor development. Results show that effectiveness of the antioxidants was highest in the case of TBHQ followed by ascorbyl palmitate + citric acid, BHA, and tocopherols. Oxidative stability of ramyon fried in rapeseed oil + 0.02% TBHQ or in 3:7 blends of rapeseed/palm oils was almost equal to that of ramyon fried in pure palm oil.

Alcoholic Beverages

Palamand and Grigsby (249) showed that upon storage, beer develops stale flavors with concomitant loss of freshness. It is now known that beer staling involves a number of reactions, of which oxidation is one of the most important. Several attempts have been made to minimize or retard beer oxidation by the use of antioxidants, but these have met with only a limited amount of success. A number of antioxidants were screened, and it was found that serine and taxifolin provide significant protection to beer against oxidative deterioration in its flavor. Further, evidence was found to the effect that these antioxidants react with the substrate rather than with O_2 as the main step in their antioxidant activity. Palamand and Grigsby (249) have also discussed the mode of action of serine and taxifolin in beer and the results obtained with model systems containing major oxidation precursors of beer.

Baetsle (250,251), after a theoretical discussion of the role of ascorbic acid (AA) in reduction of redox potentials, presented experimental data from three breweries. Dissolved and headspace O_2 was determined at various stages of manufacture: (i) in storage tanks, (ii) after filtration, (iii) after racking, (iv) immediately after bottling, and (v) 10 days after bottling. The amounts of O_2 (mg/mL) found were (i) 0–0.15; (ii) 0.25–0.90 at the filter exit, 0.20–0.65 in the filtered beer tank; (iii) no values given; (iv) 0.75–1.65; and (v) 0.15–0.40. The presence of 15 ppm AA reduced O_2 concentration (mg/bottle) from 0.467 at (iv) to 0.25 at (v). The presence of 50 ppm AA reduced O_2 content from 0.90 mg/bottle immediately after filtering to 0.55 within 1 h and to 0.25 mg/bottle 4 days after bottling. On an industrial scale,

2 g AA/hL beer preserved the color and flavor of beer during storage for 66 days. Oxygen uptake during bottling can be reduced by shortening the racking cock from 107 to 100 mm; this reduced O_2 concentrations from 0.49 to 0.25 mg/L immediately after bottling and from 0.53 to 0.22 mg/L 10 days after bottling. Recommendations for minimization of the O_2 content of beer are also given. Ascorbic acid at concentrations of 2–6 g/hL is suitable for protecting bottled beer of low O_2 content from the spoilage effect of O_2. Development of colloidal haze and variations in odor, flavor, and color are prevented or delayed, and the stability of the beer is considerably extended (252).

3.9.1 Sensory Analysis Methods

Any good sensory analysis method is always product-specific and objective-oriented. The description and quantification of various attributes is of primary importance. This can be achieved by a panel with basic training in recognition, description, and scaling of sensory quality parameters. Training of the panel requires initial recognition tests for each of the parameters followed by discriminatory tests between different levels of a specified attribute, and finally the development of the ability to scan and quantify different technical levels of sophistication, keeping in view the product-specific profile.

Antioxidant incorporation is invariably tied up with storage studies as storage stability is the specific goal of their incorporation. Hence, the analysis will be over a predetermined time frame. Nonparametric methods, which are longitudinal in nature, are applicable only to the whole set at a given time such as two-sample difference recognition tests, difference from control tests, signal detection (R index) tests, and multiple-sample intensity/quality ranking tests. This is hardly the case when a storage study is involved and the testing has to be latitudinal. Scoring methods that can provide absolute values as near as possible are more suitable for comparisons over a period of time (253,254).

The nature and appropriateness of the scoring methods are linked to the quality and integrity of the information they convey (255). The measurements are not of simple physical quantities but of sensations evoked and perceived at a psychological level in response to various physical and chemical stimuli. The scaling of these responses so that they reflect the hidden psychological processes is as complex as the human being. Hence, the scales should be constructed to reflect the sensitivity of the measuring instrument and the human being to the perceived responses to the various stimuli. Sensitivity is generally greater to off-notes, pain sensations, darker shades of color, appearance factors, structure/texture variations, and off-taste factors that are not normal or do not form part of the profile of the product. The effect of antioxidants may be apparent in one or more quality attributes.

Panel training for recognition of differences in relative quality attributes in the permissible range of edible quality is the basic requirement. The sensitivity of the panel has to be assessed closely in the required range by following a good method, and the whole process repeated periodically to monitor the performance evaluations. For greater differences, the method of ranking followed by inversions can be used to cover the entire range. The differences between consecutive sample points should be reduced gradually and fixed at a comfortable level (just noticeable difference) for the panel, which will provide the sensitivity at different levels. The feasibility of differentiation by the panel coupled with understanding of the range of change is essential for fixing the cutoff point.

Fixing the difference in quality between samples is important in sensory analysis as the just noticeable difference (JND) is not constant, changing with the concentration of the stimulus (256) and making the interval property demands difficult to meet. The three basic assumptions that must be complied with for statistical treatment of data are that the scores are independent, have equal variance, and are normally distributed. The condition of independence of scores can be ascertained by following the standard laboratory practice of scoring each sample on an individual basis, reducing to a minimum the bias arising within the panel by adopting efficient random coding methods, and conducting the evaluation sessions in taste booths, which can reduce the inter- and intrapanel bias.

Equality of variance becomes a problem when the length of the scale used by panelists varies or when the range in quality of the samples being evaluated is small and is near the endpoints of the scale. This problem can be eliminated by rigorous panel training and by including a range of quality in every set of samples. Even if the normality assumptions are only partially met, the conclusions will still be valid enough, as they are based primarily on the means and not on individual values.

Scales are tools with which the intensities of attributes are measured by the panelists (257). Several types of scales are in use—nominal, ordinal, category/interval, and ratio scales, as shown in Fig. 3.11. Ratio scales are the best suited for studies on antioxidant effects. But sometimes category/interval scales can be used, depending on the training of the panel. Unipolar scales suitable for these studies as the adjectives reflect the intensity differences from zero to strong for each specific attribute. Bipolar scales consist of hedonic root words such as acceptance/rejection, preference/no preference, and like/dislike. The adjectives used are extremely, moderately, slightly, zero/absence, to strong and the like.

Ratio scales and their modifications are structured or unstructured scales anchored with descriptors either at several points or only at their endpoints. Scales anchored at more than two points are known as structured scales; if the anchors are only at the endpoints, they are known as unstructured scales. Generally a 15-cm scale similar to the QDA scale of Stone and coworkers (258,259) has been found to be useful. The ratio scaling method of magnitude estimates is also being used for several products (260,261). An understanding of the method requires concentrated effort on the part of the panelists. The variance has been found to be larger than while working with untrained panel.

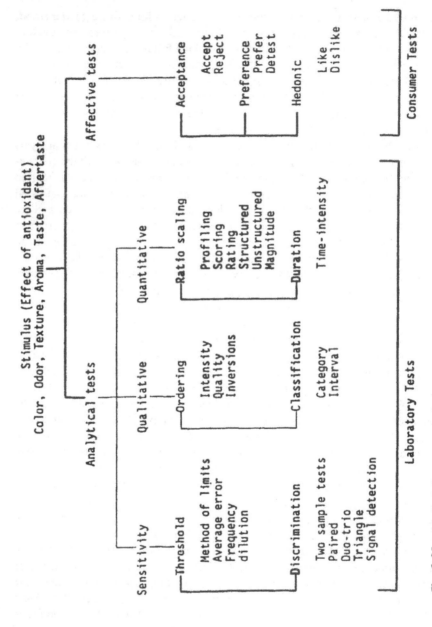

Fig. 3.11   Methods for sensory analysis.

The pathway of oxidative changes in sensory parameters is quite complex, and hence it is normally not possible to hit upon a simple cause-and-effect relationship. Changes occur simultaneously in more than one parameter. Under such situations it becomes necessary to identify the most powerful factors for discrimination through dimensionality reduction and empirical modeling for predictability.

Multivariate methods become necessary when the breakdown products lead to parallel changes in several quality attributes. Principal components analysis (PCA) and canonical discriminant analysis (CDA) are the most widely used multivariate methods (262). It is worthwhile to analyze the whole set of data as a single set so that the position of the samples and the directional differences can be explained. It has been found that multivariate methods contribute substantially to building empirical relationships and predictions as well as to tracing the pathway of changes or deteriorations occurring in the study.

Principal components analysis (PCA) is an invaluable procedure that can be applied to any multivariate data set for characterization. It can also be used to great advantage as an exploratory technique as it indicates relationships among groups of variables in a data set. PCA exhibits the relationships between objects or samples. The data matrix represents a multidimensional space, the number of dimensions being equal to the number of variables. The structure of the sample space is simple with a small number of variables and becomes more complex with increasing numbers of variables. The PCA method reduces the dimensionality through searching for linear combinations of variables that are orthogonal to each other and account for the greatest proportion of variance in the original data at each stage. Many of the variables through the data collected while conducting any experiment are likely to be collinear, in which case PCA will provide the best understanding of the system.

Canonical discriminant analysis (CDA) is a variation of the Mahalonobis D^2 method for dealing with multiple samples. The advantages of the method are that it provides estimates of the distances between samples and also maps the variables as directional vectors. The geometric representation in any two or three planes is used to visualize the position of the samples as well as the directional vectors. Based on the directional vectors, suitable nomenclature can be provided for the axes and quadrants that will depict the product quality and provide an understanding of operating characteristics for the changes in product quality.

3.9.2 Case Studies on Edible Oils

The Approach

Sensory perception of oxidative changes in oils has time and again been identified as an indispensable indicator of quality (263–266). However, specifications for odor and taste of oil continue to be limited to prescribing freedom from foreign or rancid odor, mainly due to the nonavailability of suitable methods for evaluating sensory attributes. Several workers have, however, attempted to standardize such methods (266–268). These methods have been studied for a long time along with the evaluation of storability, packaging materials, time–temperature effects, composition of the oil, effect of incorporation of antioxidants, and so on to ultimately derive the correlative inferences of physicochemical parameters with sensory responses.

Several studies have been carried out to detect oxidation at early stages through chemical tests of peroxide value, free fatty acids, anisidine value, etc. Recently the Kreis test was standardized (269) using multivariate optimization methods based on response surface methodology (270) and the optimum-seeking method of Nelder and Mead (271) and Olsson (272). Still sensory methods hold the key to ultimately deciding the quality of an oil (273–278). In a study on oxidative rancidity in groundnut oil employing chemical indices (PV, AV, FFA, and Kreis value), sensory quality, and their correlation (268), both olfactory (odor) and gustatory (flavor) responses were used. Testing was carried out on oil storage up to 18 weeks of storage. Odor had a higher correlation and higher slope than flavor with chemical tests. Among the chemical tests, the Kreis test was found more sensitive than peroxide value.

There have also been several reports on gas chromatographic techniques to identify rancidity in fats and oils and correlate the results to sensory analysis (279–284). The approaches have concentrated on direct injection of the oils or purge-trapping of volatiles using internal standards to quantify responses and correlate them with sensory values of flavor intensity or ranking collected from a trained panel. Morrison et al. (283) and Jervi et al. (281) reported satisfactory correlations ($r = 0.82$–0.90), whereas Warner et al. (280) described the attempt as futile. These groups have studied sunflower oil, soybean oil, and multiphase systems of salad dressings with edible oils. Attempts to correlate either the flavor of by-products of oils or defective notes with GC peaks would be more comprehensive.

The sensory perception of breakdown products is recorded at minute levels. The possible aldehydes originating from various unsaturated acid hydroperoxides and their flavors have been recorded by Maera (285), Farstrup and Nissen (266), and Nawar (286). The scission products of fatty acids such as aldehydes, short-chain esters, oxo-esters, dicarboxylic acids, and furans are perceived at room temperature. The following have been reported as some breakdown products of oils and fats: cis-3-hexenal and trans-2-hexanal with green beany note, trans-4-heptenal with green to tallow odor, trans-6-nonenal with hydrogenated fat odor, trans-(2E)-6-nonedienal with beany note, trans-2,6-nonedienal with cucumber note, and trans-2,4-decadienal with stale fried oil odor. Their odor thresholds, however, are very different and are well recorded as dependent on the matrix in which they are dispersed. The threshold values of these volatile aldehydes originating from oxidized n-3 polyunsaturated fatty acids are recorded to be much lower than the corresponding aldehyes from less unsaturated fatty acids (287–289). Their volatility, the next important factor in perception, is proportional to the product of vapor pressure and activity coefficients of the compound. A high molecular weight implies a low vapor pressure of the volatile compound, and the molecular interaction is affected by the solvent. It is recorded that the odor threshold in water decreases (higher sensitivity) with increasing chain length within the homologous series hexanal, nonanal, undecanal, whereas the opposite is seen when these compounds are dissolved in oil.

For improving the activity coefficient of phases, it is apparent that the pH of the solvent plays an important role. In a modification of sample presentation in testing oils, a food acid buffer system for dispersing oils is reported by Baldwin et al. (290) and Narasimhan et al. (269) in studies on soybean and palm oil. The initial tests in the authors' work used several pH levels ranging from 4 to 8 for identifying the magnitude of odor responses with 30 naive panelists. A pH of 4.5 was found to be most suitable, and the method was standardized. It was also observed that in sensory tests, either olfactory (odor) or gustatory (nasopharyngeal, flavor) responses can be used for detecting the presence of breakdown products. However, for ease of perception and less panel fatigue, olfactory testing was preferred over gustatory tests. Farstrup and Nissen (266) also recorded a preference for using odor over taste testing in their work on fish oils.

Sensory quality of edible oil is generally referred to the noticeable off-flavor or rancidity. Each oil in its unrefined state has a definite odor profile that may be desired by the consumer. Definition of quality therefore has to be linked to these odor notes or to a deviation in the refined oil. The quality changes can be traced from the first stage of "fresh" with desirable odor notes or blandness in refined oil to the second stage of "identifiable breakdown products" through the third and final stage of "objectionable" or "rancid" odor. The first important task is therefore to develop a vocabulary for describing the odor quality of oils in these possible stages and in turn for describing the grades. Several factors play a role in identifying the three stages under bias-free conditions, for example, dispersion medium, temperature of testing, and panel training for the unambiguous understanding of the descriptors. The following procedure was employed by the authors in testing edible oils and fats for their sensory quality.

Sample Preparation

A buffered medium with ions common to the food is preferred to water at pH 6.5–7 because of its increased surface activity and thus better release of low volatiles (27°C). The citrate–phosphate buffer is prepared fresh on the day of testing from 115 mL of 0.1 M citric acid (21.25 g of citric acid in 1000 mL or 1L of water) and 85 mL of 0.2 M sodium phosphate dibasic (anhydrous, 28.75 g in 1000 mL or 1L of water) made up to 4 liters with distilled water and pH adjusted to 4.5. Into a 250-mL erlenmeyer flask, 100 mL of buffer solution is measured and 10 mL of oil is pipetted or 10 g of fat is weighed. The flask is maintained at 27°C for oils and at 35–40°C for fats. Head-space volatiles are accumulated by letting the flask stand for 10–15 min. A blank is similarly prepared with only the buffer solution to enable the panel members to familiarize themselves with the background odor to be considered as "white" odor.

Test Method

A suitable parametric or nonparametric method should be chosen depending on whether it is to be used only to identify the difference from the control or to quantitate the data on a structured scale for each of the odor notes. When odor quality grading or profiling is to be carried out on a structured scale, a predetermined set of odor descriptors broadly covering the grades in the three stages of quality—first stage of fresh, second stage of deviation from fresh with other noticeable odor notes, and third stage with distinct objectionable/repulsive odor notes specific to each oil—should be the guideline. These descriptors, being specific to each oil, should be unambiguously understood by the panelists. In our studies, the descriptors were first derived from free-choice profiling for odor quality using at least 50 responses and several samples of varying quality. Among these, the terms used by more than one-third of the panelists were used in the score card for odor profile analysis. Table 3.16 gives the word descriptors derived for the five oils studied in our laboratory. Testing was carried out in an odor-free room.

Analysis of Data

The data obtained from duplicate runs of the sensory tests were analyzed, using appropriate analytical procedures, either for significance of difference or for

Table 3.16 Odor Descriptors of Edible Oils

Rice bran oil	Palm oil	Groundnut oil (raw)	Groundnut oil (refined)	Sunflower oil (refined)	Mustard oil
Oily	Nutty	Nutty	Oily	Oily	Sulfury
Sweet	Earthy	Sweet	Sweet	Sweet	Acrid/pungent
Heated	Green	Earthy	Heated	Seedy	Herbal
Brany/husky	Haylike	Green	Nutty	Heated oil	Green
Nutty	Sweet	Beany	Harsh	Harsh	Sour
Green	Cooking oil	Seedy	Chemical	Chemical	Vinegar-like
Harsh	Heated oil	Cooking oil	Rancid	Rancid	Rancid
Rancid	Harsh	Wet	Musty	Musty	
Musty	Rancid	Rancid			
	Musty	Musty			

frequency of descriptors and as mean values on the scale used, which was in turn anchored to the grades. More usefully, the information available on the quality parameters, both physicochemical and sensory, was analyzed through multivariate analysis. The principal components analysis used in these studies essentially characterizes the samples and identifies the important attributes that are responsible for their final classification, in essence, identifying the directional vectors that are important for each axis in the PCA mapping. Analysis of data collected over a period of time can throw light on the pathway of quality changes. The use of multivariate analysis becomes inevitable as the changes are complex and cannot be attributed to a single variable in the system to be correlated to the different quality levels.

Storage Studies on Palm Oil   Palm oil is one of the major edible oils of the Asian population. In addition to its use in Indonesia and Malaysia, it has recently entered the homes of other subcontinents because of its beneficial use and acceptance by the population in their daily diet. Though the refined oil is occasionally colored with a tint of yellow due to the carotenoids and is sometimes not free-flowing due to a higher proportion of the stearine fraction, the desirable odor quality and high smoke point are positive features. The oil is also valued for its low PUFA content and its inhibitory effect on tumor growth (291). Quality analysis is carried out for chemical and physical parameters (this analysis is mandatory) and also for odor quality. In an extensive study on palm oil in our laboratory (269), quality profiles for both odor (Table 3.16) and chemical parameters were traced from 0 to 180 days of storage in unit packs of 1 liter. Edible oil is prone to oxidation, with the rate of oxidation depending on several factors such as the presence of natural oxidants and metals, fatty acid composition, and available moisture and oxygen. The study consisted of quality analysis of oil stored in seven types of packaging materials with different configurations of the structure of the laminate and coextrusion of the films affecting the barrier properties with respect to moisture and oxygen. Three storage conditions of relative humidity and temperature—65% RH, 27°C; 35% RH, 45°C; and 92% RH, 38°C—were considered. The data generated were analyzed by PCA to trace the pathway of changes, followed by a study of the matrix of the underlying intercorrelations among the parameters. The important discriminating factors in the odor profile pattern and the chemical parameters were obtained. Figure 3.12 gives the projection onto the planes of I and II with the directional vectors as the attributes. The salient observation in this study was that PV correlated well with Kreis value as both optical density and Lovibond tintometric value ($r = 0.93$). However, the Kreis value recorded as absorbance at 545 nm is preferred to the measure of redness in the Lovibond tintometer, which is more subjective. Correlation of PV with FFA content was lower ($r = 0.66$). The Kreis value showed the same relationship as PV with rancid or musty notes (stage 3) ($r = 0.77$), but the early indicators of changes (stage 2) were better related to

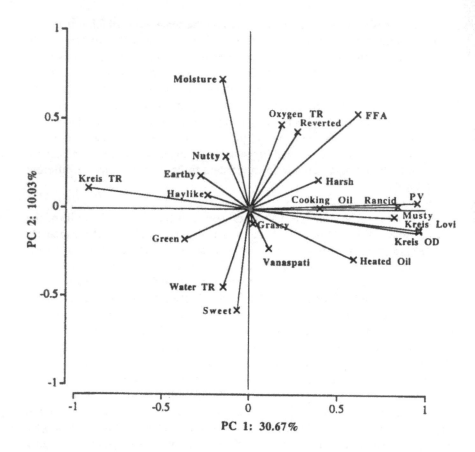

Fig. 3.12   Palm oil: PCA of attributes projected onto first two planes.

Kreis value than to PV. Each attribute is independent in the odor profile as no correlation was seen among them except for musty and rancid notes ($r = 0.71$). This indicates soundness of selection of word descriptors and use of nonsynonymous terms. For palm oil, conditions of high humidity and low temperature were found to be most detrimental.

Other Edible Oils   Addition of antioxidants to oils and their effect on quality were the subject of the next series of studies with edible oils. Refined rice bran, groundnut, mustard, and sunflower oils were studied to determine their profiles and chemical constituents. In this study, the effects of the antioxidant TBHQ at 0.02% with storage under various conditions were analyzed. Oils were stored under five humidity conditions, 11, 22, 56, 75, and 95%, and three temperatures, 27, 35, and

40°C. As in the normal scheme, odor was evaluated by a panel and physicochemical analyses for PV, FFA, Kreis value, and moisture content were carried out.

The data generated were analyzed using PCA (292–294). The effects of antioxidant are very distinctly seen in refined rice bran, sunflower, and groundnut oils, whereas in the case of mustard oil, with its natural configuration of fatty acids preventing easy oxidation, the effect or the benefit is not substantial even at accelerated conditions of storage. The observations on rice bran oil and refined groundnut oil are discussed below as examples.

Rice Bran Oil   Rice bran is one of the nonconventional sources of edible oils. From the milled paddy, oil is obtained from the 35% portion comprising the husk and bran; the remaining 65% is the milled rice. Though the yield of oil from the bran is 1–1.5%, it is a substantial source because of its volume in rice-growing countries. The steps to stabilize the oil, inactivating the lipases through heat treatment followed by acid treatment to maintain pH around 4, subject the oil to drastic treatment. This is followed by refining steps that include degumming, neutralizing, bleaching, deodorizing, and winterizing to make the oil edible. The oil at this stage has characteristic odor that is acceptable to consumers. Additionally, because of its high flashpoint and fatty acid profile, similar to those of peanut oil, its use is becoming more popular.

From the initial free-choice profiling responses, 23 descriptors were listed. From these, nine were selected because they were used by more than a third of the panelists. The final list of descriptors was derived for use in the score card from among these nine (Table 3.16). The samples were presented according to the standard method using a buffer system (pH 4.5) at 10% dispersion level. The panel examined the headspace odor profile and rated the intensity of the perceived notes on a scale from "none" to "strong." The data on distance from "none" to the "score" were computed, and the mean value was used for further analysis. The odor profile data were analyzed for frequency and mean value of each attribute to trace the changes during each sample withdrawal. Figure 3.13 gives an example of the comparative pattern of rice bran oil as initial sample and as tested at the end of 30 days of storage at 40°C and 22% RH. The data on physicochemical parameters were collected as mean value of duplicate analyses for percent moisture, PV, FFA, and Kreis value by standard methods. At the end of the study, the total integrated data of odor profile with nine descriptors and four instrumental measures were analyzed by ANOVA, and the correlations were studied. Among the odor profile notes, no significant correlations were found, indicating that each attribute is reasonably independent of the others. The highest correlation was seen between rancid and musty notes ($r = 0.62$) followed by branny/husky and nutty notes ($r = 0.34$) and oily with heated oily note ($r = 0.42$). Harsh note correlated with rancid ($r = 0.34$) and musty ($r = 0.40$) notes. Among the instrumental measures no significant correlations were seen. The Kreis test showed a negative correlation

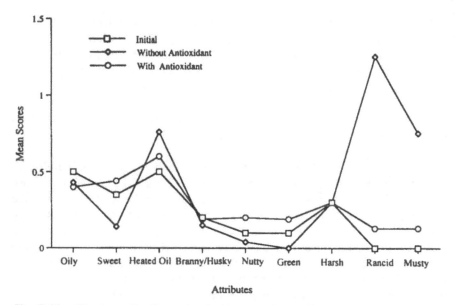

Fig. 3.13   Rice bran oil: effects of antioxidant on odor profile.

with the characteristic oily, sweet, heated oil, and nutty notes. The PV also showed a negative correlation with these odor notes but of smaller magnitude. This is important to notice, as the Kreis test, with its potential to indicate the early breakdown products, can show the quality deviation from "fresh" as observed by odor notes of nutty and green. Peroxide value was related to end notes of rancid and musty quality; FFA correlated negatively with heated oil, branny/husky, nutty; and the Kreis value also indicated the logical correlation of free fatty acids and breakdown products of fatty acids.

The integrated data, when subjected to PCA, revealed the pattern of characterization and a classification of the parameters surfaced. The basic structure of the PCA is given in Table 3.17 for five roots only, as trace percent comes down to less than 10% after the third axis. As an example, only two roots will be discussed. The cumulative trace percent reached 45% only with three roots that were worthwhile considering. The characterizing directional vectors for each root along with their respective signs are given in Table 3.18. It is evident from the table that nutty, oily, and sweet notes indicate good oil quality, whereas heated, harsh, branny/husky, and green represent the next grade of oil, and rancid and musty indicate that the oil has undergone a degree of oxidation. Among the physicochemical tests, the Kreis test measures the quality changes before the rancidity becomes apparent. Peroxide value is useful only when rancidity has progressed to more than perceptible levels. The value of FFA content lies in between that of the Kreis and PV tests. Figure

Table 3.17 Rice Bran Oil: Structure of Principal Components

Root	Eigen value	Trace %	Cumulative	df	Chi square
1	2.36025	18.16	18.16	78	454.83
2	1.97995	15.23	33.39	66	331.03
3	1.47681	11.36	44.75	55	224.80
4	1.15623	8.89	53.64	45	164.32
5	1.05236	8.10	61.74	36	131.62

3.14a gives the projection as vectors of the attributes of odor and physicochemical parameters on the first two planes along with positioning of the initial sample.

The initial sample was characterized by dominant odor notes of oily and heated oil note and lower levels of sweet and nutty notes. This stage can be considered the first stage of oil quality. This was followed by branny/husky odor, which develops with time and is indicative of second-stage quality. The mild green odor notes also appear here. The third stage consists of harsh, rancid, and musty notes, which develop during continued storage. The projections also indicate the positions of the Kreis, PV, and FFA tests on the sample plane. The Kreis test is away from the initial

Table 3.18 Rice Bran Oil: Factor Pattern for the First Three Factors[a]

Attributes	Factors		
	1	2	3
Oily	0.26	0.59	−0.50
Sweet	0.31	0.43	−0.20
Heated oil	0.18	0.73	0.06
Branny/husky	0.44	0.32	0.37
Nutty	0.44	0.43	0.06
Green	−0.10	0.38	0.22
Harsh	−0.45	0.51	0.35
Rancid	−0.72	0.29	0.13
Musty	−0.78	0.27	0.12
Kreis value	0.00	−0.13	0.51
PV	−0.30	0.23	−0.31
FFA	−0.35	−0.00	−0.69
Moisture	−0.47	0.01	−0.04

[a]Factors are linear functions of all the attributes. Underlined vectors in factors 1 and 2 have been mapped in Fig. 3.14a.

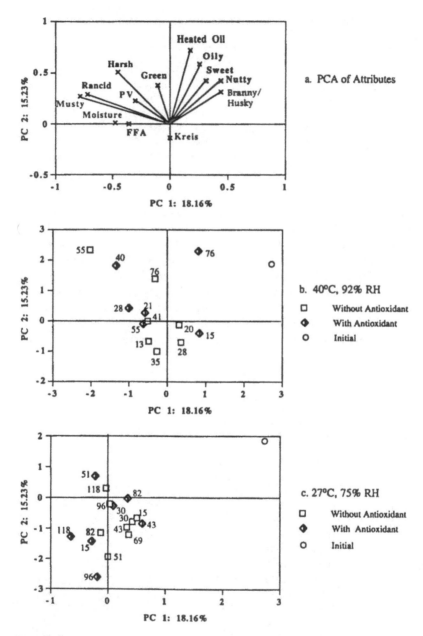

Fig. 3.14 Rice bran oil. (a) PCA of attributes projected on to first two axes; (b–f) pathway traced over the period of storage. Labels within parts b–f indicate days of storage. Interpretation of parts b–f is based on the basic structure of directional vectors indicated in part a.

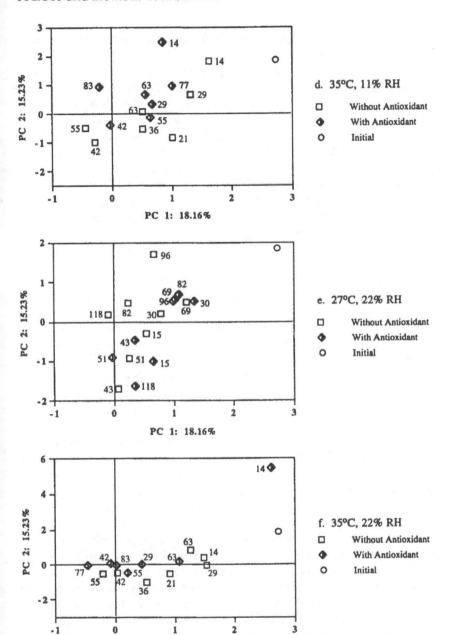

d. 35°C, 11% RH

☐   Without Antioxidant
◆   With Antioxidant
○   Initial

e. 27°C, 22% RH

☐   Without Antioxidant
◆   With Antioxidant
○   Initial

f. 35°C, 22% RH

☐   Without Antioxidant
◆   With Antioxidant
○   Initial

stage 1 location but toward stage 2, while PV follows closer to the end of the stage 3 position with rancid, musty, and harsh notes.

The analysis of the pathway of change in the overall study segregated by temperature distinctly indicated the effect on quality. The effect of humidity was more severe than that of temperature. This is demonstrated by our earlier work on palm oil also, where the effect is discussed in terms of kilocalories required to cause the breakdown of fatty acids. The presence of TBHQ prevented an early deterioration. This is distinctly indicated in the pathway of sensory quality of oil from stage 1 (initial) with characteristic oily, heated oil, and sweet nutty notes to the development of higher Kreis value and green, harsh notes, not entering the third stage of musty, rancid notes with high PV when the antioxidant is present. Without the antioxidant, the breakdown of fatty acids and subsequent development of rancid/musty notes are perceptible.

Analysis of the data based on humidity effect both in the presence and absence of TBHQ projects a different dimension. The presence of higher humidity (92%) at higher temperature is highly damaging to the oil irrespective of the presence of antioxidant (Fig. 3.14b). The initial sample quality was lost within 28 days of storage with higher FFA content and Kreis value. By 55 days the development of rancidity and musty notes with high PV was observed in both, and the presence of antioxidant was not effective in protecting the sample. A similar effect was seen in a sample stored at 75% RH, but the effect was clearer in the presence of antioxidant, which protected the oil for up to 55 days (Fig. 3.14c). Unlike oil unprotected for 36 days, the movement from initial quality (stage 1) to distinct deterioration of quality (stage 2) was swift and this clearly demonstrated the damage due to the presence of humidity.

Dry heat (40°C) with low humidity (11%) was more damaging than a lower temperature (35 or 27°C) and low humidity (Fig. 3.14d). An increase in humidity to 22% resulted in a similar trend. The change from initial stage to onset of deterioration (stage 2) is more distinctly seen at higher temperatures both with and without antioxidant. However, at lower temperature and low humidity (11 or 22%), the appearance of breakdown products and in turn, the development of green, sweet notes were delayed. The oil is not protected beyond 55 days by TBHQ (Fig. 3.14e). The presence of antioxidant retained the initial quality for up to 83 days but resulted in a sudden movement to stage 3 with high rancid musty notes and high PV. This demonstrates that the free-radical reaction is suppressed by the antioxidant, the formation of breakdown products is substantially prevented when the temperature is still low (27°C), and low humidity helps prevent the deterioration of quality. A dynamic change is noticed in oil at low humidity and temperature also, but the length of time to reach the different stages of quality is longer. Addition of antioxidant has delayed the process further, and the desirable low PV, FFA, and Kreis values and odor notes of oily, heated oil characters with low sweet and nutty notes are retained for up to 118 days. The corresponding features are retained up to 69 days only in the absence of TBHQ. The oils stored at 35°C and 22% RH (Fig. 3.14f) did not show distinct differences in the pathway with the addition of antioxidant, though some differences were observed during the initial 29 days of storage.

Refined Groundnut Oil. Groundnuts (peanuts) have been an important source of edible oils for a long time. There have been advances in techniques of extraction in terms of design of screw press, solvent extraction unit, and the refining process including physical and chemical procedures in terms of reducing capital cost, operating cost, and improving quality maintenance. Packaging and storage in either bulk or unit packs have also been constantly reassessed. Regulations and laws cover some aspects of consumer protection, especially for additives, but the dynamic state of oils demands due consideration in quality monitoring. As to sensory quality, unequivocal odor or flavor descriptors need to be established along with limits of physicochemical parameters. In our study on the effect of antioxidants on refined groundnut oil, these aspects were considered with special reference to temperature and humidity as detailed earlier.

In deriving the odor descriptors of refined groundnut oil, the panel was provided a range of samples with the facility for free-choice profiling and also a list of descriptor terms commonly applied to edible oils. From the first four sessions, descriptors were selected, and panel homogeneity and consensus were obtained. Eight descriptors were arrived at for refined groundnut oil that could cover "excellent" to "defective" oils (Table 3.16). The oils were analyzed for the physicochemical parameters moisture content, PV, FFA, and Kries value by standard methods. The samples were subjected to analysis at each withdrawal in duplicate. The odor profile was also monitored with the trained panel at each withdrawal. Figure 3.15 gives the profile pattern at the end of 60 days of storage at 40°C and 75% RH, with and without the addition of antioxidant. This intermediate plotting indicates a distinct reduction in desirable odor notes when there was no antioxidant. At the end of the study, the data for eight odor attributes and four physicochemical parameters were analyzed by ANOVA to obtain the basic correlation matrix. Among the odor descriptors, the highest correlation was seen between musty and rancid notes ($r = 0.66$) followed by oily and nutty notes ($r = 0.44$). The Kreis value significantly correlated with PV only ($r = 0.88$). These two parameters in turn correlated positively with rancid note ($r = 0.41$) and negatively with desirable notes of oily, sweet, and nutty ($r = -0.14$) to a small extent. FFA also showed a negative correlation with these desirable notes but of slightly higher magnitude ($r = -0.29$). These observations confirmed the independence of each attribute and the nonsynonymy of odor descriptors.

The characterization of refined groundnut oil for odor quality through PCA confirmed the three stages of oil quality. The basic structure of the PCA is given in Table 3.19, and Fig. 3.16a presents the projection in the plane of the first two axes along with the directional vectors and attributes identified therein. The trace percent

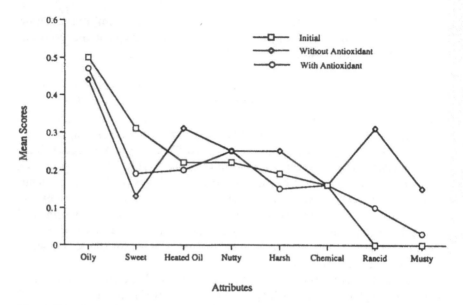

Fig. 3.15   Refined groundnut oil: effects of antioxidant on odor profile.

fell below 10% after four roots, and the cumulative trace had reached 62.78%. The characterizing directional vectors for each root along with their respective signs are given in Table 3.20. Taking into account the first two roots (cumulative trace 41.85%), the classification of attributes will be as follows: The initial sample of stage (grade) 1 is dominant with oily, nutty, and sweet notes and very low levels of harsh and heated oil notes. It is low in peroxide value, FFA, and Kreis value. In stage 2, these dominant odor notes are much subdued, with chemical and harsh notes becoming more pronounced. Development of FFA is closely related to this stage and

Table 3.19   Refined Groundnut Oil: Structure of Principal Components

Root	Eigen value	Trace %	Cumulative	df	Chi square
1	2.73450	22.79	22.79	66	651.50
2	2.28695	19.06	41.85	55	500.38
3	1.30515	10.88	52.72	45	350.26
4	1.20684	10.06	62.78	36	300.53
5	1.08039	9.00	71.78	28	245.78
6	0.77989	6.50	78.28	21	187.82
7	0.63042	5.25	83.53	15	159.29
8	0.58039	4.84	88.37	10	142.11

Table 3.20 Refined Groundnut Oil: Factor Pattern for the First Four Factors[a]

	Factors			
Attributes	1	2	3	4
Oily	0.25	0.69	0.28	0.20
Sweet	0.41	0.51	−0.25	0.00
Heated oil	−0.23	0.56	−0.01	0.48
Nutty	−0.28	0.65	0.11	0.04
Harsh	0.08	0.09	0.31	0.48
Chemical	0.26	−0.13	0.65	0.27
Rancid	−0.70	0.29	0.36	−0.32
Musty	−0.58	0.20	0.50	−0.41
Kreis value	−0.83	0.12	−0.27	0.31
PV	−0.82	0.14	−0.27	−0.27
FFA	−0.32	−0.48	0.31	0.18
Moisture	0.05	−0.63	0.10	0.41

[a]Factors are linear functions of all the attributes. Underlined vectors in factors 1 and 2 have been mapped in Fig. 3.16a.

with uptake of moisture. The third stage is distinctly identified by high rancid or musty notes and the absence of oily, sweet, or nutty notes.

The pathway of change in the study on refined groundnut oil with and without the antioxidant is indicative of the important role of temperature and humidity in determining the quality of oil. The humidity level influences the quality much more than temperature in refined groundnut oil. It was observed that the deviation from stage 1 to stage 3 was much faster in all five humidity conditions when the temperature was high (40°C) than when it was low (27°C). By the end of about 20 days, the oil at high temperature moved from stage 1, whereas at low temperature it remained at stage 1 for another 14 days, moving after 34 days. The oil at high temperature was closer to stage 3 than the oil stored at lower temperature. The effect of antioxidant is clearly seen here, where the sample did not reach stage 3. Figures 3.16b and 3.16c show the positions of these samples. Total protection, however, was not possible, even with the addition of antioxidant, when the humidity was high. Figure 3.16d shows the effect more clearly.

The storage temperature of oil in these studies was not high enough to cause serious damage to the sample. The combination of temperature with humidity and the presence or absence of TBHQ influences oil quality. It can be distinctly traced when the pathways of the two humidities, 92 and 22%, are compared across the two temperatures (Figs. 3.16b and 3.16d). Figures 3.16d and 3.16e show a comparison of similar relative humidities but different temperatures, and Figs. 3.16e and 3.16f show a comparison of different relative humidities but similar tempera-

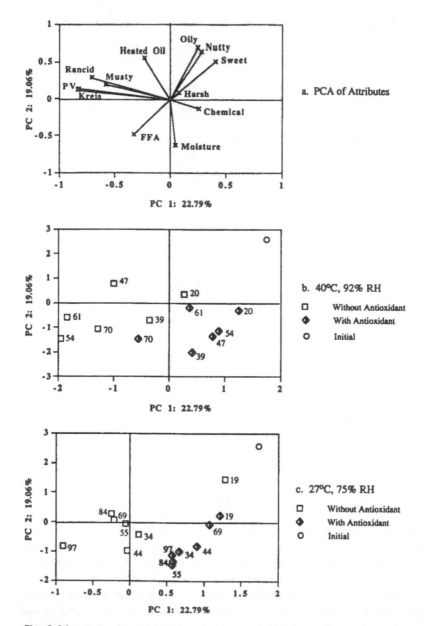

Fig. 3.16   Groundnut oil. (a) PCA of attributes projected onto first two axes; (b–f) pathway traced over the period of storage. Labels within parts b–f indicate days of storage. Interpretation of parts b–f is based on the basic structure of directional vectors indicated in part a.

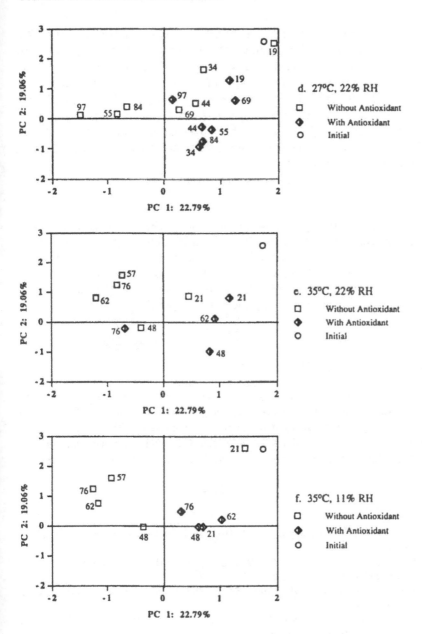

d. 27°C, 22% RH

☐   Without Antioxidant
◆   With Antioxidant
○   Initial

e. 35°C, 22% RH

☐   Without Antioxidant
◆   With Antioxidant
○   Initial

f. 35°C, 11% RH

☐   Without Antioxidant
◆   With Antioxidant
○   Initial

tures. Oil maintained at 27°C and 22% RH differed significantly from that at 27°C and 75% RH (Fig. 3.16c). The addition of TBHQ protected the sample up to 96 days and they were positioned over successive withdrawals in the first two quadrants, representing the first two stages of oil quality. Without the antioxidant, the positioning of the sample was similar only up to 44 days at 22% RH (Fig. 3.16d). The samples at higher relative humidity (75%) could retain the position in the first two quadrants for only up to 34 days when no antioxidant was used (Fig. 3.16c); addition of antioxidant did help in retaining the samples in these areas, but edging toward stage 2 of oil quality.

Conclusions

It has been observed that specific odor descriptors can be used in association with defined stages of edible oil quality. With proper panel training and data analysis, sensory evaluation is very effective for identifying early stages of oxidation and can be used as a sensitive, precise measure of antioxidant activity. In addition, it can be a valuable tool for tracing the degradation pathway in food products and the role of food antioxidants in delaying the onset of specific undesirable changes in food quality.

ACKNOWLEDGMENTS

We wish to acknowledge the willing assistance rendered by our colleagues during the preparation of this review. The permission granted by the Director of the Central Food Technological Institute to present some relevant parts of the work carried out at the Institute is also gratefully acknowledged. Our thanks are also due to Dr. D. L. Madhavi, for the substantial contributions during the preparation of this review.

REFERENCES

1. A. J. Lehman, O. G. Fitzhugh, A. A. Nelson, and G. Woodard, *Adv. Food Res.*, *3*: 197 (1951).
2. P. P. Coppen, in *Rancidity in Foods* (J. C. Allen and R. J. Hamilton, Eds.), Elsevier Applied Science, London, 1989, p. 83.
3. M. H. Gordon, in *Food Antioxidants* (B. J. F. Hudson, Ed.), Elsevier Applied Science, London, 1990, p. 1.
4. D. L. Madhavi and D. K. Salunkhe, in *Food Additive Toxicology* (J. A. Maga and A. J. Tu, Eds.), Marcel Dekker, Inc., New York, 1994, p. 89.
5. T. P. Labuza, *CRC Crit. Rev. Food Technol.*, 2: 355 (1971).
6. A. F. Bickel, E. C. Kooijman, C. La Lau, W. Roest, and P. Piet, *J. Chem. Soc., Part III, 1953*: 3211 (1953).
7. C. Karahadian and R. C. Lindsay, *J. Amer. Oil Chem. Soc.*, *65*: 1159 (1988).

8. R. E. Morse, in *Encyclopedia of Food Technology* (A. H. Johnson and M. S. Peterson, Eds.), AVI, Westport, CT, 1975.

9. R. J. Sims and J. A. Fioriti, in *CRC Handbook of Food Additives*, Vol. 1, (T. E. Furia, Ed.), CRC, Boca Raton, FL, 1980, p. 13.

10. J. D. Dziezak, *Food Technol.*, *40*(9): 94 (1986).

11. Anon., *Prepared Foods*, *160*(8): 83 (1991).

12. J. Pokorny, *Trends Food Sci. Technol.*, 2: 223 (1991).

13. J. Loliger, in *Free Radicals and Food Additives* (O. I. Arouma and B. Halliwell, Eds.), Taylor and Francis, London, 1991, p. 121.

14. M. Namiki, *Crit. Rev. Food Sci. Nutr.*, 29: 273 (1990).

15. D. E. Pratt and B. J. F. Hudson, in *Food Antioxidants* (B. J. F. Hudson, Ed.), Elsevier Applied Science, London, 1990, p. 171.

16. P. Schuler, in *Food Antioxidants* (B. J. F. Hudson, Ed.), Elsevier Applied Science, London, 1990, p. 99.

17. T. Koga and J. Terao, *J. Agric. Food Chem.*, 42: 1291 (1994).

18. T. Osawa, H. Katsuzaki, Y. Hagiwara, H. Hagiwara, and T. Shibamoto, *J. Agric. Food Chem.*, *40*: 1135 (1992).

19. H. Haraguchi, K. Hashimoto, and A. Yagi, *J. Agric. Food Chem.*, *40*: 1349 (1992).

20. T. Tsuda, K. Ohshima, S. Kawakishi, and T. Osawa, *J. Agric. Food Chem.*, 42: 248 (1994).

21. G. O. Igile, W. Oleszek, M. Jurzysta, S. Burda, M. Fafunso, and A. A. Fasanmade, *J. Agric. Food Chem.*, 42: 2445 (1994).

22. K. Wu, W. Zhang, P. B. Addis, R. J. Eple, A. M. Salih, and J. Lehrfeld, *J. Agric. Food Chem.*, *42*: 34 (1994).

23. S. Z. Dziedzic and B. J. F. Hudson, *Food Chem.*, *11*: 161 (1983).

24. E. F. Kurth and F. L. Chan, *J. Amer. Oil Chem. Soc.*, 28: 433 (1951).

25. D. E. Pratt and E. E. Miller, *J. Amer. Oil Chem. Soc.*, *61*: 1064 (1984).

26. P. D. Duh, D. B. Yeh, and G. C. Yen, *J. Amer. Oil Chem. Soc.*, *69*: 814 (1992).

27. N. Ramarathnam, T. Osawa, M. Namiki, and S. Kawakishi, *J. Agric. Food Chem.*, *37*: 316 (1989).

28. G. M. Sapers, O. Panasiuk, F. B. Talley, and R. L. Shaw, *J. Food Sci.*, *40*: 797 (1975).

29. I. A. Radaeva and N. A. Tyukavkina, USSR Patent 350,451 (1972).

30. S. E. O. Mahgoub and B. J. F. Hudson, *Food Chem.*, *16*: 97 (1985).

31. K. Herrmann, *J. Food Technol.*, *11*: 433 (1976).

32. O. Akaranta and T. O. Odozi, *Agric. Wastes*, *18*: 299 (1986).

33. M. H. Lee and R. L. Sher, *J. Chin. Agric. Chem. Soc.*, 22: 226 (1984).

34. T. Matsuzaki and Y. Hara, *Nippon Nogeikagaku Kaishi*, *59*: 129 (1985).

35. J. D. Su, T. Osawa, S. Kawakishi, and M. Namiki, *Phytochemistry*, 27: 1315 (1988).

36. T. Ariga, I. Koshiyama, and D. Fukushima, *Agric. Biol. Chem.*, *52*: 2717 (1988).

37. M. T. Meunier, E. Duroux, and P. Bastide, *Plantes Med. Phytotherapie*, 23(4): 267 (1990).

38. T. Ariga, I. Koshiyama, and D. Fukushima, U.S. Patent 4,797,421 (1989).

39. K. Igarashi, K. Takanashi, M. Makino, and T. Yasui, *J. Jpn. Soc. Food Sci. Technol.*, *36*: 852 (1989).

40. K. Igarashi, T. Yoshida, and E. Suzuki, *J. Jpn. Soc. Food Sci. Technol.*, *40*: 138 (1993).

41. H. Kozlowska, M. Naczk, F. Shahidi, and R. Zadernowski, in *Canola and Rapeseed: Production, Chemistry, Nutrition, and Processing Technology* (F. Shahidi, Ed.), Van Nostrand Reinhold, New York, 1990, p. 193.

42. S. W. Souci and E. Mergenthaler, in *Die Bestandteile der Lebensmittel* (J. Schormuller, Ed.), Springer-Verlag, Berlin, 1965, p. 1159.

43. T. Ohta, S. Yamasaki, Y. Egashira, and H. Sanada, *J. Agric. Food Chem.*, 42: 653 (1994).

44. T. Ohta, H. Takashita, K. Todoroki, K. Iwano, and T. Ohba, *J. Brew. Soc. Jpn.*, 87: 922 (1992).

45. C. W. Dubois and D. K. Tressler, *Proc. Inst. Food Technol.*, 202 (1943).

46. J. R. Chipault, G. R. Mizuno, and J. M. Hawkins, and W. O. Lundberg, *Food Res.*, 17: 46 (1952).

47. Y. Saito, Y. Kimura, and T. Sakamoto, *Eiyou to Shokuryo*, 29: 505 (1976).

48. A. Palitzsch, H. Schulze, F. Metzl, and H. Bass, *Fleischwirtzshaft*, 49: 1394 (1969).

49. A. Palitzsch, H. Schulze, G. Lotter, and A. Steidrele, *Fleischwirtzshaft*, 54: 63 (1974).

50. N. Nakatani and R. Inatani, *Agric. Biol. Chem.*, 48: 2081 (1984).

51. C. M. Houlihan, C. T. Ho, and S. S. Chang, *J. Amer. Oil Chem. Soc.*, 62: 96 (1985).

52. R. Inatani, N. Nakatani, H. Fuwa, and H. Seto, *Agric. Biol. Chem.*, 46: 1661 (1982).

53. R. Inatani, N. Nakatani, and H. Fuwa, *Agric. Biol. Chem.*, 47: 521 (1983).

54. X. C. Weng and M. H. Gordon, *J. Agric. Food Chem.*, 40: 1331 (1992).

55. K. Q. Zhang, Y. Bao, P. Wu, R. T. Rosen, and C. T. Ho, *J. Agric. Food Chem.*, 38: 1194 (1990).

56. N. Nakatani and H. Kikuzaki, *Agric. Biol. Chem.*, 51: 2727 (1987).

57. H. Kikuzaki and N. Nakatani, *Agric. Biol. Chem.*, 53: 513 (1989).

58. K. Miura and N. Nakatani, *Agric. Biol. Chem.*, 53: 3043 (1989).

59. N. Nakatani, K. Miura, and T. Inagaki, *Agric. Biol. Chem.*, 53: 1375 (1989).

60. N. Nakatani, Y. Tachibana, and H. Kikuzaki, in *Proceedings*, 4th Biennial General Meeting of the Society for Free Radical Research, Kyoto, Japan, 1988, p. 453.

61. N. Nakatani, R. Inatani, H. Ohta, and A. Nishioka, *Environ. Health Perspect.*, 67: 135 (1986).

62. R. E. Kramer, *J. Amer. Oil Chem. Soc.*, 62: 111 (1985).

63. H. Kikuzaki and N. Nakatani, *J. Food Sci.*, 58: 1407 (1993).

64. N. Nakatani and K. Ikeda, *Nippon Kasei Gakkaishi*, 79 (1984).

65. J. Burri, M. Graf, P. Lambelet, and J. Loliger, *J. Sci. Food Agric.*, 110: 153 (1989).

66. S. S. Chang, B. Ostric-Matijasevic, O. Hsieh, and A. L. Huang-Cheng-Li, *J. Food Sci.*, 42: 102 (1977).

67. U. Bracco, J. Loliger, and J. L. Viret, *J. Amer. Oil Chem. Soc.*, 58: 686 (1981).

68. K. Taguchi, K. Iwami, M. Kawabata, and F. Ibuki, *Agric. Biol. Chem.*, 52: 539 (1988).

69. F. Ibuki and K. Iwami, *Nutr. Sci. Soyprotein*, 11(1): 17 (1990).

70. G. Bertelsen, A. Ohlen, and L. H. Skibsted, *Z. Lebensm. Unters. Forsch.*, 192: 319 (1991).

71. R. E. Hayes, G. N. Bookwalter, and E. B. Bagley, *J. Food Sci.*, 42: 1527 (1977).

72. S. J. Bishov and A. S. Henick, *J. Food Sci.*, 37: 873 (1972).

73. L. Srinivas, V. K. Shalini, and M. Shylaja, *Arch. Biochem. Biophys.*, 292: 617 (1992).

74. E. A. Decker and A. D. Crum, *J. Food Sci.*, 56: 1179 (1991).

75. E. A. Decker and H. Faraji, *J. Amer. Oil Chem. Soc.*, 69: 650 (1990).

76. R. J. Marcuse, *J. Amer. Oil Chem. Soc.*, *39*: 97 (1962).
77. M. J. Taylor, T. Richardson, and R. D. Jasensky, *J. Amer. Oil Chem. Soc.*, *58*: 622 (1981).
78. A. Popovici, A. Isopescu, D. Dima, and A. D. Peterscu, *Ind. Aliment.*, *16*: 248 (1965).
79. M. M. Merzametov and L. I. Gadzhuva, *Izv. Vyssikh. Ucheb. Zaved. Pish. Tech.*, *6*: 20 (1982).
80. M. M. Merzametov and L. I. Gadzhuva, *Izv. Vyssikh. Ucheb. Zaved. Pish. Tech.*, *5*: 35 (1982).
81. H. Lingnert and C. E. Eriksson, *J. Food Process. Preserv.*, *4*: 161 (1980).
82. C. Macku and T. Shibamoto, *J. Agric. Food Chem.*, *39*: 1990 (1991).
83. J. P. Eiserich, C. Macku, and T. Shibamoto, *J. Agric. Food Chem.*, *40*: 1982 (1992).
84. J. P. Eiserich and T. Shibamoto, *J. Agric. Food Chem.*, *42*: 1060 (1994).
85. Y. Fukuda, T. Osawa, S. Kawakishi, and M. Namiki, *Nippon Shokuhin Kogyo Gakkaiashi*, *35*: 28 (1988).
86. Y. Fukuda, T. Osawa, M. Namiki, and T. Ozaki, *Agric. Biol. Chem.*, *49*: 301 (1985).
87. T. Osawa, M. Nagata, M. Namiki, and Y. Fukuda, *Agric. Biol. Chem.*, *49*: 3351 (1985).
88. Y. Fukuda, *J. Jpn. Soc. Food Sci. Technol.*, *38*: 393 (1987).
89. P. Budowski, F. G. T. Menezenes, and F. G. Doller, *J. Amer. Oil Chem. Soc.*, *27*: 377 (1950).
90. T. Kurechi, K. Kikugawa, and S. Aoshima, *Chem. Pharm. Bull.*, *29*: 2351 (1981).
91. K. Kikugawa, M. Arai, and T. Kurechi, *J. Amer. Oil Chem. Soc.*, *60*: 1528 (1983).
92. G. W. Burton and K. U. Ingold, *Science*, *224*: 569 (1984).
93. C. S. Foote and R. W. Denny, *J. Amer. Chem. Soc.*, *90*: 6233 (1968).
94. C. S. Foote, Y. C. Chang, and R. W. Denny, *J. Amer. Chem. Soc.*, *92*: 5116 (1970).
95. M. M. Mathews-Roth, T. Wilson, E. Fujumori, and N. Krinsky, *Photochem. Photobiol.*, *19*: 217 (1974).
96. A. H. Clements, R. H. Van den Engh, D. T. Frost, and K. Hoogenhout, *J. Amer. Oil Chem. Soc.*, *50*: 325 (1973).
97. R. D. Hill, V. van Leeuwen, and R. A. Wilkinson, *N.Z. J. Dairy Sci. Technol.*, *12*: 69 (1977).
98. H. Lingnert, in *Lipid Oxidation: Biological and Food Chemical Aspects* (R. Marcuse, Ed.), Conference, Goteborg, Sweden, 1989, p. 183.
99. R. Stocker, A. N. Glazer, and B. N. Ames, *Proc. Natl. Acad. Sci. USA*, *84*: 3918 (1987).
100. S. Nishibori and M. Namiki, *Nippon Kaseigaku Kaishi*, *39*: 1173 (1988).
101. Y. Endo, R. Usuki, and T. Kareda, *J. Amer. Oil Chem. Soc.*, *62*: 1387 (1985).
102. D. Boskou and I. D. Morton, *J. Sci. Food Agric.*, *27*: 928 (1976).
103. R. J. Sims, J. A. Fioriti, and M. J. Kanuk, *J. Amer. Oil Chem. Soc.*, *49*: 298 (1972).
104. T. Takagi and T. Iida, *J. Amer. Oil Chem. Soc.*, *57*: 326 (1980).
105. M. H. Gordon and P. Magos, *Food Chem.*, *10*: 141 (1983).
106. J. A. Maga, *Smoke in Food Processing*, CRC Press, Boca Raton, FL, 1988.
107. D. D. Duxbury, *Food Process, USA*, *52*(8): 54 (1991).
108. Y. Ishikawa, K. Marimoto, and T. Hamasaki, *J. Amer. Oil Chem. Soc.*, *61*: 1864 (1984).
109. T. Aoyama, Y. Nakakita, M. Nakagawa, and H. Sakai, *Agric. Biol. Chem.*, *46*: 2369 (1982).
110. K. Kikugawa, A. Kunugi, and T. Kurechi, in *Food Antioxidants* (B. J. F. Hudson, Ed.), Elsevier Applied Science, London, 1990, p. 65.

111. Y. Harano, O. Hoshino, and T. Ukita, *J. Hyg. Chem.*, *13*: 197 (1967).
112. T. Kurechi, *J. Hyg. Chem.*, *13*: 191 (1967).
113. T. Kurechi, *J. Hyg. Chem.*, *15*: 301 (1969).
114. T. Kurechi, M. Aizawa, and A. Kunugi, *J. Amer. Oil Chem. Soc.*, *60*: 1878 (1983).
115. T. Kurechi, M. Aizawa, and A. Kunugi, *J. Amer. Oil Chem. Soc.*, *60*: 1878 (1983).
116. J. Lehman and H. T. Slover, *Lipids*, *11*: 853 (1976).
117. T. Kurechi and A. Kunugi, *J. Amer. Oil Chem. Soc.*, *60*: 33 (1983).
118. T. Kurechi and T. Kato, *J. Amer. Oil Chem. Soc.*, *57*: 220 (1980).
119. J. R. Chipault, in *Autoxidation and Antioxidants*, Vol. 2 (W. O. Lundberg, Ed.), Wiley, New York, 1962, p. 477.
120. W. O. Lundberg, in *Autoxidation and Antioxidants*, Vol. 2 (W. O. Lundberg, Ed.), Wiley, New York, 1962, p. 451.
121. R. A. Ivanova, N. S. Pimenova, E. I. Kozlov, and V. F. Tsepalov, *Kinet. Katal.*, *20*: 1423 (1979).
122. T. Kurechi and T. Kato, *Chem. Pharm. Bull.*, *31*: 1772 (1983).
123. E. Niki, T. Saito, A. Kawakami, and Y. Kamiya, *J. Biol. Chem.*, *259*: 4177 (1984).
124. S. Z. Dziedzic and B. J. F. Hudson, *Food Chem.*, *14*: 45 (1984).
125. S. R. Husain, J. Tearo, and S. Matsushita, *J. Amer. Oil Chem. Soc.*, *63*: 1457 (1986).
126. B. J. F. Hudson and M. Ghavami, *Lebensm. Wiss. Technol.*, *17*: 191 (1984).
127. A. L. Tappel, *Geriatrics*, *23*: 97 (1968).
128. S. J. Bishov and A. S. Henick, *J. Food Sci.*, *40*: 345 (1975).
129. M. E. Endean, *Scientific and Technical Surveys*, British Manufacturing Industries Research Association, Surrey, UK, 1976.
130. K. Robards and S. Dilli, *Analyst*, *112*: 933 (1987).
131. S. P. Kochhar and J. B. Rossell, in *Food Antioxidants* (B. J. F. Hudson, Ed.), Elsevier Applied Science, London, 1990, p. 19.
132. A. Emmerie and C. Engel, *Rev. Trav. Chim.*, *57*: 1351 (1938).
133. *AOAC Official Methods of Analysis*, 14th ed., AOAC, Arlington, VA, 1984.
134. G. J. Bukovits and A. Lazerovich, *J. Amer. Oil Chem. Soc.*, *64*: 517 (1987).
135. A. Deschnytere, H. Deelstra, and Z. Fresenius, *Anal. Chem.*, *324*: 1 (1986).
136. N. K. Andrikopoulos, H. Brueschweiler, H. Felber, and C. Taeschler, *J. Amer. Oil Chem. Soc.*, *68*: 359 (1991).
137. B. Denis Page and C. F. Charbonneau, *J. Assoc. Off. Anal. Chem.*, *72*: 259 (1989).
138. P. Vinas, M. H. Cordoba, and C. Sanchez-Pedreno, *Food Chem.*, *42*: 241 (1991).
139. *Fed. Reg.*, *51*: 25012 (1986).
140. H. J. Kim and Y. K. Kim, *J. Food Sci.*, *51*: 1360 (1986).
141. H. J. Kim, G. Y. Park, and Y. K. Kim, *Food Technol.*, *41*: 85 (1987).
142. FAO/WHO, Joint FAO/WHO Expert Committee on Food Additives, FAO Food and Nutrition Paper No. 4, Food and Agriculture Organization of the United Nations, Rome, Italy, 1978.
143. F. Garcia Montelongo, M. J. Sanchez, J. C. Garcia Castro, and A. Hardisson, *Anal. Lett.*, *24*: 1875 (1991).
144. W. A. Behrens and R. Madere, *J. Liq. Chromatogr.*, *15*: 753 (1992).
145. ISO, ISO 5553: 1980, p. 3, Int. Standards Organization, 1980.
146. Y. Tonogai and M. Iwaida, *J. Food Protec.*, *44*: 835 (1981).

147. A. Matsunaga, A. Yamamoto, E. Mizukami, K. Kawasaki, and T. Ooizumi, *J. Jpn. Soc. Food Sci. Technol.*, *37*: 20 (1990).
148. T. Hamano, Y. Mitsuhiashi, S. Yamamoto, Y. Matsuki, Y. Tonogai, K. Nakamura, and Y. Ito, *J. Jpn. Soc. Food Sci. Technol.*, *34*: 603 (1987).
149. M. Oishi, K. Onishi, and S. Sakai, *Annu. Rep. Tokyo Metropol. Res. Lab. Publ. Health*, *34*: 211 (1983).
150. J. de Jong, A. van Polanen, and J. J. M. Driessen, *J. Chromatogr.*, *553*: 243 (1991).
151. C. Retho, D. Maitre, F. Blanchard, and L. Diep, *Ann. Falsif. l'Exper. Chim. Toxicol.*, *83*: 145 (1990).
152. *AOAC Offical Methods of Analysis*, 12th ed., AOAC, Washington, DC, 1975.
153. S. H. Ashoor and M. J. Knox, *J. Chromatogr.*, *299*: 288 (1984).
154. Y. Fujita, I. Mori, K. Fujita, and T. Tanaka, *Jpn. J. Toxicol. Environ. Health*, *33*: 56 (1987).
155. S. Tsuji, Y. Tonogai, and Y. Ito, *J. Food Prot.*, *49*: 914 (1986).
156. H. Ushiyama, M. Nishijima, K. Yasuda, H. Kamimura, S. Tabata, and T. Nishima, *J. Food Hyg. Soc. Jpn.*, *28*: 200 (1987).
157. AOAC, *AOAC Official Methods of Analysis*, 15th ed., AOAC, Arlington, VA, 1990.
158. K. D. Bos, C. Verbeek, and C. H. Peter van Eeden, *J. Agric. Food Chem.*, *39*: 1770 (1991).
159. J. O. Ragnarsson and T. P. Labuza, *Food Chem.*, *2*: 291 (1977).
160. J. B. Rossell, in *Rancidity in Foods* (J. C. Allen and R. J. Hamilton, Eds.), Elsevier Applied Science, London, 1989, p. 23.
161. E. N. Frankel, *Trends Food Sci. Technol.*, *4*: 220 (1993).
162. B. N. Stuckey, in *CRC Handbook of Food Additives*, Vol. 1 (T. E. Furia, Ed.), CRC Press, Boca Raton, FL, 1972, p. 185.
163. J. I. Gray and F. J. Monahan, *Trends Food Sci. Technol.*, *3*: 315 (1992).
164. AOCS, *American Oil Chemists' Society Official and Tentative Methods*, CD-8-53, AOCS, Chicago, IL.
165. H. Mitsuda, K. Yasumoto, and K. Iwami, *Eiyo to Shokuryo*, *19*: 210 (1966).
166. H. Kreis, *Chem. Ztg.*, *26*: 323 (1902).
167. IUPAC, *IUPAC Standard Methods for the Analysis of Oils, Fats, and Derivatives*, 7th rev. ed., Oxford, UK, 1987.
168. O. Cavazzana, *Rass. Chim.*, *32*: 135 (1980).
169. A. S. Henick, M. F. Benca, and J. H. Mitchell, *J. Amer. Oil Chem. Soc.*, *31*: 88 (1958).
170. G. L. Marco, *J. Amer. Oil Chem. Soc.*, *45*: 594 (1968).
171. B. Zhao, X. Li, R. G. He, S. J. Cheng, and X. Wenjuan, *Cell Biophys.*, *14*: 175 (1989).
172. K. Robards, A. F. Kerr, and E. Patsalides, *Analyst*, *113*: 213 (1988).
173. T. Miyazawa, K. Fujimoto, and T. Kaneda, in *Lipid Peroxidation in Biological Systems* (A. Sevanian, Ed.), AOCS Monograph, Americal Oil Chemists' Society, Champaign, IL, 1988, p. 1.
174. K. Warner and E. N. Frankel, *J. Amer. Oil Chem. Soc.*, *62*: 100 (1985).
175. S. L. Melton, *Food Technol.*, *3*(7): 105 (1983).
176. H. F. Liu, A. M. Booren, J. I. Gray, and R. L. Crackel, *J. Food Sci.*, *57*: 803 (1992).
177. A. D. Clarke, C. B. Ramsey, V. E. Hornsby, G. W. Davis, and R. D. Galyean, *Lebensm. Wissen. Technol.*, *23*: 267 (1990).

178. D. J. Buckley, J. I. Gray, A. Asghar, J. F. Price, R. L. Crackel, A. M. Booren, A. M. Pearson, and E. R. Miller, *J. Food Sci.*, *54*: 1193 (1989).

179. N. G. Marriott and P. P. Graham, *Proc. Eur. Meeting Meat Res. Workers*, No. 33, Vol. I, 4:2, 1987, p. 166.

180. D. G. Olson and R. E. Rust, *J. Food Sci.*, *38*: 251 (1973).

181. L. E. Jeremiah, *J. Food Prot. 51*: 105 (1988).

182. L. D. Titarenko, A. N. Svidovskii, and E. P. Tikhanova, *Tovarovedenie*, *24*: 10 (1992).

183. M. Oishi, K. Onishi, M. Nishijima, K. Nakagomi, and H. Nakazawa, *Jpn. J. Toxicol. Environ. Health*, *36*: 69 (1990).

184. M. Kuwahara, H. Uno, A. Fujiwara, T. Yoshikawa, and I. Uda, *Jpn. J. Food Sci. Technol.*, *18*: 64 (1971).

185. K. L. Empson, T. P. Labuza, and E. Graf, *J. Food Sci.*, *56*: 560 (1991).

186. A. M. Salih, J. F. Price, D. M. Smith, and L. E. Dawson, *J. Food Quality*, *12*: 71 (1989).

187. J. Pikul, A. Niewiarowicz, and J. Kijowski, *Fleischwirtschaft*, *63*: 960 (1983).

188. S. J. Salminen and A. L. Branen, *Kemia Kemi.*, *6*: 762 (1979).

189. M. A. Einerson and G. A. Reineccius, *J. Food Process. Preserv.*, *1*: 279 (1977).

190. A. Atkinson, R. P. Merwe, and L. G. Swart, *Agroanimalia*, *4*: 63 (1972).

191. L. E. Dawson, K. E. Stevenson, and E. Gertonson, *Poultry Sci.*, *54*: 1134 (1975).

192. U.S. Army Natick Research and Development Command, Massachusetts, Project Report No. DAAG17 73 C-0214, 1975.

193. A. J. St Angelo, M. Koohmaraie, K. L. Crippen, and J. Crouse, *J. Food Sci.*, *56*: 359 (1991).

194. V. Giaccone, T. Civera, R. M. Turi, and E. Parisi, *Fleischwirtschaft*, *71*: 1442 (1991).

195. S. Barbut and G. S. Mittal, *Meat Sci.*, *30*: 279 (1991).

196. A. Mikkelsen, G. Bertelsen, and L. H. Skibsted, *Z. Lebensm. Unters. Forsch.*, *192*: 309 (1991).

197. A. J. St Angelo, K. L. Crippen, H. P. Dupuy, and C. James, Jr., *J. Food Sci.*, *55*: 1501 (1990).

198. T. L. Wheeler, S. C. Seideman, G. W. Davis, and T. L. Rolan, *J. Food Sci.*, *55*: 1274 (1990).

199. G. C. Arganosa, *Diss. Abstr. Int.*, B: 49(5): 1454: Order No. DA 8811391, 1988, p. 266.

200. R. L. Crackel, J. I. Gray, A. M. Booren, A. M. Pearson, and D. J. Buckley, *J. Food Sci.*, *53*: 656 (1988).

201. T. F. Mann, *Diss. Abstr. Int.* B: 48(3): 609 610: Order No. DA 8712676, 1987, p. 116.

202. J. P. Roozen, *Food Chem.*, *24*: 167 (1987).

203. T. M. Dessouki, A. A. El Dashlouty, M. M. El Ebzary, and H. A. Heikal, *Agric. Res. Rev.*, *58*: 311 (1980).

204. M. F. Chastain, D. L. Huffman, W. H. Hsieh, and J. C. Cordray, *J. Food Sci.*, *47*: 1779 (1982).

205. S. J. Riet and M. M. van de Hard, *J. Amer. Dietetic Assoc.*, *75*: 556 (1979).

206. L. W. Haymon, E. Brotsky, W. E. Danner, C. W. Everson, and P. A. Hammes, *J. Food Sci.*, *41*: 417 (1976).

207. R. C. Benedict, E. D. Strange, and C. E. Swift, *J. Agric. Food Chem.*, *23*: 167 (1975).

208. S. O. Toemek, A. Saygin Guemueskesen, and M. Serdaroglu, *Chem. Mikrobiol. Technol. Lebens.*, *13*: 15 (1991).

209. M. Asahara, *Bull. Jpn. Soc. Sci. Fish.*, *53*(9): 1617 (1987).
210. E. H. Lee, S. Y. Cho, Y. J. Cha, J. K. Jeon, and S. K. Kim, *Bull. Korean Fish. Soc.*, *14*: 201 (1981).
211. J. J. Licciardello, E. M. Ravesi, and M. G. Allsup, *Marine Fish. Rev.*, *Natl. Oceanic Atmos. Admin.*, *44*: 15 (1982).
212. A. Biggar, N. A. M. Eskin, and M. Vaisey, *J. Fish. Res. Board Can.*, *32*: 227 (1975).
213. M. H. Chahine and R. F. MacNeill, *J. Amer. Oil Chem. Soc.*, *51*: 37 (1974).
214. R. N. Farragut, U.S. Dept. of Commerce, Natl. Oceanic and Atmospheric Administration Tech. Rep. NMFS SSRF 650 iv, 12, 1972.
215. I. I. Lapshin and I. I. Teplitsyna, *Rybn. Khoz.*, *45*: 64 (1969).
216. A. R. Patel, *Diss. Abstr. Int.*, B: 36(5): 2143 2144, Order No. 75 23773, 1975.
217. G. S. Sindhu, M. A. Brown, and A. R. Johnson, *J. Dairy Res.*, *42*: 185 (1975).
218. E. G. Hammond, *Amer. Dairy Rev.*, *32*: 40 (1970).
219. D. B. Min, D. B. Ticknor, S. H. Lee, and G. A. Reineccius, *J. Food Sci.*, *55*: 401 (1990).
220. G. Hall and H. Lingnert, *J. Food Quality*, *7*: 131 (1984).
221. I. A. Radaeva, L. S. Dmitrieva, and E. A. Bekhova, *XIX Int. Dairy Congr.*, *USSR, 1E*: 618 (1974).
222. J. Abbot, *J. Soc. Dairy Technol.*, *24*: 182 (1971).
223. W. Kuzio, *Confect. Prod.*, *40*: 387 (1974).
224. B. K. Wadhwa, S. Kaur, and M. K. Jain, *Indian J. Dairy Sci.*, *44*: 119 (1991).
225. C. N. Rao, B. V. R. Rao, T. J. Rao, and G. R. R. M. Rao, *Asian J. Dairy Res.*, *3*: 127 (1984).
226. P. N. Thakar, P. S. Prajapati, A. J. Pandya, K. G. Upadhya, and S. H. Vyas, *Gujarat Agric. Univ. Res. J.*, *9*: 40 (1984).
227. S. Mattsson and S. Johansson, 18th International Dairy Congress, Sydney, Australia, *1E*: 226 (1970).
228. Y. K. Jha, S. Singh, and S. Singh, *Indian J. Dairy Sci.*, *30*: 1 (1977).
229. C. Kanno, K. Yamaguchi, and T. Tsugo, *Agric. Biol. Chem.*, *34*: 1652 (1970).
230. C. B. Watkins, J. E. Harman, and G. Hopkirk, *N. Z. J. Exper. Agric.*, *16*: 55 (1988).
231. K. Kimura, H. Nishimura, I. Iwata, and J. Mizutani, *Agric. Biol. Chem.*, *47*: 1661 (1983).
232. H. J. Bielig and U. Feier, *Chem. Mikrobiol. Technol. Lebens.*, *5*: 93 (1976).
233. C. C. King, Jr., *Diss. Abstr. Int.*, B: 47: 1347: Order No. DA 8614972, p. 160, 1986.
234. C. G. Cavaletto and H. Y. Yamamoto, *J. Food Sci.*, *36*: 81 (1971).
235. A. N. Svidovskii, S. N. Ishmakova, and E. N. Budyak, *Tovarovedenie*, *24*: 54 (1991).
236. M. Aoyama, H. Kanematsu, I. Niiya, M. Tsukamoto, S. Tokairin, and T. Matsumoto, *Yukagaku*, *40*: 202 (1991).
237. P. S. Yan and P. J. White, *J. Agric. Food Chem.*, *38*: 1904 (1990).
238. B. Tokarska, Z. J. Hawrysh, and M. T. Clandinin, *Can. Inst. Food Sci. Technol. J.*, *19*: 130 (1986).
239. K. Warner, E. N. Frankel, J. M. Snyder, and W. L. Porter, *J. Food Sci.*, *51*: 703 (1986).
240. D. B. Min and J. Wen, *J. Food Sci.*, *48*: 1172 (1983).
241. T. L. Mounts, K. A. Warner, G. R. List, J. P. Fredrich, and S. Koritala, *J. Amer. Oil Chem. Soc.*, *55*: 345 (1978).
242. J. Slowikowska, A. Jakubowski, K. Pilat, and Z. Rudzka, *Zesz. Problem. Postep. Nauk Roln.* *136*: 235 (1973).

243. D. Ohyabu, A. Mukai, and S. Ohta, *J. Jpn. Oil Chem. Soc.*, *28*: 317 (1979).
244. P. Baardseth, *Food Add. Contam.*, *6*: 210 (1989).
245. H. C. Lee and B. P. Klein, *Food Chem.*, *32*: 151 (1989).
246. D. O'Beirne and A. Ballantyne, *Int. J. Food Sci. Technol.*, *22*: 515 (1987).
247. W. A. Gould and S. Yusof, Research Circular, Ohio Agric. Res. Develop. Center, Columbus, No. 271, 1982, p. 33.
248. Y. B. Park, H. K. Park, and D. H. Kim, *Korean J. Food Sci. Technol.*, *21*: 468 (1989).
249. S. R. Palamand and J. H. Grigsby, *Abstr. Papers Amer. Chem. Soc.*, *168*: AGFD 42 (1974).
250. G. Baetsle, *Fermentation*, *70*: 23 (1974).
251. G. Baetsle, *Fermentation*, *70*: 83 (1974).
252. W. Postel, *Brauwissenschaft*, *25*: 196 (1972).
253. H. Stone and J. L. Sidel, *Sensory Evaluation Practices*, Academic Press, New York, 1985.
254. M. O'Mahony, *Sensory Analysis of Foods*, Marcel Dekker, New York, 1986.
255. C. Gay and R. Mead, *J. Sensory Stud.*, *7*: 205 (1992).
256. S. Stevens and E. H. Galanter, *J. Exp. Psychol.*, *54*: 377 (1957).
257. D. G. Land and R. Shepherd, in *Sensory Analysis of Foods* (J. R. Piggot, Ed.), Elsevier Applied Sciences, London, 1984, p. 155.
258. H. Stone, J. Sidel, S. Oliver, A. Woolsey, and R. C. Singleton, *Food Technol.*, *28*: 24 (1974).
259. H. Stone, J. L. Sidel, and J. Bloomquist, *Cereal Foods World*, *25*: 642 (1980).
260. S. S. Stevens, in *Handbook of Perception*, Vol. II, *Psychophysical Judgement and Measurement* (E. C. Carterette and M. P. Friedman, Eds.), Academic Press, New York, 1974, p. 361.
261. H. R. Moskowitz, *J. Food Quality*, *3*: 195 (1977).
262. W. W. Cooley and P. R. Lohnes, *Multivariate Data Analysis*, John Wiley and Sons, New York, 1971.
263. G. Hoffman, *Chem. Ind.*, *1970*: 729 (1970).
264. J. I. Gray, *J. Amer. Oil Chem. Soc.*, *55*: 539 (1978).
265. D. B. Josephson, R. C. Lindsay, and G. Olafsdottir, in *Seafood Quality Determination* (D. E. Kramer and J. Liston, Eds.), Elsevier, Amsterdam, 1987, p. 27.
266. K. Farstrup and J. A. Nissen, in *Proceedings of the VI International Flavor Conference, Greece* (G. Charalambous, Ed.), Elsevier, Amsterdam, 1989, p. 897.
267. *AOCS Recommended Practice Cg -2-83*, Flavor Panel Evaluation of Vegetable Oils, AOCS, Champaign, IL.
268. S. Narasimhan, K. G. Raghuveer, C. Armugham, K. K. Bhat, and D. P. Sen, *J. Food Sci. Technol.*, *23*: 273 (1986).
269. S. Narasimhan, G. K. Sarvamangala, A. K. Vasantkumar, N. Chand, and D. Rajalakshmi, *J. Amer Oil Chem. Soc.*, *71*:1267 (1994).
270. A. I. Khuri and J. A. Cornell, *Response Surfaces: Design and Analysis*, Marcel Dekker, New York, 1987.
271. A. Nelder and R. Mead, *Computer J.*, *7*: 308 (1965).
272. P. A. Olsson, *J. Qual. Technol.*, *6*: 53 (1974).
273. J. C. Allen, in *Recent Advances in Chemistry and Technology of Fats and Oils* (R. J. Hamilton and A. Bhati, Eds.), Elsevier, New York, 1987, p. 31.

274. H. A. Moser, H. J. Dutton, C. D. Evans, and J. C. Crown, *Food Technol.*, *4*: 105 (1950).
275. H. W. Jackson, *J. Amer. Oil Chem. Soc.*, *58*: 227 (1981).
276. C. D. Evans, *J. Amer. Oil Chem. Soc.*, *32*: 596 (1955).
277. S. S. Arya, S. Ramanujam, and P. K. Vijayaraaghavan, *J. Amer. Oil Chem. Soc.*, *46*: 28 (1969).
278. N. A. M. Eskin and C. Frenkel, *J. Amer. Oil Chem. Soc.*, *53*: 746 (1976).
279. K. Warner, C. D. Evans, G. R. List, H. P. Dupuy, J. I. Wadsworth, and Goheen, *J. Amer. Oil Chem. Soc.*, *55*: 252 (1978).
280. K. Warner, E. N. Frankel, J. M. Snyder, and W. L. Porter, *J. Food Sci.*, *51*: 703 (1986).
281. P. K. Jervi, G. D. Lee, D. R. Krickson, and E. A. Butkus, *J. Amer. Oil Chem. Soc.*, *48*: 121 (1971).
282. H. M. Blumenthal, J. R. Tront, and S. S. Chang, *J. Amer. Oil Chem. Soc.*, *53*: 496 (1976).
283. W. H. Morrison III, B. G. Lyon, and J. A. Robertson, *J. Amer. Oil Chem. Soc.*, *58*: 23 (1981).
284. J. A. Fioriti, M. J. Kanuk, and R. J. Sims, *J. Amer. Oil Chem. Soc.*, *54*: 219 (1977).
285. M. L. Maera, in *Fats and Oils: Chemistry and Technology* (R. J. Hamilton and A. Batti, Eds.), Elsevier Applied Science, London, 1980, p. 193.
286. W. W. Nawar, *ACS Symp. Ser. 409*: 94 (1989).
287. E. N. Frankel, *Prog. Lipid Res. 22*: 1 (1982).
288. E. N. Frankel, in *Flavor Chemistry of Fats and Oils* (D. B. Min and T. H. Sponser, Eds.), Amer. Oil Chem. Soc., Champaign, IL, 1985, p. 1.
289. H. W. S. Chan, *Autoxidation of Unsaturated Lipids*, Academic Press, London, 1987.
290. R. E. Baldwin, M. R. Cloninger, and R. C. Lindsay, *J. Food Sci.*, *38*: 528 (1973).
291. Kalyanasundaram, *Asian Food J.*, *4*: 3 (1989).
292. CFTRI Annual Progress Report, Dept. of Sensory Analysis and Statistical Services, Central Food Technological Research Institute, Mysore, India, 1988/89, p. 169.
293. CFTRI Annual Progress Report, Dept. of Sensory Analysis and Statistical Services, Central Food Technological Research Institute, Mysore, India, 1989/90, p. 186.
294. CFTRI Annual Progress Report, Dept. of Sensory Analysis and Statistical Services, Central Food Technological Research Institute, Mysore, India, 1993/94, p. 14.

4

Technological Aspects of Food Antioxidants

D. L. Madhavi
University of Illinois, Urbana, Illinois

R. S. Singhal and P. R. Kulkarni
University of Bombay, Bombay, Maharashtra, India

4.1 INTRODUCTION

Application of antioxidants offers an effective, economical means of delaying the onset of oxidative degradation in food products. Consequently, the food products have longer shelf life and remain acceptable longer without loss of nutritional value and sensory characteristics. Thus antioxidants have become indispensable entities in food manufacturing. Oxidation in food products can also be delayed by other means such as vacuum packaging, packaging under inert gas, low-temperature storage, and the use of packaging materials to exclude air and light. However, many of these methods may not be economical, and many have limited applications because of the nature of the food product.

Antioxidants can be effectively employed in a wide range of food products as they are relatively easy to incorporate and have a broad range of activities that can withstand the various food processing operations. One of the major advantages of the use of antioxidants is that the food manufacturer will be in a position to select the best possible antioxidant from a wide array of compounds or design new combinations and concentrations to suit the needs of specific food products. The use of antioxidants also expands the choice of ingredients for the food technologist and aids in the development of novel food products.

The proper and effective use of antioxidants is largely dependent on an understanding of their chemistry, mode of action, and functions in model and food

159

systems. Many of these compounds have other important functions or are employed in a variety of food applications apart from their role as antioxidants. For example, the major phenolic antioxidants BHA, BHT, and TBHQ have antimicrobial properties. Similarly, oxygen scavengers, ascorbic acid, sulfites, and phosphates have antimicrobial properties. Phosphates are employed in food processing for a variety of functions such as pH stabilization, emulsification, and acidification. The chelators citric acid and tartaric acid are used mainly as food acidulants. Lecithin is a major emulsifying agent in food processing.

This chapter gives an overview of the methods of preparation, properties, and food applications of some of the commonly used primary antioxidants, oxygen scavengers, and chelators. In addition, some of the synthetic and natural compounds that have a potential for large-scale application are described. However, this chapter emphasizes only the antioxidant properties of the various compounds. At present both synthetic and natural antioxidants are being employed in a variety of food products such as fats, oils, dairy products, cereals, fried foods, baked foods, essential oils, meat products, confectioneries, fruits, and vegetables. An attempt has been made to discuss the food applications under specific food categories wherever possible.

4.2   TECHNOLOGICAL ASPECTS OF PRIMARY ANTIOXIDANTS

Primary antioxidants include polyhydroxy phenolic and hindered phenolic compounds, which function mainly as electron or hydrogen donors. Other functional groups containing nitrogen or sulfur are also effective. They function by converting the free radicals to more stable products, thus interrupting the free-radical chain reaction. The hindered phenols have electron-donating substituents such as *tertiary*-butyl groups in the ortho or para position to the –OH group, which makes them more stable to high temperatures encountered in food processing. The hindered phenolic compounds generally have carry-through properties and remain effective in fried or baked products prepared with fats and oils in which they are present.

4.2.1   Phenols

Gallates

The Gallates used as food antioxidants include propyl, octyl, and dodecyl esters of gallic acid (3,4,5-trihydroxybenzoic acid) (Fig. 4.1). Propyl gallate is one of the most widely used food antioxidants and is a component of many commercial antioxidant mixtures. It is less soluble in fats and oils than BHA and BHT and has significant solubility in water. The dodecyl and octyl esters are more soluble than propyl gallate in fats and oils. In general, the gallates do not have significant

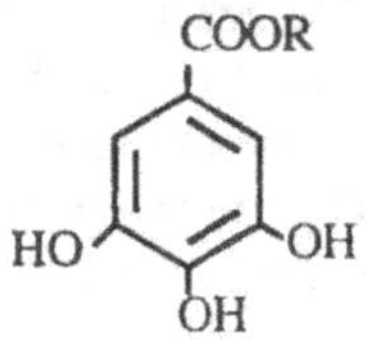

R : C_3H_7 Propyl gallate
    C_8H_{17} Octyl gallate
    $C_{12}H_{25}$ Dodecyl gallate

Fig. 4.1 Gallates.

carry-through properties. Of the three gallates, propyl gallate is the most sensitive to heat and undergoes degradation at frying temperatures. The temperature stability improves with increases in molecular weight; octyl and dodecyl gallates are more stable to heat and possess better carry-through properties. Propyl gallate also forms purple or violet complexes with iron ions, resulting in discoloration of the food product. Hence, propyl gallate is always used in combination with a metal chelator such as citric acid. This discoloration is not a problem in anhydrous fats and oils or when octyl or dodecyl gallates are used (1). The physical properties of the gallates are presented in Table 4.1.

Preparation. Propyl gallate is prepared by the esterification of gallic acid with excess isopropyl alcohol in the presence of anhydrous hydrogen chloride as a catalyst. A suspension of 25 g of gallic acid in 105 mL of anhydrous isopropyl

Table 4.1 Selected Physical Properties of the Gallates

Physical properties	Propyl gallate	Octyl gallate	Dodecyl gallate
Molecular weight	212	282.2	338.2
Physical appearance	White crystalline powder	Crystalline powder	Crystalline powder
Boiling point	Decomposes above 148°C		
Melting point (°C)	146–148	93.7–94.9	96.3–96.8
Solubility (25°C)			
Water	<1%	Insoluble	Insoluble
Vegetable oils	1–2%	Soluble	Soluble
Ethyl alcohol	>60%		
Propylene glycol	55%		
Glyceryl monooleate	5%		

Source: Refs. 5 and 35.

alcohol is cooled and saturated with dry hydrogen chloride. The reaction mixture is refluxed after standing overnight, and the solvent is evaporated after 24 h. The residue is suspended in water and extracted with ether. The extract is treated with barium carbonate, clarified, and evaporated to dryness. The crude product is extracted with chloroform, and crystalline propyl gallate is obtained on cooling the extract (3).

Preparation of the octyl and dodecyl esters requires refluxing in relatively high boiling, polar, inert solvents such a o-dichlorobenzene, anisole, phenetole, or nitrobenzene. Highest yields are obtained by using solvent mixtures in which the nitrobenzene is present in a 1:1 molar ratio with respect to gallic acid and one of the other solvents makes up the bulk of the solvent mixture. A typical preparation involves refluxing 0.3 mol of anhydrous gallic acid and 0.6 mol of n-octyl alcohol or n-dodecyl alcohol in 5.8 mol of anisole and 0.3 mol of nitrobenzene in the presence of 2.5 g of naphthalene- β-sulfonic acid as a catalyst for 24 h. The solvent mixture and a considerable amount of unreacted alcohol are removed by steam distillation. The product is dissolved in benzene and washed with water to remove excess gallic acid and the catalyst, and the crude ester is precipitated by the addition of petroleum ether. The pure esters are obtained by recrystallization from a benzene–petroleum ether mixture (4).

Food Applications. The various food uses of the gallates have been described by Maddox (1), and in the WHO Food Additive Series (5). Table 4.2 presents the levels of gallates commonly used in food products. The gallates are extremely effective antioxidants, especially for the stabilization of anhydrous fats and oils.

Table 4.2 Levels of Gallates Used in Practice in Various Food Products[a]

Product	Level (%)
Animal fats	0.001–0.01, all gallates
Vegetable oils	0.001–0.02, all gallates
Whole milk powder	0.005–0.01, mainly DG
Margarine	0.001–0.01, all gallates
Bakery products	0.01–0.04, mainly DG[b]
Cereals	0.003, all gallates
Chewing gum base	0.1, mainly PG
Candy	0.01, all gallates[b]

DG = dodecyl gallate; PG = propyl gallate.
[a]Combinations with BHA, BHT, and citric acid are frequently used.
[b]Calculated on the fat.
Source: Ref. 5.

The antioxidant activity of the three esters is approximately the same on an equimolar basis. The gallates show optimum concentrations for antioxidant activity and may act as prooxidants when used at higher levels (6). They function synergistically with primary antioxidants such as BHA, tocopherols, and TBHQ, and secondary antioxidants such as ascorbyl palmitate, ascorbic acid, and citric acid.

Fats and oils. The gallates are very effective in animal fats at the 0.001–0.01% level. Propyl gallate (PG) provides better oxidative stability to lard and rendered poultry fat than BHA and BHT. A combination of PG and BHA gives the best all-around antioxidant effect in lard and carry-through protection in foods prepared with lard such as pastry, potato chips, and crackers (7,8). Fish oils are difficult to stabilize because they are highly unsaturated. However, a combination of dodecyl gallate (DG) and BHT substantially increases the storage life of herring oil stored at room temperature and at 40°C (1). Propyl gallate has been shown to retard the oxidation of fats in frozen Atlantic fish (9). Table 4.3 presents the antioxidant properties of PG in various fats. Propyl gallate is effective in stabilizing soybean oil, cottonseed oil, palm oil, and hydrogenated vegetable oils. With refined oils, the addition of gallates or gallate-containing mixtures is beneficial in vegetable oils that are less highly unsaturated and have a fairly low natural tocopherol content such as peanut oil, coconut oil, palm oil, and olive oil (1). Addition of propyl gallate to crude oils results in higher oxidative stability of the oil after refining, bleaching, and deodorization, with better response to further antioxidant treatment (13). Butter

Table 4.3 Antioxidant Activity of Propyl Gallate in Fats and Oils

Substrate	Propyl gallate conc. (%)	Stability[a] (h)	Ref.
Butterfat	0.005	22 (9)	10
	0.02	78 (11)	10
Lard	0.05	135 (6)	11
	0.1	145 (6)	11
Cottonseed oil	0.01	10 (8)	11
	0.05	45 (12)	12
Soybean oil	0.05	54 (8)	11
Hydrogenated vegetable oil	0.01	138 (78)	12
	0.05	294 (78)	12

[a]The stabilities listed in this column were obtained in various ways. Comparison between references should not be made without consulting the original publications. Stabilities of control fats are given in parentheses.
Source: Ref. 6. With permission from John Wiley & Sons, New York.

prepared from acid cream is protected from peroxide formation by the use of octyl and dodecyl gallates at 0.01–0.03%, but the flavor stability is not improved. Dodecyl gallate at 0.01% and a mixture of 0.005% DG with 0.01% BHA were effective in suppressing the formation of peroxides in margarine. The mixture is effective at influencing the organoleptic quality of margarine. BHA retards the development of a "cardboard" taste, and DG retards development of an "old" taste (5). The gallates are effective in essential oils. A combination of 0.05% PG + 0.05% BHA + 0.05% BHT has been shown to be effective in the stabilization of orange oils, and 0.1% DG is a suitable antioxidant for lemon oil (1).

Meat products. Propyl gallate and BHA offer substantial protection to fresh beef pigments and lipids, and the antioxidant effect is carried through into the cooked product. A combination of PG with BHA and ascorbyl palmitate or ascorbic acid prolongs the shelf life of ground beef (14,15). PG also prolongs the keeping quality of chicken and turkey meat (16).

Baked and fried products. Propyl gallate provides maximum oxidative stability to frying fats and oils (17). Addition of PG and citric acid enhances the carry-through effect of BHA in fried foods (18). Similarly, addition of PG and citric acid to the fried product increases the stability of the product, which compensates for the loss of phenolic antioxidants during frying (19). In baked products, a combination of BHA and DG is effective. Ottaway and Coppock (20) demonstrated the synergism between these two antioxidants in biscuits prepared from a range of animal fats, vegetable oils, and margarine. In a Dutch shortbread made with butter, a combination of 0.02% of DG and 0.01% BHT was very effective, especially in terms of flavor stability (21).

Confectioneries and nuts. Propyl gallate either alone or in combination with BHA and citric acid is effective in retarding oxidation in butter creams used for the preparation of candies (22). A combination of PG, BHA, and BHT is effective in increasing the oxidative stability of shelled walnuts (23).

Milk products. In spray-dried and foam-dried whole milk powder, DG at the 0.02% level showed very effective results, improving shelf life by up to threefold. Propyl gallate at 0.01% also stabilizes spray-dried whole milk satisfactorily. Gallates do not significantly improve the shelf life of gas-packed whole milk powders (24–27).

Trihydroxybutyrophenone

Trihydroxybutyrophenone (2,4,5,-trihydroxybutyrophenone) (THBP) (Fig. 4.2) is a substituted phenol used mainly in food packaging materials. THBP is a tan-colored powder slightly soluble in water and soluble in fats and oils. It produces a tan to brown color in the presence of organic metal ions that can be inhibited by the addition of citric acid or other chelating agents (28). THBP is used at 0.02%

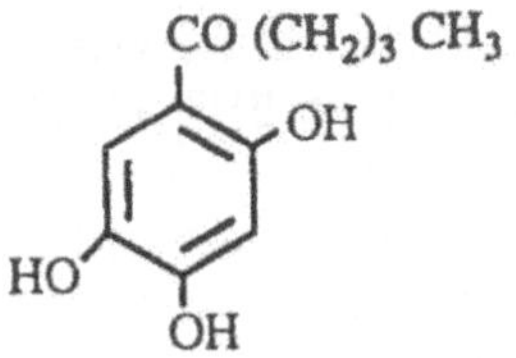

Fig. 4.2 Trihydroxybutyrophenone.

based on the fat content, and the migration of up to 0.005% from food packaging materials is permitted (29). Its physical properties are presented in Table 4.4.

Preparation. Trihydroxybutyrophenone is prepared by nuclear acylations of 1,2,4-trihydroxybenzene by Friedel–Crafts-type catalysts or Fries rearrangements of phenolic esters (30,31). In this process, 33 g of AlCl₃ is added to a solution of 12.6 g of 1,2,4-trihydroxybenzene in 200 mL of nitrobenzene. The mixture is cooled to 25°C, 16 g of butyric anhydride is added, and the mixture is heated to 60°C for 45 min. After cooling, 150 mL of ice cold 10% HCl is added. The mixture is steam-distilled to remove nitrobenzene. The residue yields 12 g (61%) of pure THBP.

Food Applications. The food applications of THBP have been described by Stuckey and Gearhart (28). THBP is about equal to or slightly better than PG in producing initial stabilities in lard and equal to PG in cottonseed oil. It is better than BHT and PG under deep fat frying conditions but about equal to BHA as tested with potato chips fried in cottonseed oil. It is not as good as BHA and BHT in carry-through properties in baked products such as crackers. Studies by Augustin and Berry (32) showed that THBP is as effective as PG in stabilizing refined,

Table 4.4 Selected Physical Properties of Trihydroxybutyrophenone

Molecular weight	196
Physical appearance	Tan powder
Melting point (°C)	149–153
Solubility	
Water	0.5% (50°C)
Vegetable oils	1.5% (20°C)
Lard	2.5% (50°C)
Ethyl alcohol	30% (20°C)
Propylene glycol	30% (20°C)

Source: Ref. 28.

bleached, and deodorized palm olein. The order of effectiveness of the various antioxidants tested in this study is TBHQ > PG ~ THBP > DLTDP > BHT > BHA. At the 0.3–0.5% level, THBP is highly effective in stabilizing vitamin A, which in general reacts poorly to antioxidant treatment. It is highly effective in stabilizing yellow grease at the 0.01–0.02% level in combination with citric acid. THBP is also an excellent antioxidant for paraffin wax, being approximately three times as active as BHT, which is used almost exclusively for preventing the breakdown in paraffin waxes. At 0.005%, THBP had a lower activity than a combination of BHA and BHT at similar levels in breakfast cereals.

Nordihydroguaiaretic Acid

Nordihydroguaiaretic acid [NDGA; 2,3-dimethyl-1,4-bis(3,4-dihydroxyphenyl)-butane] (Fig. 4.3), a naturally occurring polyhydroxyphenolic antioxidant, was widely used in food products in the 1950s and 1960s. NDGA is a major constituent of the resinous exudate from the creosote bush (*Larrea divaricata*), which grows in the southwestern United States and in Mexico. It is a grayish-white crystalline solid soluble in dilute alkaline solutions, fats, and oils. NDGA is no longer used in food products because of its toxicological properties. Its physical properties are presented in Table 4.5.

Preparation. The preparation, applications, and toxicological properties of NDGA have been reviewed by Oliveto (33).

Extraction from natural sources. The plant material, either dry or slightly dried including stems, leaves, and buds, was extracted with 5% sodium hydroxide containing 1.5% sodium hydrosulfite, a reducing agent. The reducing agent was necessary to protect the orthohydroquinone nuclei of the NDGA from oxidation during the alkali extraction. The aqueous solution was neutralized with 50% hydrochloric acid to form a flocculant precipitate. The precipitate was dissolved in methyl alcohol. The solution was acidulated with glacial acetic acid, and NDGA

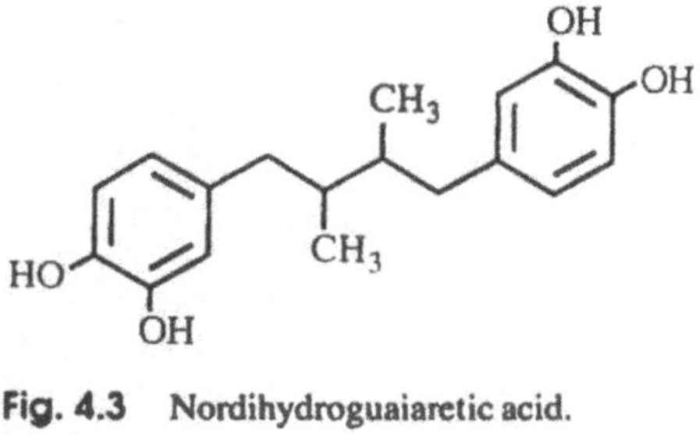

Fig. 4.3   Nordihydroguaiaretic acid.

Table 4.5 Selected Physical Properties of
Nordihydroguaiaretic Acid

Molecular weight	302.36
Physical appearance	Tan powder
Melting point (°C)	184–185
Solubility	Slightly soluble in hot water
	Soluble in dilute alkali
	Fats and oils 0.5–0.7% (20°C)
	Soluble in ethanol, propylene glycol

Source: Ref. 6.

was extracted by adding ethyl ether to the solution. The ether layer was washed with water to separate out the remainder of the alcohol. The ether layer was then extracted with 22 portions of an aqueous 5% potassium hydroxide containing 2.5% sodium hydrosulfite. Each portion was acidulated with HCl to precipitate the NDGA, and the ether was removed using a steam bath. Portions 7–18 yielded crystalline NDGA with 95% purity. When dry and powdered, the NDGA had a faint yellow color. To remove all traces of color, the crystalline product was suspended in glacial acetic acid. The acid was filtered off, and the decolorized pure NDGA was washed with water and dried (34).

Chemical synthesis. In addition to isolation from natural sources, NDGA can be obtained by chemical synthesis. The chemical synthesis involves the alkylation of 3,4-dimethoxypropiophenone with its α-bromo derivative to yield the racemic 2,3-diveratroylbutane in liquid ammonia. Acid-catalyzed cyclodehydration followed by hydrogenation produces NDGA tetramethyl ether, which can be readily converted to NDGA by treatment with HBr. NDGA can also be prepared by the hydrogenation and subsequent demethylation of the guaiaretic acid dimethyl ether, a constituent of gum guaiac (33).

Food Applications. Like other phenolic antioxidants, NDGA is more effective in animal fats than in vegetable oils. NDGA resembles PG in forming a colored complex with iron salts and does not carry through into baked or fried food products (6,35). Alone or in combination, NDGA was effective in a variety of food products in addition to fats and oils such as dairy products, margarine, meats, baked goods, and confections. It was also used in pharmaceutical preparations, particularly those containing vitamin A (33).

Hydroquinone

Hydroquinone (HQ; 1,4-dihydroxybenzene) (Fig. 4.4) was the first antioxidant to be used for the stabilization of vitamin A in fish oils in the 1930s (36). However,

Fig. 4.4 Hydroquinone.

it was not approved as a food grade antioxidant because of the results of toxicological studies. Its physical properties are presented in Table 4.6.

Preparation. The preparation, properties, uses, and environmental and health risks of HQ have been reviewed by Devillers et al. (37). Three major processes are used commercially to prepare this compound. The first method is based on the oxidation of aniline with manganese dioxide and sulfuric acid, followed by reduction with iron dust and water. The second method involves alkylation of benzene with propylene to produce a mixture of diisopropylbenzene isomers, from which, in a first step, a para isomer is isolated. This is oxidized with oxygen to produce the corresponding dihydroperoxide, which is treated with an acid to produce hydroquinone and acetone. Finally, the oxidation of phenol with hydrogen peroxide can be used to produce a mixture of products from which both hydroquinone and catechol can be isolated.

Hydroquinone is used as a stabilizer for polymers and motor fuels and oils and as a source for the production of its mono- and dialkyl ethers. It is also used in very small quantities in dermatological preparations as a depigmenting agent.

4.2.2 "Hindered" Phenols

Butylated Hydroxyanisole

Butylated hydroxyanisole (BHA; *tert*-butyl-4-hydroxyanisole) is one of the extensively used food antioxidants. Commercial BHA is a mixture of two isomers,

Table 4.6 Selected Physical Properties of Hydroquinone

Molecular weight	110.11
Physical appearance	Light tan to light gray crystals
Melting point (°C)	173–174
Solubility	Water 7.0% (25°C)
	Highly soluble in alcohol, ether

Source: Ref. 2.

2-*tert*-butyl-4-hydroxyanisole (2-BHA) and 3-*tert*-butyl-4-hydroxyanisole (3-BHA), and contains 90% of the 3-isomer (Fig. 4.5). Both isomers have a phenolic odor. BHA is highly soluble in fats and oils and insoluble in water. It has a low melting point and is steam volatile at frying temperatures. However, the residual BHA exhibits considerable carry-through properties in baked and fried foods. BHA may also develop a pink color in the presence of alkaline metals in the food product. The physical properties of BHA are listed in Table 4.7.

Preparation. Butylated hydroxyanisole is prepared by alkylating *p*-methoxyphenol with isobutylene or *tert*-butyl alcohol in the presence of phosphoric acid as a catalyst. In this process a mixture of *p*-methoxyphenol (124 g), 85% phosphoric acid (300 mL), and hexane (1 liter) is heated to 50°C with stirring, and isobutylene (56 g) or *tert*-butyl alcohol (73 g) is added over a period of 1.5 h, after which the alkylated products are recovered by neutralization and vacuum distillation. The presence of a solvent such as hexane or heptane increases the yield of the monoalkylated phenol and reduces the formation of the dialkylated derivative (38). BHA obtained by this method is a mixture of 3-BHA (43.7%) and 2-BHA (56.3%). 3- BHA is further concentrated and separated from the 2-isomer by the method of Rosenwald (39). This process involves dissolving the mixture in a water-immiscible solvent such as pentane and extracting the solution with an aqueous solution of sodium hydroxide having about 0.1–0.5 mole equivalents of the phenols. On evaporation, the water-immiscible solvent phase yields mainly the 3-isomer.

Young and Rodgers (40) developed a method wherein the product contains a major proportion of the 3-isomer, which involves reacting TBHQ with dimethyl sulfate in the presence of comminuted zinc dust in an inert atmosphere. In this process, 332 g of TBHQ and 1 g of zinc dust are slurried with water in an inert nitrogen atmosphere, the temperature is increased to reflux, and 85 g of sodium hydroxide is added. Dimethyl sulfate (140 g) is added over a 45-min period, and

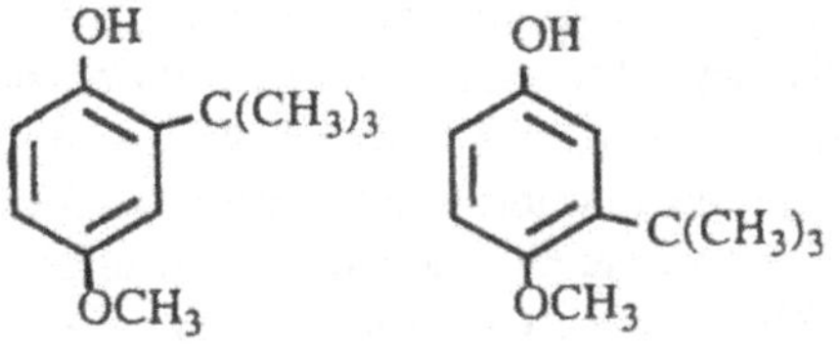

Fig. 4.5   Butylated hydroxyanisole. (a) 3-BHA; (b) 2-BHA.

Table 4.7 Selected Physical Properties of
Butylated Hydroxyanisole

Molecular weight	180
Physical appearance	White, waxy tablets
Boiling point (°C)	264–270
Melting point (°C)	48–65
Solubility (25°C)	
Water	Insoluble
Fats and oils	30–50%
Ethyl alcohol	>50%
Propylene glycol	70%
Glyceryl monooleate	50%

Source: Ref. 35.

the reactants are maintained under reflux conditions for 18 h. The reaction mixture is acidified after cooling and extracted with benzene. After the benzene extract is washed with warm water and evaporated, crude BHA is obtained as a viscous liquid or a low-melting solid. Fractional distillation results in 271 g of the product containing 79.4% 3-BHA and 17.6% 2-BHA.

Food Applications. Butylated hydroxyanisole is used in a wide variety of products such as fats, oils, fat-containing foods, and food packaging materials. Table 4.8 lists the levels of BHA commonly used in food products. The antioxidant activity of the 3-isomer is greater than that of the 2-isomer. Commercial BHA containing a high proportion of the 3-isomer is nearly as effective as pure 3-BHA mainly because of a slight synergism between the two isomers (41,42). The antioxidant activity increases with increases in concentration up to 0.02% and remains approximately constant at higher levels (42). BHA also functions synergistically with other primary antioxidants and synergists like the gallates, tocopherols, BHT, TBHQ, thiodipropionic acid, citric acid, and phosphoric acid.

Fats and oils. In animal fats and shortenings, if used alone BHA is less effective than BHT or the gallates, but its effectiveness increases with the addition of synergists. BHA alone improved the oxidative stability of lard from 4 h to 16 h, and the stability was increased to 36 h with the addition of citric acid. Various synergists such as phosphoric acid, lecithin, and methionine had similar effects (42). Mixtures of BHA with gallates or BHT and a chelator such as citric acid are more effective and advantageous if the fats are to be used for baking or frying. In general, BHA or BHT provide only a small degree of protection in frying fats and oils but have a carry-through effect in the fried product (17). BHA is less effective in vegetable oils than in animal fats and is slightly more effective in

Table 4.8 Levels of BHA Used in Practice
in Food Products[a]

Product	Level (%)
Animal fats	0.001 – 0.01
Vegetable oils	0.002 – 0.02
Bakery products	0.01 – 0.04[b]
Cereals	0.005 – 0.02
Dehydrated mashed potatoes	0.001
Essential oils	0.01 – 0.1
Chewing gum base	Up to 0.1
Candy	Up to 0.1[b]
Food packaging materials	0.02 – 0.1

[a]Combinations with BHA, gallates, and citric acid are
frequently used.
[b]Calculated on the fat.
Source: Ref. 5.

hydrogenated vegetable oils (Table 4.9). Butylated hydroxyanisole has been suggested as an antioxidant for crude vegetable oils at 0.02%. Such treatments would result in better keeping quality after refining (13). In margarine, a mixture of 0.01% BHA and 0.005% dodecyl gallate is more effective than BHA alone (21,43). BHA is widely used in stabilizing essential oils such as d-limonene, orange oil, lemon oil, lime oil, and myrcene, an aliphatic unsaturated terpene found in many essential oils (44–46).

   Baked and fried foods and cereal products. The most important property of BHA is its ability to remain active in baked or fried foods. Table 4.10 presents the comparative carry-through properties of BHA and BHT in various products prepared with lard. BHA is stable at pH values above 7.0, which contributes to its stability in baked foods (6). In low-fat foods such as cereal products, especially breakfast cereals, mashed potato, and cake mixes, BHA finds widespread use. BHA and combinations of BHA, BHT, and propyl gallate increase the stabilities of wheat germ meal, brown rice, rice bran, and dry breakfast cereals (47). The volatility of BHA and BHT is an advantageous property in low-fat foods. Small quantities of BHA or BHT added to the potato or cereal slurry before cooking or drying results in dispersion by volatilization or steam distillation, resulting in the protection of the product during processing and subsequent storage. Such products are also stabilized by the addition of relatively high concentrations of the antioxidant to the inner waxed liner of the packaging material. The antioxidant comes into contact with the fat portion of the product more efficiently on vaporization. Another approach has been to spray an emulsion of the antioxidant onto the finished product

Table 4.9　Antioxidant Activity of BHA in Fats and Oils

Substrate	BHA conc. (%)	Stability (h)
Lard	0	4
	0.01	19
	0.05	20
	0.1	21
Cottonseed oil	0	7
	0.01	9
	0.05	9
	0.1	9
Hydrogenated cottonseed oil	0	121
	0.01	108
	0.05	158
	0.10	172

Source: Compiled from Chipault (6). With permission from John Wiley & Sons, New York.

immediately prior to packaging (5). In dehydrated potato products, BHA and BHT are effective in retarding oxidative rancidity (48,49).

Meat products. Butylated hydroxyanisole is effective at stabilizing the fresh pigment of raw beef and inhibits lipid oxidation at 0.01% (50). Combinations of BHA and BHT are effective in mechanically deboned carp, ground turkey meat, Chinese fried pork fiber, and frozen bacon slices (51–54). In combination with sodium tripolyphosphate and ascorbic acid, BHA retards the development of rancidity in frozen restructured pork chops (55). It is effective in freeze-dried beef, fowl, pork, and fish (56). A combination of BHA and ascorbic acid has been reported to be effective in retarding lipid and pigment oxidation in raw ground beef

Table 4.10　Carry-Through Properties of BHA and BHT at 0.01% in Baked and Fried Foods—Oven Stability at 63°C in Days

Lard sample	Pastry			Crackers			Potato chips		
	Control	BHA	BHT	Control	BHA	BHT	Control	BHA	BHT
1	3.5	50	17	5.5	39	38	3	36	29
2	5	50	30	5	42	37	2	24	11.5
3	10	55	40	9	37	25	3	34	11
4	7	44	34	5	35	34	1.5	25	15

Source: Ref. 69.

for up to 8 days of refrigerator storage in oxygen-permeable film (15). With or without a synergist like citric acid, BHA retards discoloration and rancidity in various dry sausages (57).

Milk products. Butylated hydroxyanisole is slightly effective in increasing the shelf life of foam-dried whole milk powder at 0.01% and has no effect in gas-packed whole milk powder (27,58). BHA markedly improves the keeping quality of processed cheese stored under varying conditions at 5 or 20°C (5).

Spices, nuts, and confectioneries. Butylated hydroxyanisole is effective in stabilizing the color or paprika and cayenne peppers (59). Shelled and ground nuts are particularly vulnerable to oxidation. A combination of BHA and BHT is highly effective in stabilizing shelled walnuts, ground pecans, and peanuts (60). BHA is effective in stabilizing the fresh aroma and flavor of roasted and salted peanuts when added to the cooking oil and the salt (61). BHA is effective in stabilizing a variety of shelled nuts when incorporated in an edible protective coating. BHA extends the shelf life of roasted macadamia nuts and almonds (62–64). Addition of BHA or a combination of BHA, PG, and citric acid inhibits oxidation in candies when added to the butter used for preparing the candies (22).

Butylated Hydroxytoluene

Butylated hydroxytoluene (BHT; 2,6-di-*tert*-butyl-*p*-cresol or 2,6,di-*tert*-butyl-4-methylphenol) (Fig. 4.6) is also extensively used in food products and is similar to BHA in many of its properties. However, BHT is not as effective as BHA mainly because of the presence of two *tert*-butyl groups, which offer greater steric hindrance than BHA (65). BHT is also highly soluble in fats and oils and insoluble in water. It is insoluble in propylene glycol and moderately soluble in glyceryl monooleate, solvents used for commercial antioxidant formulations. BHT is a white crystalline solid with a faint phenolic odor. It is more steam-volatile than BHA and has less carry-through properties than BHA. BHT also forms a yellow color in the presence of iron ions in food products or packaging materials. Its physical properties are presented in Table 4.11.

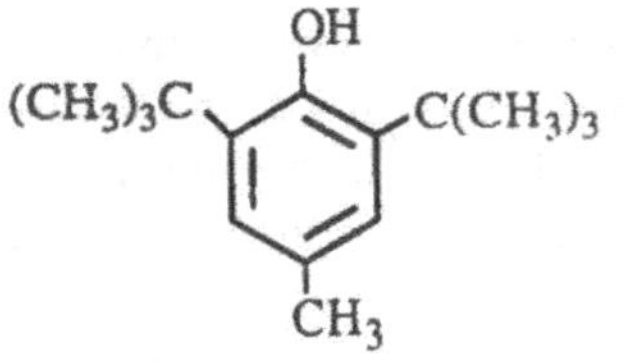

Fig. 4.6   Buylated hydroxytoluene.

Table 4.11 Selected Physical Properties of
Butylated Hydroxytoluene

Molecular weight	220
Physical appearance	White, granular crystals
Boiling point (°C)	265
Melting point (°C)	69.7
Solubility (25°C)	
Water	Insoluble
Fats and oils	25–40%
Ethyl alcohol	25%
Propylene glycol	Insoluble
Glyceryl monooleate	15%

Source: Ref. 35.

Preparation. The methods of preparation for BHT have been reviewed by Voges (66). Preparation of BHT involves reaction of pure anhydrous *p*-cresol with 2 mol or slightly more than 2 mol of isobutane in the presence of 2 wt % concentrated sulfuric acid as a catalyst at 70°C and at an isobutane partial pressure of about 0.1 MPa for about 5 h. The reaction mixture is washed with water at 70°C to remove the acid, and pure BHT is obtained by crystallization and recrystallization of the crude product from an ethanol–water mixture. Replacement of sulfuric acid with tetraphosphoric acid, methane disulfonic acid, or methane trisulfonic acid has been reported to improve product yield, color, and odor (67,68). Another method involves condensation of 2,6-di-*tert*-butylphenol with formaldehyde to yield 4,4′-methylenebis(2,6-di-*tert*-butylphenol), which is subsequently heated to 200°C with NaOH and methanol to give BHT in 80% yield.

Food Applications. Butylated hydroxytoluene resembles BHA in many of its antioxidant properties and is used in similar food applications. However, BHT is less effective than BHA and has less carry-through properties. The various food uses of BHT are described by Dugan et al. (69) and in the WHO Food Additive Series (5). Table 4.12 presents the levels of BHT commonly used in food products. BHT functions synergistically with BHA, TBHQ, and chelators like citric acid and has no synergistic activity with propyl gallate. BHT does not have an optimum concentration, and the stability of fats to which it is added continues to increase with concentration, although the rate of increase is less at higher levels. At concentrations higher than 0.02%, BHT imparts a phenolic odor to the fats (6,43,69).

Fats and oils. Butylated hydroxytoluene is more effective than BHA for stabilizing animal fats and is used at concentrations of 0.005–0.02%. Table 4.13

Table 4.12 Levels of BHT Used in Practice in Food Products[a]

Product	Level (%)
Animal fats	0.001–0.01
vegetable oils	0.002–0.02
Bakery products	0.01 –0.04[b]
Cereals	0.005–0.02
Dehydrated mashed potatoes	0.001
Essential oils	0.01 –0.1
Chewing gum base	Up to 0.1
Food packaging materials	0.02 –0.1

[a]Combinations with BHA, gallates, and citric acid are frequently used.
[b]Calculated on the fat.
Source: Ref. 5.

presents the antioxidant properties of BHT in lard. BHT is particularly effective if the lard is packed in paper containers or is otherwise in direct contact with paper. In mixtures of BHA, BHT, gallates, and citric acid, BHT is used at levels of 0.001–0.01%. BHT is found to be better than BHA or a BHA–BHT combination for stabilizing ghee (70). Like BHA, BHT provides very little protection to the frying fats and oils but has a carry-through effect in the fried product (17,21). Like BHA, BHT is less effective in vegetable oils and has no significant effect on the storage stability of margarine (21,71). BHT is also widely used in stabilizing

Table 4.13 Antioxidant Activity of BHT and BHT + BHA in Lard

Antioxidant	Concentration (%)	AOM stability[a] (h)
Control	0	11
BHT	0.005	36
BHT	0.01	53
BHT	0.02	64
BHA	0.02	54
BHT + BHA	0.005 + 0.01	80
BHT + BHA	0.01 + 0.01	102

[a]AOM = active oxygen method.
Source: Ref. 69.

essential oils such as *d*-limonene, orange oil, lemon oil, lime oil, and myrcene, an aliphatic unsaturated terpene found in many essential oils (44–46).

Baked and fried foods and cereal products. In baked or fried products, the carry-through properties of BHT are not as high as those of BHA (Table 4.10). In various cereal products and other low-fat foods, BHT is as effective as BHA and is widely used (47–49).

Meat products. Butylated hydroxytoluene is also effective in retarding the hematin-catalyzed oxidation of lard at 37°C. Combinations of BHA and BHT are effective in mechanically deboned carp, ground turkey meat, Chinese fried pork fiber, and frozen bacon slices (51–54). Like BHA, BHT is effective in freeze-dried beef, fowl, pork, and fish (56). BHT is also effective in retarding discoloration and rancidity in various dry sausages (57).

Milk products. At 0.008%, BHT has been used to stabilize whole milk. However, a combination of BHT and DG is more effective than BHT alone. Milk reconstituted from powder with BHT may show some phenolic flavor, particularly on heating (43).

Nuts and confectioneries. A combination of BHA and BHT is highly effective at stabilizing shelled walnuts, ground pecans, and peanuts (60). Like BHA, BHT is effective in stabilizing a variety of shelled nuts when incorporated in an edible protective coating. It is also effective in extending the shelf life of roasted macadamia nuts and almonds (62–64). BHA is widely used in retarding flavor loss, off-flavor development, toughness, and brittleness due to oxidation in chewing gums (65).

tert-*Butyl Hydroquinone*

tert-Butyl hydroquinone (TBHQ) (Fig. 4.7) was approved for food use in 1972. TBHQ is very effective in stabilizing fats and oils, especially polyunsaturated crude vegetable oils. The two para hydroxyl groups are believed to be responsible for its antioxidant activity (8). TBHQ is stable to high temperatures and is less steam-volatile than BHA and BHT. TBHQ is a white or light tan crystalline powder that is

Fig. 4.7   *tert*-Butyl hydroquinone.

slightly soluble in water and does not form a complex with iron or copper. It has significant solubilities in a rather wide range of fats and oils and solvents (8). The physical properties of TBHQ are presented in Table 4.14.

Preparation. *tert*-Butyl hydroquinone is prepared by reacting equimolar proportions of hydroquinone and *tert*-butyl alcohol in the presence of phosphoric acid as catalyst. In this process a mixture of 300 mL of toluene, 110 g of hydroquinone, and 400 mL of 85% phosphoric acid is heated to 92°C, and 74 g of *tert*-butyl alcohol is introduced over a 30-min period. After the addition, the hot toluene layer is separated and subjected to steam distillation to remove the solvent. The aqueous residue is filtered hot to isolate di-*tert*-butyl hydroquinone. The filtrate is cooled to bring about crystallization of TBHQ at 29.8% yield (40).

Another method involves reacting equimolar proportions of hydroquinone and isobutylene in the presence of phosphoric acid. A mixture of 147 g of hydroquinone, 250 g of 85% phosphoric acid, and 500 mL of xylene is heated to 105°C, and 55 g of isobutylene is added with agitation over a 1-h period. The supernatant xylene layer is separated and allowed to cool to bring about crystallization of TBHQ containing small amounts of hydroquinone and di-*tert*-butyl hydroquinone. TBHQ is further purified by recrystallization from hot water (40).

Food Applications. The activity of TBHQ is equivalent to or greater than that of BHA, BHT, or PG. TBHQ does not contribute to the color and odor of fats and oils. Its solubility characteristics are similar to those of BHA and superior to those of BHT or PG. It has no carry-through effects in baked products but carries through in frying to impart stability to the fried product (8). TBHQ functions synergistically with other antioxidants and chelators such as PG, BHT, BHA, tocopherols, ascorbyl palmitate, citric acid, and EDTA. The most interesting property of TBHQ is perhaps its effectiveness in various oils and fats where other phenolic antioxidants are

Table 4.14 Selected Physical Properties of *tert*-Butyl Hydroquinone

Molecular weight	166.22
Physical appearance	White to light tan crystals
Melting point (°C)	126.5 – 128.5
Solubility (25°C)	
Water	<1%
Fats and oils	5 – 10%
Ethyl alcohol	60%
Propylene glycol	30%
Glyceryl monooleate	10%

Source: Ref. 8.

relatively ineffective. The addition of citric acid generally enhances the activity of
TBHQ, and the combination is used in vegetable oils, shortenings, and animal fats.

Fats and oils. *tert*-Butyl hydroquinone is equivalent to BHA and possibly more
effective than BHT or PG in increasing the oxidative stability of lard as measured
by the active oxygen method (Table 4.15). It is also effective in retarding oxidative
rancidity in lard when incorporated into the packaging material (35). In rendered
poultry fat, TBHQ is more effective than BHA, BHT, or PG. At 0.02%, TBHQ
increased the oxidative stability of rendered poultry fat from 5 to 56 h, whereas
BHA, BHT, and PG at 0.02% increased it to 18, 20, and 30 h, respectively (8).
TBHQ is more effective in increasing the oxidative stability of crude whale oil than
BHA or PG (72). It is also effective in stabilizing the highly unsaturated mackerel
skin lipids (73).

The potency of TBHQ is equivalent to or greater than that of BHA, BHT, or PG
in stabilizing a variety of crude and refined vegetable oils. It is especially effective
in cottonseed, soybean, and safflower oils (Table 4.16), where other phenolic
antioxidants like BHA, BHT, and PG have little or no activity (8,13). TBHQ is also
effective in hydrogenated soybean oil and liquid canola shortenings (74,75).
Studies by Sherwin and Luckadoo (13) indicate that vegetable oils protected with
TBHQ during storage in the crude form might yield refined, bleached, and

Table 4.15 Comparison of TBHQ with BHA,
BHT, and PG in Lard by the Active Oxygen
Method

Treatment	AOM stability (hours to reach 20 meq peroxide content)
Control	4
0.005 TBHQ	23
0.010 TBHQ	38
0.020 TBHQ	55
0.005 BHA	27
0.010 BHA	36
0.020 BHA	42
0.005 BHT	12
0.010 BHT	18
0.020 BHT	33
0.005 PG	12
0.010 PG	20
0.020 PG	42

Source: Ref. 8.

Table 4.16 Comparison of TBHQ with BHA, BHT, and PG in Vegetable Oils by the Active Oxygen Method

Treatment	AOM stability (hours to reach 70 meq peroxide content)		
	Cottonseed oil	Soybean oil	Safflower oil
Control	9	11	6
0.010 TBHQ	24	29	40
0.020 TBHQ	34	41	29
0.030 TBHQ	42	53	77
0.010 BHA	9	12	8
0.020 BHA	9	10	6
0.030 BHA	9	10	7
0.010 BHT	10	12	10
0.020 BHT	11	13	7
0.030 BHT	13	15	8
0.010 PG	19	21	13
0.020 PG	30	26	10
0.030 PG	37	31	12

Source: Ref. 8.

deodorized oils with somewhat higher initial oxidative stability and with better response to further antioxidant treatment. Combinations of TBHQ, citric acid, and ascorbyl palmitate are highly effective in soybean oil and bint oil (50% sunflower oil + 50% cottonseed oil) (76,77). TBHQ is more effective in crude, refined, bleached, and deodorized palm olein than PG, BHT, and BHA (32,78,79). Studies by Sherwin (80) and Huffaker (81) indicate that TBHQ treatment has the potential to replace hydrogenation of vegetable oils. A combination of TBHQ and PG is reported to be highly effective in peanut oil prepared from unroasted and roasted peanuts (82). A combination of TBHQ and tocopherols is effective in margarine (83). At 0.02%, TBHQ is effective in increasing the oxidative stability of olive oil from 7 to 12 h. Addition of citric acid increased the stability to 58 h (65). TBHQ is also more effective than BHA and PG in retarding oxidative rancidity of lemon oil, orange oil, and peppermint oil (35).

Baked and fried foods and cereal products. Studies by Sherwin and Thompson (8) indicated that TBHQ, like PG, does not survive the processing conditions involved in the preparation of pastry and crackers. Hence, TBHQ does not carry through into the baked product, and BHA and BHT are more effective. However, unlike BHA and BHT, TBHQ has significant carry-through effects in frying operations. Sherwin and Thompson (8) reported that in potato chips prepared with cottonseed, soybean, and safflower oils stabilized with 0.02% TBHQ, BHA, BHT, or PG, TBHQ was significantly more effective than PG in enhancing the oxidative stability of the chips, while BHA and BHT were virtually ineffective. The efficacy of TBHQ has also been reported in other fried products such as deep-fried instant noodles (84,85), fish crackers (86), tapioca chips (87), and banana chips (88). In breakfast cereals such as corn flakes, wheat flakes, and oat cereal, TBHQ is similar to BHA and BHT in protecting the oxidizable lipids when added directly to the product (8). TBHQ is as effective as BHA when incorporated in waxed glassine liners in packaging materials used for breakfast cereals (35).

Meat products. *tert*-Butyl hydroquinone is effective in retarding the flavor deterioration in refrigerated and frozen beef patties (89). A combination of TBHQ and BHA is effective in retarding both discoloration and flavor deterioration in restructured pork or beef steaks (90). Other foods in which TBHQ is effective include minced carp (91) and fish (73).

Tocopherols

The tocopherols comprise a group of chemically related tocols and tocotrienols that are widely distributed in plant tissues, especially in nuts, vegetable oils, fruits, and vegetables. Some of the rich sources are wheat germ, corn, sunflower seed, rapeseed, soybean oil, alfalfa, and lettuce. Tables 4.17–4.19 present the content of tocopherols in some selected cereal grains, foods of animal origin, and vegetable oils. All the compounds exhibit antioxidant and vitamin E activity. However, they have limited carry-through properties and are less effective than BHA, BHT, and PG. The tocols and tocotrienols include α, β, γ, and δ homologs. The basic structural unit in all the homologs is a 6-chromanol ring with a phytol side chain (Fig. 4.8). The homologs differ in the number of methyl groups bound to the aromatic ring. The tocotrienols differ from tocols in having an unsaturated side chain, with double bonds at the 3'-, 7'-, and 11'-positions.

The tocols are generally referred to as tocopherols. The antioxidant activity of the α, β, γ, and δ homologs varies with the degree of hindrance and temperature. At physiological conditions around 37 °C, the antioxidant activity is in the order $\alpha > \beta > \gamma > \delta$, similar to the biological activity. At higher temperatures of 50–100°C, a reversal in the order of antioxidant activity is observed: $\delta > \gamma > \beta > \alpha$. α-Tocopherol is the most abundant of all the tocopherols, and its biological activity is twice that of the β and γ homologs and 100 times that of the δ homolog (93). α-Tocopheryl acetate, the stable form of α-tocopherol, is not an antioxidant because the active hydroxyl group is protected. However, under certain conditions, for example, in acidic aqueous systems, tocopherol is released by slow hydrolysis of the tocopheryl acetate (94). Figure 4.9 presents the antioxidant activity of some tocopherols and their degradation products (95).

Tocopherols are pale yellow, clear, viscous, oily substances, insoluble in water

Table 4.17 Approximate Tocopherol Content of Vegetable Oils (mg/kg)

Vegetable oils	Tocopherols				Tocotrienols			
	α	β	γ	δ	α	β	γ	δ
Coconut	5– 10	—	5	5	5	Trace	1– 20	—
Cottonseed	40– 560	—	270–410	0	—	—	—	0
Maize grain	60– 260	0	400–900	1– 50	—	0	0–240	0
Maize germ	300– 430	1– 20	450–790	5– 60	—	—	—	—
Olive	1– 240	0	0	0	—	—	—	—
Palm	180– 260	trace	320	70	120–150	20– 40	260–300	70
Peanut	80– 330	—	130–590	10– 20	—	—	—	—
Rapeseed	180– 280	—	380–590	10– 20	—	—	—	—
Safflower	340– 450	—	70–190	230–240	—	—	—	—
Soybean	30– 120	0– 20	250–930	50–450	0	0	0	—
Sunflower	350– 700	20– 40	10– 50	1– 10	—	—	—	—
Walnut	560	—	590	450	—	—	—	—
Wheat germ	560–1200	660–810	260	270	20– 90	80–190	—	—

Source: Ref. 94. With permission from Elsevier Publishers.

Table 4.18 Approximate Tocopherol Content of Cereal
Grains

Product	Tocopherol (mg/100 g)			α-Tocotrienol (mg/100g)
	α	β	γ	
Whole yellow corn	1.5	—	5.1	0.5
Yellow corn mean	0.4	—	0.9	—
Whole wheat	0.9	2.1	—	0.1
Wheat flour	0.1	1.2	—	—
Whole oats	1.5	—	0.05	0.3
Oat meal	1.3	—	0.05	0.3
Whole rice	0.4	—	0.4	—
Milled rice	0.1	—	0.3	—

Source: Ref. 92.

and soluble in fats and oils. The physical properties of α-tocopherol are listed in
Table 4.20.

α-Tocopherol and its acetate have also been chemically synthesized. Synthetic
products are designated as *dl*-α-tocopherol and *dl*- α-tocopheryl acetate, which are
actually mixtures of the four racemates. A synthetic water-soluble form of α-to-

Table 4.19 Approximate α-Tocopherol
Content (mg/kg) of Selected Foods of Animal
Origin

Product	α-Tocopherol (mg/kg)
Beef	6
Chicken	4
Pork	5
Lard	12
Cod	2
Halibut	9
Shrimp	7
Milk, spring	0.2
Milk, autumn	1.1
Butter	10–33
Eggs	5–11

Source: Ref. 94. With permission from Elsevier Publishers.

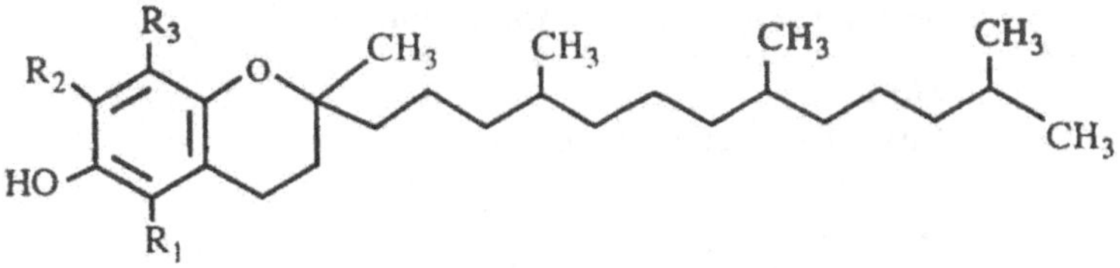

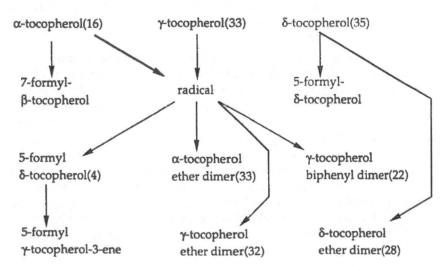

Fig. 4.8 Tocopherols.

	R_1	R_2	R_3	
	CH_3	CH_3	CH_3	α-tocopherol
	CH_3	H	CH_3	β-tocopherol
	H	CH_3	CH_3	γ-tocopherol
	H	H	CH_3	δ-tocopherol

copherol, *d*-α-tocopheryl polyethylene glycol 1000 succinate (TPGS), has been prepared by the esterification of *d*-α-tocopheryl acid succinate with polyethylene glycol having an average molecular weight of 1000. It is a pale yellow waxy substance that provides 260 mg of *d*-α-tocopherol per gram and forms a clear solution in water at concentrations up to 20% (96).

Fig. 4.9 Antioxidant activity of some tocopherols and their decomposition products in lard. Length of induction period (days) is given in parentheses. (Control without antioxidants, 2 days.) [From Ishikawa and Yiki (95).]

Table 4.20 Selected Physical Properties of α-Tocopherol

Molecular weight	430.72
Physical appearance	Pale yellow, clear, viscous, oily substance
Boiling point (°C)	200–220
Solubility	Insoluble in water; freely soluble in vegetable oils and organic solvents

Source: Ref. 2.

Preparation. The commercial methods of preparation of α-tocopherol and its acetate have been reviewed by Schuler (94), Herting (97), and Kasparek (98). Tocopherols are obtained either by extraction from natural sources or by chemical synthesis.

Extraction from natural sources. The major natural source is the sludge obtained in the deodorization of vegetable oils and fats. Besides the tocopherols, such distillates may also contain sterols, free fatty acids, and triglycerides. Separation of tocopherols is achieved by several methods: by esterification with a lower alcohol, washing, and vacuum distillation; by saponification; or by fractional liquid–liquid extraction. Further purification is by molecular distillation, extraction, or crystallization or by a combination of these processes. The concentrate so obtained contains relatively high amounts of γ- and δ-tocopherols. They are converted to α-tocopherol by methylation followed by acetylation to obtain the relatively stable α-tocopheryl acetate (94).

Chemical synthesis. The chemical synthesis of α-tocopherol is based on the condensation of 2,3,5-trimethylhydroquinone with phytol, isophytol, or phytyl halogenides (Fig. 4.10). The condensation is carried out in acetic acid or in an inert solvent such as benzene with an acidic catalyst such as zinc chloride, formic acid, or boron trifluoride ethyl etherate. Earlier, natural δ-phytol or isophytol were used, and the resulting product was a mixture of the two isomers. At present, synthetic isophytol is used and the product all-rac- α-tocopherol is a racemic mixture of eight stereoisomers. The crude product is purified by vacuum distillation. The all-rac forms of β-, γ-, and δ-tocopherols can be synthesized by the same method. Instead of trimethylquinoline, the appropriate dimethylquinoline or monomethylquinoline is used. 2,5-Dimethylquinoline yields all-rac-β-tocopherol, 2,3-dimethylquinoline yields all-rac-γ-tocopherol, and methyl hydroquinone yields all-rac-δ-tocopherol (94,97,98).

Food Applications. The food applications of tocopherols have been reviewed by Klaui (99) and Schuler (94) and in the WHO Food Additives Series (5). Tocopherols are considered the major natural antioxidants in vegetable and animal

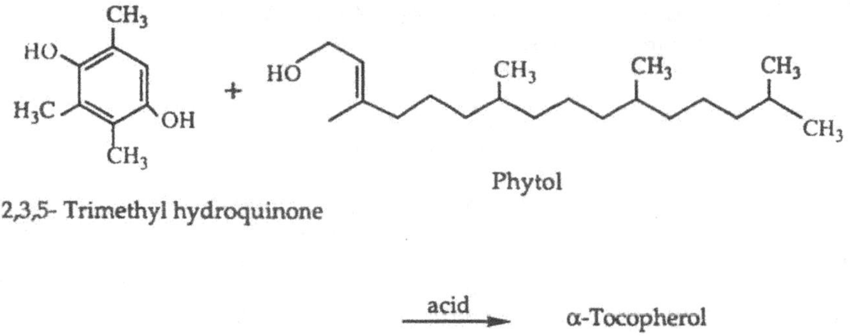

Fig. 4.10   Chemical synthesis of α-tocopherol.

fats. They are relatively weak antioxidants compared to the synthetic phenolic antioxidants and have limited carry-through properties. At the concentrations necessary to give adequate protection to low-stability fats, they usually contribute an undesirable foreign flavor of their own (6). The tocopherols function synergistically with other antioxidants such as BHA, TBHQ, ascorbyl palmitate, lecithin, amino acids, ascorbic acid, and various spice extracts. The use of tocopherols is limited to fats, oils, and fat-containing foods.

Fats and oils. The tocopherols are effective in stabilizing lard and other animal fats. α-Tocopherol is most effective in lard at intermediate concentrations (e.g., 0.01%) and low temperatures (e.g., 20°C). At higher temperatures (e.g., 97°C), γ-tocopherol is most active in stabilizing lard (100,101).

The tocopherols, however, are more effective in combination with other anti-oxidants or synergists such as ascorbyl palmitate, ascorbic acid, lecithin, or citric acid than when used alone. The shelf life of lard has been doubled by adding a mixture consisting of α-tocopherol (0.001%), ascorbyl palmitate (0.0025%), and citric acid (0.0005%) in storage tests under commercial conditions (5). A combination of 0.02% dl-α-tocopherol or dl-γ-tocopherol with ascorbyl palmitate is effective in stabilizing various animal fats such as chicken fat, pork fat, and beef fat (102). Studies by Pongracz (103) showed that α-tocotrienol is more effective than α-tocopherol in lard and that the activity is further enhanced by the addition of ascorbyl palmitate.

Antioxidant mixtures containing 100 ppm α-tocopherol + 500 ppm ascorbyl palmitate + 500 ppm lecithin (mixture A) or 100 ppm α-tocopherol + 500 ppm ascorbyl palmitate + 500 ppm lecithin + 100 ppm octyl gallate (mixture G) were highly effective in stabilizing lard at room temperature and lard heated at 150°C (104,105) (Tables 4.21 and 4.22). The antioxidant mixtures A and G were also more highly effective in stabilizing butterfat stored at 80°C than α-tocopherol (106). Incorporation of 100 ppm of dl-α-tocopherol and 500 ppm of ascorbyl palmitate

Table 4.21 Antioxidant Efficacy of Antioxidant Mixtures A and G at Room Temperature in Various Oils[a]

Oil or fat	Peroxide value after 180 days at room temperature			
	BHA + BHT, (0.1 g each/kg)	A (2 g/kg)	G (2 g/kg)	Control
Peanut oil	—	0	0	14.2
Sunflower oil	20.5	1.6	1.1	26.4
Soybean oil	—	0.8	0.3	19.8
Lard	—	0	0	5.4

[a]Mixture A: ascorbyl palmitate + tocopherol + lecithin. Mixture G: mixture A + octyl gallate.
Source: Ref. 99.

in beef tallow decreases the formation of oxidized cholesterol derivatives during heat treatment (107). The shelf life of hydrogenated whale oil under commercial storage conditions was increased from 1 to 2 years with a mixture of α-or γ-tocopherol (0.008%) and ascorbyl palmitate (0.0024%) (108).

In vegetable oils that contain endogenous tocopherols, the addition of a weaker homolog (e.g., α-tocopherol) to a product that already contains a stronger homolog (e.g., γ-tocopherol) is not beneficial. The antioxidant mixtures A and G have been more effective in vegetable oils than tocopherols. In peanut, sunflower, soybean, rapeseed, and palm oils, the antioxidant mixtures A and G are highly effective both

Table 4.22 Antioxidant Efficacy of Antioxidant Mixtures A and G in Heated Oils (150°C)—Percentage of Lost Glycerides[a]

Oil or fat	Control		A (0.2%)		G (0.2%)	
	3 days	4 days	3 days	4 days	3 days	4 days
Lard	9.1	13.3	3.5	6.6	0	0
Soybean oil	9.0	13.3	—	11.8	—	8.2
Groundnut oil	7.3	9.2	3.5	2.5	1.2	0.7
Sunflower oil	21.6	49.3	11.5	47.7	5.9	40.6
Rapeseed oil	8.3	15.6	6.9	11.6	6.3	7.8
Palm oil	8.6	13.9	6.8	9.2	4.4	6.2

[a]Mixture A: ascorbyl palmitate + tocopherol + lecithin. Mixture G: mixture A + octyl gallate

in oils stored at room temperatures and in oils heated at 150°C (104,105) (Tables 4.21 and 4.22).

In peppermint, combinations of α-tocopherol with lecithin or ascorbyl palmitate are as effective as BHA in retarding oxidation (109). γ-Tocopherol is effective in retarding oxidation of *d*-limonene, and its antioxidant activity is similar to that of BHA or BHT, whereas α-tocopherol is a prooxidant under certain conditions. Hence, the behavior of mixed tocopherols in orange oil depends on the content of α-*tocopherol* in comparison with that of γ- and δ-tocopherols (109).

Baked and fried products. The tocopherols have carry-through properties in baked and fried products. Crackers, pastry, and potato chips prepared with lard treated with 0.01–0.1% tocopherol, either alone or in combination with BHA (0.01%), were appreciably more resistant to rancidity than control samples (6,110). In another study, Pongracz et al. (109) compared the effectiveness of various antioxidants including α- and γ-tocopherols in the stabilization of a commercial frying oil and the fried product. After repeated fryings of 35 batches of potato chips, γ-tocopherol was more effective than α-tocopherol and was comparable to TBHQ and octyl gallate in protecting the oil and the fried product.

Vitamin A and carotenoids. The tocopherols are effective in stabilizing natural vitamin A in a methyl linoleate model system and in cod-liver oil. γ-Tocopherol is more potent than α-*tocopherol* (109,111). The tocopherols also stabilize carotenoids in model systems and in an orange drink. In a model system containing 150 mg of pure β-carotene per kilogram of vaseline, 500 ppm of γ-*tocopherol* or 2000 ppm of Ronoxan, a commercially available synergistic mixture of 5% *dl*- α-tocopherol, 25% ascorbyl palmitate, and 70% lecithin, are highly effective. After 3 months of storage exposed to daylight, the control sample showed 33% loss of β-carotene, whereas samples with γ-tocopherol or Ronoxan showed 0% loss, and samples with α-tocopherol showed 7% loss of β-carotene. In an orange drink, β-carotene is better protected by α-tocopherol, and apocarotenol is better protected by γ-tocopherol (109).

Meat products. A combination of α-tocopherol, ascorbyl palmitate, and citric acid is highly effective in retarding the formation of white spots in uncured pork sausages (94). In salami, γ-tocopherol is effective in retarding lipid oxidation and the formation of rancid off-flavors but is ineffective in improving the color stability (109). The tocopherols are effective in enhancing color stability and retarding lipid oxidation and off-flavor development in beef, pork, turkey, and chicken meat if the animals are fed a diet rich in tocopherols prior to slaughter (112). They are effective in fish preserves with ascorbates (113,114).

Nuts and confectioneries. A combination of α-tocopherol and ascorbyl palmitate is effective in retarding lipid oxidation and rancidity development in whole or broken nuts such as walnuts, almonds, and hazelnuts. The antioxidant mixture dissolved in 96% ethanol is applied either by a spray technique or by dipping the nuts. The tocopherols in combination with ascorbyl palmitate or synergistic mixtures including lecithin or citric acid are effective in the protection of the lipid phase in toffees and caramels. The α- and γ-tocopherols are effective in improving keeping quality of chewing gum base. In one of the studies, α-tocopherol was as effective as BHT in protecting two types of chewing gums (94,109).

Gum Guaiac

Gum guaiac, a resinous exudate from the wood of *Guajacum officinale* L. or *G. sanctum* L., is a naturally occurring antioxidant. Commercial gum guaiac is a mixture of phenols consisting of α- and β-guaiaconic acids, guaiacic acid, guaiaretic acid, vanillin, guaiac yellow, and guaiac saponins. α-Guaiaconic acid is the most effective compound. Gum guaiac is in the form of brown or greenish-brown irregular lumps that are insoluble in water and sparingly soluble in fats and oils.

The antioxidant properties of gum guaiac have been described by Grettie (115) and Chipault (6). Gum guaiac was the first antioxidant to be approved for food use. It was used for the stabilization of lard in combination with phosphoric acid for a number of years. Gum guaiac is more effective in animal fats than in vegetable oils. It is stable to higher temperatures and has carry-through properties (116). Gum guaiac has been reported to add flavors and odors to shortenings and is no longer being used in food products.

Ionox Compounds

The Ionox compounds are derivatives of BHT and include, in increasing order of structural complexity, Ionox-100 (2,6-di-*tert*-butyl-4-hydroxymethyl phenol), Ionox-201 [di-(3,5-di-*tert*-butyl-4-hydroxybenzyl ether], Ionox-220 [di-(3,5-di-*tert*-butyl-4-hydroxyphenyl)methane], Ionox-312 [2,4,6-tri-(3',5'-di-*tert*-butyl-4-hydroxybenzyl)phenol], and Ionox-330 [2,4,6-tri-(3',5'-di-*tert*-butyl-4- hydroxybenzyl)mesitylene] (Fig. 4.11). Ionox-100 is approved for food use, and Ionox-330 is approved for incorporation into food-packaging materials.

Preparation. Ionox-100 is prepared by the method of Rocklin and Morris as described by Wright et al. (117). In this process a solution of potassium *tert*-butoxide (3.5 mg) in *tert*-butyl alcohol (1 mL) is added to a solution of 2,6-di-*tert*-butylphenol (1.2 g) and paraformaldehyde (100 mg) in *tert*-butyl alcohol (4 g) under N_2 at 0°C, and the mixture is maintained at 12°C for 40 min. The catalyst is inactivated with CO_2, and the solution is neutralized by the addition of 2 mL of aqueous 10% NH_4Cl. After evaporation of the solvent, the residue is adsorbed onto a column of acid-washed alumina, and Ionox-100 is eluted with light petroleum–diethyl ether (7:3, v/v).

Ionox-330 is prepared by the alkylation of mesitylene (1,3,5-trimethylbenzene)

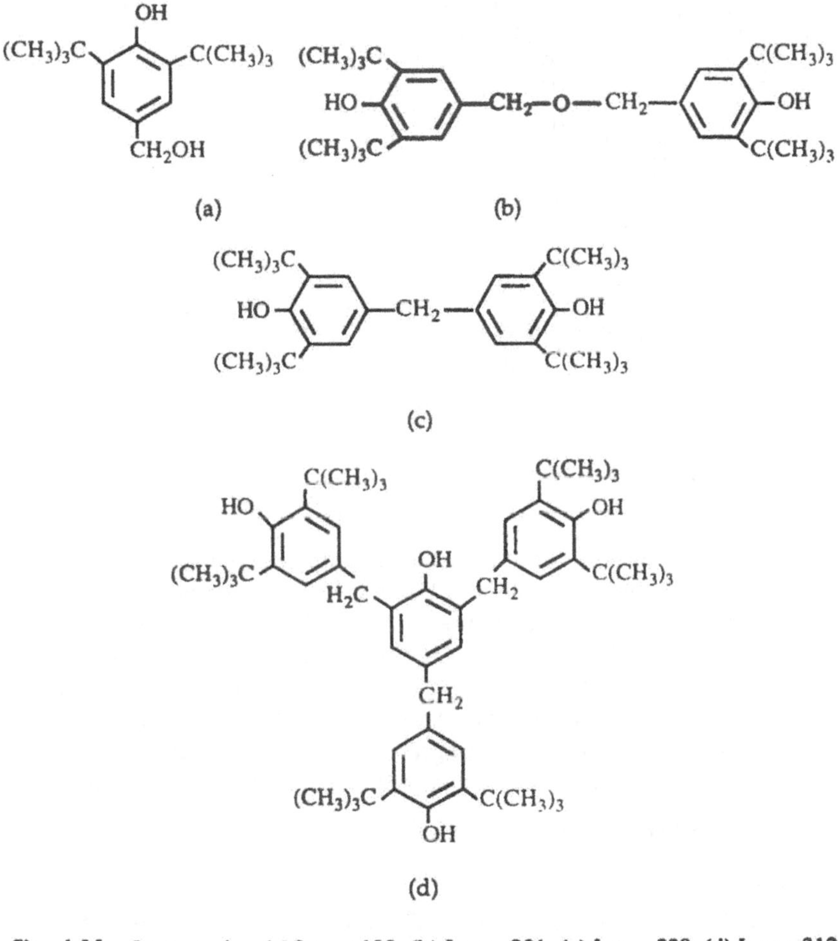

Fig. 4.11   Ionox series. (a) Ionox-100; (b) Ionox-201; (c) Ionox-220; (d) Ionox-312; (e) Ionox-330.

(0.4 mol) with 3,5-di-*tert*-butyl-4-hydroxybenzyl alcohol (0.1 mol) in methylene chloride at 4°C in the presence of 80% sulfuric acid as a catalyst. Ionox-330 is obtained after recrystallization from isopentane (118).

Food Applications. Ionox-100 has been available since 1967. The objective in changing the BHT structure to include a hydroxymethyl group rather than a simple methyl group was to decrease its volatility and help in the stabilization of oils during frying, but in most applications this small structural change had little effect on antioxidant effectiveness. As a consequence, this material has not been used commercially to any extent (35).

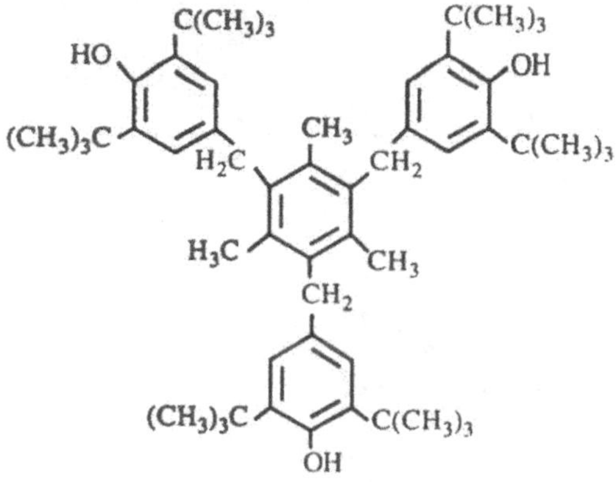

(e)

Fig. 4.11   (*Continued*)

Ionox-330 was evaluated as a stabilizer for polypropylene and safflower oil (118). The polypropylene samples containing 0.25% or 0.5% of the compound were tested by heat aging, outdoor exposure, and Fadeometer (Table 4.23). Ionox-330 increased the heat stability of the polypropylene by a factor of about 70 and the weather stability by a factor of about 6 in the heat aging test, and, in the Fadeometer test, it more than doubled the ultraviolet stability of the polypropylene sample. In sunflower oil, the number of days to rancidity measured by a gravimetric procedure was enhanced from 5.75 days in the control to 9 days in the samples treated with 0.02% Ionox-330.

4.2.3   Miscellaneous Primary Anitoxidants

Ethoxyquin

Ethoxyquin (EQ; 6-ethoxy-1,2-dihydro-2,2,4-trimethylquinoline) (Fig. 4.12) is widely used as a feed antioxidant, particularly in dehydrated alfalfa as a protective

Table 4.23   Antioxidant Activity of Ionox-330 in Polypropylene

Additive (% wt)	Heat aging (days)	Exposure (weeks)	Fadeometer (h)
None	<0.5	~0.5	~30
0.25	30	3.5	78
0.5	37	3.0	78

Source: Ref. 118.

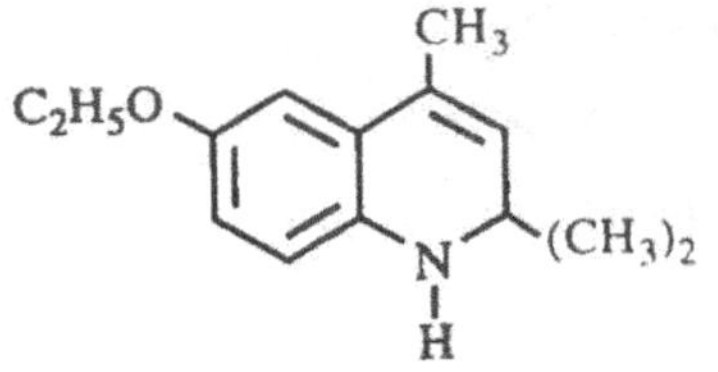

Fig. 4.12 Ethoxyquin.

agent for carotenoids (119–121). Ethoxyquin nitroxide, the stable free radical formed upon the oxidation of EQ, is more effective than EQ (122,123). The physical properties of EQ are listed in Table 4.24.

Preparation. Ethoxyquin is synthesized via *p*-phenetidine–acetone or aniline–*p*- phenetidine–mesityl oxide condensation products (124) (Fig. 4.13). In the first method, *p*-phenetidine acetone anil was prepared by the reaction of acetone with *p*-phenetidine at 135–140°C for 13 h. Ethoxyquin was obtained (85% yield) by heating the anil with excess *p*-phenetidine and toluenesulfonic acid at 130–140°C for 4 h (Fig. 4.13). In the second method, *p*-phenetidine mesityl oxide anil or aniline mesityl oxide anil was prepared by heating the corresponding amine with mesityl oxide in benzene at 104°C for 24 h. Ethoxyquin was obtained in similar yields by reaction with *p*-phenetidine and toluenesulfonic acid as before.

Food Applications. Ethoxyquin has proved to be the most active and usable compound for the stabilization of carotene in forage products. It affords protection at a level of 0.015% in alfalfa meal (119). Ethoxyquin was found to be comparable to Endox, a commercial formulation of BHA, phosphoric acid, and EDTA, in inhibiting the degradation of asthaxanthin during storage at 4°C (125). Ethoxyquin is effective in controlling superficial scald on apples and pears. This could be due to the antioxidant action of ethoxyquin and/or inhibition of conjugated triene biosynthesis, which is highly correlated to the incidence of superficial scalds

Table 4.24 Selected Physical Properties of Ethoxyquin

Molecular weight	217.3
Physical appearance	Yellow liquid
Boiling point (°C)	123—125
Solubility (25°C)	Insoluble in water; soluble in organic solvents

Source: Ref. 2.

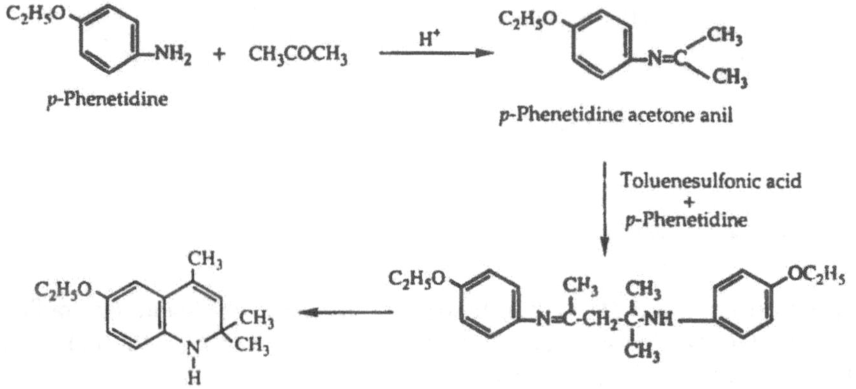

Fig. 4.13   Synthesis of ethoxyquin.

(126–129). Ethoxyquin is effective in protecting the pigments of paprika (130) and crawfish meal (131). In crawfish meal, ethoxyquin was found to be more effective than BHA in pigment retention (Table 4.25). Ethoxyquin is also an effective antioxidant for fish oils, fish meals, and squalene (122,132,133). Ethoxyquin nitroxide and quinone imine, the oxidation products of ethoxyquin, are potent antioxidants. It has been postulated that in fish oils the oxidation product quinone imine may play a more significant role than ethoxyquin (Table 4.26). Ethoxyquin is also a potent inhibitor of nitrosamine formation in bacon at levels as low as 20 ppm (135,136). Lecithin works synergistically with ethoxyquin in preventing quality changes in fats and oils (137).

Table 4.25   Pigment Retention in Crawfish Meal

Antioxidant	% Concn.	% Pigment retained		
		1 wk	2 wk	3 wk
BHA	0	57.6	43.6	33.8
	0.1	66.5	53.9	50.7
	0.5	84.1	74.5	64.0
	1.0	81.3	78.0	68.7
Ethoxyquin	0	57.8	40.0	31.5
	0.1	83.0	82.2	70.2
	0.5	90.2	85.5	74.6
	1.0	88.5	85.9	80.4

Source: Ref. 131.

Table 4.26 Effect of Ethoxyquin and Quinone Imine in Mackerel Oil

Antioxidant (0.1%)	Oxygen uptake (μmol/g)			
	100 h	300 h	500 h	1000 h
None	10	158	211	555
Ethoxyquin	10	43	51	66
Quinone imine	0	5	8	13

Source: Ref. 134.

Anoxomer

Anoxomer is a polymeric, nonabsorbable antioxidant developed by Weinshenker et al. (138). It is not a single molecular mass compound but is a distribution of molecular masses centered about a "peak" at 4000 daltons. Anoxomer is an off-white powder that is insoluble in water and propylene glycol and highly soluble in organic solvents, fats, and oils. It is stable to frying temperatures of 190°C for over a period of 6 h and provides carry-through protection to the fried product. Anoxomer adds no color, odor, or taste to the oils under standard use conditions (139). Table 4.27 presents the physical properties of Anoxomer.

Preparation. Anoxomer is prepared by the condensation of divinylbenzene, hydroxyanisole, *tert*-butyl hydroquinone, *tert*-butyl phenol, and bisphenol under ortho alkylation conditions (138).

Food Applications. The antioxidant properties of Anoxomer have been described by Weinshenker (139). Anoxomer is active in animal fats and vegetable

Table 4.27 Selected Physical Properties of Anoxomer

Molecular mass	~4000 daltons
Physical appearance	Off-white powder
Solubility (25°C)	
Water	Insoluble
Fats and oils	20%
Ethyl alcohol	>20%
Propylene glycol	Insoluble
Glyceryl monooleate	10%

Source: Ref. 139.

oils such as lard, tallow, soybean, cottonseed, peanut, corn, and safflower oils. It does not show a prooxidant effect even at higher concentrations, and, unlike that of other antioxidants, its antioxidant activity increases with concentration. It is highly effective in stabilizing rendered chicken fat at 27 and 45°C at increasing concentrations. The fat is considered rancid when the peroxide value reaches a value of 10 meq/kg. At 27°C, addition of Anoxomer at 100, 200, 500, and 800 ppm increased the number of weeks to reach a peroxide value of 10 meq/kg from 4 weeks in untreated sample to 16, 24, 34, and >36 weeks, respectively. At the 800 ppm Anoxomer level, the peroxide value was only 5.1 meq/kg at 36 weeks. At 45°C, the samples containing 200, 500, and 800 ppm Anoxomer were protected against rancidity for 14 weeks. In a mackerel skin lipid system, Anoxomer was better than BHA and BHT and less active than TBHQ at the 0.05–0.1% level (140). In capelin fish oils, Anoxomer is effective at 0.02% (141).

Anoxomer is also highly effective in increasing the life of frying fats at 300–1000 ppm levels. A deep-frying fat formulation containing 20% soy oil, 70% liquid palm oil, and 10% hardened palm oil containing 300 ppm of Anoxomer was stable after 17 days of repetitive fryings with a peroxide value of only 1.4 meq/kg. Anoxomer was effective in controlling the formation of aldehydes as secondary breakdown products of peroxides in frying fats (142,143). In unstabilized natural lemon essence oil, Anoxomer was effective at concentrations of 1000–5000 in protecting against oxidative degradation (144).

Anoxomer also has carry-through properties in fried and baked products. The carry-through effect has been observed in potato chips prepared from hydrogenated soybean oil containing 1000 ppm Anoxomer and in sugar cookies prepared from lard containing 20–600 ppm Anoxomer. Anoxomer at 200 ppm was effective in retarding rancidity in cooked and raw ground turkey meat. It was also highly effective in styrene-butadiene bubble gum base at 1000 ppm. A combination of Anoxomer with calcium stearate or distearyl dithiopropionate was effective in polymer systems such as polypropylene and polyisoprene (139).

Trolox-C

Trolox-C (6-hydroxy-2,5,7,8-tetramethyl chroman-2-carboxylic acid) (Fig. 4.14) is a synthetic phenolic antioxidant developed by Scott et al. (145). It resembles α-tocopherol structurally except for the replacement of the hydrocarbon side chain by a COOH group. Trolox-C is a colorless, tasteless solid, slightly soluble in water and sparingly soluble in oils. It is stable for 2 months at room temperature (22°C) and at 45°C In vegetable oils and animal fats, Trolox-C was found to have two to four times the activity of BHA and BHT and was more active than α-tocopherol, propyl gallate, and ascorbyl palmitate (146,147). However, Trolox-C is not being used commercially at present. Its physical properties are presented in Table 4.28.

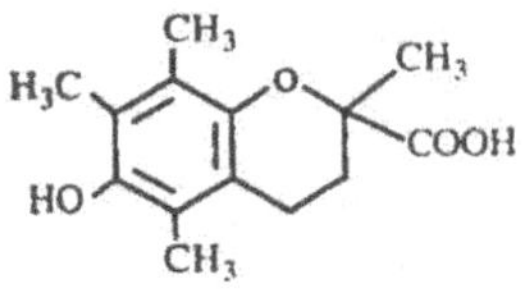

Fig. 4.14 Trolox-C.

Preparation. The preparation of Trolox-C has been described by Scott et al. (145). The process involves the condensation of trimethyl hydroquinone and methyl vinyl ketone in a methanol–trimethyl orthoformate–sulfuric acid mixture to obtain the ketal 6-hydroxy-2-methoxy-2,5,7,8- tetramethylchroman. The ketal can be converted to Trolox-C by two routes.

In the first route, the ketal is hydrolyzed to the hemiketal with concentrated HCl in acetone to form 2,6-dihydroxy-2,5,7,8-tetramethylchroman. The hemiketal is acetylated with acetic anhydride in pyridine to form 4-(2,5-diacetoxy-3,4,6-trimethylphenyl)butan-2-one. The diacetate is converted to the cyanohydrin by reacting with potassium cyanide in dimethylformamide to obtain 2-cyano-4-(2,5-diacetoxy-3,4,6-trimethylphenyl)butan-2-ol. Warming the cyanohydrin in concentrated HCl causes acetate and nitrile hydrolysis and ring closure to give Trolox-C in 66% overall yield (Fig. 4.15).

In the second route, the ketal is first acetylated to the acetate, which upon acid hydrolysis yields the hemiketal. The hemiketal is converted to the cyanohydrin. The cyanohydrin is converted to the hydroxy ester by reaction with anhydrous hydrogen chloride in methanol followed by hydrolysis of the intermediate iminium salt. Trolox-C is obtained by saponification followed by acid-catalyzed cyclization of the hydroxy ester.

Table 4.28 Selected Physical Properties of Trolox-C

Molecular weight	250.32
Physical appearance	White to light yellow powder
Melting point (°C)	189–195
Solubility (25°C)	
Water	0.053%
Fats and oils	0.17–0.7%
Ethyl alcohol	16%
Propylene glycol	1.6%

Source: Ref. 146.

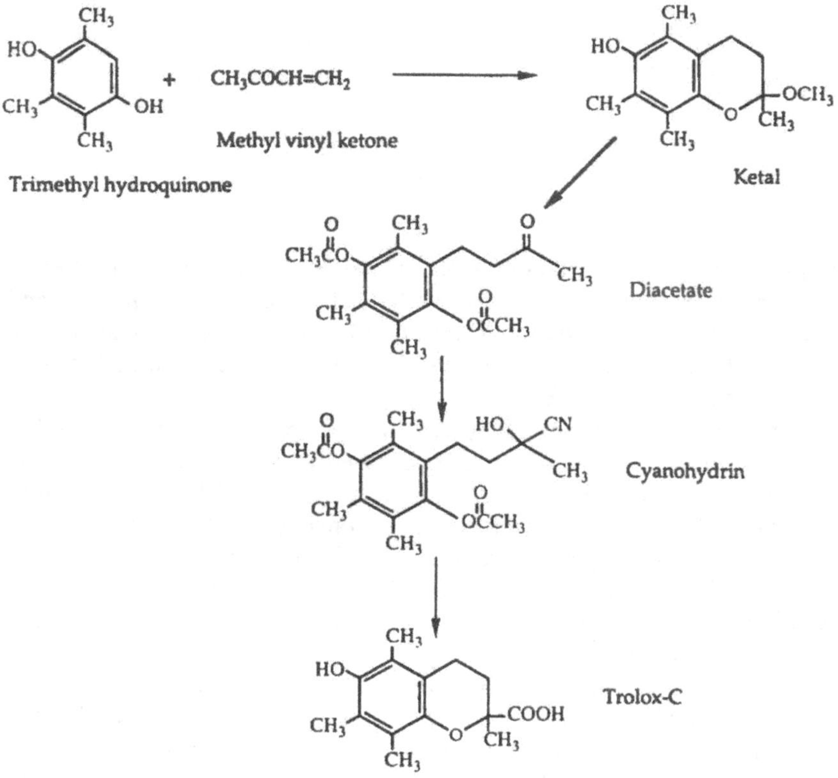

Fig. 4.15 Synthesis of Trolox-C.

Food Applications. The antioxidant properties of Trolox-C have been described by Cort et al. (146,147). Trolox-C is two to four times as active as BHA, BHT, PG, and ascorbyl palmitate in thin-layer tests using soybean oil and chicken fat. It is also more active than α- or γ-tocopherol. Lard treated with Trolox-C displayed the greatest stability and required 38 days to reach a peroxide value of 70 meq/kg at 45°C. Addition of ascorbyl palmitate or ascorbic acid further enhanced the stability to 60 and 62 days, respectively. In chicken and pork fat tested in thin layers, Trolox-C was more effective than THBQ and the tocopherols. Trolox-C was more effective at the 0.02% level than BHA, BHT, and PG in a series of vegetable oils including corn, peanut, sunflower, and safflower oils at 45°C (Table 4.29). It is better than TBHQ in corn and safflower oils, equivalent to TBHQ in sunflower oil, and less effective than TBHQ in soybean oil. It is not as effective at a higher temperature of 98°C. However, addition of ascorbyl palmitate significantly enhances the activity of Trolox-C at both 45 and 98°C. Trolox-C also functions

Table 4.29 Antioxidant Activity of Trolox-C in
Vegetable Oils at 45°C—Days to Reach a Peroxide
Value of 70 meq/kg[a]

Antioxidant (0.02%)	Substrate oil			
	Corn	Peanut	Sunflower	Safflower
Trolox-C	25	34	27	39
TBHQ	24	30	27	33
PG	21	26	19	16
BHT	13	15	9	10
BHA	15	15	8	8
None	12	15	6	6

[a]Thin layer test results.
Source: Ref. 146.

synergistically with TDPA, amino acids, ascorbic acid, and other natural compounds such as quercetin, curcumin, ferulic acid, and caffeic acid. Trolox-C in combination with ascorbic acid is highly effective in stabilizing crude palm oil and olive oil. It is comparable to BHA in retarding oxidation in orange oil, lemon oil, and grapefruit oil. Troloxylamino acids derived by a covalent attachment of amino acids to Trolox-C are even more active than Trolox-C in vegetable oils (148).

4.3 TECHNOLOGICAL ASPECTS OF SYNERGISTIC OR SECONDARY ANTIOXIDANTS

Secondary or synergistic antioxidants function by different mechanisms. They may function as electron or hydrogen donors to primary antioxidant radicals, thereby regenerating the primary antioxidant. They also provide an acidic environment that enhances the stability of primary antioxidants. Oxygen scavengers such as ascorbic acid react with free oxygen and remove it in a closed system. The chelating agents remove prooxidant metals, thus preventing metal-catalyzed oxidations.

4.3.1 Oxygen Scavengers

Sulfites

Sulfites represent a group of compounds, including sulfur dioxide (SO_2), sodium sulfite (Na_2SO_3), sodium bisulfite ($NaHSO_3$), and sodium metabisulfite ($Na_2S_2O_5$), that are commonly used as antimicrobial agents. They have weak antioxidant properties and are used as oxygen scavengers. In general, sulfites are added in excess (1.5 mol/0.5 mol oxygen) to ensure complete and accelerated removal of oxygen. Sulfites are highly effective in preventing browning in fruits and vegetables. However, their use is restricted mainly because of reports of adverse allergic-type reactions in asthmatic individuals that were attributed to the consumption of sulfite-containing foods. Table 4.30 presents the physical properties of sulfites.

Preparation. The various methods of manufacturing sulfur dioxide and sulfur compounds have been described by Canning (149) and Weil (150).

Sulfur dioxide. Sulfur dioxide is obtained by burning sulfur in air or oxygen. Oxidation of sulfides in the roasting of sulfide minerals also results in sulfur dioxide. Sulfur dioxide is obtained in a relatively pure state by the reduction of sulfuric acid with copper or by the treatment of sulfites or hydrogen sulfites with strong acids.

Sodium sulfite. In a typical manufacturing process, a solution of sodium bicarbonate is allowed to percolate downward through a series of absorption towers through which sulfur dioxide is passed countercurrently. The solution leaving the towers is mainly sodium bisulfite containing 27% sulfur dioxide. Sodium sulfite is prepared by reacting the solution with aqueous sodium carbonate or sodium hydroxide. Anhydrous sodium sulfite (96–99% pure) is crystallized above 40°C. Sodium sulfite is a colorless crystalline substance that oxidizes to the sulfate on exposure to air.

Sodium bisulfite. Sodium bisulfite is produced by saturating a solution of sodium carbonate with sulfur dioxide. Sodium bisulfite is isolated as a white powder after precipitation with alcohol. It oxidizes to the sulfate on exposure to air. The bisulfite of commerce consists chiefly of sodium metabisulfite. Commercial aqueous solutions contain 26–77% sulfur dioxide.

Sodium metabisulfite. Sodium metabisulfite is prepared by reacting aqueous sodium hydroxide, sodium carbonate, or sodium sulfite solution with sulfur diox-

Table 4.30 Selected Physical Properties of SO_2 and Sulfites

Physical properties	Sulfur dioxide	Sodium sulfite	Sodium bisulfite	Sodium metabisulfite
Molecular weight	64.07	126.06	104.07	190.13
Appearance	Colorless gas	Colorless crystals	White powder	White crystals
Odor	Strong suffocating		SO_2 odor	SO_2 odor
Density	1.5 (liquid)		1.48	
Solubility (25°C)				
Water	8.5%	Freely soluble	Freely soluble	Freely soluble
Ethanol	25%	Insoluble	Soluble	Slightly soluble

Source: Ref. 2.

ide. Sodium metabisulfite is a white granular solid that is freely soluble in water and glycerol and slightly soluble in ethyl alcohol. It is stable when kept protected from air and moisture.

Food Applications. Sulfites are effective in preventing enzymatic browning and preserve freshness in raw packaged or unpackaged fruits and vegetables. They are also effective in dehydrated fruits and vegetables, soups, fruit juices, and beer. Sulfites react with molecular oxygen, forming the corresponding sulfates. SO_2 and sulfites inhibit numerous enzymes including polyphenol oxidase, lipoxygenase, and ascorbic oxidase. The mechanism of action of sulfites in preventing browning reactions is not known but very likely involves several different types of actions. Sulfites may directly inhibit the enzyme, interact with intermediates of the reaction and prevent their participation in reactions leading to the formation of the brown pigments, or act as reducing agents promoting the formation of phenols from quinones (151). They are also effective in inhibiting non-enzymatic browning by reacting with carbonyl intermediates, thereby preventing their participation in reactions leading to the formation of brown pigments (152).

Sulfites are used in prepeeled potatoes, sliced potatoes, cut apples, and other fruits supplied to the baking industry, fresh mushrooms, and table grapes. Sulfites also prevent ascorbic acid oxidation by ascorbic acid oxidase and other enzymes. They are effective in preventing off-flavor formation in dehydrated peas and other vegetables mainly by inhibiting the enzyme lipoxygenase. Sulfites prevent the oxidation of essential oils and carotenoids. In beer, SO_2 inhibits oxidative changes that are considered undesirable to flavor development (151).

The use of sulfites in fresh meats, meat products, and fish is generally prohibited in the United States on the grounds that they could restore a bright color and appearance of freshness to faded meat. In some European countries, they are permitted in sausages, sausage meats, shrimp, and dried fish. In the United States, they are allowed in shrimp to prevent black spot formation (153).

Ascorbic Acid

L-Ascorbic acid (vitamin C) (3-keto-*L*-glucofuranolactone) (Fig. 4.16) and sodium ascorbate are used as oxygen scavengers and synergists in a wide variety of food products. Fruits and vegetables are valuable sources of ascorbic acid (Table 4.31). The physical properties of ascorbic acid and sodium ascorbate are presented in Table 4.32.

Preparation. The methods of preparation of ascorbic acid have been described by Jaffe (154). Commercially, ascorbic acid is prepared from D-glucose by the method of Reichstein and Grussner (155) with some modifications. Figure 4.17 presents some of the major reactions in the synthesis of ascorbic acid. This process

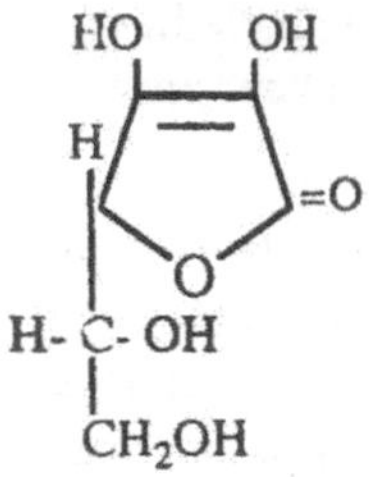

Fig. 4.16 Ascorbic acid.

includes inversion of the glucose chain so that the C-1 of D-glucose becomes C-6 of L-ascorbic acid. D-Glucose is quantitatively hydrogeneated to D-sorbitol at elevated temperature and pressure in the presence of nickel catalyst, and the yield is >97%. Sterile D-sorbitol solution is oxidized on fermentation with *Acetobacter suboxydans* to L-sorbose (>90% yield). The L- sorbose is isolated by crystallization. The process can be carried out on a large scale in either batch or continuous modes.

L-Sorbose, on treatment with acetone and sulfuric acid as a catalyst, is converted to a mixture of 2,3-*O*- isopropylidene-α-sorbose and 2,3:4,6- isopropylidene-α-L-sorbofuranose. The solution is neutralized, and the acetone is removed by distillation. The products are extracted with toluene. Catalysts such as ferric chloride or ferric bromide can be used in place of sulfuric acid.

2,3-4,6-Isopropylidene or L-sorbofuranose can be oxidized at elevated temperatures in dilute sodium hydroxide to 2,3:4,6-bis-*O*-isopropylidene-2-oxo-L-gulonic acid. Sodium hypochlorite is the oxidant, and >90% yield has been obtained with catalytic amounts of nickel chloride. Oxidation can also be carried out by the direct electrochemical method in alkaline solution or by catalytic oxidation with oxygen in the presence of platinum or palladium.

Several methods are followed for the direct cyclization of 2,3:4,6-bis-*O*-isopropylidene-2-oxo-L-gulonic acid. One of the methods involves treatment of the acid with hydrochloric acid gas in a water-free chloroform–ethanol mixture. At the end of the reaction, crude L-ascorbic acid (>80% yield) is filtered and purified by recrystallization from dilute ethanol.

Food Applications. Ascorbic acid is used as an antioxidant in many food products, including processed fruits, vegetables, meat, fish, dairy products, soft drinks, and beverages. Table 4.33 presents the levels used in practice. The food applications of ascorbic acid and sodium ascorbate have been reviewed by a number of investigators (94,156,157). The function of ascorbic acid as an antioxidant in food systems is to scavenge oxygen and thereby prevent oxidation of oxygen-sensitive food constituents, to shift the redox potential of the system to a reducing range, to regenerate phenolic or fat-soluble antioxidants, to maintain sulfhydryl groups in −SH form, to act synergistically with chelating agents, and/or

Table 4.31 Content of L-Ascorbic Acid in Representative Foods

Food substance	L-Ascorbic acid (mg/100 g)
Vegetables	
Asparagus	15– 30
Brussel sprouts, broccoli	90–150
Cabbage	30– 60
Cauliflower	60– 80
Carrot	9
Kale	120–180
Leek	15– 30
Onion	10– 30
Peas, beans	10– 30
Parsely	170
Peppers	125–200
Potatoes	10– 30
Spinach	50– 90
Tomatoes	20– 33
Fruits	
Apples	10– 30
Bananas	10
Grapefruit	40
Guava	300
Hawthrone berries	160–180
Oranges, lemons	50
Peaches	7– 14
Pineapples	17
Rose hips	1000
Strawberries	40– 90
Meat, fish, and milk	
Meat, beef, pork, fish	≤ 2
Liver, kidney	10– 40
Cow's milk	1– 2

Source: Ref. 154.

to reduce undesirable oxidation products (157). In such systems ascorbic acid is oxidized to dehydroascorbic acid. For the initiation of oxygen scavenging, the C-2 position must be unsubstituted to form the free radical.

Fats and oils. Ascorbic acid functions synergistically with phenolic antioxidants such as BHA, PG, and the tocopherols in retarding oxidation in fats and oils.

Table 4.32 Selected Physical Properties of Ascorbic Acid and Sodium Ascorbate

Physical properties	Ascorbic acid	Sodium ascorbate
Molecular weight	176.13	198.11
Physical appearance	White crystalline powder	White to slightly yellowish crystalline powder
Melting point (°C)	190–192	Approx. 200
Taste	Acidic	Soapy
Solubility (% at 25°C)		
Water	33	89
Ethanol	2	Insoluble
Vegetable oil	Insoluble	Insoluble

Source: Ref. 94.

Addition of ascorbic acid at 0.025–0.1% to lard containing α-tocopherol at 75°C significantly enhances the induction period (158). Combinations of BHA and ascorbic acid are effective in inhibiting hematin-catalyzed oxidation of lard emulsion at 37°C. Such systems are used for the evaluation of antioxidants for use in meat and meat products (159). In aqueous fat systems, particularly in the absence of phenolic antioxidants and in the presence of copper, ascorbic acid may act as a prooxidant due to the formation of a copper–ascorbic acid complex. The prooxidant effects have been reported in several cases, including lard, butter, and frozen cream (160–162). The addition of EDTA or polyphosphates, however, eliminates the prooxidant effects (163,164).

In vegetable oils containing natural tocopherols, ascorbic acid functions as a synergist (165). A combination of α-tocopherol, ascorbyl palmitate, lecithin, and ascorbic acid was found to be effective in vegetable oils such as soybean, sunflower, and peanut oils (103). Popovici et al. (166) reported that a combination of histidine and ascorbic acid is highly effective in corn oil. Ascorbic acid in the presence of other antioxidants such as BHA and PG is effective in retarding the development of rancidity in hydrogenated almond oil, hydrogenated whale oil, cacao butter, and margarine (167).

Fruits and vegetables. Ascorbic acid is highly effective in preventing browning in processed fruits and vegetables. The browning reactions are mainly due to the oxidation of phenolic compounds by polyphenol oxidases resulting in the formation of orthoquinones. On polymerization with other phenolic compounds, the orthoquinones form irreversible brown or purple compounds. Ascorbic acid prevents browning by removing the oxygen, reducing the orthoquinones to the orthophenolic forms, and also possibly by inhibiting the polyphenol oxidases.

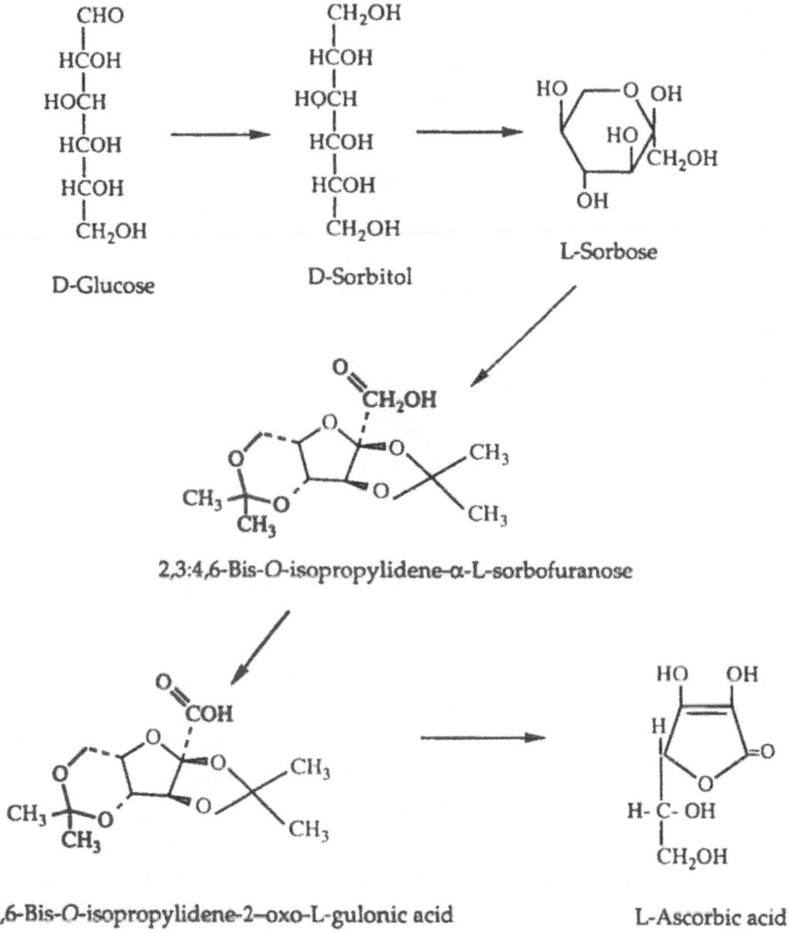

Fig. 4.17 Synthesis of L-ascorbic acid.

Ascorbic acid is effective when added to frozen and canned fruits and fruit juices. Most varieties of frozen peaches, apricots, pears, plums, nectarines, bananas, and apples, which are low in ascorbic acid content, readily discolor and develop off-flavor when the frozen cut or pitted fruit is thawed. This enzymatic oxidation can be prevented by the addition of ascorbic acid. Fruits with intermediate levels of ascorbic acid such as strawberries, cherries, and pineapple also benefit by the addition of ascorbic acid. Ascorbic acid is effective in packed-in-syrup canned apples, pears, figs, and grapes. Ascorbic acid is used during processing and storage of various fruit juices. In the preparation of opalescent juice concentrates, addition

Table 4.33 Levels of Ascorbic Acid Used
in Practice in Food Products

Product	Level (%)
Fruit juices	0.005 – 0.02
Soft drink	0.005 – 0.03
Citrus oils	0.01
Wine	0.005 – 0.015
Beer	0.002 – 0.006
Frozen fruits	0.03 – 0.045
Canned fruits	0.025 – 0.04
Canned vegetables	0.1
Fresh meat	0.02 – 0.05
Cured meat	0.02 – 0.05
Milk powder	0.02 – 0.2

Source: Ref. 5.

of ascorbic acid to apple, pear, grape, and tomato pulp improves their color and taste (157).

Ascorbic acid is used to prevent discolorations in various processed vegetables such as peeled potatoes, sauerkraut in brine, artichoke hearts, canned mushrooms, carrots, beets, and cauliflower. In canned mushrooms, the addition of a chelator such as EDTA or citric acid with ascorbic acid was found to be beneficial. A combination of propyl gallate and ascorbic acid is effective in preserving the pungency of horseradish powder (157).

Meat products. Ascorbic acid is effective in preventing discoloration of fresh and cured meat products and rancidity in cooked meat products. Addition of ascorbic acid maintains the desirable red color of fresh meat and ground meat held under refrigeration. The surface treatment of fresh meat either by spraying or dipping in ascorbic acid solution delays surface discoloration. Addition of about 200 mg/kg ascorbic acid to hamburgers or uncured sausages delays the formation of metmyoglobin by 1 or 2 days. However, the use of ascorbic acid in fresh meat is prohibited in many countries (157).

In addition to preserving color, ascorbic acid also retards lipid oxidation in fresh meat. In frozen meat, sodium ascorbate preserves the fresh red color for up to 6 weeks (157). A combination of ascorbic acid and polyphosphates retards the development of rancidity in cooked pork (168). In frozen cured cooked pork, a combination of sodium polyphosphate and ascorbic acid retards the development of salt-catalyzed lipid oxidation for more than 1 year of storage (169). A combination of ascorbic acid, BHA, and sodium tripolyphosphate has been reported to be effective in protecting restructured pork chops during frozen storage (55). Ascorbic acid in combination with BHA or PG is effective in retarding lipid and pigment oxidation in raw ground beef for up to 8 days of refrigerator storage in oxygen-permeable membrane (15).

In cured meats, ascorbic acid catalyzes the formation of nitrosomyoglobin; increases the yield of nitric oxide, thus reducing the level of residual nitrite; and protects the cured meat pigment nitrosomyochrome from oxidation. The effectiveness of ascorbic acid is improved by the addition of chelators like EDTA or polyphosphates. Ascorbic acid treatment has been used in various cured meat products such as sliced bacon, corned beef, ham, bologna, and wieners. Ascorbates also prevent the formation of nitrosamines in cured meat products (156,157,170). Ascorbic acid or erythorbic acid or their sodium salts are required to be added to cured bacon in the United States (170).

Ascorbates are effective in various fish products as inhibitors of oxidative rancidity when used alone or in combination with citric acid or nicotinamide. Ascorbic acid delays the development of rancidity and the rusting or yellowing of filleted freshwater and marine fish. Ascorbic acid treatment increases the frozen shelf life of freshwater lake herring fillets from about 3 months to at least 7 months. Other applications include frozen anchovies, oysters, sturgeon, and scallops. Ascorbic acid is also effective in fresh, iced, salted, and canned fish (157).

Dairy products. Ascorbic acid delays the development of oxidized flavor in whole milk. Addition of a combination of 200–250 mg of ascorbic acid and 100 mg of sodium citrate per liter of milk before spray drying protects the lipids, flavor, and vitamins A and D even after prolonged storage. Ascorbic acid improves the flavor stability of nonaerated sweetened cream in storage for up to 3 months. Ascorbic acid is highly effective as an antioxidant and antihydrolytic agent in butter. It also prevents the development of a reddish coloration that sometimes appears in ripening cheese stored on wooden boards (157).

Carbonated beverages, beer, and wine. Ascorbic acid is employed as an oxygen scavenger in bottled and canned carbonated beverages to prevent the deterioration of flavor and color of the beverages. In bottled beverages it protects the carotenoids from fading on exposure to sunlight. Ascorbic acid prevents the oxidation haze, changes in flavor, development of wildness, and darkening or discoloring in beer. It also improves the taste, flavor, and clarity of wine and helps to stabilize the oxidation–reduction potential (157).

Ascorbyl Palmitate

Ascorbyl palmitate (6-*O*-palmitoyl-L-ascorbate) (Fig. 4.18) is highly effective as an oxygen scavenger and a synergist. It is a white or yellowish-white crystalline powder with a slight citrus-like odor. Ascorbyl palmitate is more soluble in oils and solvents than ascorbic acid. However, its solubility is still very low compared with that of other antioxidants, and it is usually necessary to use ascorbyl palmitate

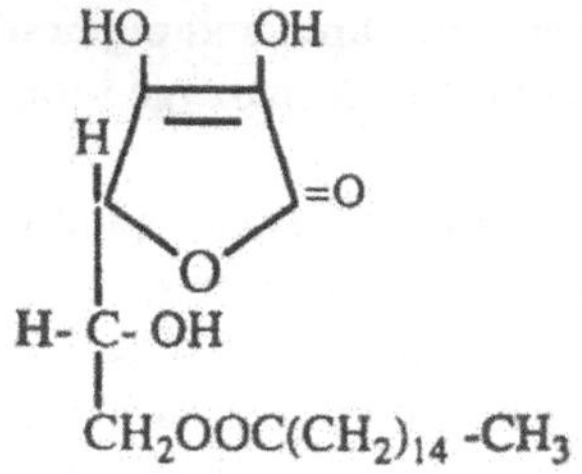

HO   OH

H
$\begin{array}{c} \\ \\ \end{array}$ =O

H- C- OH

CH$_2$OOC(CH$_2$)$_{14}$ -CH$_3$

Fig. 4.18   Ascorbyl palmitate.

in combination with a solubilizing agent such as a monoglyceride (171). Table 4.34 presents the physical properties of ascorbyl palmitate.

Preparation. Ascorbyl palmitate is prepared by reacting equimolar quantities of ascorbic acid and palmitic acid or its methyl ester in the presence of an excess of 95–100% sulfuric acid at room temperature for 16–24 h (172,173).

Food Applications. Ascorbyl palmitate functions synergistically with other antioxidants, especially tocopherols. Table 4.35 presents the levels of ascorbyl palmitate used in food products. Commercial antioxidant mixtures containing ascorbyl palmitate, *dl*-α-tocopherol, and lecithin are available. In such mixtures lecithin functions as a solvent for ascorbyl palmitate and also as a synergist. Ascorbyl palmitate can be weighed directly into oils, dissolved in ethanol and added to the oils, or dissolved in a special oil such as decaglycerol octaoleate. With the last method, solubility of 0.05% in oils has been achieved. Increasing antioxidant activity has been demonstrated from 0.003–5.0% levels (102).

Ascorbyl palmitate is highly effective in synergistic mixtures with tocopherols,

Table 4.34   Selected Physical Properties of Ascorbyl Palmitate

Molecular weight	414.55
Physical appearance	White or yellowish-white powder
Melting point (°C)	Approx. 113
Taste	Soapy
Solubility (25°C)	
Water	0.0002%
Vegetable oils	0.03–0.12%
Ethyl alcohol	12.5%
Propylene glycol	0.48%
Glycerol	0.01%

Source: Ref. 102.

Table 4.35 Levels of Ascorbyl Palmitate
Used in Practice in Food Products

Product	Level (%)
Animal fats	0.01 −0.02
Vegetable oils	0.01 −0.02
Butter	0.001−0.02
Whole milk powder	0.01 −0.05

Source: Ref. 5.

lecithin, citric acid, and octyl gallate in protecting animal fats, vegetable oils, vitamin A, carotenoids, essential oils, nuts, and confectionery products. The various applications of ascorbyl palmitate in synergistic mixtures have been described under the food applications of tocopherols.

Fats and oils. In animal fats, combinations of ascorbyl palmitate, tocopherols, and lecithin are effective in retarding oxidation reactions. In chicken fat, beef fat, and pork fat, combinations of *dl*-α-tocopherol or *dl*-γ-tocopherol and ascorbyl palmitate are quite effective (102). Studies by Wom Park and Addis (107) have shown that incorporation of 500 ppm of ascorbyl palmitate and 100 ppm of *dl*-α-tocopherol in beef tallow decreases the formation of oxidized cholesterol derivatives during heat treatment.

Ascorbyl palmitate is highly effective either alone or in combination with tocopherols and lecithin in vegetable oils. Ascorbyl palmitate enhances the shelf life of most of the vegetable oils at 0.01%. It is more effective at 0.01% than BHA and BHT at 0.02% in soybean oil. Table 4.36 presents the comparative antioxidant activities of ascorbyl palmitate, BHA, BHT, and PG in safflower, sunflower, peanut, and corn oils. Here again, ascorbyl palmitate is more effective than BHA and BHT at 0.01%. Ascorbyl palmitate in combination with PG and TDPA gave extended protection to all four oils (102). It is effective in stabilizing olive oil for up to 6 months at 0.02% level (167).

Ascorbyl palmitate is also highly effective in protecting the deep-frying fats and oils and the fried product. Studies by Gwo et al. (174) showed that addition of 0.02% ascorbyl palmitate reduces the color development of frying fat (animal fat/vegetable oil shortening) and partially hydrogenated soybean oil in simulation studies. It also reduces peroxide values and the development of conjugated diene hydroperoxides and their subsequent degradation to volatile compounds such as decanal and 2,4-decadienal. Similar effects have been reported in a commercial french fry fat and peanut oil used for frying chicken parts. The recovery of ascorbyl palmitate from the fats over a 10-day frying period was 96% for partially hydro-genated soybean oil and 90–96% for an animal fat or vegetable oil shortening. This

Table 4.36 Antioxidant Activity of Ascorbyl Palmitate in
Vegetable Oils at 45°C—Days to Reach a Peroxide Value of 70
meq/kg

Antioxidant[a]	Substrate oil			
	Safflower	Sunflower	Peanut	Corn
None	7	6	15	12
AP (0.01%)	11	10	26	21
AP+PG+TDPA[a] (0.01% each)	25	22	46	31
BHA (0.02%)	8	8	15	15
BHT (0.02%)	10	9	15	13
PG (0.02%)	16	19	26	21

[a]AP = ascorbyl palmitate, TDPA = thiodipropionic acid.
Source: Ref. 102.

indicates that ascorbyl palmitate is potentially still active, even at these conditions
of high heat, light, and long frying times. Ascorbyl palmitate also protects the fried
product. In a comparative study on the efficacy of several antioxidants in potato
chips made with cottonseed oil treated with the different antioxidants, ascorbyl
palmitate was better than BHT (Table 4.37) (102).

Dairy products. In foam-dried whole milk stored in air packs, ascorbyl palmitate
at 0.5% showed a stabilizing effect for 6 months. Ascorbyl palmitate extended the

Table 4.37 Stability of Potato Chips Cooked
in Cottonseed Oil with Various Antioxidants at
45°C

Antioxidant[a] (0.02%)	Days to reach rancidity
None	10
BHA	10
AP	15
BHT	14
PG	24
DLTDP	14

[a]AP = ascorbyl palmitate; DLTDP = dilauryl
thiodipropionate.
Source: Ref. 102.

keeping quality of spray-dried whole milk by 50–70% at 20, 37, and 47°C. Citric acid at 0.01% enhanced the action of ascorbyl palmitate. However, ascorbyl palmitate was inferior to the gallates in these applications (26, 175).

Erythorbic Acid

Erythorbic acid (3-keto-D-glucofuranolactone) (Fig. 4.19) and its sodium salt are strong reducing agents and function by reducing molecular oxygen. Erythorbic acid is the D isomer of ascorbic acid. It is not biologically active and does not occur naturally in food products. Its physical properties are presented in Table 4.38.

Preparation. Erythorbic acid is prepared by reacting 2-keto-D-gluconate with sodium methoxide. In this process, the gluconate is obtained by oxidizing potassium diacetone-3- ketogluconate. The ester is dissolved in methanol, metallic sodium is added, and the mixture is warmed to precipitate the sodium salt of erythorbic acid. The free acid is obtained by treating the sodium salt with H_2SO_4 in the presence of methanol or acetone and concentrating the filtrate in vacuo. Erythorbic acid is purified by crystallization from dioxane (176).

Food Applications. Many investigators have used erythorbic acid interchangeably with ascorbic acid in food products such as juices, canned fruits, and meat products. A combination of erythorbic acid and citric acid has been suggested as an alternative to sulfites. This combination is used to retard oxidative rancidity and discoloration in frozen seafood, salad vegetables, cole slaw, and apples. Erythorbic acid and sodium erythorbate retard oxidative deterioration in frozen fruits at levels of 150–200 ppm. USDA regulations permit the use of erythorbates only in conjunction with curing agents in cured pork and beef cuts, cured comminuted meat products, cured poultry, and cured comminuted poultry products (177). Table 4.39 presents the levels of erythorbic acid used in food products.

Erythorbic acid is generally used interchangeably with ascorbic acid on the

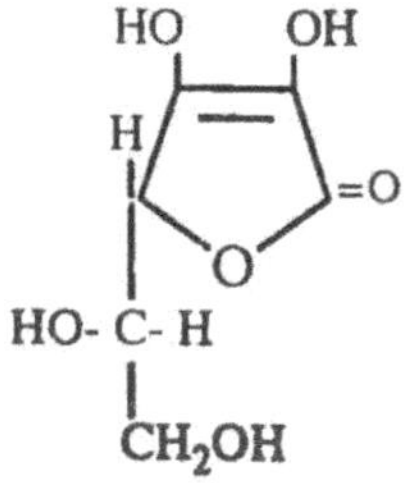

Fig. 4.19   Erythorbic acid.

Table 4.38 Selected Physical Properties of Erythorbic Acid and Sodium Erythorbate

Physical properties	Erythorbic acid	Sodium erythorbate
Molecular weight	176.13	216.12
Melting point (°C)	164–169	~200
Taste	Acidic	Soapy
Solubility in water at 25°C (%)	40	16

Source: Ref. 5.

assumption that the two isomers have similar antioxidant properties. However, in a critical review on the comparative properties of ascorbic acid and erythorbic acid, Borenstein (178) concluded that ascorbic acid is a superior antioxidant in food applications, particularly in processes involving heat. Erythorbic acid oxidizes more rapidly than ascorbic acid in food products and in model systems (179,180). Studies by Hope (181) showed that in canned apple halves, ascorbic acid consumed 96% of the available oxygen whereas erythorbic acid consumed only 75% of the available oxygen. Sixty-six percent of the ascorbic acid was oxidized in processing, versus 90% of the erythorbates. In another study involving enzymatic browning of frozen peaches, Reyes and Luh (182) observed that ascorbic acid was more stable than erythorbic acid. The inference from this work is that erythorbic acid would be consumed 30–50% faster than ascorbic acid in peaches, and hence browning would commence earlier in the erythorbate treatments. The conclusions drawn from these studies can be extended to other products, especially the color stability of cured meat, which depends on the residual reducing agents. The difference in the oxidative stability of ascorbic acid versus erythorbic acid should favor a better color shelf life for cured meats processed with ascorbic acid or sodium ascorbate (178).

Table 4.39 Levels of Erythorbic Acid Used in Practice in Food Products

Product	Level (%)
Fruit juices	0.005–0.02
Beer	0.002–0.006
Frozen fruits	0.03 –0.045
Canned fruits	0.025–0.04
Cured meat	0.02 –0.05

Source: Ref. 5.

4.3.2 Chelating Agents

Phosphoric Acid and Phosphates

Phosphoric acid and its derivatives are widely used as chelators, antimicrobials, emulsion stabilizers, and anticaking agents. Phosphoric acid also functions as a synergist with primary antioxidants in vegetable oils. Phosphoric acid occurs widely in nature and plays an important part in metabolism. The phosphates used as chelators include sodium and potassium orthophosphates (Na_2HPO_4, NaH_2PO_4, K_2HPO_4, KH_2PO_4), pyrophosphates ($Na_4P_2O_7$, $Na_2P_2H_2O_7$), sodium tripolyphosphate ($Na_3P_5O_{10}$), and sodium polyphosphates ($NaPO_3)_{n=4-15}$, of which the pyrophosphates and tripolyphosphate are the most effective. Phosphates form very stable soluble complexes with the transition metal ions, such as copper, iron, and nickel. Table 4.40 presents data on the physical properties of phosphoric acid.

Preparation. The methods of preparation of phosphoric acid and phosphates have been described by Hudson and Dolan (183). Phosphoric acid is produced commercially by either the wet process or the electric furnace process. The wet process acid, produced directly from phosphatic ores, is characterized by relatively high production volume, low cost, and low purity. It is used worldwide for fertilizer manufacture. Furnace and thermal acid, manufactured from elemental phosphorus, is more expensive and considerably more pure than the wet process acid. It is produced in much smaller quantities and is used almost exclusively for applications requiring high purity.

Wet process. In this method, phosphoric acid is manufactured by digestion of phosphate rock (apatite forms) with sulfuric acid. The acid is separated from the resultant calcium sulfate slurry by filtration. Chemical precipitation and solvent extraction are the main methods for purifying wet process acid, although crystallization and ion exchange have also been used.

Furnace process. In the manufacture of phosphoric acid from elemental phosphorus, the phosphate rock is reduced in an electric or blast furnace to elementary phosphorus, which is burned in excess air to form phosphorus pentoxide. The phosphorus pentoxide is hydrated to form phosphoric acid.

Table 4.40 Selected Physical Properties of Phosphoric Acid

Molecular weight	98.0
Physical appearance	Clear syrupy liquid
Melting point (°C)	42
Solubility at 25°C	Miscible with water, ethyl alcohol

Source: Ref. 2.

The product is purified further with hydrogen sulfite to remove arsenic. Oxidation and hydration are carried out in either one or two steps. Because of its purity, the furnace grade phosphoric acid is often preferred for food applications.

In phosphate salts, any or all three acidic protons of phosphoric acid are replaced by cationic species. Both mono- and disodium phosphates are prepared commercially by neutralization of phosphoric acid with sodium carbonate or hydroxide. Crystals of a specific hydrate are obtained by evaporation of the resultant solution within the temperature range over which the hydrate is stable. The condensed phosphates such as polyphosphoric acids, pyrophosphates, and polyphosphates are derived from mono- and diphosphates by condensation processes involving loss of water.

Food Applications. Phosphates are widely used in food processing. The numerous reports and reviews that have appeared on the subject have been compiled in an exhaustive review on the various food applications of phosphates by Ellinger (184). Phosphoric acid and phosphates function as synergists with primary antioxidants in fats and oils. Phosphates are highly effective as chelating agents in the preservation of fruit and vegetable products, dairy products, meat products, and seafoods.

Fats and oils. Phosphoric acid and phosphates function as synergists with BHA, BHT, PG, and other primary antioxidants in stabilizing fats and oils. In vegetable oils phosphoric acid may acts as a synergist for tocopherols. Phosphoric acid, however, is sparingly soluble in oils and fats and may produce discoloration at high temperatures. Pyrophosphates, tripolyphosphates, and hexametaphosphates have synergistic effects with primary antioxidants whereas orthophosphates are ineffective. Table 4.41 presents the synergistic effects of phosphates with various primary antioxidants in aqueous lard system. A combination of glycine and phosphoric acid or phosphates has been reported to improve the stability of dry fat systems against oxidative rancidity (5,184).

Dairy products. A combination of sodium polyphosphate with ascorbic acid and/or α-tocopherol is effective in stabilizing the flavor of sterilized cream (186). Phosphates are also effective in stabilizing butter against oxidation (187). Sodium hexametaphosphate is reported to be effective in inhibiting Maillard-type discoloration products in sweetened condensed milk during an 8-month shelf life test (188).

Fruit and vegetable products. Phosphates are effective in stabilizing fruit and vegetable products against discoloration and development of off-flavors. Deobold et al. (189) reported that combinations of sodium acid pyrophosphate or citric acid with a mixture of BHA and BHT retard the development of off-flavors in sweet potato flakes. Trisodium pyrophosphate is reported to extend the shelf life of dried vegetables when they are dipped in the solution of phosphate before being blanched

Table 4.41 Synergistic Effects of Phosphates with Primary Antioxidants in Aqueous Lard System

Phosphates[a]	Control	Days to turn rancid at 45°C[b]		
		Tocopherol (0.005%)	BHA (0.005%)	PG (0.005%)
Buffer	2	4	22	11
DSP	2	7	43	7
STP	3	12	48	18
SHMP	7	—	57	28
IMP	8	—	64	28

[a]Concentration, 0.1% in pH 7.5 borate buffer. DSP = disodium orthophosphate; STP = sodium tripolyphosphate; SHMP = sodium hexametaphosphate; IMP = insoluble metaphosphate or Maddrell's salt.
[b]Time required for half-bleaching of carotene.
Source: Ref. 185.

and dried (190). Sodium hexametaphosphate and metaphosphoric acid have been reported to prevent the oxidation of ascorbic acid in model systems and in sugared and nonsugared citrus juices (191,192). Disodium phosphate in combination with sodium sulfate and sodium bisulfite protects the flavor and texture of frozen mushrooms (193).

Polyphosphates, from pyrophosphates through the long-chain potassium and sodium metaphosphates, are effective in preserving the natural red color of tomatoes, ketchup, all types of red berries and their preserves, and juice products (194). Sodium acid pyrophosphate is highly effective in preventing the after-cooking darkening of potatoes. Sodium acid pyrophosphate or a combination of sodium acid pyrophosphate and trisodium pyrophosphate inhibits the discoloration of sweet potato flakes and of canned and frozen sweet potatoes (195,196). Phosphates are also effective in preventing browning in apple slices and preserve their texture and flavor. A commercial process has been developed for efficient treatment of apple slices with phosphates (197,198).

Meat products. Phosphates are widely used in the preservation of meat products. Phosphates are effective in protecting the red color of various fresh meats and cured meat products and in preventing the development of off-flavor. The subject has been reviewed by Ellinger (184). Addition of phosphates retards the oxidation of oxymyoglobin to the brown metmyoglobin mainly by maintaining a pH of about 6.5–6.8 (199,200). In fresh whole hams, infusion of a combination of sodium hexametaphosphate and lactic acid significantly improved the pH and the color (201). Pyrophosphate, tripolyphosphate, and hexametaphosphate in combination with ascorbic acid are effective in protecting cooked pork against the development of oxidative rancidity (168). The combination is also effective in protecting refrigerated and frozen roast beef slices against the development of rancid odors and flavors (202). In ground meat products, the red color is preserved for 9–12 days by a combination of monosodium phosphate, ascorbic acid, and sodium propionate (203). Phosphates are also effective in preserving the color and flavor of various fresh, cooked, and refrigerated poultry and fish products (184).

In cured meat products, the pH of the system seems to be highly important in determining the effects of phosphates on the color of cured meat. According to Hall (199), pH values of 6.5–7.0 are necessary to protect the discoloration of frankfurters, hams, and other cured products when polyphosphates having chain lengths of 3 or longer are used. Brissey (200) maintained the pH of curing salt pickles with disodium phosphate at 6.6–6.8. Combinations of disodium phosphate and sodium hexametaphosphate or of disodium phosphate and ascorbic acid increased the color shelf life of bologna up to 70% if it was held at refrigerated temperatures when exposed to fluorescent lights (204). In canned baked hams, the addition of either tripolyphosphate or hexametaphosphate significantly increases the color retention and prevents fat oxidation (205).

Ethylenediaminetetraacetic Acid

Ethylenediaminetetraacetic acid (EDTA) (Fig. 4.20) and its disodium salt (Na_2EDTA) and calcium disodium salt ($CaNa_2EDTA$) are widely used as chelating agents in food products. They form stable complexes with many di- or polyvalent metal ions. EDTA is a white crystalline powder moderately soluble in water and insoluble in organic solvents. The disodium and calcium disodium salts are more soluble in water. Table 4.42 presents the physical properties of EDTA, $CaNa_2EDTA$, and Na_2EDTA.

Preparation. The various methods of preparation were compiled by Hart (206). The original commercial synthesis of EDTA was from ethylenediamine, chloracetic acid, and caustic soda to form the tetrasodium salt. However, the material produced in this way was generally contaminated with sodium chloride and had to be purified for commercial purposes. This process is no longer used commercially. Today, the two principal manufacturing processes for EDTA and

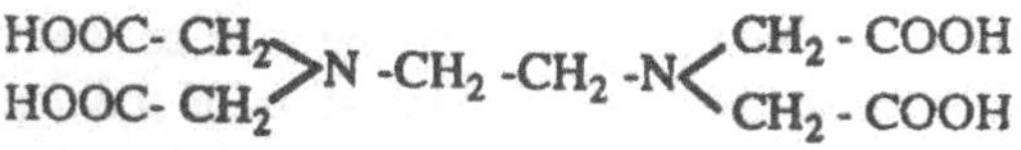

$$\text{HOOC- }CH_2\diagdown N\text{ -}CH_2\text{ -}CH_2\text{ -N}\diagup^{CH_2\text{ - COOH}}_{CH_2\text{ - COOH}}$$

Fig. 4.20   Ethylenediaminetetraacetic acid.

Table 4.42 Selected Physical Properties of EDTA and Its Salts

Physical properties	EDTA	Na$_2$ EDTA	CaNa$_2$ EDTA
Molecular weight	292.24	336.21	374.28
Physical appearance	White powder	White powder	White powder
Solubility			
Water (25°C)	0.05%	Soluble	Soluble
Organic solvents	Insoluble	Insoluble	Insoluble

Source: Ref. 2.

related chelating agents are both based on cyanomethylation of the parent polyamine.

1. The most widely used method is the alkaline cyanomethylation of ethylenediamine by means of sodium cyanide and formaldehyde:

   $$H_2NCH_2CH_2NH_2 + 4CH_2O + 4NaCN + 4H_2O \rightarrow$$
   $$(NaOOCCH_2)_2NCH_2CH_2N(CH_2COONa)_2 + 4NH_3$$

   This method offers high yields (>90%) of the chelating agent. The principal by-product is ammonia, which is continuously boiled off during the reaction. However, some of the ammonia is cyanomethylated to yield salts of nitrilotriacetic acid, of N- carboxymethylglycine, and of glycine. In addition, glycolic acid salts are formed from the reaction of sodium cyanide and formaldehyde. These impurities are not detrimental to most chelating applications.

2. The second commercial method for producing EDTA is the two- step Singer synthesis. In this process, the cyanomethylation step is separate from the hydrolysis. Hydrogen cyanide and formaldehyde react with ethylenediamine to form insoluble ethylenedinitrilotetraacetonitrile (EDTN), 2,2′,2″,2‴-(1,2-ethanediyldinitrilo)tetrakisacetonitrile in high yield (>96%). The intermediate nitrile is separated, washed, and subsequently hydrolyzed with sodium hydroxide to tetrasodium EDTA, with liberation of the by-product, ammonia. Carrying out the synthesis in two steps eliminates most of the impurities and yields a pure form of the chelating agent.

   $$H_2NCH_2CH_2NH_2 + 4CH_2O + 4HCN \rightarrow (NCCH_2)_2NCH_2CH_2N$$
   $$(CH_2CN)_2 + 4H_2O$$
   $$(NCCH_2)_2NCH_2CH_2N(CH_2CN)_2 + 4NaOH \rightarrow$$
   $$(NaOOCCH_2)_2NCH_2CH_2N(CH_2COONa)_2 + 4NH_3$$

Food Applications. The food applications of EDTA were reviewed by Furia (207,208). EDTA has a high stability constant compared to those of other chelating agents and forms stable metal complexes with a greater variety of metals than other chelating agents. The pH of the food system also plays an important role in the metal-complexing functions of EDTA. EDTA becomes increasingly dissociated as the pH rises, and the quantity of metal complexed also increases. The chelation of calcium by EDTA is best accomplished above pH 8.5, and that of zinc, above pH 6.0. Below pH 3.0, EDTA will not chelate calcium. At pH <2.5, only metals having high ionic charges can form stable complexes such as ferric EDTA. Frequently, hydroxyl ions also compete with chelating agents for metal ions. For example, EDTA will prevent the precipitation of ferric hydroxide by forming the soluble ferric EDTA complex only if the hydroxyl ion concentration is kept low. It will, however, form stable complexes with ferrous ions in highly alkaline media.

Although it is not usually possible to "selectively" chelate a metal ion, the introduction of a second chelating agent can provide some degree of selectivity. For example, in a system containing calcium and iron, EDTA could begin chelating calcium only after first chelating iron. If salicylic acid were now introduced, two complexes would be formed—an iron salicylate complex and a calcium–EDTA complex. The net result is the "selective" chelation of calcium by EDTA in the presence of iron. When small differences exist between the stability constants of two metal complexes of EDTA, both metals will be partially chelated as with magnesium and strontium (207,208).

Fats and oils. EDTA functions synergistically with BHA, BHT, PG, ascorbic acid, and isoascorbic acid in stabilizing fats and oils. EDTA has been successfully used to extract metallic hydrogenation catalysts from fats and oils, improve the oxidative stability of partial glyceride fatty esters prepared by ester interchange, inhibit the copper-catalyzed autoxidation of linoleic acid, stabilize soybean oil, inhibit the rancidity of roasted nuts, inhibit the autoxidation of essential oils, and retain good flavor in oleomargarine, lard, salad dressings, and sandwich spreads. Though EDTA and its salts are poorly soluble in fats and oils, only small quantities are necessary for maximum activity. In a study on the stabilization of lard, Na_2EDTA was found to be more effective than citric acid in conjunction with BHT. Na_2EDTA showed a 43% improvement over citric acid. In emulsion systems, EDTA and its salts are readily incorporated (207,208).

Fruit and vegetable products. Ethylenediaminetetraacetic acid is highly effective in preventing discoloration and the development of off-flavors in various canned and frozen vegetables. The addition of EDTA before blanching and/or retorting prevents undesirable color changes. Na_2EDTA is effective in various products such as canned whole-kernel and cream-style corn, canned sliced beets, and canned legumes (e.g., lima beans, pinto beans, chickpeas, lima beans, red-eyed beans) (208). It prevents the surface darkening of a variety of canned vegetables such as sweet potatoes, yams, cauliflower, eggplant, asparagus, Brussels sprouts, and turnips (209). EDTA is highly effective in preventing discoloration in a variety of potato products. Na_2EDTA increases the shelf life of raw, prepeeled potatoes and prevents after-cooking darkening (210,211). It is also effective in preventing after-cooking darkening, poor texture, and blistering of frozen french fries (209,212). The greening of potato tubers exposed to fluorescent light is reduced by spraying them with an aqueous wax emulsion containing 0.1–5% of CaNa2EDTA (213,214). EDTA functions synergistically with ascorbic acid in preventing browning in canned apple sections (181). It also prevents the pink discoloration in canned pears (215). EDTA is used for the prevention of discoloration of freeze-dried banana slices added to ready-to-eat cereals (216).

Vitamin stabilization. EDTA is effective in preventing metal-catalyzed oxidation of vitamins. It inhibits the copper-catalyzed oxidation of ascorbic acid. EDTA is used in stabilizing ascorbic acid in liquid formulations or concentrated stock solutions and in fruit juices such as tomato, grapefruit, cranberry, and black currant juices. EDTA also stabilizes carotene; vitamins A, D, E, K, and B12; folic acid; and thiamine (208).

Dairy products. EDTA is effective in preventing off-flavors due to copper-catalyzed oxidation of milk. Since the copper concentration in off-flavored milk is seldom higher than 1.0 ppm, about 5.0 ppm EDTA, Na2EDTA, or CaNa2EDTA is usually effective (217–219). EDTA prevents the after-cooking greening of scrambled eggs prepared from acidified and glucose-free whole egg powders (220).

Beverages. EDTA is effective in inhibiting preformed casse and metal-catalyzed oxidative changes in wines. EDTA prevents copper casse in white table wines by lowering the redox potential. The gushing or wildness and chill haze in beer caused by trace metal contaminants is effectively prevented by EDTA. Also, EDTA protects beer stabilizers like ascorbic acid from oxidation in the presence of metal contaminants. In nondistilled vinegar, precipitation of proteins or metallic tannates and cloudiness are inhibited by EDTA (207,208).

In canned carbonated beverages, EDTA promotes flavor retention, reduces discoloration and turbidity on mixing with alcoholic beverages, and reduces the incidence of corrosion. A combination of EDTA and ascorbic acid inhibits fading of FDC dyes on exposure to light. In the case of FDC Yellow No. 6, approximately 50% of the original color strength was retained by the addition of EDTA after a 90-min exposure to direct sunlight (207).

Meat products. EDTA retards nitric oxide hemoglobin formation in cured meats (221). A combination of EDTA with ascorbic acid and nitrous oxide or sodium nitrite significantly accelerates the curing of cooked sausage products (222). Na2EDTA aids in more efficient curing of various meat products including ham, bacon, and frankfurters by magnesium salts by solubilizing magnesium (223). A surface spray of 0.05–0.1% of Na2EDTA in combination with some nitrite inhibits the gray surface discoloration of meat (e.g., pet foods, in slack-filled containers). A combination of ascorbic acid and Na2EDTA stabilizes the color and flavor of ground beef (224).

Ethylenediaminetetraacetic acid is also effective in inhibiting discoloration, off-flavor, and odor in various fish products. The blue-green and grayish discoloration of canned, frozen, iced, and fresh shellfish is prevented by solutions of EDTA and alum. EDTA-alum treatment of canned shrimp prevents discoloration and maintains firmness of the pack (225). EDTA is also effective in preventing the conversion of odorless trimethylamine oxide in fish fillets to trimethylamine. Haddock fillets treated with EDTA solutions showed an extended storage life for 11 days (226). In salmon, 100–500 ppm of Na_2EDTA is effective in retarding odor development due to oxidative rancidity and discoloration (227).

Tartaric Acid

L(+)-Tartaric acid (2,3-dihydrosuccinic acid) (Fig. 4.21) occurs widely in nature in many fruits and vegetables and is classified as a fruit acid. It is a dihydroxydicarboxylic acid with two chiral centers and exists as the dextro- and levorotatory acids, the meso form, and the racemic mixture. The commercial product is the natural dextrorotatory form, L(+)-tartaric acid. L-Tartaric acid occurs in grapes as its acid potassium salt and is obtained as a by-product of wine making. Tartaric acid is an effective chelator and an acidulant. Table 4.43 presents the physical properties of tartaric acid.

Preparation. L(+)-Tartaric acid and its acid potassium salt (cream of tartar) are obtained as by-products of wine making. The acid potassium salt is first converted to calcium tartrate by reacting with calcium hydroxide, which is then hydrolyzed to tartaric acid and calcium sulfate. Crude tartrates are recoverable from various raw materials (228):

1. The press cakes from grape juice (i.e., unfermented "marcs" or partly fermented "pomace") are boiled with water, and alcohol, if present, is distilled. The hot mash is settled and decanted, and the clear liquor is cooled to crystallize. The recovered high-test crude cream of tartar (vinaccia) has an 85–90% cream of tartar content.
2. Lees, which are the dried sediments in the wine fermentation vats, consist

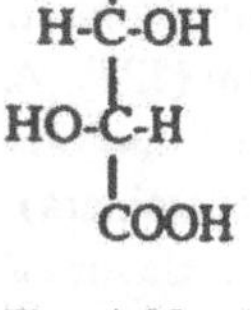

```
      COOH
        |
   H-C-OH
        |
  HO-C-H
        |
      COOH
```

Fig. 4.21   Tartaric acid.

Table 4.43 Selected Physical Properties of Tartaric Acid

Molecular weight	150.09
Physical appearance	White powder
Melting point (°C)	168–170 (anhydrous)
Solubility at 25°C	Highly soluble in water and ethyl alcohol; soluble in glycerol

Source: Ref. 2.

of yeast cells, pectinaceous substances, and tartars. Their content of total tartaric acid equivalent ranges from 16 to 40%.

3. Argols, the crystalline crusts that form in the vats in the secondary fermentation period, contain more than 40% tartaric acid; they are high in potassium hydrogen tartrate and low in the calcium salt.

It is usually advantageous to combine the manufacture of tartaric acid, cream of tartar, and the Rochelle salt in one plant. This permits the most favorable disposition of the mother liquors from the three processes. The following chemical reactions are involved.

1. Formation of calcium tartrate from crude potassium tartrate:

$$2KHC_4H_4O_6 + Ca(OH)_2 \rightarrow 2CaC_4H_4O_6 + K_2SO_4 + 2H_2O$$

2. Formation of tartaric acid from calcium tartrate:

$$CaC_4H_4O_6 + H_2SO_4 \rightarrow H_2C_4H_4O_6 + CaSO_4$$

3. Formation of the Rochelle salt from argols:

$$2KHC_4H_4O_6 + Na_2CO_3 \rightarrow 2KNaC_4H_4O_6 + CO_2 + H_2O$$

4. Formation of cream of tartar from tartaric acid and Rochelle salt liquors:

$$2H_6C_4O_6 + 2KNaC_4H_4O_6 + K_2SO_4 \rightarrow 4KHC_4H_4O_6 + Na_2SO_4$$

Food Applications. Tartaric acid is effective as a metal chelator or a synergist with other antioxidants and is comparable to citric acid in its activity. Hartman (229) reported that tartaric acid is effective in stabilizing commercial edible beef and mutton tallow and lard. Dutton et al. (230) reported that tartaric acid improves the oxidative stability and retards the flavor deterioration of soybean oil. The use of tartaric acid as a stabilizing agent in ground dry spices, milk products, and cheese has been patented (231–233).

Citric Acid

Citric acid (2-hydroxy-1,2,3-propanetricarboxylic acid) (Fig. 4.22) is a natural constituent in plants and animals and an intermediate in the Krebs cycle. It is also

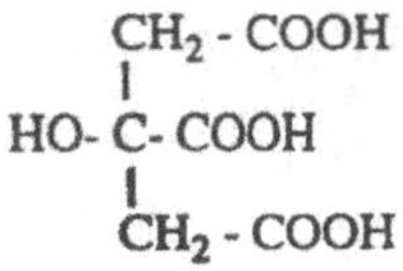

$$\begin{array}{c} CH_2 - COOH \\ | \\ HO- C- COOH \\ | \\ CH_2 - COOH \end{array}$$

Fig. 4.22   Citric acid.

one of the most versatile and most widely used organic acids in foods and pharmaceuticals. Citric acid is widely distributed in plants, microbes, and animals. Table 4.44 presents the citric acid content of some fruits and vegetables. Citric acid is a white, odorless solid highly soluble in water. Table 4.45 lists the physical properties of citric acid.

Preparation. Citric acid is obtained mainly from fermentation processes. Either a surface or submerged fermentation process is used. The various methods of preparation have been described by Bouchard and Merritt (234).

Table 4.44   Citric Acid Content of Some Fruits and Vegetables

Source	Citric acid/citrate (%)
Fruits	
Lemons	4.0 −8.0
Grapefruit	1.2 −2.1
Tangerines	0.9 −1.2
Oranges	0.6 −1.0
Red currants	0.7 −1.3
Black currants	1.5 −3.0
Raspberries	1.0 −3.0
Gooseberies	1.0
Strawberries	0.6 −0.8
Apples	0.008
Vegetables	
Potatoes	0.3 −0.5
Tomatoes	0.25
Asparagus	0.08 −0.2
Turnips	0.05 −1.1
Peas	0.05
Butternut squash	0.007−0.025
Corn kernels	0.02
Lettuce	0.016
Eggplant	0.01

Source: Ref. 234.

Table 4.45 Selected Physical Properties of
Citric Acid

Molecular weight	192.12
Physical appearance	White crystalline powder
Melting point (°C)	153 (Anhydrous)
Solubility at 20°C	
Water	59.2%
Organic solvents	Insoluble

Source: Ref. 2.

Surface process. The microbial production of citric acid on a commercial scale was first begun by Pfizer Inc. in 1923. The process uses certain strains of *Aspergillus niger*, which when grown on the surface of a sucrose and salts solution produces significant amounts of citric acid. Variations of this surface culture technique still account for a substantial portion of the world's production of citric acid.

The surface process involves inoculating sugar solutions along with sources of assimilable nitrogen, phosphate, magnesium, and various trace minerals with spores of *A. niger* in shallow aluminum or stainless steel pans. The mold grows on the solution surface, producing a rubbery, convoluted mycelial mass. Air is circulated over the surface to provide oxygen and to control the temperature of evaporative cooling.

Because of its relatively low cost, molasses is often the preferred source of sugar for this fermentation process. However, since it is a by-product of sugar refining, molasses varies considerably in composition, and not all types are suitable for citric acid production. Beet molasses is generally preferred to cane molasses.

Submerged process. In this process, *A. niger* is grown dispersed throughout a liquid medium. The fermentation vessel usually consists of a sterilizable tank with a capacity of thousands of gallons and equipped with a mechanical agitator and a means of introducing sterile air. The inoculum is transferred aseptically to the production fermentor containing the proper media. Citric acid can also be produced by a submerged process using a certain species of yeast (*Candida guilliermondii*). The process, patented by Miles Laboratories, requires shorter fermentation times than *A. niger* fermentation.

Citric acid is generally recovered from the fermented aqueous solutions by first separating out the microorganisms, generally by rotary filtration or centrifugation, and then precipitating the citrate ion as the insoluble calcium salt. If fermented solutions are of sufficiently high purity, it may be possible to recover crude citric acid by direct crystallization; however, commercial success with this method has not been reported.

The recovery of citric acid via calcium salt precipitation is a highly complex process. The clarified fermentation broth is treated with a calcium hydroxide slurry in two stages. In the first stage, oxalic acid when present is removed as a by-product, and in the second stage the citrate is precipitated. The citrate is formed by adding a lime slurry to obtain a neutral pH. After filtration, the resulting calcium citrate is acidified with sulfuric acid, which converts the salt to calcium sulfate and citric acid. The calcium sulfate is filtered and washed free of citric acid solution. The clear citric acid solution then undergoes a series of crystallization steps to achieve the physical separation of citric acid from the remaining trace impurities. The finished citric acid is dried and sifted in conventional rotary drying equipment. Because anhydrous citric acid is hygroscopic, care must be taken to reach the final moisture specification during drying and to avoid storage under high-temperature, high-humidity conditions. The citric acid yield from fermentation processes ranges from 50 to 70% based on the sugar conversion.

Food Applications. Citric acid is widely used as an acidulant and a chelator in many food products. It is used with both primary and oxygen scavengers at levels of 0.1–0.3%. It is a constituent of many commercial antioxidant formulations. Citric acid chelates metal ions at levels of 0.005–0.2% in fats and oils (177). Table 4.46 presents the efficacy of 0.01% citric acid in stabilizing soybean oil containing various prooxidant metal salts. Unlike EDTA, citric acid and citrates are readily solubilized in fats and oils from propylene glycol solution concentrates.

Fats and oils. In lard, citric acid increases the effectiveness of BHA. BHA increases the oxidative stability of lard from 4 to 16 h, which is further increased to 36 h with the addition of citric acid (42). Citric acid is widely used in oil

Table 4.46 Effects of Citric Acid on the Oxidative Stability of Soybean Oil

	Peroxide value (AOM conditions, 8 h)	
Salt (ppm)	With citric acid	Without citric acid
Control	10.7	46.6
Ferric chloride (3.0)	125	293
Manganese chloride (3.0)	13.1	85.4
Cobalt chloride (3.0)	9.1	239
Chromium chloride (3.0)	17.8	153
Copper chloride (3.0)	291	294
Copper chloride (0.3)	44.4	288

Source: Ref. 230.

refineries, where it is added during deodorization, the final step of the refining process. Small quantities of an aqueous solution of the acid are sucked into the oil, preferably during the cooling-down process (235). Citric acid functions as a synergist with other antioxidants in various vegetable oils. Some linoleate-containing oils such as soybean oils, crambe, mustard oil, and rape oil are effectively stabilized by the addition of 0.01% citric acid (5). A combination of 0.02% TBHQ and 0.01% citric acid increased the oxidative stability of cottonseed oil from 10 to 30 h. The oxidative stability of corn oil was increased from 6 to 26 h by a combination of 0.015% BHA, 0.005% PG, and 0.003% citric acid (65). In olive oil, the addition of citric acid increased the effectiveness of TBHQ. *tert*-Butylhydroquinone increases the oxidative stability of olive oil from 7 to 12 h. The stability was enhanced from 12 to 58 h on addition of 0.01% citric acid with 0.02% TBHQ (65). In crude palm oil, TBHQ and citric acid treatment significantly reduced oxidation and improved bleachability (236). Citric acid is used as a stabilizing agent in the synthesis of rearranged fats for shortenings, in oleomargarine and similar spreads, and in various salad dressings (237).

A combination of 0.02% BHA and 0.01% citric acid is effective in stabilizing yellow grease and feed-grade tallow. Studies on the effect of various antioxidants on the keeping quality of yellow grease have shown that citric acid improves the effectiveness of BHA, BHT, and PG (7,65).

Fruit and vegetable products. Citric acid has found considerable use in combination with antioxidants in processing fresh-frozen fruits. In frozen fruits, citric acid protects ascorbic acid from metal-catalyzed oxidation and also neutralizes the residual lye left from peeling operations, which would tend to destroy ascorbates. Mixtures of citric acid and erythorbic acid are used to retard the browning of bananas. In canned apples, a combination of citric acid and ascorbic acid is effective in preventing browning. The pink discoloration of canned pears is significantly inhibited by the addition of citric acid. Citric acid prevents discoloration of canned sliced beets. Citric acid inhibits after-cooking discoloration in potatoes. However, it is not as effective as EDTA or the pyrophosphates. Citric acid also prevents the discoloration of onions (208,237).

Meat products. Citric acid in combination with tocopherols and ascorbyl palmitate is effective in retarding oxidative rancidity in restructured beef patties (238). A combination of citric acid and BHA increased the oxidative stability of pork sausage from less than 5 h to 12 h (65). Citric acid retards the formation of nitric oxide-hemoglobin in cured meats. In dry-cured hams, a combination of BHA, BHT, and citric acid showed a significant reduction in oxidative rancidity and preserved the flavor (239). Combinations of citric acid with phenolic antioxidants prevent rancidity in frozen creamed turkey (240).

Combinations of citric acid with BHA, tocopherols, ascorbic acid, or propyl gallate are effective in various seafoods. In canned shellfish products, the addition of 0.7% citric acid to a 1% salt and 1–2% sugar brine prevents discoloration (241). Combinations of citric acid with BHA or PG is effective in controlling oxidative flavor changes of frozen minced Atlantic whiting for up to 8 weeks (242). In dried herrings, a combination of ascorbic acid and citric acid retards the oxidation and hydrolytic processes of fatty and nitrogen fractions and retains their appropriate organoleptic properties (243). A mixture of citric acid and ascorbic acid is used as a dip for oily fish to prevent surface tissue from becoming brownish and gummy, a condition known as rusting. A similar dip is used for shrimp. Dilute citric acid solutions have been suggested for preventing black spot formation in shrimp. Citric acid solutions are also used in the canning of crab meat to inhibit discoloration and the development of odors and off-flavors. Similar treatment is suitable for the canning or freezing of lobster meat, scallops, and oysters (237).

Other uses. Citric acid aids in flavor retention of stored beverages. It is used in wine, cider, and nondistilled vinegar to prevent casse. Citric acid is used in the stabilization of various vitamins. It is also incorporated with primary antioxidants in various candies to prevent oxidation of nuts and other ingredients. A combination of BHA, PG, and citric acid inhibits oxidation in candies when added to the butter used for preparing the candies. Citric acid improves the keeping quality of whole milk powder stabilized by ascorbyl palmitate or dodecyl gallate. It is also used as a stabilizer for spices and in wrapping material with other antioxidants (22,175, 208,237).

Citrate Esters

Esters of citric acid, for example, stearyl and isopropyl citrates, are also effective as synergists. The esters are more soluble than citric acid in oils and fats. To disperse the monoesters in oils and fats, hydrophilic coupling agents such as alkyl and alkylene diesters and triesters of citric acid, aliphatic alcohols having at least six carbon atoms, monocarboxylic acids having at least 10 carbon atoms, and monoglycerides of monocarboxylic acids in which the acyl groups have at least 10 carbon atoms are used (244). The esters are used in many commercial formulations.

Monoisopropyl citrate is a viscous, colorless syrup that exhibits some crystallization upon standing. The commercial product contains approximately 27% w/w of monoisopropyl citrate, 9% w/w diisopropyl citrate, 2% w/w triisopropyl citrate, and 62% by weight of a mixture of mono- and diglycerides, which is an oil-miscible semisolid material. Stearyl citrate is a solid. The commercial product, which melts at approximately 58°C, contains 12.5% monostearyl citrate, 75% distearyl citrate, and 12.5% tristearyl citrate. Isopropyl citrate is soluble in water and ethanol and is dispersible with some difficulty in oils. Stearyl citrate is readily soluble in oils. A mixture of mono- and diglycerides of a vegetable oil is used as a vehicle for dissolving isopropyl citrate (5).

Preparation. Monoisopropyl citrate is prepared by refluxing equal parts of citric acid and anhydrous isopropyl alcohol for 118 h at 92°C. The alcohol and most of the water are removed by evaporation in vacuo. The residue is taken up in ethyl ether, and sufficient low-boiling petroleum ether is added to form a precipitate. The precipitate is treated by this procedure several times, and the solvent is finally removed by heating at 130°C with stirring. The product is a mixture of about 90% monoester and 10% diester (245).

Stearyl citrate is prepared by heating an excess of anhydrous citric acid with stearyl alcohol in dry pyridine. The mixture is heated for 4–6 h at about 100°C and then held at 40–50°C for 14–20 h. It is then poured into a water–ice mixture containing sufficient H_2SO_4 to combine with pyridine to form pyridine sulfate. The aqueous mixture is extracted with ethyl ether, and sufficient petroleum ether is added to precipitate the monoester (244).

Food Applications. Studies by Vahlteich et al. (246) showed that to be effective as metal deactivators the esters of citric acid must have two free carboxyl groups. The di- and triesters are completely inactive. The monoisopropyl and stearyl citrates function as chelators and as inhibitors of flavor defects in commercially refined and deodorized edible oils and fats containing traces of metal contaminants. The esters are equivalent in their effectiveness on an equimolar basis. The use of esters alleviates some of the disadvantages associated with the use of citric acid in oils undergoing processing. Citric acid is sparingly soluble in oils. A maximum of only about 0.005% citric acid can be dissolved in oils. Also, during the deodorization process, which involves high temperatures, citric acid gives rise to various degradation products of known and unknown composition. The concentration of free citric acid in such treated oils is therefore variable and unknown.

Vahlteich et al. (246) reported that in a model system consisting of methyl esters of cottonseed salad oil fatty acids, isopropyl citrate is highly effective in retarding oxidation. Addition of tocopherol results in marked stabilization against peroxide formation. The esters are also highly effective at the 0.02% level in commercial corn and soybean salad oils, which contain approximately 0.08% natural tocopherols. The esters are also effective in extending the relative shelf life of limpid soybean oil by about 50% and that of hydrogenated soybean oil by more than 100–200% (Table 4.47). A combination of butylated hydroxyanisole, propyl gallate, and isopropyl citrate is highly effective in stabilizing a vegetable shortening under commercial processing conditions (246). The esters are effective in the stabilization of nut meats and cheese (247,248). They prevent the development of off-flavors in fried potato chips and french-fried potatoes. The esters are effective when used alone in vegetable oils and in combination with tocopherols in animal fats used in the preparation of the fried products (249).

Table 4.47 Performance of Isopropyl and Stearyl Citrates in Stabilizing the Flavor of Soybean Oil Stored Under Air at 35°C

Oil	Additive[a]	Relative shelf life (wk)[b]	Overall flavor performance[c]
Salad oil	None	5	215
	0.02% IC	7	325
	0.15% SC	7.5	425
Margarine oil	None	5	165
	0.02% IC	14+	550+
	0.15% SC	14+	700+
Shortening oil	None	7	315
	0.02% IC	14+	625+
	0.02% SC	14+	600+

[a]IC = Isopropyl citrate; SC = stearyl citrate.
[b]Number of weeks scored above "fair".
[c]Relative shelf life × degree of acceptability.
Source: Ref. 246.

Phytic Acid

Phytic acid (myoinositol hexaphosphoric acid) (Fig. 4.23) occurs widely in nature as a mixture of calcium, magnesium, and potassium salts. Phytic acid constitutes 1–5% by weight of most plant seeds like cereals and oilseeds and typically accounts for 60–90% of the total phosphorus. Some of the important physiological functions of phytic acid include storage of phosphorus, high-energy phosphate groups, and cations, and protection of seeds against oxidative damage during storage (250). Phytic acid is a strong chelating agent. At low pH it precipitates iron quantitatively. At higher pH it forms insoluble complexes with all other polyvalent cations. Phytic

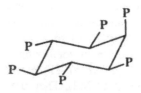

P : H_2PO_4

Fig. 4.23 Phytic acid.

acid is used widely as a food additive in Japan. It is a syrupy straw-colored liquid that decomposes on heating. Table 4.48 presents the physical properties of phytic acid.

Preparation. Phytic acid is prepared by extraction from various plant sources with dilute acid, followed by precipitation using alkalis, bicarbonates, or alcohol and further purification by chromatographic or chemical methods. The various methods have been described by Graf (250).

Phytic acid can be extracted from cereals, bran, glutens, plant embryos, and seeds using aqueous H_2SO_4. It is then precipitated with ammonia gas and purified by ion-exchange chromatography on Amberlite IR-120 and Amberlite IR-45. Similarly, phytic acid can be produced from aqueous HCl extracts of rice bran and other wastes of the food industry, precipitated with $NaHCO_3$, and purified by extraction with ether from a 0.1 N HCl solution. In another procedure, the HCl extract of rice bran is decolorized and then treated with $Ca(OH)_2$ and $Mg(OH)_2$, and its pH is raised from 3.4 to 7.0 with NaOH to give a phytin-like salt of phytic acid in high yield; 23 g of phytin is also obtained from 250 g of defatted rice bran by simply treating the decolorized HCl extract with ethanol and drying the precipitate. Ammonium chloride has been employed for the isolation of phytin from cottonseed grist and from the mother liquor of wastes following the production of edible proteins from cottonseed oil cake. Aluminum phytate is prepared by adding $AlCl_3$ to aqueous H_2SO_4 extracts of defatted rice bran and adjusting the pH to 3.0–4.0. Two examples of successful industrial production of phytic acid include the development of continuous extraction of phytin from cottonseed oil cake with 0.57% HNO_3 and the large-scale isolation of pure phytic acid from raw materials such as meal, rice bran, and other by-products of the food industry (250).

Most of the above preparations consist of phytin-like salts containing non-stoichiometric amounts of mono- and divalent cations. These phytins can be easily converted into a number of well-defined salts such as Na_{12}-phytate, $CaNaH_9$-phytate, and $MgKH_9$-phytate and recrystallized from ethanol or methanol. Free phytic acid is prepared by passing phytin over a cation-exchange resin in the H^+ form. However, all crystallization attempts have been unsuccessful, which neces-

Table 4.48 Selected Physical Properties of Phytic Acid

Molecular weight	660.08
Physical appearance	Syrupy, straw-colored liquid
Solubility (25°C)	Miscible with water, 95% ethyl alcohol, glycerol

Source: Ref. 2.

sitates its storage as a solution. Phytic acid is stored as a concentrated solution of one of the above salts or as a starch powder prepared by spray-drying or vacuum-drying an aqueous mixture of phytic acid and dextrin. In the United States, commercially available phytic acid is prepared from corn and sold as a hydrated dodecasodium salt (250).

Food Applications. The antioxidant functions of phytic acid have been described by Graf and Eaton (251). Phytic acid has a relatively high binding affinity for iron and is a potent inhibitor of iron-mediated hydroxyl radical formation. In the absence of phytate, Fe^{3+} is completely insoluble at neutral or higher pH due to the formation of large polynuclear iron hydroxide complexes. High concentrations of phytic acid solubilize iron in the form of $Fe(III)_1$-phytate, whereas low concentrations precipitate iron as $Fe(III)_3$- and $Fe(III)_4$-phytate complexes. Chelation of iron by phytic acid prevents the formation of polynuclear aggregates even at iron concentrations as low as 6 nM. The six coordination sites of trivalent iron are occupied by water and hydroxide ions in aqueous solutions. Most chelating agents displace five of these ligands and form a pentadentate chelate with H_2O occupying the sixth coordination site. Phytic acid, however, is unique in occupying six coordination sites and displacing all of the coordination water in the $Fe(III)_1$ complex. Iron-catalyzed formation of hydroxyl radical requires the availability of at least one reactive iron coordination site as well as iron solubility. Phytic acid preserves the solubility and makes the metal totally unreactive. In molar phytate-to-iron ratios of 0.25 and above, the superoxide-driven generation of hydroxyl radical is completely blocked.

Phytic acid appears to be one of the most effective agents for the inhibition of iron-mediated lipid oxidation in food products. In meat products, it presumably removes myoglobin-derived iron from negatively charged phospholipids and prevents their autoxidation and off-flavor formation (251). Studies by Empson et al. (252) indicate that phytic acid substantially inhibits oxygen uptake, malondialdehyde formation, and warmed-over flavor development in cooked minced chicken breasts stored in sealed oxygen-impermeable containers at 4°C. Similar effects in cooked ground beef have been reported by St. Angelo et al. (253). In a soybean oil-in-water emulsion system, the oxidation was significantly decreased by the addition of 1 M phytic acid. Phytic acid also significantly reduces the iron-catalyzed oxidation and partially inhibits the copper-mediated oxidation of ascorbic acid in model systems (252).

Lecithin

Phospholipids or phosphatides occur in all living organisms and make up 1–2% of many crude oils and animal fats. The term *lecithin* was initially used for crude fractions of phospholipids obtained from vegetable oils or fats. At present it is used to refer to phosphatidylcholine (1,2-diacylglycero-3-phosphorylcholine). It is also used to describe soybean lecithin, which is composed of soybean phosphatides and soybean oil (254). Chemically, phosphatidylcholine consists of a glyceride in which two of the hydroxyl groups are esterified with fatty acids and the third is esterified with a phosphoric acid that in turn is bound to a molecule of choline. In phosphatidylethanolamine, the choline is replaced by ethanolamine (6). The structures of the major phosphatides present in soybean lecithin are presented in Figure 4.24. The commercial source of lecithin is mainly soybeans.

Crude lecithin, purified phosphatidylcholine, and phosphatidylethanolamine function as synergists and chelators and may also bring about decomposition of hydroperoxides. They are particularly useful as surfactants and emulsifying agents. Phosphatidylcholine (PC) and phosphatidyl ethanolamine (PE) are potent synergists with a number of primary antioxidants and oxygen scavengers at elevated temperatures above 80°C. Phosphatidylserine is less effective, and phosphatidylinositol has no synergistic effect (255).

Physical Properties. Some of the parameters that characterize commercial lecithin are acetone-insoluble matter, acid value, moisture content, hexane-insoluble matter, color, consistency, and clarity. Commercial lecithin consists of about 66% phospholipids, the other major components being the triglycerides. Lecithin is classified as plastic or fluid according to its viscosity. Each of these is further classified into natural, bleached, or unbleached grades based on color. Only the unbleached grades are permitted as food additives. Commercial lecithin can also be produced as a powder by alcohol or acetone extractions. It is soluble in oils and is hydrated in water with the formation of an emulsion (254). Table 4.49 presents the specifications for commercial soybean lecithin.

Phosphatidylcholine is a waxy mass when its acid value is about 20 and is a pourable thick liquid when its acid value is around 30. It is yellow to brownish in color. It forms a colloidal suspension in water and is readily soluble in ethanol, sparingly soluble in acetone, and practically insoluble in cold fats and oils. Phosphatidylethanolamine is a yellow amorphous substance that is practically insoluble in water and acetone and slightly soluble in ethanol (2).

Preparation. Commercial soybean lecithin is prepared from crude soybean oil obtained from mature soybean seeds by compression. The crude oil contains about 2–3% of the phosphatides. The process involves addition of water (1–2.5 vol %) to the crude oil to hydrate the phosphatides. The water–oil mixture is then heated at 70°C for 30–60 min in a pipeline agitator that ensures thorough mixing. The oil-insoluble lecithin fraction, a wet gum known as lecithin hydrate, is separated by centrifugation. The process is called degumming. To obtain a lighter colored product, hydrogen peroxide and benzoyl peroxide are used to bleach the lecithin. The bleached gum contains 25–50% water. It is dried by thin-film vacuum drying at a temperature of 80–105°C and a pressure of 3.3–4 kPa. The crude lecithin is further purified by fractionation with various solvents (254).

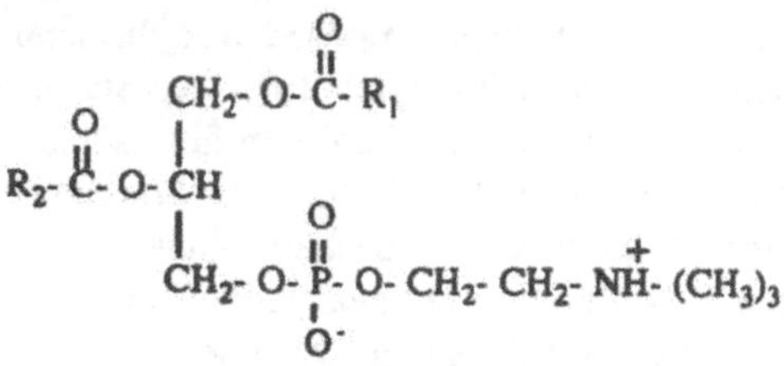

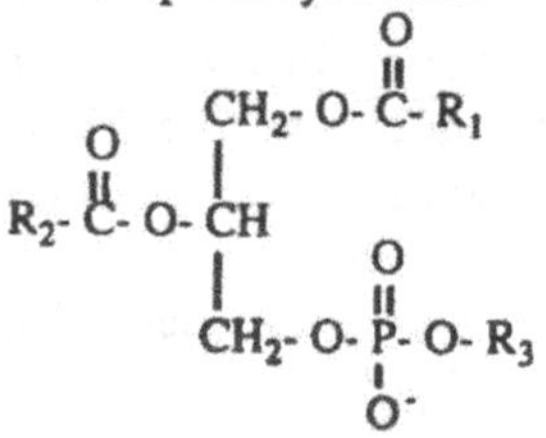

$$\underset{\underset{O^-}{\overset{\displaystyle O}{\overset{\|}{\underset{R_2\text{-}\overset{\displaystyle O}{\overset{\|}{C}}\text{-}O\text{-}CH}{\quad}}}}{\overset{CH_2\text{-}O\text{-}\overset{\displaystyle O}{\overset{\|}{C}}\text{-}R_1}{\Big|}}}$$

CH₂- O- C- R₁ ... R₂- C- O- CH ... CH₂- O- P- O- CH₂- CH₂- NH- (CH₃)₃

Phosphatidyl choline

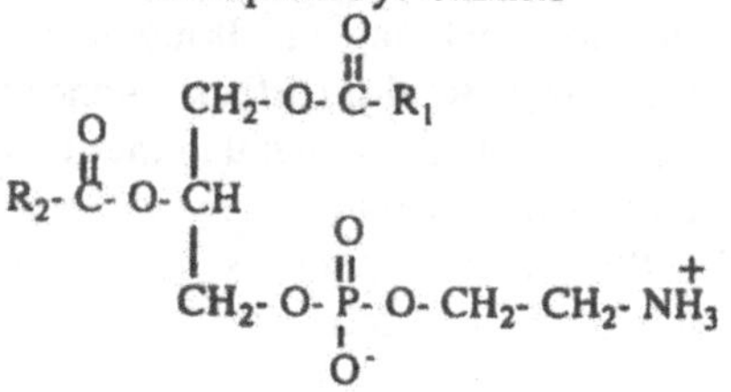

CH₂- O- C- R₁ ... R₂- C- O- CH ... CH₂- O- P- O- CH₂- CH₂- NH₃

Phosphatidyl ethanolamine

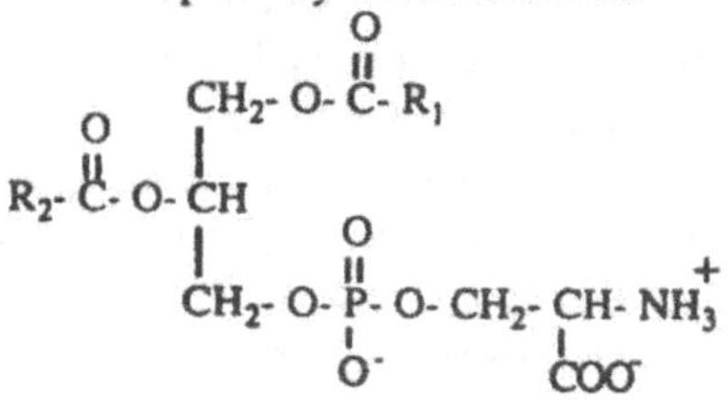

CH₂- O- C- R₁ ... R₂- C- O- CH ... CH₂- O- P- O- CH₂- CH- NH₃ ... COO⁻

Phosphatidyl serine

CH₂- O- C- R₁ ... R₂- C- O- CH ... CH₂- O- P- O- R₃

Phosphatidyl inositol

Fig. 4.24 Major phosphatides in soybean lecithin. R_1, R_2 are fatty acid residues; R_3 represents inositol.

Table 4.49 Commercial Soybean Lecithin Specifications

	Grade		
Analysis	Natural	Bleached	Double-bleached
Fluid lecithin			
Acetone-insoluble (% min)	62	62	62
Moisture (% max)	1	1	1
Benzene-insoluble (% max)	0.3	0.3	0.3
Acid value (max)	32	32	32
Color (max)	10	7	4
Viscosity at 25°C (max)	15	15	15
Plastic lecithin			
Acetone-insoluble (% min)	65	65	65
Moisture (% max)	1	1	1
Benzene-insoluble (% max)	0.3	0.3	0.3
Acid value (max)	30	30	30
Color (max)	10	7	4
Penetration (mm, max)	22	22	22

Source: Ref. 254.

Most of the triglycerides and fatty acids can be separated from crude lecithin by acetone fractionation to give oil-free lecithin powders. Acetone-fractionated lecithin has greater hydrophilicity and greater emulsifying capacity than the crude form. Lecithin can also be fractionated further by alcohol extraction. Phosphatidylcholine is concentrated in the alcohol-soluble fraction, and the alcohol-insoluble fraction becomes rich in the hydrophobic phosphatidylinositol. Phosphatidylethanolamine is evenly divided between the alcohol-soluble and -insoluble fractions. Lecithin containing large amounts of PC is obtained by adsorption and distribution chromatography. Products containing up to 95% PC are produced for pharmaceutical use by this method (254). A process has also been developed for the large-scale separation of PE from other phosphatides. The alcohol-insoluble fraction is dissolved in water-saturated chloroform and precipitated with methanol by warming to 40–45°C and cooling to room temperature. The precipitate is redissolved and reprecipitated five times in the same way. The methanol solution, which is enriched with PE, is purified by precipitating the impurities with 30% lead acetate solution. Excess lead acetate is removed by passing carbon dioxide through the solution. The methanol is evaporated, and PE is extracted with hexane. The product obtained on evaporation of hexane is 85.6% PE (256).

Food Applications. The phospholipids function synergistically with primary antioxidants, especially the tocopherols, in stabilizing various fats and oils.

Lecithin also functions as a carrier for antioxidants such as PG in commercial mixtures. Commercial mixtures containing lecithin, α-tocopherol, and ascorbyl palmitate or ascorbic acid are available. In such mixtures, ascorbic acid regenerates tocopherol mediated by lecithin. However, the mechanism is not understood. The various applications of the synergistic mixtures have been described here under the food applications of tocopherols.

Studies by Hudson and his associates have shown that in lard and soybean oil at elevated temperatures of 80–140°C, PE greatly enhances the activity of various primary antioxidants in edible oils with limited stability. Phosphatidylcholine and phosphatidylserine are also effective, and phosphaltidylinositol is without synergistic effect. In these studies, PE was shown to act synergistically with various polyhydroxyflavones, flavanones, and polyhydroxyisoflavones, tocopherols, and some of their derivatives (255,257–259). Phosphatidylethanolamine is particularly effective with α-tocopherol, 3,4-dihydroxychalcone, and BHA in stabilizing lard. The synergistic activity is small or negligible below 80°C. At 120°C, the synergistic effect increases progressively as the concentration of the synergist increases from 0.025 to 0.25%. At a given level of synergist, its effect is proportionately greater at low rather than high levels or primary antioxidant (255). Table 4.50 presents the synergistic effects of PE with various primary antioxidants. In commercially refined unhydrogenated soybean oil, a combination of tocopherols and PE or PI is effective in increasing the stability of the oil (260). Bhatia et al. (261) reported that

Table 4.50 Effects of Concentration of Dipalmitoyl Phosphatidylethanolamine (DPE) on Its Synergistic Action in Cooperation with Primary Antioxidants Added to Lard at 120°C

Primary antioxidant (0.025%)	Induction period (h)						Percent synergistic efficiency[a]		
	% DPE						% DPE		
	Nil	0.025	0.05	0.1	0.25	0.025	0.05	0.1	0.25
None	0.3	0.45	0.5	0.65	1.7	—	—	—	—
PG	12.7	9.1	11.1	13.1	21.9	−43	−17	0	+36
dl-α-Tocopherol	4.5	5.9	7.1	10.8	20.7	+22	+35	+57	+78
BHA	2.6	3.5	4.8	5.0	12.6	+23	+56	+44	+70
BHT	0.7	0.9	1.1	1.5	2.1	+8	+25	+37	0
TBHQ	3.9	3.9	7.1	7.7	9.4	−4	+42	+45	+45

[a] % Synergistic efficiency = $100 \left[(I_M - I_L) - (I_A - I_L) - (I_S - I_L)\right]/(I_M - I_L)$, where I_L = induction period of the substrate, I_A = induction period of the substrate + primary antioxidant, I_S = induction period of the substrate + synergist, and I_M = induction period of the substrate + primary antioxidant + synergist.
Source: Ref. 255.

PE is more effective than PC and PA in preventing the autoxidation of butter fat. Studies by Kashima et al. (262) showed that the phospholipids are also effective at lower temperatures. In refined perilla oil and tocopherol-enriched perilla oil, PE and PS markedly enhanced oxidative stability at 37°C. A mixture of PG, lecithin, and ascorbyl palmitate stabilizes crude corn oil (263). In menhaden oil held at 50°C, a combination of ethoxyquin and lecithin was highly effective in protecting against oxidative rancidity (264). A mixture of PG, ascorbic acid, and lecithin is used with nitrite in the curing of meat (265).

4.3.3 Secondary Antioxidants

Thiodipropionic Acid and Thiodipropionates

Thiodipropionic acid [TDPA; 3,3'-thiobis(propanoic acid)] and its dilauryl and distearyl esters (Fig. 4.25) function as synergists and chelators. They are also effective in decomposing alkyl hydroperoxides into more stable compounds and terminating peracid oxidations. TDPA and its derivatives are approved for use in edible fats and oils and other food products. However, they are not being used to any great extent in food products (266). TDPA is obtained from hot water as nacreous crystals with a melting point of 123–128°C. One gram dissolves in 26.9 mL of water at 26°C. It is freely soluble in hot water, ethyl alcohol, and acetone (2).

Preparation. Thiodipropionic acid is prepared by the reaction of β-propiolactone with ethyl thiohydracrylate, a mercaptan. Theoretically, 1 mol of the β-propiolactone is required for each mole of the mercapto radical present in the mercaptan for the reaction. β-Propiolactone is obtained economically from ketene and formaldehyde. In this process, an aqueous solution of the mercaptan salt is prepared by adding 20.1 parts of ethyl thiohydrate and 6 parts of sodium hydroxide to 50 parts of water and cooling the solution to 3°C. β-Propiolactone (10.8 parts) is then added with stirring and sufficient cooling to maintain the temperature of the solution at 3–10°C. The reaction mixture is allowed to stand for about 3 h to

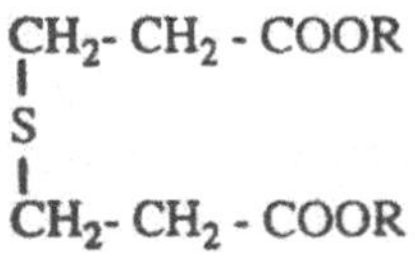

$$CH_2\text{-}CH_2\text{-}COOR$$
$$|$$
$$S$$
$$|$$
$$CH_2\text{-}CH_2\text{-}COOR$$

R : - H   thiodipropionic acid

- $(CH_2)_{11}$ -CH_3 dilauryl ester

- $(CH_2)_{17}$ -CH_3 distearyl ester

Fig. 4.25   Thiodipropionic acid and esters.

complete the reaction, and then an additional 6 parts of sodium hydroxide is added, and the mixture is refluxed for 3 h to hydrolyze the ester linkage present in the product. The mixture is acidified with concentrated hydrochloric acid and extracted with ether. TDPA is obtained as a solid on evaporating the ether (267).

Food Applications. The thiodipropionates are generally regarded as synergists rather than primary antioxidants. Thiodipropionic acid functions as a metal chelator and decomposes alkyl hydroperoxides into more stable compounds. TDPA and dilauryl thiodipropionate (DLTDP) also inactivate perioxy acids, which are capable of oxidizing aldehydes, ketones, and olefins (266,268–271). TDPA functions synergistically with BHA and ascorbyl palmitate.

Thiodipropionic acid is effective in stabilizing lard, and a strong synergistic effect has been observed between TDPA and BHA. Addition of phosphoric acid reduces the effectiveness of TDPA either alone or in combination with BHA (42). In safflower oil, TDPA (0.05%) and DLTDP (0.072%) extend the storage period from 21 days at 43°C to 193 and 113 days, respectively (41). A combination of 0.01% each of ascorbyl palmitate, PG, and TDPA is more effective in protecting safflower, sunflower, peanut, and corn oils than either PG or ascorbyl palmitate used alone. The synergistic mixture is also more effective than BHA and BHT (102) (Table 4.36). In another study comparing the effectiveness of various antioxidants in refined, bleached, and deodorized palm olein, Augustin and Berry (32) reported that DLTDP at 200 ppm is effective both at room temperature and at 60°C in improving the oxidative stability of the oil. The order of effectiveness is TBHQ > PG ~ THBP > DLTDP > BHT > BHA as measured by peroxide value and conjugated diene formation. DLTDP at 0.02% is effective in stabilizing potato chips made with cottonseed oil treated with the antioxidant. The results indicate that DLTDP is comparable to BHT and ascorbyl palmitate and is better than BHA (102) (Table 4.37).

4.3.4 Miscellaneous Antioxidants

Sodium Nitrate and Sodium Nitrite

Sodium nitrate ($NaNO_3$) and sodium nitrite ($NaNO_2$) have been used in meat curing for centuries. Sodium nitrate has no antioxidant activity but becomes functional on reduction to nitrite. The important functions of sodium nitrite include stabilization of meat color, texture improvement, development of the characteristic cured meat flavor, elimination of the problem of warmed-over flavor, and antimicrobial activity. Nitrites probably function as metal chelators, may form nitroso compounds that have antioxidant properties, and convert the heme proteins into stable nitric oxide forms (272). Table 4.51 presents the physical properties of sodium nitrate and sodium nitrite.

Preparation. Sodium nitrate. Sodium nitrate occurs in nature in deposits associated with sodium and potassium chlorides, potassium nitrate, sodium

Table 4.51 Selected Physical Properties of Sodium Nitrate and Sodium Nitrite

Physical properties	Sodium nitrate	Sodium nitrite
Molecular weight	85.01	69.0
Physical appearance	White powder	White or slightly yellow powder
Melting point (°C)	308	271
Solubility (25°C)		
Water	1 g/1.1 mL	1 g/1.5 mL
Alcohol	1 g/125 mL	Slight

Source: Ref. 2.

sulfate, and other salts. Commercial sodium nitrate is obtained from caliche, the ore mainly found in Chile. Sodium nitrate is extracted with water or a brine solution followed by fractional crystallization. A modified Guggenheim process is used for commercial production. The ore is crushed, and the coarse fraction is separated from the fines. The coarse fraction is leached countercurrently with water at around 40°C in vats. From the weak sodium nitrate brines, sodium sulfate and sodium iodate are removed by fractional crystallization and extraction. When the concentration of sodium nitrate reaches 420–440 g/liter, which is more than 50% of the dissolved solids, it is recovered from the brine by crystallization. The fines, which form 20–25% of the crushed product, are leached separately with a solution containing 150–200 g/liter of sodium nitrate at 40–50°C in a filter plant. The resulting solution, which contains 350 g/liter of sodium nitrate, is mixed with the leaching solution from the vats for further processing. The process recovers 80–85 wt % of the sodium nitrate in the ore. Synthetic sodium nitrate is produced by the neutralization of nitric acid with sodium carbonate or sodium hydroxide (273).

Sodium nitrite. Sodium nitrite is prepared by the reaction of oxides of nitrogen with aqueous sodium hydroxide or sodium carbonate solutions. Nitrogen gas can be produced by several processes, including the arc process of nitrogen fixation and the air oxidation of ammonia, or as a waste gas in the production of nitric acid and oxalic acid. To maximize the formation of nitrite and minimize the formation of nitrate, it is common to use a mixture with an excess of nitric oxide over nitrogen dioxide as well as to adjust the process conditions. Thus, a solution of sodium nitrite with a low concentration of nitrate is prepared by rapidly cooling a gas mixture containing, in parts by volume, 0–10 parts O_2, 2–12 parts NO_x, and 15–20 parts inert gas from about 900°C to 300–350°C and then dispersing it under the surface of an aqueous sodium hydroxide solution at 30–120°C. The process produces a dilute solution, which after evaporation and drying typically gives a product containing 99 wt % $NaNO_2$ (274).

Food Applications. Nitrites are widely used in various meat products such as bacon, corned beef, frankfurters, ham, luncheon meats, fermented sausages, shelf-stable canned cured meats, and perishable canned cured meat. Nitrites are also used in various fish and poultry products. Nitrites are converted in meat to nitrous acid, which is further reduced to nitric oxide. Nitric oxide converts myoglobin to nitrosylmyoglobin, the deep red pigment in uncooked cured meats. On heating, nitrosylmyoglobin forms the stable, cured, cooked, pink meat pigment nitrosylmyochrome. Nitrosylmyochrome is similar to nitrosylmyoglobin with a denatured globin moiety. Addition of ascorbic acid enhances the formation of nitric oxide and protects the cured meat pigments from oxidation. Nitrite may function as an antioxidant by different mechanisms: by converting heme proteins to their catalytically inactive and stable nitric oxide heme proteins, by chelating trace metals present in meat, by stabilizing lipids per se in the muscle against oxidation, and by the formation of nitroso compounds that possess antioxidant properties (272).

The addition of low levels of sodium nitrite (20 mg/kg) significantly inhibits lipid oxidation in various cooked meats such as fish, chicken, pork, and beef, while 50 mg/kg shows a highly significant effect. Sodium nitrite at the 200 mg/kg level caused a 17-fold reduction in TBA values in fish and 12-fold reductions in chicken, pork, and beef (272). Similar results were reported by Fooladi et al. (275) in chicken, pork, and beef at nitrite levels of 156 mg/kg. Sodium nitrite at 10 mg/kg also inhibits light-catalyzed oxidation in an emulsified unsaturated fatty acid system (276). Sodium nitrite inhibits lipid oxidation catalyzed by iron, copper, cobalt, or myoglobin in heated water extracts of pork muscle (272). Similar observations were reported in a linoleic acid model system containing prooxidant metals (276). Nitrosylmyoglobin is also effective in inhibiting lipid oxidation catalyzed by iron, copper, cobalt, or myoglobin in heated water extracts of pork muscle. The antioxidant effects of nitrosylmyoglobin were demonstrated in linoleate and β-carotene-linoleate aqueous model systems by Kanner et al. (277). Nitrosylmyoglobin may function as a quencher of free radicals, thus terminating the free radical chain reaction (272,277). Nitrite also causes significant reduction in nonheme iron in heated aqueous extracts of muscles (272). This could be mainly because of the stabilizing effects of heat on nitrosylmyoglobin to form the stable complex nitrosomyochrome, thus preventing denaturation of myoglobin and concomitant release of free iron (272,278).

Amino Acids

Amino acids function as primary antioxidants, synergists, and chelators. Their antioxidant properties depend on their concentration and the pH of the medium

(279). Glycine, methionine, tryptophan, histidine, proline, and lysine have been found to be effective in fats and oils. The methods of preparation of some of the amino acids are compiled in Table 4.52.

Food Applications. The antioxidant properties of amino acids have been studied in model systems and also in food products. Marcuse (279) compared the antioxidative effects of several amino acids in herring oil emulsions containing about 0.01 M tocopherol and in a linoleic acid model system. In the case of herring oil, 10 of the 11 amino acids tested had antioxidative effects of varying degree. The effect was strongest with histidine. The results obtained for autoxidation of linoleic acid show that amino acids may have an antioxidative effect in the absence of tocopherols or other primary antioxidants. The effect is considerably dependent on pH. In the case of alanine and histidine, the effect is relatively small at pH 7.5 and strong at pH 9.5. With increasing concentration of amino acid, the antioxidative effect decreases, and it may become prooxidative at pH 7.5. The amino acids also show a synergistic effect with α-tocopherol at pH 7.5. At pH 9.5, the strong effect of amino acid addition is only slightly reinforced by tocopherol (279). Various troloxyl amino acids formed by a covalent attachment of the amino acids with Trolox-C have been found to have better antioxidant properties than Trolox-C in a linoleate emulsion system (148). Amino acids such as threonine, histidine, and aspartic acid function synergistically with ferulic acid (281). Similarly, the antioxidant activity of methionine is considerably enhanced by quercetin (282).

Of the various amino acids, glycine has been listed as a GRAS substance for addition to fats and oils up to 0.01%. Olcott and Kuta (283) reported that pro-

Table 4.52 Preparation of Amino Acids

Amino acid	Preparation
Methionine	Casein digestion with pancreatin, industrial synthesis with β-methylmercapto propinaldehyde
Tryptophan	From casein; from β-indolylaldehyde and hippuric acid; from hydantoin, 3-indoleacetonitrile, and α-ketoglutaric acid phenylhydrazone
Lysine	Acid hydrolysis of casein, fibrin, or blood corpuscle paste
Glycine	From gelatin or from silk fibroin
Proline	From wheat gliadin or from gelatin
Histidine	From blood corpuscles; from glyoxaline and hippuric acid: from 4-chloromethylglyoxaline and ethylchloromalonate

Source: Compiled from *Merck Index* (2).

line is effective in vegetable oils, probably acting in conjunction with natural antioxidants. For example, samples of a refined olive oil containing 0, 0.025, and 0.05% added proline had induction periods at 60°C of 4, 12, and 23 days, respectively. In sardine oil, proline at 0.02% is equivalent to BHA and almost five times as effective at 0.1% (284). The nitroxide of proline is effective in menhaden oil (285). Methionine is reported to be antioxidative in rapeseed, soy, and olive oils (286–290). Cystine is effective in soy, rice, and corn oils as well as in lard and tallow (290,291). Popovici et al. (166) reported that a combination of histidine and ascorbic acid is highly effective in corn oil. Histidine also functions synergistically with α-tocopherol in inhibiting oxidation of herring oil (292).

Amino acids function as antioxidants in milk fat, characteristically delaying the induction period (293) (Table 4.53). They are effective in anhydrous milk fat or ghee, where they delay both hydrolytic and oxidative rancidity (294). Merzametov and Gadzhuva (295) reported that combinations of tryptophan at 0.2% and lysine at 0.15 and 0.2% are highly effective in butterfat. Similarly, combinations of cystine with BHT or tocopherols are highly effective in milk fat (296,297). Methionine has been proposed as a synergist in dairy products (298).

Amino acids are also effective in meat products and in preventing browning reactions in fruits. S-Nitrosocysteine, a compound generated during curing of meat, acts as an antioxidant in ground turkey meat (299). The meat of antarctic krill is effective in retarding oxidation in canned and preserved fish products. The antioxidant principle has been identified as a tocopherol and a mixture of free amino acids (300). A combination of cysteine and ascorbic acid inhibits browning in apple slices and juice (301).

Table 4.53 Effect of Amino Acids on Induction Period (IP) of Milk Fat at 95°C[a]

Treatments	IP (h)
None	<10
Cysteine	>150
Tryptophan	<150
Lysine	>150
Alanine	120
Serine	90
Histidine	70
Tyrosine	65

[a]10 mg of amino acid in 2 mL of hexane was added to 200 mg of milk fat.
Source: Ref. 293.

Spices

Spices form potential sources of natural antioxidants. They are effective in a number of food products, including confectioneries, fats, oils, meat products, and baked goods. Chipault and coworkers (302,303) investigated the antioxidant properties of 78 samples representing 36 spices in fats. Ground spices and extracts were used for the studies. Allspice, cloves, sage, oregano, rosemary, and thyme were found to be active in all types of fats examined. The antioxidant compounds in spices are mainly phenolic. However, spice extracts are limited in use mainly because of their color, odor, and taste. Numerous studies have been conducted to identify the specific active principles and develop methods to obtain bland, odorless, and tasteless products.

In rosemary (*Rosmarinus officinalis* L. family Labiatae) leaf extract, approximately 90% of the antioxidant activity has been attributed to carnosol, a phenolic diterpene. Other active compounds are rosmaridiphenol, rosmarinic acid, carnosic acid, rosmanol, isorosmanol, and epirosmanol (Fig. 4.26) (304–307). The same compounds have been isolated in sage (*Salvia officinalis* L. family Labiatae) (307).

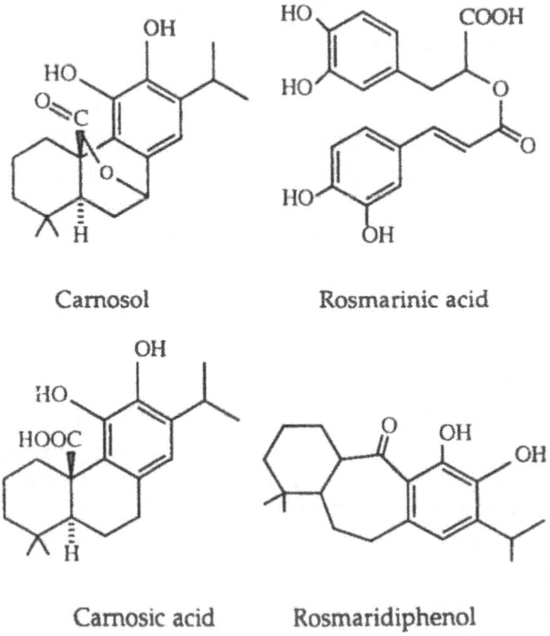

Carnosol

Rosmarinic acid

Carnosic acid   Rosmaridiphenol

Fig. 4.26   Some antioxidant compounds from rosemary.

Commercial methods have been developed for the preparation of odorless and colorless antioxidant compounds from rosemary and sage (308,309).

Preparation of Antioxidants from Rosemary and Sage. The process of Chang et al. (308) involves extraction of the crude antioxidant with ethyl ether under refluxing conditions. The solvent is removed, and the crude material is washed with water several times, dissolved in methanol, and bleached with activated carbon by stirring at 60°C for 15 min. The purified material is a light-colored fine powder and has a little odor of the original spice. An odorless and tasteless product is obtained by vacuum distillation or molecular distillation of a suspension of the purified antioxidant in cottonseed oil.

The process of Bracco et al. (309) involves premilling the material to a particle size of about 2 mm and whirl-sieving to collect particles smaller than 600 μ. The particles are suspended in peanut oil, and the moisture content is maintained at 2–4%. The lipid suspension is microionized in a ball mill. The microionization of the spice in edible oil is the main step of the process, which ruptures the cell wall to free the antioxidant from the protoplasm, permits intimate contact between the lipid phase and the microionized protoplasm, and yields a lipid suspension for the molecular distillation step. The finely dispersed antioxidant components have a molecular weight range lower than that of the natural triglycerides of the oil used and can therefore be physically separated by molecular distillation on either a fall-film or centrifugal system. In the fall-film system, a two-stage process is used. The lipid suspension at 80–90°C is passed through the first distillation unit (140°C, pressure 0.3×10^{-2} mm, condenser at 80–100°C) to eliminate the volatile components, which are then cold trapped at −196°C. The second distillation is at a higher temperature of 205–215°C with a condenser temperature of 80–110°C and pressure lower than 1×10^{-3} mm to obtain the antioxidant compounds. In the centrifugal method, the microionized suspension is spread by centrifugal force in a thin film across the face of the rotor. The antioxidant fraction evaporates and condenses on the internal water-cooled leaf condenser.

Food Applications. The antioxidant activity of various ground spices, extracts, and purified compounds has been reported. Chipault and coworkers (302,303,310) studied the antioxidative activity of ground spices in lard, ground pork, two types of mayonnaise, and a french dressing. Allspice, cloves, sage, oregano, rosemary, and thyme increased the stability of all the fat substrates in which they were tested, but the relative effectiveness of the spices varied with the substrate. Compared to other spices, sage and rosemary were highly effective in lard. Cloves were highly effective in simple oil–water emulsions and in ground pork. Oregano was most effective in mayonnaise and french dressing. Oregano, rosemary, sage, and thyme were effective in pie crust. Dubois and Tressler (311) reported that black pepper, sage, mace, and ginger retard the development of rancidity in frozen ground pork and beef. Black pepper, red pepper, and sage extend the keeping quality of frozen pork sausage (312). In soybean, linseed, olive, and sesame oils, various spices were more effective at 0.0003% than BHA at 0.02% (Table 4.54) (313).

Petroleum ether–soluble, ethanol-soluble, and water-soluble fractions from various spices have also been tested for their antioxidant activity in lard. In general, the soluble fractions have lower activity than the ground spices (302,314). The antioxidative effectiveness of rosemary extract is comparable to that of BHA and BHT in precooked dried wheat flakes, potato flakes, and frozen ground pork (304). Table 4.55 compares the antioxidant activity of rosemary extract with that of other primary antioxidants in lard.

The antioxidant activities of purified total antioxidants or specific compounds have been reported. Purified, odorless, tasteless rosemary antioxidant is effective in lard, chicken fat, sunflower oil, corn oil, and soybean oil. It inhibits flavor reversion in soybean oil. Ascorbic acid functions synergistically with rosemary antioxidant. The purified antioxidant is effective in fried products such as potato chips as it has a lower volatility and better stability than BHA and BHT at higher temperatures (308). In chicken fat, purified rosemary and sage antioxidants show antioxidant activity comparable to that of BHA and BHT (309).

Carnosic acid, carnosol, rosmanol, epirosmanol, isorosmanol, rosmaridiphenol, and rosmariquinone are some of the active components isolated from rosemary antioxidant (305,306,315–317). Like other phenolic antioxidants, carnosic acid is a potent antioxidant for animal fats. Synergism between ascorbyl palmitate and carnosic acid has been demonstrated (109). Rosmanol showed stronger activity

Table 4.54 Antioxidant Activity of Spices in Fats and Oils

	Antioxidant index[a]			
Spice (0.0003 %)	Soybean oil	Linseed oil	Olive oil	Sesame oil
Fennel	2.3	2.2	8.0	5.1
Ginger	1.9	2.6	7.8	3.8
Capsicum	2.0	2.6	7.3	5.4
Cloves	5.1	6.0	23.8	8.5
Garlic	2.3	2.2	6.6	7.0
Turmeric	2.0	2.1	8.5	3.3
BHA (0.02%)	1.1	1.1	2.5	1.1
Control	9.5	3.7	16.0	8.7

[a]Antioxidant index $= \dfrac{\text{time for treated sample to reach peroxide value} \times 100}{\text{time for control sample to reach peroxide value} \times 100}$

Source: Ref. 313.

Table 4.55 Rosemary Extract Compared with
Other Antioxidants in Lard

Antioxidant, 0.02%	AOM stability (h)	Antioxidant index
Control	26	—
Rosemary extract	156	6.0
BHA	68	2.6
BHT	98	3.7
DG	162	6.2

Source: Ref. 318.

than BHA in lard as well as in a water–oil emulsion system (315,316).
Rosmaridiphenol and rosmariquinone are reported to be as effective as BHA and
BHT (305,306).

Vanillin, used mainly as a flavoring agent, also has potent antioxidant properties. Pyenson and Tracy (319,320) reported that vanillin is effective as an antioxidant in dried whipped cream and in ice cream mixes. Studies by Burri et al.
(321,322) and Arouma et al. (323) showed that vanillin is effective in dried rice
flakes. Addition of 200 or 500 ppm vanillin to dried rice flakes reduced the
formation of pentane in a 3-month storage study at 20°C. Vanillin is also effective
in shortenings (324).

Flavonoids

Flavonoids include a variety of phenolic compounds with a C_6–C_3–C_6 chemical
structure (Fig. 4.27). They occur widely in common edible fruits, leaves, seeds, and
other parts of food plants as either glycosides or aglycones. The subgroups are
classified based on the substitution pattern of the C ring and the position of the B
ring. The major subgroups are flavonols, anthocyanins, flavones, isoflavones,
catechins, proanthocyanidins, and aurones. Chalcones, flavanones, leucoanthocyanins, and dihydroflavonols are the intermediates in the flavonoid biosynthetic
pathway. Other related compounds are cinnamic acids, some of which are the
precursors of the flavonoids. Many of these compounds have antioxidant properties
in model systems.

The structure–activity relationships of various flavonoids were reviewed by
Pratt and Hudson (325). In general, flavonoids function as primary antioxidants,
chelators, and superoxide anion scavengers. The aglycones are more active than
the glycosides. The antioxidant activity is determined by the position and degree
of hydroxylation of the B ring. The presence of hydroxyl groups at the 3′-, 4′-, and
5′-positions on the B ring enhances the antioxidant activity compared to that of a

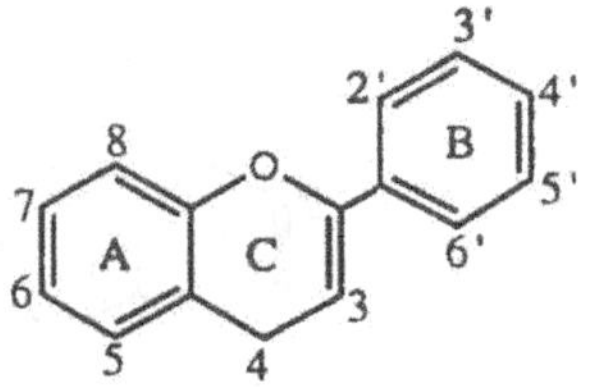

C6-C3-C6 configuration of flavonoids

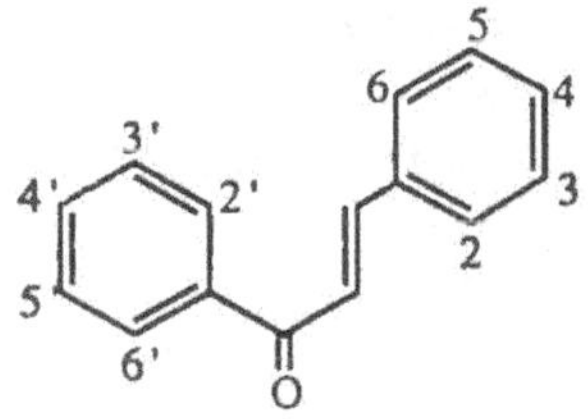

Chalcones
Butein 2'=4'=3=4=OH
Okanin 2'=3'=4'=3=4=OH

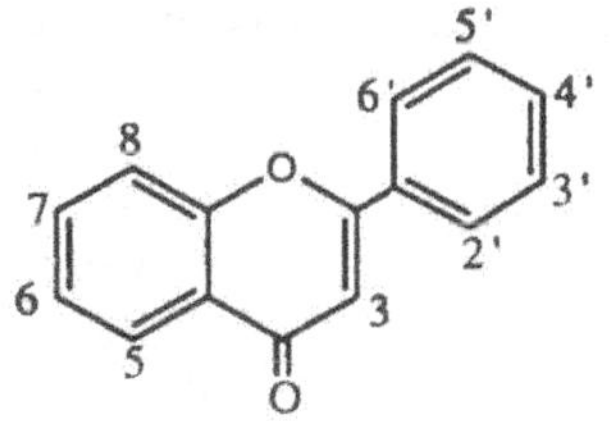

Flavones
Luteolin 5=7=3'=4'=OH
Isovitexin 4'=5=7=OH, 6=Glucose

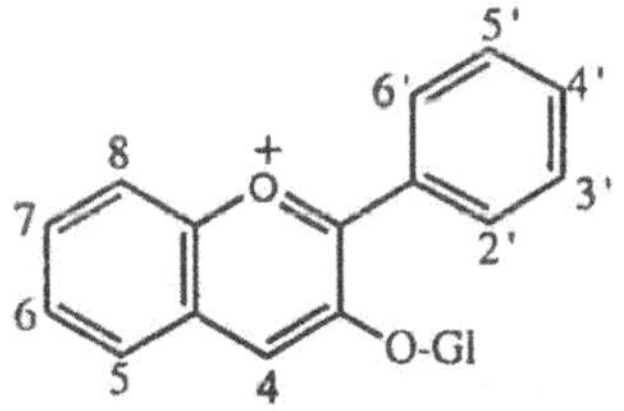

Anthocyanins

Isoflavones

Cyanidin-3-glucoside 5'=4'=5=7=OH
Malvidin-3-glucoside 5=7=4'=OH, 3'=5'=OCH3

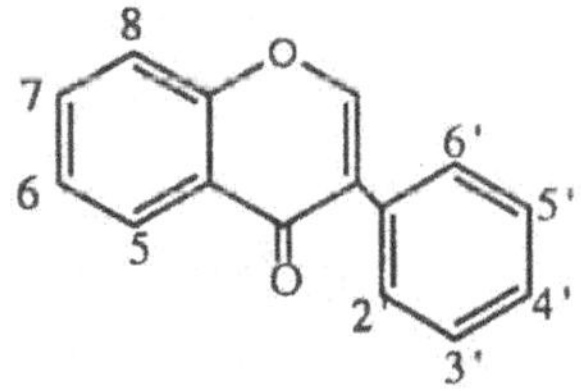

Daidzein 7=4'=OH
Genistein 5=7=4'=OH

Fig. 4.27 Some antioxidant flavonoids and related compounds.

single hydroxyl group (326–328). In addition, the presence of a 3-hydroxyl group and the 2–3 double bond on the C ring also appears to have an effect on the antioxidant properties (329,330). Some of the effective compounds are quercetin, myricetin, robinetin, quercetagetin, gossypetin, catechins, and proanthocyanidins. Intermediates such as dihydroquercetin, eriodictyol, chalcones like butein and okanin, and dihydrochalcones are very effective as antioxidants. Phenolic acids

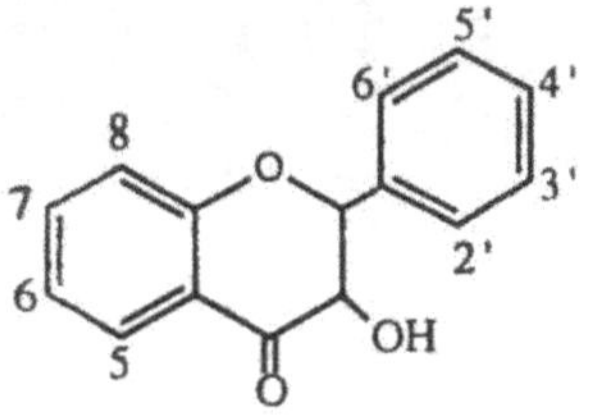

Dihydroflavonols
Dihydroquercetin 3=5=7=3'=4'=OH

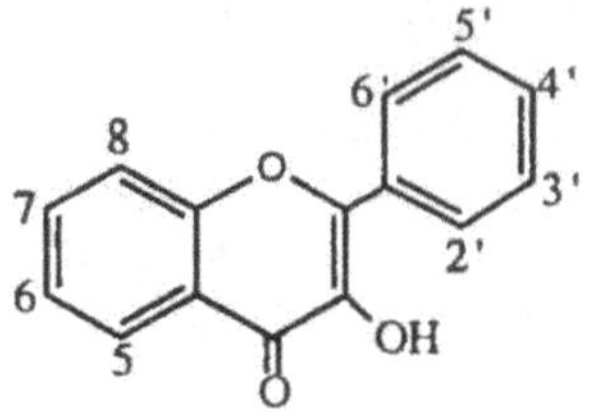

Flavonols
Quercetin 3=5=7=3'=4'=OH
Myricetin 3=5=7=3'=4'=5'=OH
Gossypetin 3=5=7=8=4'=5'=OH

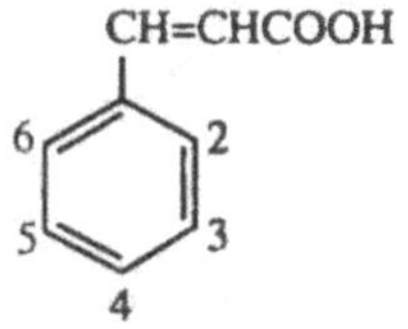

Cinnamic acids
Ferulic acid 4=OH, 3=OCH3
Caffeic acid 3=4=OH

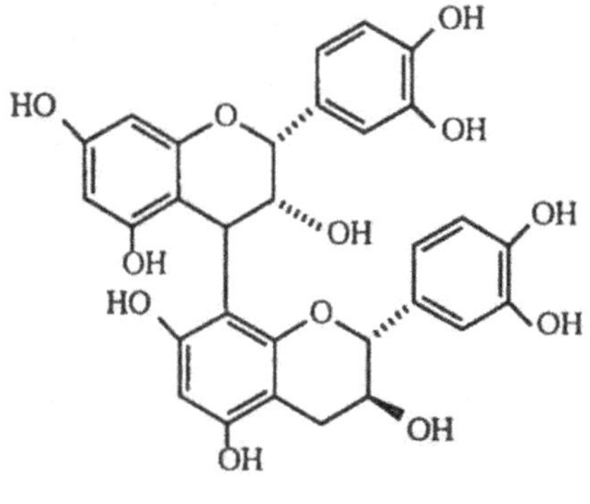

Procyanidin B-1

Fig. 4.27 (*Continued*)

such as caffeic acid and chlorogenic acid also have significant antioxidant activity. However, flavonoids are not being used as antioxidants in food products at present.

Preparation. Flavonoid-rich fruits, oilseeds, and whole bark form potential commercial sources of flavonoids and related compounds. Fruits and vegetables, for example, grapes, blueberries, cranberries, and red cabbage, are rich sources of anthocyanins. In general, the flavonoids are extracted from their sources with alcohol, and compounds of interest are isolated by chromatography and recrystallization. The chemical synthesis of various anthocyanidins, flavonols, and flavones has been reported (331,332).

Commercially, anthocyanins from grapes, cranberries, roselle, and cabbage are available as aqueous extracts or powders. Grape wine pomace and cranberry presscake have been used as raw materials. The various extraction methods have been reviewed by Markakis (333). The anthocyanins are extracted with water, water containing SO_2, or acidified alcohols. In grape wine pomace, the SO_2 extract contains purer anthocyanin than the water extract; also these anthocyanins are more stable as colorants for soft drinks than the anthocyanins from water extract. Extraction with methanol or ethanol containing a small amount of mineral acid (1% HCl or less) is commonly used. If acylated anthocyanins are present as in grapes, an organic acid such as tartaric acid, acetic acid, or formic acid is used. Extraction of anthocyanins from cranberry presscake was reported by Chiriboga and Francis (334,335). The presscake is treated several times with methanol containing 0.03% HCl, the methanol in the extract is evaporated under vacuum, and the aqueous concentrate is passed through a cation-exchange resin (Amberlite CG-50). The resin absorbs the pigment and allows many impurities to be washed off the column with water. The anthocyanins along with other flavonoids are eluted with methanol containing 0.0001% HCl, and the eluate is freeze-dried. A similar extraction procedure has been applied to miracle fruit, blueberries, and cabbage. The concentrates are obtained after evaporation of the organic solvent under reduced pressure, thus minimizing exposure to oxygen and high temperatures. The concentrates may be marketed at room temperature or frozen, or they may be spray-dried or freeze-dried (333).

The whole bark or the cork portion of Douglas fir forms a potential commercial source of dihydroquercetin, which on oxidation yields quercetin. Several procedures have been developed for large-scale preparation of dihydroquercetin from Douglas fir bark. The bark may be extracted with hot water and the concentrated solution extracted with ethyl ether to yield crude dihydroquercetin, which is subsequently purified by recrystallization; it may be directly extracted with a solvent such as constant-boiling alcohol–benzene or methyl ethyl ketone–water to yield an extract that is subsequently separated into a crude water-soluble dihydroquercetin fraction that may be further purified; or it may be extracted with benzene or chlorinated solvents such as trichloroethylene to remove waxes and then extracted with diethyl ether to give a crude dihydroquercetin fraction that may be further purified by recrystallization from hot water (336).

Food Applications. The antioxidative activity of various flavonoids has been studied in model systems and in food products. The 3,4-dihydroxychalcones butein and okanin are potent antioxidants in lard. Butein at 0.02% is twice as active as quercetin or α-tocopherol and about six times as active as BHT in lard. 3,4-Dihydroxychalcones are also more effective than the corresponding flavanones and cinnamic acids (255,307). Dihydroquercetin, a dihydroflavonol, has the same antioxidant activity as quercetin. It is more active than the flavanones naringenin, eriodictyol, and hesperetin in a stripped corn oil system and lard (325). Dihydroquercetin is effective in lard, cottonseed oil, and butter oil (337). However, it has low solubility in fats and oils and has no carry-through properties in fried foods.

The antioxidant activity of various flavonols and flavones has been reported. Quercetin is effective in inhibiting copper-catalyzed oxidation of lard by forming a weak chelate with copper (338). A combination of Tenox-5 (25% BHA + 25% BHT) and quercetin is more effective in extending the shelf life of potato flakes than Tenox-5 (48). Quercetin is also effective in dry milk products (339). In an ethyl linoleate model system, quercetin is more effective than PG on an equimolar basis. However, myricetin and robinetin are more potent than quercetin in stripped corn oil. This could be due to the presence of an additional hydroxyl group at the 5'-position in myricetin and robinetin (325). Flavones like luteolin and isovitexin are also effective in model systems (257,340).

Isoflavones from soybeans are effective in fish preserves in conjunction with tocopherols and phospholipids (341). The isoflavones genistein and prunetin are effective in lard and show considerable synergism with phosphatidylethanolamine (257). However, in comparison with the flavonols, flavones, flavanones, and chalcones, the isoflavones show a relatively low order of antioxidant activity (325).

The antioxidant activity of phenolic acids and their esters is again dependent on the degree of hydroxylation. Of the various phenolic acids, chlorogenic acid and caffeic acid are more effective in stripped corn oil than quinic acid, coumaric acid, and ferulic acid (325). Sato et al. (342) reported that caffeic acid is effective in preventing quality deterioration in sake. Proanthocyanidin dimers having a higher antioxidant activity than α-tocopherol in a linoleic acid–β-carotene–water system have been reported (343). Proanthocyanidins from fruits such as grapes, black currants, and bilberries are effective as superoxide anion scavengers and inhibitors of lipid oxidation (344). Proanthocyanidins are effective in salad oil, frying oil, and lard (345). Anthocyanins from various sources, including grapes, bilberries, black currants, and eggplant, have antioxidant activity in model systems (344,346, 347). The antioxidant activity of catechins has been described here in the section on tea extracts.

Vitamin A

Vitamin A or all-*trans*-retinol [3,7-dimethyl-9-(2,6,6-trimethyl-1-cyclohexen-1-yl] (Fig. 4.28) occurs widely in all animal tissues and is stored in the liver as the palmitate. Some of the sources of vitamin A are eggs, milk, butter, and fish. Table 4.56 presents the physical properties of vitamin A.

Preparation. Vitamin A can be extracted from natural sources such as fish oils where it occurs in esterified form. However, commercial vitamin A is produced mostly by chemical synthesis. The synthetic methods that are commercially exploited for the production of vitamin A are described by Isler and Kienzle (348). The synthetic methods in general involve the key intermediate β-ionone [4-(2,6,6-trimethyl-1-cyclohexen-1-yl)-3-buten-2-one], a well-known commercial product used in the perfume industry that is synthesized mainly from acetone.

β-Ionone can be converted to vitamin A by various means. The two major processes used commercially are those of Hoffman-La Roche and Badische Anilin-und-Soda-Fabrik (BASF). In the Roche process, β-ionone is converted to a C_{14} aldehyde that is further condensed with (Z)-3-methyl-2-penten-4-yn-1-ol in a Grignard reaction. Partial hydrogenation and rearrangement leads to vitamin A acetate. In the BASF process, β-ionol is rearranged to a C_{15} phosphonium salt, which then, in a Wittig reaction with 3-formylcrotonyl acetate, yields vitamin A directly. The structures of the important intermediates are presented in Fig. 4.29.

Food Applications. Vitamin A has very limited use as an antioxidant. It is easily oxidized on exposure to air and light and may become prooxidant. However, vitamin A has antioxidant effects in fats and oils in the dark and inhibits the formation of free acids in vegetable oils. Studies have shown that it is effective in pork lard at 100 mg/kg. In a 6-month storage study at 16–27°C and relative humidity of 48–72%, the acid and peroxide values remained at acceptable levels for 4 months, compared to 2 months for the control, with relatively little loss in the sensory quality throughout storage (349).

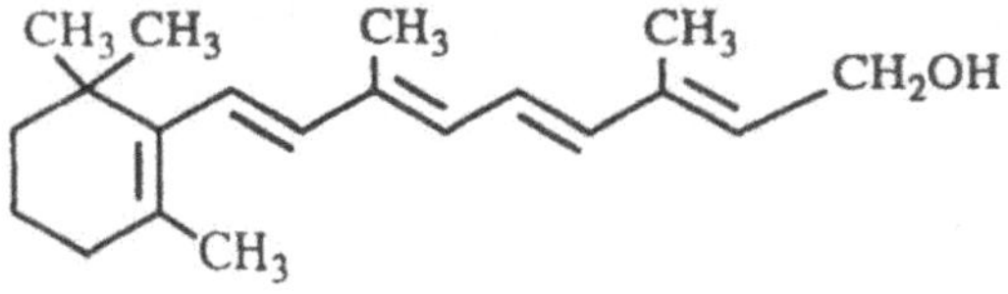

Fig. 4.28 Vitamin A.

Table 4.56 Selected Physical Properties of Vitamin A

Molecular weight	286.44
Physical appearance	Yellow prisms
Melting point (°C)	62–64
Solubility at 25°C	Practically insoluble in water or glycerol; soluble in ethyl alcohol; soluble in fats and oils

Source: Ref. 2.

β-*Carotene*

β-Carotene (provitamin A) (Fig. 4.30) occurs widely in all plants. Carrots, green leafy vegetables, and tropical fruits such as papayas and mangoes are rich sources of β-carotene. Table 4.57 presents the physical properties of β-carotene. β-Carotene is used mainly as a food colorant. Structurally, β-carotene consists of two β-ionone rings and an 18-carbon polyene side chain. β-Carotene and related carotenoids function as singlet oxygen quenchers and free-radical trapping agents. The antioxidant action is limited to low oxygen partial pressures less than 150 mm Hg, and at higher oxygen pressures β-carotene may become prooxidant (350,351).

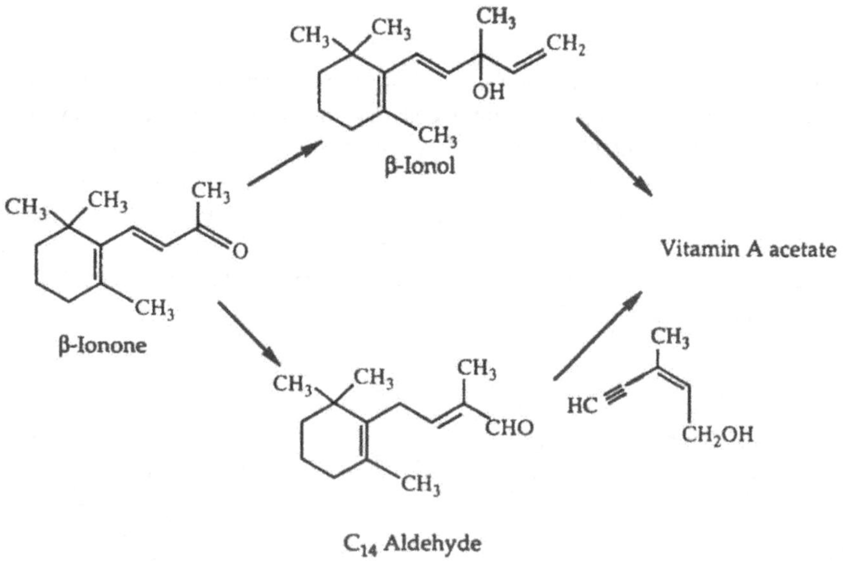

Fig. 4.29 Intermediates in vitamin A synthesis.

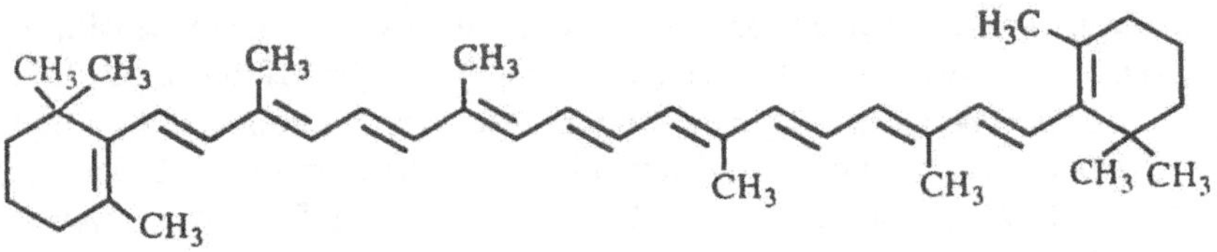

Fig. 4.30 β-Carotene.

β-Carotene is insoluble in water and soluble in fats and oils. Because of the conjugated double-bond structure of the molecule, β-carotene is sensitive to oxidative decomposition on exposure to air. On exposure to air, the absorption maximum drops to 25% of its initial value in 6 weeks at 20°C either in darkness or in light. β-Carotene is also sensitive to light, temperature, and acidic conditions. It is almost completely destroyed at 45°C in air after 6 weeks. Contact with acids in the presence of oxygen causes rapid deterioration accelerated by heat. However, the stability improves on dissolution or suspension in vegetable oils.

Preparation. The commercial synthesis of β-carotene is closely related to that of vitamin A. The starting material is β-ionone, which is converted to the C_{14} aldehyde as in vitamin synthesis. The aldehyde is converted to β-carotene in two different ways. In the first method the C_{14} aldehyde is converted to a C_{19} aldehyde by a chain-lengthening process. Two moles of the C_{19} aldehyde β-apo-14′-carotenol and 1 mol of acetylene are condensed in a Grignard reaction to form the C_{40} diol. The 15,15′-didehydro-β-carotene, upon partial hydrogenation and isomerization, yields β-carotene. Another method involves Wittig condensations of either 2 mol of a C_{15} phosphonium salt with a C_{10} dialdehyde or a C_{20} phosphonium salt with retinal (348,352).

Food Applications. β-Carotene inhibits the light-catalyzed oxidation of fats and oils such as butter, butterfat, and corn, coconut, rapeseed, groundnut, soybean, and olive oils (353–355). In another study on the quenching mechanism of β-carotene,

Table 4.57 Selected Physical Properties of β-Carotene

Molecular weight	536.85
Physical appearance	Deep purple to red crystals
Melting point (°C)	176–183
Solubility at 25°C	Practically insoluble in water or glycerol; sparingly soluble in ethyl alcohol; soluble in fats and oils

Source: Ref. 2.

Lee and Min (356) observed that β-carotene inhibits chlorophyll-sensitized photooxidation of soybean oil by quenching singlet oxygen. Terao et al. (357) reported that tocopherols extend the inhibitory effects of β-carotene in a methyl linoleate model system and in soybean oil. A combination of 5–10 ppm of β-carotene and 100 ppm of citric acid is effective in inhibiting flavor deterioration initiated by light in soybean oil (358). Studies by Clements et al. (359) showed that β-carotene at 0.46 ppm reduces the rise in peroxide value of soybean oil after 6 h at 20°C under photooxidizing conditions from 36 meq/kg for the control to 30 meq/kg for the sample containing β-carotene. β-Carotene is also effective in inhibiting lipid oxidation initiated by xanthine oxidase (360).

Tea Extracts

Tea leaves are from the evergreen plant *Camellia sinensis*, family Camelliaceae. The leaves are a rich source of polyphenols, mainly catechins, a group of flavonoids. The leaves are processed in different ways to produce green and black teas. The leaves are steamed, rolled, and dried to produce green tea. The steaming inactivates the polyphenol oxidase systems, thereby preventing the oxidation of polyphenols. To produce black tea, the leaves are partially dried, crushed, fermented for a few hours, and dried. During the fermentation process, the catechins are oxidized by a polyphenol oxidase present in the tea shoots, resulting in the formation of theaflavin, theaflavin gallate, and thearubigin (361). Both green and black tea extracts have antioxidant properties. The extracts contain epicatechin (EC), epigallocatechin (EGC), epicatechin gallate (ECg), and epigallocatechin gallate (EGCg) and gallocatechins, of which EGCg forms the major component (Fig. 4.31). Table 4.58 presents the catechin composition of green tea. The fermentation process reduces the concentration of EGC and EGCg in black tea. The catechins are insoluble in oil and soluble in alcohol.

Preparation. Extraction of the antioxidant principles from green tea and black tea leaves has been reported. Crude catechin extract is obtained by treating the green tea leaves with ethanol or acetone (363). For the preparation of a crude catechin powder, the leaves are extracted in water and the extract is spray-dried. The water-soluble green tea powder is dissolved in hot water. The aqueous extract is washed with chloroform and extracted with ethyl acetate. The ethyl acetate layer is evaporated, and the concentrated solution is freeze-dried to yield a crude powder containing >90% catechins. The four main compounds are isolated from the crude powder by liquid chromatography (364). Black tea leaves are extracted with water at 120–210°C. The extract contains 0.005–1.5% by weight tea solids, of which 5% is polyphenols. The extract can be reextracted with a water-immiscible organic solvent and treated with tannase to produce a superior antioxidant (365,366).

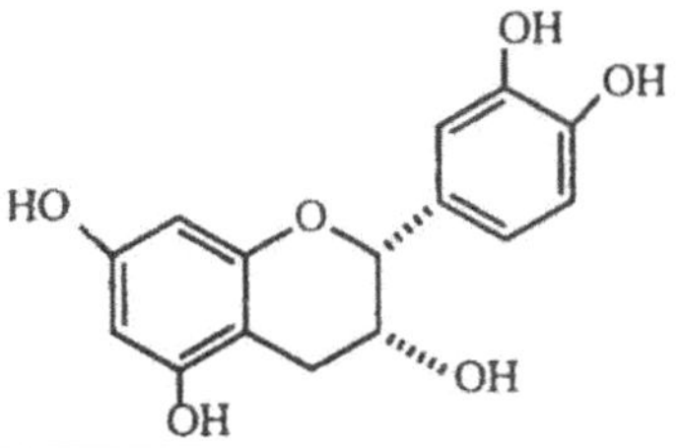

(-)-Epicatechin

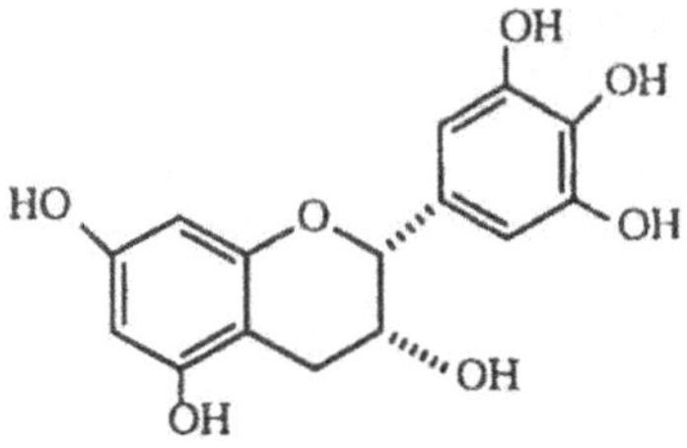

(-)-Epigallocatechin

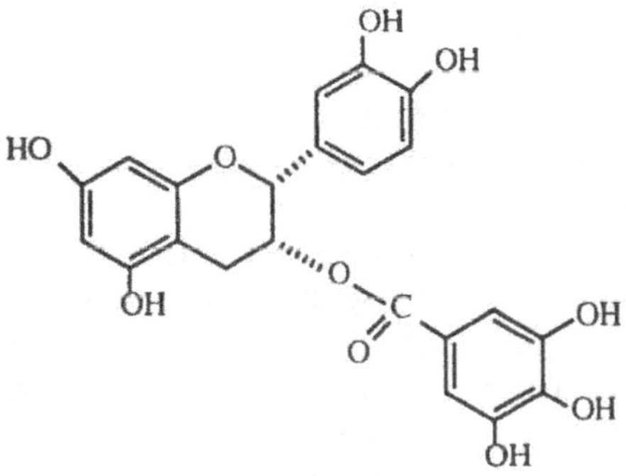

(-)-Epicatechin gallate

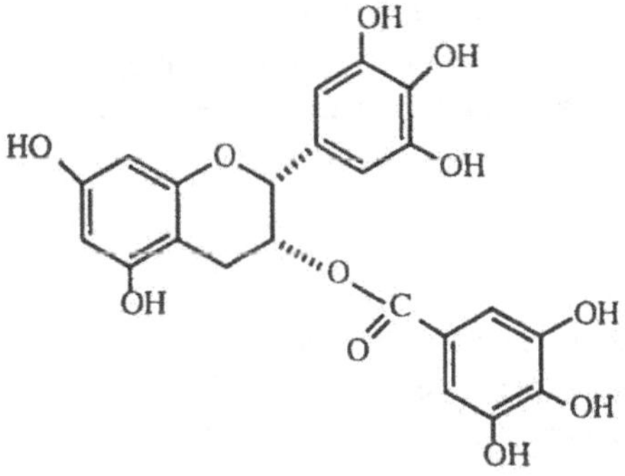

(-)-Epigallocatechin gallate

Fig. 4.31 Some antioxidant compounds from green tea.

Table 4.58 Composition of Catechins in Green Tea

Compound	Absolute %	Relative %
(+)-Gallocatechin	1.44	1.6
(–)-Epigallocatechin	17.57	19.3
(–)-Epicatechin	5.81	6.4
(–)-Epigallocatechin gallate	53.90	59.1
(–)-Epicatechin gallate	12.51	13.7

Source: Ref. 362.

Food Applications. Studies by Matsuzaki and Hara (364) showed that the antioxidant activity of the crude catechin is stronger than that of BHA or *dl*-tocopherol in lard and in salad oil. The antioxidant activity of the four main compounds has also been tested in lard. On an equimolar basis, the order of activity is $EC < EC_g < EGC < EGC_g$. This result implies that the presence of a hydroxy group at C-5′ increases the antioxidant activity. EGC_g is effective in both oil and aqueous systems, as shown by studies in a linoleic acid–water system. The ethanol and acetone extracts of green tea are highly effective in soybean, corn, palm, and groundnut oils and lard (363). The acetone extracts from roasted tea have shown significant antioxidant activity in biscuit formulations, whereas the water extracts have a prooxidant effect (367). EGCg also functions synergistically with ascorbic acid, α-tocopherol, citric acid, and tartaric acid in lard (364). Applications of tea antioxidants in frying oils, potato flakes, meat emulsions, mayonnaise, margarine, frozen fish, margarine, precooked cereals, chicken fat, pork, and cheese have been patented (368–370).

Zinc

Zinc is an essential micronutrient and functions as a biological antioxidant. Some of the dietary sources of zinc are meat, seafoods, wheat bran, germ, brewer's yeast, and whole grains. Table 4.59 presents the physical properties of zinc.

Preparation. The sources and methods of manufacture of zinc and zinc compounds have been reviewed by Lloyd and Showak (371) and Lloyd (372). The major source of zinc in nature is the ore sphalerite, which consists of ZnS with 67% zinc. The sulfide is converted to the oxide by roasting at 900°C followed by sintering to complete the roast and eliminate the volatile materials. The crude zinc oxide concentrate is subjected to leaching in dilute sulfuric acid to remove contaminants like iron, which forms an important step in purification. Zinc is obtained by electrowinning or by pyrometallurgical processes in the Imperial

Table 4.59 Selected Physical Properties of Zinc

Atomic weight	65.38
Physical appearance	Bluish-white metal
Melting point (°C)	419.5
Boiling point (°C)	908

Source: Ref. 2.

Smelting Furnace wherein zinc oxide is reduced with carbon monoxide at 1000–1050°C.

Zinc oxide is prepared by the vaporization of metallic zinc and oxidation of the vapors with preheated air. The medicinal grade contains 99.5% or more zinc oxide. Zinc chloride is prepared by the reaction of aqueous hydrochloric acid with crude zinc oxide. The solution is purified by precipitation of the heavy metals and other contaminants with zinc powder, partial neutralization, and oxidation with chlorine. Zinc chloride is obtained by crystallization. Zinc sulfate is formed by the reaction of zinc oxide with sulfuric acid. The heptahydrate formed is converted to the monohydrate by dehydration at 100°C. Very pure zinc sulfate is formed during the preparation of zinc by electrowinning. Zinc stearate is prepared by the reaction of zinc chloride with stearic acid (2,372).

Food Applications. Zinc functions as an antioxidant at the cellular level. It is a component of the active site of the cytosolic superoxide dismutase in eukaryotes. Zinc also inhibits lipid oxidation at the membrane level. Zinc is a redox-inactive metal that functions by displacing the redox-active metal ions like copper and iron from their biological binding sites. This markedly reduces the site-specific damage exerted by the redox-active ions in combination with superoxide radicals. Zinc is a component of over-the-counter antioxidant preparations and multivitamin–multimineral formulations. Zinc chloride, sulfate, stearate, and oxide are some of the forms in which zinc is incorporated into supplemental preparations.

Selenium

Selenium is an essential micronutrient and also a biological antioxidant. Dietary sources of selenium include vegetables like cabbage, celery, and radish, brewer's yeast, fish, whole grains, and meat. Selenium is present in organic form in these sources, which is the major nutritional form for animals and humans. Table 4.60 presents the physical properties of selenium.

Preparation. The various methods of preparation of elemental selenium are reviewed by Elkin (373). The principal sources of selenium are metal sulfide

Table 4.60 Selected Physical Properties of Selenium and Sodium Selenite

Physical properties	Selenium	Sodium selenite
Atomic/molecular weight	78.96	172.95
Physical appearance	(a) Red or crystalline	Tetragonal prisms
	(b) Gray or metallic	
	(c) Amorphous	
Melting point (°C)	<200°C	
Solubility at 25°C	Insoluble in water and alcohol	Freely soluble in water; insoluble in alcohol

Source: Ref. 2.

deposits that are mined primarily for copper, zinc, nickel, and silver. Most of the selenium is obtained as a by-product of precious metal recovery from electrolytic copper refinery slimes. Selenium occurs in slimes mainly as $CuAgSe$, Ag_2Se, and $Cu_{2-x}Se_x$, where $x < 1$. Selenium is converted to a water-soluble form, followed by reduction to the elemental state. The most important methods of achieving this are smelting with soda ash, roasting with soda ash, direct oxidation, and roasting with sulfuric acid. Selenium recovery varies between 80 and 95%. The crude selenium is purified by distillation at 310–400°C. Sodium selenite is prepared by evaporating an aqueous solution of sodium hydroxide and selenous acid between 60 and 100°C or by heating a mixture of sodium chloride and selenium oxide (2).

Food Applications. Selenium functions as an antioxidant at the cellular level. It is necessary for the synthesis and activity of glutathione peroxidase, which catalyzes the reduction of hydrogen peroxide and organic hydroperoxides by reducing equivalents derived from glutathione. The activity of the enzyme is dependent on the presence of four atoms of selenium at the active site of the enzyme.

Selenium and vitamin E show a synergistic effect in depressing lipid peroxide formation in vivo and in vitro in an NADPH-dependent microsomal lipid peroxidation system. Selenium also exerts a protective effect against UV irradiation, carcinogenesis, and aging. At present, selenium is incorporated in many over-the-counter antioxidant preparations and multivitamin–multimineral formulations. Sodium selenite is the commonly used supplemental form of selenium.

REFERENCES

1. D. N. Maddox, *Flavors,* May/June: 117 (1976).
2. *The Merck Index,* 9th ed., Merck and Co., Inc., Rahway, New Jersey, 1976.
3. W. G. Christiansen, *J. Amer. Chem. Soc.,* 48: 1358 (1926).

4. W. C. Ault, J. K. Weil, G. C. Nutting, and J. C. Cowan, *J. Amer. Chem. Soc.*, 69: 2003 (1947).
5. WHO, WHO Food Additive Series, No. 3, World Health Organization, Geneva, Switzerland, 1972.
6. J. R. Chipault, in *Autoxidation and Antioxidants*, Vol. 2 (W. O. Lundberg, Ed.), Wiley, New York, 1962, p. 477.
7. W. M. Gearhart and B. N. Stuckey, *J. Amer. Oil Chem. Soc.*, 32: 386 (1955).
8. E. R. Sherwin and J. W. Thompson, *Food Technol.*, 21(6): 912 (1967).
9. A. E. Tseluiko, *Nauch. Issled. Inst. Morsk. Ryb, Khoz. Okeanogr.*, 29: 229 (1970).
10. C. H. Lea, *J. Soc. Chem. Ind. (Lond.)*, 63: 107 (1944).
11. J. W. Higgins and H. C. Black, *Oil Soap*, 21: 277 (1944).
12. S. G. Morris, L. A. Kraekel, D. Hammer, J. S. Myers, and R. W. Riemenschneider, *J. Amer. Oil Chem. Soc.*, 24: 309 (1947).
13. E. R. Sherwin and B. M. Luckadoo, *J. Amer. Oil Chem. Soc.*, 47: 19 (1970).
14. B. E. Greene, *J. Amer. Oil Chem. Soc.*, 48: 637 (1971).
15. B. E. Greene, I. M. Hsia, and M. W. Zipser, *J. Food Sci.*, 36:940 (1971).
16. M. Jacobson and H. H. Koehler, *J. Agric. Food Chem.*, 18: 1069 (1970).
17. R. J. Sims and H. D. Stahl, *Baker's Digest*, October: 50 (1970).
18. J. E. Magoffin and R. W. Bentz, *J. Amer. Oil Chem. Soc.*, 26: 687 (1949).
19. L. Sair and L. A. Hall, *Food Technol.*, 5(2): 69 (1951).
20. F. J. H. Ottaway and J. B. M. Coppock, *J. Sci. Food Agric.*, 5: 294 (1958).
21. F. D. Tollenaar and H. J. Vos, *J. Amer. Oil Chem. Soc.*, 35: 448 (1958).
22. J. H. Mitchell, Jr. and A. S. Henick, in *Autoxidation and Antioxidants*, Vol. 2 (W. O. Lundberg, Ed.), Wiley, New York, 1962, p. 477.
23. L. B. Rockland, D. M. Swarthout, and R. A. Johnson, *Food Techol.*, 15(3): 112 (1961).
24. J. Abbot and R. Waite, *J. Dairy Res.*, 29: 55 (1962).
25. J. Abbot and R. Waite, *J. Dairy Res.*, 32: 143 (1965).
26. A. Tamsma, T. J. Mucha, and M. J. Pallansch, *J. Dairy Sci.*, 46: 114 (1963).
27. J. W. Pette and J. A. B. Smith, *Int. Dairy Fed.*, II-Doc. 14 (1963).
28. B. N. Stuckey and W. M. Gearhart, *Food Technol.*, 11: 676 (1957).
29. R. J. Lewis, Sr., *Food Additives Handbook*, Van Nostrand Reinhold, New York, 1989.
30. A. Bell, M. B. Knowles, and C. E. Tholstrup, U.S. Patent 2,759,828 (1956).
31. A. Bell, M. B. Knowles, and C. E. Tholstrup, U.S. Patent 2,848,345 (1958).
32. M. A. Augustin and S. K. Berry, *J. Amer. Oil Chem. Soc.*, 60: 1520 (1983).
33. E. P. Oliveto, *Chem. Ind.*, 17: 677 (1972).
34. O. Gisvold, U.S. Patent 2,382,475 (1945).
35. R. J. Sims and J. A. Fioriti, in *CRC Handbook of Food Additives*, Vol. 1 (T. E. Furia, Ed.), CRC Press, Boca Raton, FL, 1980, p. 13.
36. O. S. Privett, in *Autoxidation and Antioxidants*, Vol. 2 (W. O. Lundberg, Ed.), Wiley, New York, 1962, p. 985.
37. J. Devillers, P. Boule, P. Vasseur, P. Prevot, R. Steiman, F. Seigle-Murandi, J. L. Benoit-Guyod, M. Nendza, C. Grioni, D. Dive, and P. Chambon, *Ecotoxicol. Environ. Safety*, 19: 327 (1990).
38. R. H. Rosenwald, U.S. Patent 2,470,902 (1949).
39. R. H. Rosenwald, U.S. Patent 2,459,540 (1949).
40. D. S. Young and G. F. Rodgers, U.S. Patent 2,722,556 (1955).

41.   J. W. Thompson and E. R. Sherwin, *J. Amer. Oil Chem. Soc.*, *43*: 683 (1966).
42.   H. R. Kraybill, L. R. Dugan, Jr., B. W. Beadle, F. C. Vibrans, V. Swartz, and H. Rezabek, *J. Amer. Oil Chem. Soc.*, *26*: 449 (1949).
43.   F. D. Tollenaar ad H. J. Vos, *Fette, Seifen, Anstrichum.*, *58*: 112 (1956).
44.   R. Roos and J. P. Ostendorf, *Perfum. Essent. Oil Rec.*, *58*: 25 (1967).
45.   W. M. Gearhart, B. N. Stuckey, and E. R. Sherwin, *Food Technol.*, 2: 260 (1957).
46.   I. I. Rusoff and J. L. Common, U.S. Patent 2,657, 997 (1951).
47.   B. N. Stuckey, *Food Technol.*, *9*(11): 585 (1955).
48.   G. M. Sapers, O. Panasiuk, and F. B. Talley, *J. Food Sci.*, *40*: 797 (1975).
49.   R. M. Stephenson, T. Sano, and P. R. Harris, *Food Technol.*, *12*(11): 622 (1958).
50.   B. E. Greene, *J. Food Sci.*, *110* (1969).
51.   Y. M. Naidu, *Diss. Abstr. Int.*, *B44*(12):3581: Order No. DA 8407219, (1984), p. 204.
52.   V. M. Olson and W. J. Stadelman, *Poultry Sci.*, *59*(12): 2733 (1980).
53.   T. Y. Hou, *J. Chin. Agric. Chem. Soc.*, *18*: 53 (1980).
54.   L. E. Jeremiah, *J. Food Prot.*, *51*: 105 (1988).
55.   N. G. Marriott and P. P. Graham, *Proc. Eur. Meeting Meat Res. Workers*, *33*(1): 4:2, 1987, p. 166.
56.   D. E. Mook and P. L. McRoberts, U.S. Patent 3,459,561 (1969).
57.   W. F. Douglas and W. E. Phalen, U.S. Patent 2,901,354 (1959).
58.   A. Hadorn and K. Zuercher, *Mitt. Geb. Lebensm. Unters.*, *57*: 127 (1966).
59.   J. G. Lease and E. J. Lease, *Food Technol.*, *16*: 104 (1962).
60.   D. M. Swarthout, R. A. Johnson, and S. DeWitte, *Food Technol.*, *12*(11): 599 (1958).
61.   S. R. Cecil and J. G. Woodroof, *Food Ind.*, 23 February: 81 (1951).
62.   C. G. Cavaletto and H. Y. Yamamoto, *J. Food Sci.*, *36*: 81 (1971).
63.   N. E. Harris, D. E. Westcott, and A. S. Henick, *J. Food Sci.*, *37*: 824 (1972).
64.   R. Shea, *Food Proc.*, *26*: 148 (1965).
65.   E. R. Sherwin, in *Food Additives* (A. L. Branen, P. M. Davidson, and S. Salminen, Eds.), Marcel Dekker, New York, 1990, p. 139.
66.   H. Voges, in *Ullmann's Encyclopedia of Industrial Chemistry*, 5th ed., Vol. A19, VCH, Weinheim, FRG, 1985, p. 313.
67.   G. H. Stillson, U.S. Patent 2,428,745 (1947).
68.   W. V. McConnell and H. E. Davis, U.S. Patent 3,082,258 (1963).
69.   L. R. Dugan, Jr., L. Marx, C. E. Weir, and H. R. Kraybill, *Amer. Meat Inst. Found. Bull.*, No. 18, 1954.
70.   C. N. Rao, B. V. R. Roa, T. J. Rao, and G. R. R. M. Rao, *Asian J. Dairy Res.*, *3*: 127 (1984).
71.   S. Paul and A. Roylance, *J. Amer. Oil Chem. Soc.*, *39*: 163 (1962).
72.   M. H. Chahine and R. F. MacNeill, *J. Amer. Oil Chem. Soc.*, *51*: 37 (1974).
73.   P. J. Ke, D. M. Nash, and R. G. Ackman, *J. Amer. Oil Chem. Soc.*, *54*: 417 (1977).
74.   T. L. Mounts, K. A. Warner, and G. R. List, *J. Amer. Oil Chem. Soc.*, *58*: 792 (1981).
75.   Z. J. Hawrysh, P. J. Shand, C. Lin, B. Tokarska, and R. T. Hardin, *J. Amer. Oil Chem. Soc.*, *67*: 585 (1990).
76.   M. C. F. Toledo, W. Esteves, and V. E. M. Hartmann, *Cienc. Technol. Aliment.*, *5*(1): 1 (1985).
77.   M. M. Ahmad, S. Al-Hakim, and A. A. Y. Shehata, *Fette- Seifen. Anstrichm.* 85: 479 (1983).

78. J. E. Huffaker, *J. Amer. Oil Chem. Soc.*, *59*: 381 (1982).
79. C. W. Fritsch, V. E. Weiss, and R. H. Anderson, *J. Amer. Oil Chem. Soc.*, *52*: 517 (1975).
80. E. R. Sherwin, *J. Amer. Oil Chem. Soc.*, *53*: 430 (1976).
81. J. E. Huffaker, *Snack Food*, October, 1976.
82. T. Y. Shiao, J. J. Huang, and R. L. Chang, *J. Chin. Agric. Chem. Soc.*, *27*: 299 (1989).
83. M. Aoyama, T. Maruyama, H. Kanematsu, I. Niiya, M. Tsukamoto, S. Tokairin, and T. Matsumoto, *Yukagaku*, *35*(6): 449 (1986).
84. Y. B. Park, H. K. Park, and D. H. Kim, *Korean J. Food Sci. Technol.*, *21*: 468 (1989).
85. K. L. Rho, P. A. Seib, O. K. Chung, and D. S. Chung, *J. Amer. Oil Chem. Soc.*, *63*: 251 (1986).
86. S. Ahmad and M. A. Augustin, *J. Sci. Food Agric.*, *36*: 393 (1985).
87. M. A. Augustin and S. K. Berry, *J. Amer. Oil Chem. Soc.*, *61*: 873 (1984).
88. R. Rasit and M. A. Augustin, *Pertanika*, *5*(1): 119 (1982).
89. S. J. Van de Reit and M. M. Hard, *J. Amer. Diet. Assoc.*, *75*: 556 (1979).
90. M. F. Chastain, D. L. Huffman, W. H. Hseih, and J. C. Cordray, *J. Food Sci.*, *52*: 1779 (1982).
91. J. Mai and J. E. Kinsella, *J. Sci. Food Agric.*, *32*: 293 (1981).
92. D. C. Herting and E. J. E. Durry, *J. Agric. Food Chem.*, *17*: 785 (1969).
93. F. C. Johnson, *CRC Crit. Rev. Food Technol.*, 2: 267 (1971).
94. P. Schuler, in *Food Antioxidants* (B. J. F. Hudson, Ed.), Elsevier Applied Science, London, 1990, p. 99.
95. Y. Ishikawa and E. Yuki, *Agric. Biol. Chem.*, *39*: 851 (1975).
96. W. J. Krasavage and C. J. Terhaar, *J. Agric. Food Chem.*, *25*: 273 (1977).
97. D. C. Herting, in *Krik-Othmer Encyclopedia of Chemical Technology*, Vol. 24, Wiley-Interscience, New York, 1978, p. 214.
98. S. Kasparek, in *Vitamin E* (L. J. Machlin, Ed.), Marcel Dekker, New York, 1980, p. 7.
99. H. Klaui, *Flavors*, July/August: 165 (1976).
100. R. M. Parkhurst, W. A. Skinner, and P. A. Sturm, *J. Amer. Oil Chem. Soc.*, *45*: 641 (1968).
101. W. Heiman and H. von Pezold, *Fette-Seifen Anstrichm.*, *59*: 330 (1957).
102. W. M. Cort, *J. Amer. Oil Chem. Soc.*, *51*: 321 (1974).
103. G. Pongracz (1975), Unpublished, cited in Ref. 99.
104. G. Pongracz (1972), Unpublished, cited in Ref. 99.
105. K. G. Berger, *Chem. Ind.*, *March*: 194 (1975).
106. G. Pongracz, *Int. J. Vit. Nutr. Res.*, *43*: 525 (1973).
107. S. Won Park and P. D. Addis, *J. Food Sci.*, *51*: 1380 (1986).
108. J. Baltes, *Wiss. Veroff, Deut. Ges. Ernaehr.*, *16*: 169 (1967).
109. G. Pongracz, F. Kracher, and P. Schuler (1978–1987), Unpublished, cited in Ref. 94.
110. L. R. Dugan, Jr. and H. R. Kraybill, *J. Amer. Oil Chem. Soc.*, *33*: 527 (1956).
111. G. Pongracz, *Fette-Seifen Anstrichm.*, *86*: 455 (1984).
112. W. L. Marusich, in *Vitamin E* (L. J. Machlin, Ed.), Marcel Dekker, New York, 1980, p. 445.
113. M. Asahara, Y. Matsuzaki, and S. Matsumori, *Nippon Shokuhin Kogyo Gakkaishi*, *22*: 467 (1975).

114. H. S. Olcott, in *Fish in Nutrition* (H. Eirik, Ed.), Fishing News Ltd., London, 1962.
115. D. P. Grettie, *Oil Soap, 10*: 126 (1933).
116. J. W. Higgins and H. C. Black, *Oil Soap, 21*: 277 (1944).
117. A. S. Wright, D. A. A. Akintonwa, R. S. Crowne, and D. E. Hathway, *Biochem. J.,* 97: 303 (1965).
118. A. L. Rocklin and J. L. Van Winkle, U.S. Patent 3,026,264 (1962).
119. E. M. Bickoff, A. L. Livingston, J. Guggolz, and C. R. Thompson, *J. Agric. Food Chem., 2*: 1229 (1954).
120. J. Van der Veen and H. S. Olcott, *Poultry Sci., 43*: 616 (1964).
121. R. E. Knowles, A. L. Livingston, J. W. Nelson, and G. O. Kohler, *J. Agric. Food Chem., 16*: 985 (1968).
122. J. T. Weil, J. Van der Veen, and H. S. Olcott, *Nature, 219*: 168 (1968).
123. J. S. Lin and H. S. Olcott, *J. Agric. Food Chem., 23*: 798 (1975).
124. C. C. Tung, *Tetrahedron, 19*: 1685 (1963).
125. H. M. Chen and S. P. Meyers, *J. Agric. Food Chem., 30*: 469 (1982).
126. R. B. H. Wills, G. Hopkirk, and K. J. Scott, *J. Amer. Soc. Hort. Sci., 106*(5): 569 (1981).
127. C. R. Little, H. J. Taylor, and I. D. Peggie, *Sci. Hort., 13*: 315 (1980).
128. A. I. Strangev and V. K. Todorov, *Refrig. Sci. Technol., 84*: 264 (1982).
129. P. M. Chen, D. M. Varga, E. A. Mielke, T. J. Facteau, and S. R. Drake, *J. Food Sci., 55*: 171 (1990).
130. P. Vinas, M. H. Cordoba, and C. Sanchez-Pedreno, *Food Chem., 42*: 241 (1991).
131. S. P. Meyers and D. Bligh, J. Agric. Food Chem., *29*: 505 (1981).
132. H. S. Olcott, *J. Amer. Oil Chem. Soc., 35*: 597 (1958).
133. A. Atkinson, R. P. Van de Merwe, and L. G. Swart, *Agroanimalia, 4*: 63 (1972).
134. S. Thorisson, F. Gunstone, and R. Hardy, *J. Amer. Oil Chem. Soc., 69*: 806 (1992).
135. K. R. Bharucha, C. K. Cross, and L. J. Rubin, *J. Agric. Food Chem., 33*: 834 (1985).
136. K. R. Bharucha, C. K. Cross, and L. J. Rubin, *J. Agric. Food Chem., 35*: 915 (1987).
137. Y. Takagi, T. Ishizawa, and T. Iida, *J. Jpn. Oil Chem. Soc., 28*: 548 (1979).
138. N. M. Weinshenker, L. A. Bunes, and R. Davis, U.S. Patent 3,996,199 (1976).
139. N. M. Weinshenker, *Food Technol., 34*(11): 40 (1980).
140. P. J. Ke, E. Cervantes, and C. Robles-Martinez, *J. Sci. Food Agric., 34*: 1154 (1983).
141. J. K. Kaitaranta, *J. Amer. Oil Chem. Soc., 69*: 810 (1992).
142. U. Holm, *Abstr. Int. Soc. Fat Res. Congr.*, Goteborg, Sweden, June 1972.
143. G. R. List, C. D. Evans, W. F. Kwolek, K. Warner, B. K. Boundy, and J. C. Cowan, *J. Amer. Oil Chem. Soc., 51*: 17 (1974).
144. A. R. Cooper, D. D. Matzinger, and T. E. Furia, *J. Amer. Oil Chem. Soc., 56*: 1 (1979).
145. J. W. Scott, W. M. Cort, H. Harley, D. R. Parrish, and G. Saucy, *J. Amer. Oil Chem. Soc., 51*: 250 (1974).
146. W. M. Cort, J. W. Scott, M. Araujo, W. J. Mergens, M. A. Cannalonga, M. Osadca, H. Harley, D. R. Parrish, and W. R. Pool, *J. Amer. Oil Chem. Soc., 52*: 174 (1975).
147. W. M. Cort, J. W. Scott, and J. H. Harley, *Food Technol., 29*(11): 46 (1975).
148. M. J. Taylor, T. Richardson, and R. D. Jasensky, *J. Amer. Oil Chem. Soc., 58*: 622 (1981).
149. T. F. Canning, in *Kirk-Othmer Encyclopedia of Chemical Technology*, Vol. 21, Wiley-Interscience, New York, 1978, p. 245.

150. E. D. Weil, In *Kirk-Othmer Encyclopedia of Chemical Technology*, Vol. 22, Wiley-Interscience, New York, 1978, p. 107.
151. S. T. Taylor, N. A. Higley, and R. K. Bush, *Adv. Food Res.*, *30*: 1 (1986).
152. B. L. Wedzicha, *Int. J. Food Sci. Technol.*, *22*: 433 (1987).
153. D. F. Chichester and F. W. Tanner, in *CRC Handbook of Food Additives*, Vol. 1 (T. E. Furia, Ed.), CRC Press, Boca Raton, FL 1972, p. 115.
154. G. M. Jaffe, in *Kirk-Othmer Encyclopedia of Chemical Technology*, Vol. 24, Wiley-Interscience, New York, 1978, p. 8.
155. T. Reichstein and A. Grussner, *Helv. Chim. Acta*, *17*: 311 (1934).
156. J. C. Bauernfeind, *Adv. Food Res.*, *4*: 359 (1953).
157. J. C. Bauernfeind and D. M. Pinkert, *Adv. Food Res.*, *18*: 219 (1970).
158. O. S. Privett and F. W. Quackenbush, *J. Amer. Oil Chem. Soc.*, *31*: 321 (1954).
159. Y. T. Lew and A. L. Tappel, *Food Technol.*, *10*: 285 (1956).
160. B. M. Watts and R. Wong, *Arch. Biochem.*, *30*: 110 (1951).
161. T. Sabalitschka, *Milchwissenschaft*, *8*: 300 (1953).
162. A. J. Gelpi, E. W. Bryant, and L. L. Russof, *J. Dairy Sci.*, *38*: 197 (1955).
163. G. G. Kelley and B. M. Watts, *Food Res.*, *22*: 308 (1957).
164. B. M. Watts, *Food Technol.*, *10*: 101 (1956).
165. C. Golumbic and H. A. Mattill, *J. Amer. Chem. Soc.*, *63*: 1279 (1941).
166. A. Popovici, A. Isopescu, D. Dima, and A. D. Petrescu, *Ind. Aliment. Prod. Veg.*, *16*: 248 (1965).
167. G. Cerutti, *Olii Miner. Grassi Saponi, Calari Vernici*, *33*: 25 (1956).
168. M. J. Tims and B. M. Watts, *Food Technol.*, *12*: 240 (1958).
169. M. W. Zipser and B. M. Watts, *J. Agric. Food Chem.*, *15*: 80 (1967).
170. M. D. Ranken, in *Rancidity in Foods* (J. C. Allen and R. J. Hamilton, Eds.), Elsevier Applied Science, London, 1989, p. 225.
171. P. P. Coppen, in *Rancidity in Foods* (J. C. Allen and R. J. Hamilton, Eds.), Elsevier Applied Science, London, 1989, p. 83.
172. H. Nickels and A. Hackenberger, Ger. Offen. DE 3,308,922 (1984).
173. J. Delga and R. Boulu, *Ann. Pharm. Franc.*, *15*: 691 (1957).
174. Y.-Y. Gwo, G. J. Flick, Jr., H. P. Dupuy, R. L. Ory, and W. L. Baran, *J. Amer. Oil Chem. Soc.*, *62*: 1666 (1985).
175. J. Abbot, *18th Int. Dairy Congr., Sydney, Australia*, *IE*: 464 (1970).
176. K. Maurer and B. Schiedt, *Ber.*, *67B*: 1239 (1934).
177. J. D. Dziezak, *Food Technol.*, *40*(9): 94 (1986).
178. B. Borenstein, *Food Technol.*, *19*(11): 115 (1965).
179. F. J. Yourga, W. B. Esselen, and C. R. Fellers, *Food Res.*, *9*: 188 (1944).
180. W. B. Esselen, Jr., J. J. Powers, and R. Woodward, *Ind. Eng. Chem.*, *37*: 295 (1945).
181. G. W. Hope, *Food Technol.*, *15*: 548 (1961).
182. P. Reyes and B. S. Luh, *Food Technol.*, *16*: 116 (1962).
183. R. B. Hudson and M. J. Dolan, in *Kirk-Othmer Encyclopedia of Chemical Technology*, Vol. 17, Wiley-Interscience, New York, 1978, p. 426.
184. R. H. Ellinger, in *CRC Handbook of Food Additives*, Vol. 1 (T. E. Furia, Ed.), CRC Press, Boca Raton, FL, 1972, p. 617.
185. B. T. Lehmann and B. M. Watts, *J. Amer. Oil Chem. Soc.*, *28*: 475 (1951).
186. H. K. Wilson and E. O. Herreid, *J. Dairy Sci.*, *52*: 1229 (1969).

187. A. M. Malkov, A. I. Trippel, and G. A. Kirichkova, *Sb. Tr. Leningr. Inst. Sov. Torgovli.*, *1964*(12): 112 (1964).
188. S. Shtal'berg and I. Radaeva, *Proc. 17th Int. Dairy Congr.*, *Munich*, 5: 153 (1966).
189. H. J. Deobold, T. A. McLemore, N. R. Bertoniere, and J. A. Martinez, *Food Technol.*, *18*: 1970 (1964).
190. A. G. Maggi, Netherlands Patent Appl. 6,600,754 (1966).
191. S. Hanada, Y. Yokoo, E. Suzuki, N. Yamaguchi, and T. Yoshida, *Eiyo Shokuryo*, *11*: 360 (1959).
192. W. Feldheim and J. Seidemann, *Fruchtsaft-Ind.*, 7: 166 (1962).
193. C. C. Molsberry, U.S. Patent 3,342,610 (1967).
194. G. O. Hall, U.S. Patent 2,478,266 (1949).
195. M. W. Hoover, *Food Technol.*, *17*: 636 (1963).
196. *Stauffer Chemicals for Modern Potato Processing*, Food Ind. Release No. 5, Stauffer Chemical Company, Westport, Connecticut.
197. H. R. Bolin, F. S. Nury, and B. J. Finkle, *Baker's Dig.*, *38*(3): 46 (1964).
198. R. F. Nelson and B. J. Finkle, *Phytochemistry*, *3*: 321 (1964).
199. G. O. Hall, U.S. Patent 2,513,094 (1950).
200. G. E. Brissey, U.S. Patent 2,596,067 (1952).
201. L. D. Kamstra and R. L. Saffle, *Food Technol.*, *13*: 652 (1959).
202. P. Chang, M. T. Younathan, and B. M. Watts, *Food Technol.*, *15*: 168 (1961).
203. A. L. Savich and C. E. Jansen, U.S. Patent 2,830,907 (1958).
204. G. D. Wilson, *Proceedings*, Research Report to Industry, American Meat Institute Foundation, 1954, p. 32.
205. B. M. Watts, *Proceedings*, 9th Research Conference, American Meat Institute, 1957, p. 61.
206. J. R. Hart, in *Ullmann's Encyclopedia of Industrial Chemistry*, 5th ed., Vol A10, Weinheim, FRG, 1985, p. 95.
207. T. E. Furia, *Food Technol.*, *18*(12): 50 (1964).
208. T. E. Furia, in *CRC Handbook of Food Additives*, Vol. 1 (T. E. Furia, Ed.), CRC Press, Boca Raton, FL, 1972, p. 271.
209. C. R. Fellers and E. L. Morin, U.S. Patent 3,049,427 (1962).
210. W. S. Greig and O. Smith, *Amer. Potato J.*, *32*: 1 (1955).
211. M. L. Hunsader and F. Hanning, *Food Res.*, *23*: 269 (1958).
212. R. U. Makower and S. Schwimmer, *Biochim. Biophys. Acta*, *14*: 156 (1954).
213. D. D. Gull and F. M. Isenberg, *Proc. Amer. Soc. Hort. Sci.*, *71*: 446 (1958).
214. E. D. Kitzke, U.S. Patent 3,051,578 (1962).
215. B. S. Luh, S. J. Leonard, and D. D. Patel, *Food Technol.*, *14*: 53 (1960).
216. R. B. Guyer and F. B. Erickson, *Food Technol.*, *8*: 165 (1954).
217. L. R. Arrington and W. A. Krienke, *J. Dairy Sci.*, *37*: 819 (1954).
218. R. L. King and W. L. Dunkley, *J. Dairy Sci.*, *42*: 897 (1959).
219. P. F. Pierpoint, G. M. Trout, and C. M. Stine, *J. Dairy Sci.*, *46*: 1044 (1963).
220. L. T. Kline, T. Sonoda, H. L. Hanson, and J. H. Mitchell, *Food Technol.*, 7: 456 (1953).
221. T. J. Weiss, F. Green, and B. Watts, *Food Res.*, *18*: 11 (1953).
222. B. Borenstein, U.S. Patent 3,386,836 (1968).
223. I. F. Levy, U.S. Patent 3,003,883 (1961).

224. H. M. Caldwell, M. A. Glidden, G. G. Kelley, and M. Mangel, *Food Res.*, *25*: 131 (1960).
225. K. Ladenburg, U.S. Patent 2,868,655 (1959).
226. H. E. Power, R. Sinclair, and K. Savagaon, *J. Fish. Res. Board Can.*, *25*: 2071 (1968).
227. J. W. Boyd and B. A. Southcott, *J. Fish. Res. Board Can.*, *25*: 1753 (1968).
228. E. S. Berger, in *Kirk-Othmer Encyclopedia of Chemical Technology, Vol. 13*, Wiley-Interscience, New York, 1978, p. 111.
229. L. Hartman, *J. Sci. Food Agric.*, *4*: 430 (1953).
230. H. J. Dutton, A. W. Schwab, H. A. Moser, and J. C. Cowan, *J. Amer. Oil Chem. Soc.*, *25*: 385 (1948).
231. K. Buchholz, Ger. Patent 1,000,226 (1957).
232. J. A. Benckiser, Ger. Patent 938,581 (1956).
233. Borden Company, Brit. Patent 886, 519 (1962).
234. E. F. Bouchard and E. G. Merritt, in *Kirk-Othmer Encyclopedia of Chemical Technology*, Vol. 6, Wiley- Interscience, New York, 1978, p. 150.
235. A. J. C. Anderson, *Refining of Oils and Fats for Edible Purposes*, 2nd ed., Pergamon Press, London, 1962.
236. J. E. Huffaker, *Food Eng.*, 50 (May): 122 (1978).
237. W. H. Gardner, in *CRC Handbook of Food Additives*, Vol. 1 (T. E. Furia, Ed.), CRC Press, Boca Raton, FL, 1972, p. 225.
238. R. L. Crackel, J. I. Gray, A. M. Booren, A. M. Pearson, and D. J. Buckley, *J. Food Sci.*, *53*: 656 (1988).
239. D. G. Olson and R. E. Rust, *J. Food Sci.*, *38*: 251 (1973).
240. H. Lineweaver, J. D. Anderson, and H. L. Hanson, *Food Technol.*, *6*: 1 (1952).
241. J. J. Ganucheau, U.S. Patent 2,448,970 (1948).
242. J. J. Licciardello, E. M. Ravesi, and M. G. Allsup, *Marine Fish. Rev.*, *Natl. Ocean. Atmos. Admin.*, *44*(8): 15 (1982).
243. I. I. Lapshin and I. I. Teplitsyna, *Rybn. Khoz.*, *45*(2): 64 (1969).
244. H. W. Vahlteich, R. H. Neal, and C. M. Gooding, U.S. Patent 2,485,632 (1949).
245. R. H. Neal, C. M. Gooding, and H. W. Vahlteich, U.S. Patent 2,485,631 (1949).
246. H. W. Vahlteich, C. M. Gooding, C. F. Brown, and D. Melnick, *Food Technol.*, *1*: 6 (1954).
247. R. H. Neal, H. W. Vahlteich, and C. M. Gooding, U.S. Patent 2,485,636 (1949).
248. C. M. Gooding, R. H. Neal, and H. W. Vahlteich, U.S. Patent 2,485,637 (1949).
249. R. H. Neal, C. M. Gooding, and H. W. Vahlteich, U.S. Patent 2,485,635 (1949).
250. E. Graf, *J. Amer. Oil Chem. Soc.*, *60*: 1861 (1983).
251. E. Graf and J. W. Eaton, *Free Rad. Biol. Med.*, *8*: 61 (1990).
252. K. L. Empson, T. P. Labuza, and E. Graf, *J. Food Sci.*, *56*: 560 (1991).
253. A. J. St. Angelo, J. R. Vercellotti, H. P. Dupuy, and A. M. Spanier, *Food Technol.*, *42*(6): 133 (1988).
254. H. Tanno, in *Ullmann's Encyclopedia of Industrial Chemistry*, 5th ed., Vol. A15, VCH, Weinheim, 1985, p. 293.
255. S. Z. Dziedzic and B. J. F. Hudson, *J. Amer. Oil Chem. Soc.*, *61*: 1042 (1984).
256. C. R. Scholfield and H. J. Dutton, U.S. Patent 2,801,255 (1957).
257. B. J. F. Hudson and J. I. Lewis, *Food Chem.*, *10*: 111 (1983).
258. B. J. F. Hudson and M. Ghavami, *Lebensm. Wiss. Technol.*, *17*: 191 (1984).

259. S. Z. Dziedzic and B. J. F. Hudson, *Food Chem.*, *11*: 161 (1983).
260. D. H. Hildebrand, *J. Amer. Oil Chem. Soc.*, *61: 552 (1984)*.
261. I. S. Bhatia, N. Kaur, and P. S. Sukhija, *J. Sci. Food Agric.*, *29*: 747 (1978).
262. M. Kashima, G. S. Cha, Y. Isoda, J. Hirano, and T. Miyazawa, *J. Amer. Oil Chem. Soc.*, *68*: 119 (1991).
263. L. A. Hall and L. L. Gershbein, U.S. Patent 2,464,927 (1949).
264. H. S. Olcott and J. Van der Veen, *J. Food Sci.*, *28*: 313 (1963).
265. L. A. Hall, U.S. Patent 2,772,169 (1956).
266. C. Karahadian and R. C. Lindsay, *J. Amer. Oil Chem. Soc.*, *65*: 1159 (1988).
267. T. L. Gresham and F. W. Shaver, U.S. Patent 2,449,992 (1948).
268. A. W. Schwab, H. A. Moser, R. S. Curley, and C. D. Evans, *J. Amer. Oil Chem. Soc.*, *30*: 413 (1953).
269. L. M. Hill, E. G. Hammond, A. F. Carlin, and R. G. Seals, *J. Dairy Sci.*, *52*: 888 (1969).
270. H. H. Szmant, in *Organic Sulfur Compounds* (N. Kharasch, Ed.), Pergamon Press, New York, 1961, p. 154.
271. D. Swern, in *Organic Peroxides* (D. Swern, Ed.), Wiley, New York, 1971, p. 355.
272. P. A. Morrissey and J. Z. Tichivangana, *Meat Sci.*, *14*: 175 (1985).
273. S. Maya and M. Laborde, in *Kirk-Othmer Encyclopedia of Chemical Technology*, Vol. 21, Wiley-Interscience, New York, 1978, p. 228.
274. J. Kraljic, in *Kirk-Othmer Encyclopedia of Chemical Technology*, Vol. 21, Wiley-Interscience, New York, 1978, p. 240.
275. M. H. Fooladi, A. M. Pearson, T. H. Coleman, and R. A. Merkel, *Food Chem.*, *4*: 283 (1979).
276. D. S. MacDonald, J. I. Gray, and L. N. Gibbins, *J. Food Sci.*, *45*: 893 (1980).
277. J. Kanner, I. Ben-Gera, and S. Berman, *Lipids*, *15*: 944 (1980).
278. J. O. Igene, J. A. King, A. M. Pearson, and I. D. Morton, *J. Food Prot.*, *44*: 302 (1979).
279. R. J. Marcuse, *J. Amer. Oil Chem. Soc.*, *39*: 97 (1962).
280. M. J. Taylor, T. Richardson, and R. D. Jasensky, *J. Amer. Oil Chem. Soc.*, *58*: 622 (1981).
281. T. Okada, K. Nakagawa, and N. Yamaguchi, *J. Jpn. Soc. Food Sci. Nutr.*, *29*: 305 (1982).
282. Y. Takizawa, A. Ootani, Y. Makabe, and T. Mitsuhashi, *J. Jpn. Oil Chem. Soc.*, *29*: 199 (1980).
283. H. S. Olcott and E. J. Kuta, *Nature*, *183*: 1812 (1959).
284. G. D. Revankar, *J. Food Sci. Technol.*, *64*: 10 (1974).
285. J. S. Lin, T. C. Tom, and H. S. Olcott, *J. Agric. Food Chem.*, *22*: 526 (1974).
286. K. F. Mattil, L. J. Filer, and H. E. Logenecker, *Oil Soap*, *21*: 160 (1944).
287. Z. Kwapniewski, A. Rutkowski, and J. Sliwick, *Riv. Ital. Sostanze Grasse*, *39*: 190 (1962).
288. M. I. Soboleva, I. V. Sirokhman, and A. V. Troyan, *Izv. Vyssh. Ucheb. Zaved. Pishch. Tekhnol.*, *1971*(1):35 (1971).
289. Z. Kwapniewski and J. Silwick, *Probl. Postep. Nauk Roln.*, *53*: 365 (1965).
290. J. Silwick and J. Siechowski, *Riv. Ital. Sostanze Grasse*, *47*: 73 (1970).
291. H. Enei, S. Kawaski, and S. Okamura, U.S. Patent 3,585,223 (1971).
292. R. Marcuse, *Fette-Seifen Anstrichm.*, *63*: 547 (1961).
293. Z. Y. Chen and W. W. Nawar, *J. Amer. Oil Chem. Soc.*, *68*: 47 (1991).

294. S. Gupta, P. S. Sukhija, and I. S. Bhatia, *Indian J. Dairy Sci.*, *30*: 319 (1977).

295. M. M. Merzametov and L. I. Gadzhuva, *J. Amer. Oil Chem. Soc.*, *81*: 4979 (1976).

296. M. M. Merzametov and L. I. Gadzhuva, *Izv. Vyssh. Ucheb. Zaved. Pish. Tech.*, *6*: 20 (1982).

297. M. M. Merzametov and L. I. Gadzhuva, *Izv. Vyssh. Ucheb. Zaved. Pish. Tech.*, *5*: 35 (1982).

298. J. W. Stull, E. O. Herreid, and P. H. Tracy, *J. Dairy Sci.*, *34*: 187 (1951).

299. J. Kanner and B. J. Juven, *J. Food Sci.*, *45*: 1105 (1980).

300. A. Seher and D. Loschner, *Fette-Seifen Anstrichm.*, *87*(11): 454 (1985).

301. J. R. L. Walker and C. E. S. Reddish, *J. Sci. Food Agric.*, *15*: 902 (1964).

302. J. R. Chipault, G. R. Mizuno, and J. M. Hawkins, *Food Res.*, *17*: 46 (1952).

303. J. R. Chipault, G. R. Mizuno, and W. O. Lundberg, *Food Technol.*, *10*: 209 (1956).

304. J. Loliger, in *Free Radicals and Food Additives* (O. I. Arouma and B. Halliwell, Eds.), Taylor and Francis, London, 1991, p. 121.

305. C. M. Houlihan, C. T. Hou, and S. S. Chang, *J. Amer. Oil Chem. Soc.*, *61*: 1036 (1984).

306. C. M. Houlihan, C. T. Hou, and S. S. Chang, *J. Amer. Oil Chem. Soc.*, *62*: 96 (1985).

307. M. Namiki, *Crit. Rev. Food Sci. Nutr.*, *29*(4): 273 (1990).

308. S. S. Chang, B. Ostric-Matijasevic, O. Hsieh, and A. L. Huang-Cheng-Li, *J. Food Sci.*, *42*: 102 (1977).

309. U. Bracco, J. Loliger, and J. L. Viret, *J. Amer. Oil Chem. Soc.*, *58*: 686 (1981).

310. J. R. Chipault, G. R. Mizuno, and J. M. Hawkins, *Food Res.*, *20*: 443 (1955).

311. C. W. Dubois and D. K. Tressler, *Proc. Inst. Food Technol.*, 202 (1943).

312. I. D. A. Atkinson, S. R. Cecil, J. G. Woodroof, and E. Shelor, *Food Ind.*, *19*: 1198 (1947).

313. T. Hirahara and H. Takai, *Jpn. J. Nutr.*, *32*: 1 (1974).

314. Y. Watanabe and Y. Ayono, *J. Jpn. Soc. Food Nutr.*, *27*: 181 (1974).

315. R. Inatani, N. Nakatani, H. Fuwa, and H. Seto, *Agric. Biol. Chem.*, *46*: 1661 (1982).

316. R. Inatani, N. Nakatani, and H. Fuwa, *Agric. Biol. Chem.*, *47*: 521 (1983).

317. N. Nakatani and R. Inatani, *Agric. Biol. Chem.*, *48*: 2081 (1984).

318. B. Ostric-Matijasevic, *Rev. Fr. Corps Gras*, (8–9): 443 (1963).

319. H. Pyenson and P. H. Tracy, *J. Dairy Sci.*, *31*: 539 (1940).

320. H. Pyenson and P. H. Tracy, *J. Dairy Sci.*, *33*: 315 (1950).

321. J. Burri, M. Graf, P. Lambelet, and J. Loliger, Eur. Patent 340500 (1988).

322. J. Burri, M. Graf, P. Lambelet, and J. Loliger, *J. Sci. Food Agric.*, *110*: 153 (1989).

323. O. I. Arouma, P. J. Evans, H. Kaur, L. Sutcliffe, and B. Halliwell, *Free Rad. Res. Commun.*, *10*: 143 (1990).

324. K. M. Narayanan, N. S. Kapur, G. S. Bains, and D. S. Bhatia, *Food Sci.*, *6*: 245 (1957).

325. D. E. Pratt and B. J. F. Hudson, in *Food Antioxidants* (B. J. F. Hudson, Ed.), Elsevier Applied Science, London, 1990, p. 171.

326. W. Heimann and F. Reiff, *Fatt Seifen*, *55*: 451 (1957).

327. A. Letan, *J. Food Sci.*, *31*: 518 (1966).

328. A. Letan, *J. Food Sci.*, *31*: 395 (1966).

329. A. C. Mehta and T. R. Seshadri, *J. Sci. Ind. Res.*, *18B*: 24 (1959).

330. D. L. Crawford, R. O. Sinnhuber, and H. Aft, *J. Food Sci.*, *26*: 139 (1962).

331. K. Hayashi, in *The Chemistry of Flavonoid Compounds* (T. A. Geissman, Ed.), Pergamon Press, London, 1962, p. 248.

332. J. Gripenberg, in *The Chemistry of Flavonoid Compounds* (T. A. Geissman, Ed.), Pergamon Press, London, 1962, p. 406.

333. P. Markakis, in *Anthocyanins as Food Colors* (P. Markakis, Ed.), Academic Press, New York, 1982, p. 245.

334. C. D. Chiriboga and F. J. Francis, *J. Amer. Soc. Hort. Sci.*, *95*: 233 (1970).

335. C. D. Chiriboga and F. J. Francis, *J. Food Sci.*, *38*: 464 (1973).

336. H. L. Hergert, in The Chemistry of Flavonoid Compounds (T. A. Geissman, Ed.), Pergamon Press, London, 1962, p. 553.

337. E. F. Kurth and F. L. Chan, *J. Amer. Oil Chem. Soc.*, *28*: 433 (1951).

338. S. E. O. Mahgoub and B. J. F. Hudson, *Food Chem.*, *16*: 97 (1985).

339. I. A. Radaeva and N. A. Tyukavkina, USSR Patent 350,451 (1972).

340. T. Osawa, H. Katsuzaki, Y. Hagiwara, H. Hagiwara, and T. Shibamoto, *J. Agric. Food Chem.*, *40*: 1135 (1992).

341. V. V. Lisitski, G. I. Kas'yanov, and V. I. Bessarabov, *Rybn. Khoz.*, *11*: 61 (1984).

342. M. Sato, K. Nakamura, and M. Tadenuma, Jpn. Patent 7,241,040 (1972).

343. T. Ariga, I. Koshiyama, and D. Fukushima, *Agric. Biol. Chem.*, *52*: 2717 (1988).

344. M. T. Meunier, E. Duroux, and P. Bastide, *Plant. Med. Phytotherapie.*, *23*(4): 267 (1990).

345. T. Ariga, I. Koshiyama, and D. Fukushima, U.S. Patent 4,797,421 (1989).

346. K. Igarashi, K. Takanashi, M. Makino, and T. Yasui, *J. Jpn. Soc. Food Sci. Technol.*, *36*: 852 (1989).

347. K. Igarashi, T. Yoshida, and E. Suzuki, *J. Jpn. Soc. Food Sci. Technol.*, *40*: 138 (1993).

348. O. Isler and F. Kienzle, in *Kirk-Othmer Encyclopedia of Chemical Technology*, Vol. 24, Wiley-Interscience, New York, 1978, p. 140.

349. L. D. Titarenko, A. N. Svidovski, and E. P. Tikhanova, *Tovarovedenie*, *24*: 10 (1992).

350. M. H. Gordon, in *Food Antioxidants* (B. J. F. Hudson, Ed.), Elsevier Applied Science, London, 1990, p. 1.

351. C. S. Foote and R. W. Denny, *J. Amer. Chem. Soc.*, *90*: 6233 (1968).

352. H. Klaui and J. C. Bauernfeind, in *Carotenoids as Colorants and Vitamin A Precursors* (J. C. Bauernfeind, Ed.), Academic Press, New York, 1981, p. 47.

353. A. Sattar, J. M. deMan, and J. C. Alexander, *J. Amer. Oil Chem. Soc.*, *53*: 473 (1976).

354. A. Kiritsakis and L. R. Dugan, *J. Amer. Oil Chem. Soc.*, *62*: 892 (1985).

355. R. S. Tsai, J. J. Huang, and R. L. Chang, *J. Chin. Agric. Chem. Soc.*, *27*: 233 (1989).

356. E. C. Lee and D. B. Min, *J. Food Sci.*, *53*: 1984 (1988).

357. J. Terao, R. Yamaguchi, H. Murakami, and S. Matsushita, *J. Food Proc. Preserv.*, *4*: 79 (1980).

358. K. Warner and E. N. Frankel, *J. Amer. Oil Chem. Soc.*, *64*: 213 (1987).

359. A. H. Clements, R. H. Van den Engh, D. T. Frost, and K. Hoogenhout, *J. Amer. Oil Chem. Soc.*, *50*: 325 (1973).

360. E. W. Kellogg *III and I.* Fridovich, *J. Biol. Chem.*, *250*: 8812 (1975).

361. E. A. H. Roberts, in *The Chemistry of Flavonoid Compounds* (T. A. Geissman, Ed.), Pergamon Press, London, 1962, p. 468.

362. Food Research Laboratories, Mitsui Norin Co., Ltd. Fujieda, Japan, 1991.

363. M. H. Lee and R. L. Sher, *J. Chin. Agric. Chem. Soc.*, *22*: 226 (1984).

364. T. Matsuzaki and Y. Hara, *Nippon Nogeikagaku Kaishi*, *59*: 129 (1985).

365. J. Mai, L. J. Chambers, and R. E. McDonald, U.K. Patent Appl. GB 2151123A (1985).

366. J. Mai, L. J. Chambers, and R. E. McDonald, Eur. Patent Appl. EP 0169936A1 (1986).
367. M. Oyaizu, *J. Jpn. Soc. Cold Preserv. Food, 14*: 144 (1988).
368. J. Mai, L. J. Chambers, and R. E. McDonald, U.S. Patent 4,891,231 (1989).
369. J. Mai, L. J. Chambers, and R. E. McDonald, U.S. Patent 4,925,681 (1990).
370. I. V. Sirokhman, *Tovarovedenie, 18*: 38 (1985).
371. T. B. Lloyd and W. Showak, in *Kirk-Othmer Encyclopedia of Chemical Technology*, Vol. 24, Wiley-Interscience, New York, 1978, p. 807.
372. T. B. Loyd, in *Kirk-Othmer Encyclopedia of Chemical Technology*, Vol. 24, Wiley-Interscience, New York, 1978, p. 851.
373. E. M. Elkin, in *Kirk-Othmer Encyclopedia of Chemical Technology*, Vol. 20, Wiley-Interscience, New York, 1978, p. 575.

5

Toxicological Aspects of Food Antioxidants

D. L. Madhavi
University of Illinois, Urbana, Illinois

D. K. Salunkhe
Utah State University, Logan, Utah

5.1 INTRODUCTION

Toxicological studies are crucial in determining the safety of an antioxidant for food use and also in determining the acceptable daily intake (ADI) levels. Compounds like nordihydroguaiaretic acid (NDGA) have been banned, and the ADIs for widely used antioxidants such as butylated hydroxyanisole (BHA), butylated hydroxytoluene (BHT), and the gallates have changed over the years mainly because of their toxicological effects in various species. A number of synthetic and natural compounds with very effective antioxidative properties have not been approved for food use because of insufficient toxicological information. The present chapter highlights the toxicological aspects such as acute toxicity, short-term and long-term studies, biochemical changes, teratogenicity, mutagenicity, and carcinogenicity of some of the antioxidants used in the food industry. The ADIs allocated by the Joint FAO/WHO Expert Committee on Food Additives (JECFA) are included wherever possible.

5.2 TOXICOLOGICAL ASPECTS OF PRIMARY ANTIOXIDANTS

5.2.1 Phenols

Gallates

The gallate group comprises the propyl, octyl, and dodecyl esters of gallic acid (3,4,5-trihydroxybenzoic acid). Of the three, propyl gallate (PG) is more effective as an antioxidant and is widely used in stabilizing animal fats and vegetable oils.

It is also effective in meat products, spices, and snacks. In 1980, JECFA allocated a group ADI of 0–0.2 mg/kg bw. But later studies indicated that octyl and dodecyl gallates may have an adverse effect on reproduction. Hence, in a reevaluation in 1987, an ADI of 0–2.5 mg/kg bw was allocated to PG and no ADIs were allocated to OG and DG because of insufficient toxicological information (1).

TOXICOLOGICAL STUDIES

Absorption, Metabolism, and Excretion.   In rats, nearly 70% of an oral dose of PG was absorbed in the gastrointestinal tract, OG and DG were absorbed to a lesser extent. All three esters were hydrolyzed to gallic acid and free alcohol. The free alcohols were metabolized through the Krebs cycle. Gallic acid was methylated to yield 4-*O*-methyl gallic acid, which was the main metabolite found in urine either free or conjugated with glucuronic acid. Small quantities of gallic acid were also excreted as glucuronide as well as in free form. Significant amounts of unchanged esters were excreted in the feces. Similar urinary metabolites were present in rabbits in addition to small quantities of pyrogallol (Fig. 5.1). In pigs, the metabolism was similar to that observed in rats (2–4).

Acute Toxicity.   The acute oral LD_{50} of PG in rats was 3600–3800 mg/kg bw and in mice 2000–3500 mg/kg bw (5,6). The acute oral LD_{50} of OG in rats was 4700 mg/kg bw, and for DG, 6500 mg/kg bw (7).

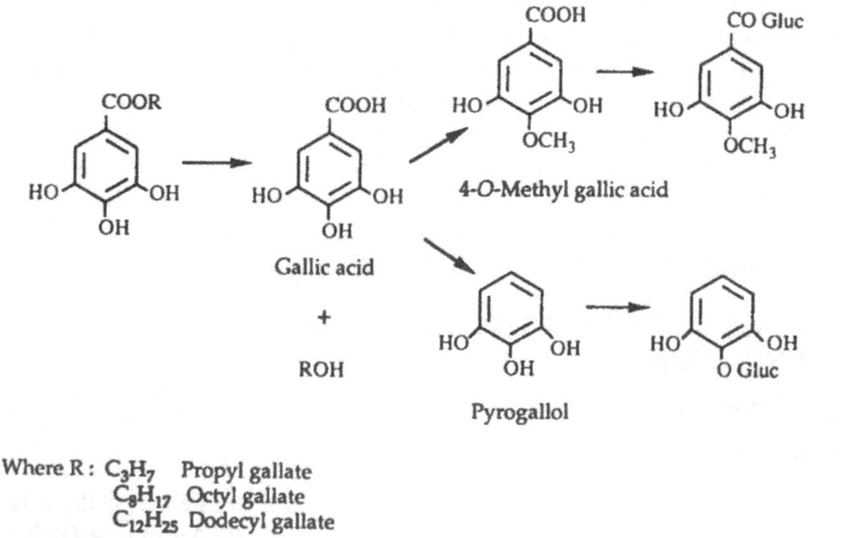

Where R : C_3H_7 Propyl gallate
C_8H_{17} Octyl gallate
$C_{12}H_{25}$ Dodecyl gallate

Fig. 5.1   Major metabolites of the gallates. (From Ref. 4.)

Short-Term Studies.   In rats fed PG at dose levels of 0.1, 0.25, 0.3, and 0.5% for 6 weeks, no adverse changes were observed in body weight, liver weight, or total liver lipid content (8). Higher levels of 1.17 and 2.34% in the diet caused a reduction in weight gain, retarded growth, and about 40% deaths accompanied by renal damage in the first month. No pathological changes were observed in the survivors (5). In a 4-week study in rats, at a dose level of 2.5%, PG produced growth retardation, anemia, hyperplasia in the outer kidney medulla, and an increase in the activity of cytoplasmic and microsomal hepatic drug-metabolizing enzymes. An increase in enzyme activity was also observed at the 0.5% level (9). In mice fed 0.5 or 1% PG for 3 months, no toxic effects were observed (10). Propyl gallate had no adverse effects in dogs at the 0.0117% level for over a period of 12 months (5). In rats fed 0.2% dodecyl gallate for 70 days, no changes in body weight were observed (11). However, in weanling rats given 2.5 and 5% DG, 100% mortality was observed within 10 and 7 days, respectively (12). In a 13-week study in rats fed 0.1, 0.25, and 0.5% octyl gallate, no deleterious changes were observed in body weight or food consumption, and hematological studies, urinalysis, and histopathological examination showed no compound-related effects (13). In guinea pigs, 0.2% of PG, OG, or DG had no demonstrable ill effects (14).

Long-Term and Carcinogenicity Studies.   In a 2-year study, rats fed 5% PG showed reduced food intake, growth retardation, and patchy hyperplasia of the stomach (6). Orten et al. (5) observed, in addition to retarded growth, lower hemoglobin levels and mottled kidneys in a 2-year study in rats at dose levels of 1.17 or 2.34%. In mice fed 0.5 and 1% PG for 21 months, no adverse effects on food intake, body weight, survival time, or hematological response were observed (Table 5.1)

Table 5.1   Relative Organ Weights and Terminal Body Weights of Mice Fed Diets Containing Propyl Gallate for 21 Months

Dietary level (%)	Number of animals	Terminal body weight (g)	Relative organ weights (g/100 g body weight)			
			Liver	Spleen	Kidneys	Adrenals
Males						
0.0	32	35.0 ± 0.9	5.89 ± 0.15	0.35 ± 0.07	1.74 ± 0.04	0.015 ± 0.001
0.5	16	36.5 ± 1.5	6.25 ± 0.30	0.35 ± 0.06	1.76 ± 0.05	0.017 ± 0.002
1.0	33	34.4 ± 0.8	6.41 ± 0.17	0.23 ± 0.02	1.68 ± 0.03	0.013 ± 0.001
Females						
0.0	34	27.5 ± 0.7	5.58 ± 0.19	0.39 ± 0.04	1.38 ± 0.04	0.014 ± 0.001
0.5	18	26.4 ± 1.3	6.01 ± 0.26	0.40 ± 0.04	1.54 ± 0.05	0.020 ± 0.002
1.0	32	27.4 ± 0.8	6.32 ± 0.25	0.41 ± 0.04	1.46 ± 0.05	0.015 ± 0.001

Source: Ref. 10.

(10). Propyl gallate at levels of 0.6 or 1.2% resulted in a dose-related reduction in body weight in both sexes in a 103-week chronic toxicity and carcinogenicity study in F344 rats and B6C3F1 mice. Male rats showed an increased incidence of hepatic vacuolization and suppurative inflammation of the prostate gland. At lower dose levels, tumors at several anatomical sites including the thyroid gland, pancreas, and adrenal glands were observed at a higher incidence than the control in males. Also, malignant lymphoma occurred with a positive trend in males. In females, incidences of uterine polyps and mammary gland adenomas were observed (Table 5.2) (15). However, the control group in this experiment showed lower tumor incidence than the historical control groups in the same laboratory. Comparison of the experimental

Table 5.2 Tumor Incidence in F344 Rats Given Propyl Gallate in the Diet for 103 Weeks

Tumor	Tumor incidence[a] at PG dietary levels (%)		
	0	0.6	1.2
Males			
Adrenal gland			
Phaeochromocytoma	4/50 (8)	13/48 (27)	88/50 (16)
Pancreas			
Islet-cell adenoma	0/50 (0)	8/50 (16)	2/50 (4)
Prepupital gland			
Adenoma	0/50 (0)	5/50 (10)	0/50 (0)
Adenoma, adenocarcinoma, or carcinoma	1/50 (2)	8/50 (16)	0/50 (0)
Thyroid gland			
Follicular cell adenoma or carcinoma[b]	0/50 (0)	0/50 (0)	3/50 (6)
Hematopoietic system			
Leukemia	16/50 (32)	7/50 (14)	6/50 (12)
Females			
Uterus			
Endometrial stromal polyps	6/50 (12)	8/50 (16)	13/50 (26)
Mammary gland			
Fibroadenoma[c]	11/50 (22)	2/50 (4)	5/50 (10)
Adenoma[b]	0/50 (0)	0/50 (0)	3/50 (6)

[a]Number of tumor-bearing animals divided by number examined at the site (percentage in parentheses).
[b]Significant positive trend ($P < 0.05$).
[c]Significant negative trend ($P < 0.05$).

results with historical controls indicated that the results were not significant. Hence, it was concluded that under the conditions of this study, PG was not carcinogenic to mice or rats. PG was found to have a protective effect against dimethylbenz[a]-anthracene-induced mammary tumors in rats (16,17).

Van Esch (2) observed no deleterious effects in rats fed 0.035, 0.2, or 0.5% OG or DG in a 2-year study. In the same study, Van Esch reported that DG had no adverse effects in pigs at the 0.2% level for 13 weeks. In rats administered DG intragastrically in doses of 10, 50, or 250 mg/kg bw for 2 years, the highest dose level resulted in increased mortality, decrease in the weight of spleen, and morphological changes in the liver, kidneys, and spleen. The dose of 50 mg/kg bw was also toxic. Ten milligrams per kilogram body weight was regarded as the threshold dose for rats (18).

Reproduction. Propyl gallate was without any adverse effects in a three-generation study in rats fed 0.03, 0.2, and 0.5% in the diet (2). In a teratogenicity study, Tanaka et al. (19) observed rats fed PG at dose levels of 0.4, 1, and 2.5% from day 0 to day 20 of pregnancy and found maternal toxicity and slight retardation of fetal development at the higher dose levels but observed no teratogenic effects. At dose levels of 2.5, 12, 54, or 250 mg/kg bw per day, PG had no adverse effects on organogenesis in rabbits (20). It was reported to prevent the occurrence of fetal abnormalities caused by vitamin E deficiency and hydroxyurea in rats and rabbits (21, 22).

No adverse effects were observed in two three-generation studies in rats fed OG at dose levels of 0.035–0.5% (2,7). In another two-generation study in rats at dose levels of 0.1 and 0.5%, a marked reduction in weaning survival index and body weights was observed at the 0.5% level. A dose-dependent reduction in the number of implantation sites and a reduction in the number of corpora lutea were observed (9). Dodecyl gallate had no adverse effects at a dose level of 0.2% in a three-generation study in rats (7). At a higher dose level of 0.5%, Van Esch (2) observed loss of litters due to underfeeding and a slight hypochromic anemia at the 0.2% level.

Skin Toxicity. The gallates have been reported to cause allergic contact dermatitis. In three reports from Switzerland, gallates in margarine were said to have caused occupational hand dermatitis (23–25). Van Ketel (26) reported one case of occupational hand and face dermatitis from peanut butter containing OG. In recent years a greater number of cases have been reported mainly because of topically applied products such as lip balm and body lotion in which the gallates are being used to a greater extent than other antioxidants. Sensitization did not occur when there was oral exposure. In a recent study Hausen and Beyer (27) studied the sensitizing capacity of the three gallates using a modern sensitization procedure in guinea pigs (Table 5.3). It was found that PG and OG were moderate sensitizers whereas DG was the strongest sensitizer. Hausen and Beyer (27) also

Table 5.3 Cases and Sources of Gallate
Allergy[a]

Source of sensitization	Responsible sensitizer[b]	No. of cases
Antibiotic ointment	PG	3
Antipsoriatic cream	PG	1
Baby lotion	PG	5
Body lotion	PG	6
Chicken fat (dripping)	OG	1
Cosmetic cream	PG	2
Cosmetic cream	DG, OG	3
Deodorant	PG	1
Leg ulcer ointment	PG	2
Lipstick, lip balm	PG	2
Moisturizing cream	PG	1
Peanut butter	OG	1
Washing powder	DG	4

[a]Original references in Hausen and Beyer (27).
[b]PG = propyl gallate; OG = octyl gallate; DG = dodecyl gallate.
Source: Ref. 27.

observed that an increase in the side-chain length closely correlated with an increase in the sensitizing potential.

Mutagenicity.   Propyl gallate has been reported to be negative in a number of mutagenicity tests. In the Ames test using four tester strains of *Salmonella typhimurium* with or without metabolic activation, PG was negative (28). It was negative in in vitro tests for chromosomal aberrations that used human embryonic lung cells (29) and was a weak inducer of sister chromatid exchange in Chinese hamster ovary cells (NTP unpublished data cited in Ref. 15). In vivo tests for chromosomal abnormalities in rat bone marrow cells, a dominant lethal study in rats, and a micronucleus test in mice were negative (29,30). Propyl gallate had no clastogenic effect in rats and mice (31). It inhibited the mutagenic activity of benzopyrene metabolites, pyrolysis products of albumin, and aflatoxin B in *S. typhimurium* TA98 (28,32,33). No information is available on the mutagenic potential of either octyl or dodecyl gallate.

Hydroquinone

Hydroquinone (HQ; 1,4-dihydroxybenzene) was proposed for use as a food antioxidant in the 1940s, and extensive studies were conducted to determine the toxicological properties of HQ and its oxidation products quinhydrone and quinone (Fig. 5.2). Hydroquinone was found to have a high level of toxicity.

TOXICOLOGICAL STUDIES

Hydroquinone is rapidly absorbed from the gastrointestinal tract and eliminated in urine as sulfate and glucuronide conjugate (34). Approximate oral LD_{50} in mg/kg bw in rats was 320; in mice, 400; in guinea pigs, 550; in pigeons, 300; in cats, 70; in dogs, 200; and in rats, for quinhydrone 225, and for quinone 225. Symptoms of HQ poisoning developed 30–90 min after oral administration. Some of the symptoms observed were hyperexcitability, tremors, convulsions, and salivation in dogs and cats and incoordination of the hind limbs in dogs. Death occurred within a few hours. Oral administration of HQ at 100 mg/kg bw in dogs and 70 mg/kg bw in cats produced mild to severe swelling of the area around the eye, nictitating membrane, and the upper lip. In dogs receiving lower doses of 25 or 50 mg/kg for 4 months, a slight involvement of the eye was observed (35). In a chronic toxicity test, Carlson and Brewer (36) reported that in rats fed 0.5% HQ for 2 years and in dogs given 100 mg/kg bw for 26 weeks, no adverse effects were observed on body weight or growth rate and no pathological changes were observed. In another 2-year feeding study [FDA (1950), unpublished data cited in Ref. 6], HQ at the 2% level increased the incidence of chronic gastrointestinal ulcerations and renal tumors in

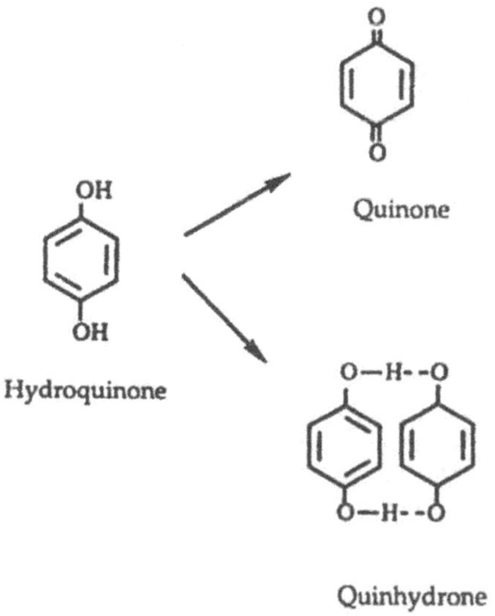

Fig. 5.2   Metabolites of hydroquinone.

rats. In lifetime feeding studies in rats, HQ at levels 100 times that proposed for human consumption had a significant retarding effect on growth rate. On the basis of toxicological data, HQ was considered a harmful and deleterious substance for addition to food (6). At present HQ is used as a stabilizer in polymers and in very small quantities in dermatological preparations as a bleaching agent.

Trihydroxybutyrophenone

Trihydroxybutyrophenone (THBP; 2,4,5-trihydroxybutyrophenone) is a substituted phenol listed as GRAS by the FDA for addition to foods and food packaging materials. THBP is effective in stabilizing vitamin A, lard, paraffin wax, mineral oil, and peanut oil (37). However, THBP is being used only in food packaging materials.

TOXICOLOGICAL STUDIES

The absorption, metabolism, and excretion of THBP has been studied in rats and dogs (38). Nearly 75% of a single oral dose of THBP was absorbed. In rats given single oral doses of 400 mg/kg bw, 6.3% of the dose was excreted as unchanged THBP in the urine and 1.7% in the feces, 23% as glucuronides and 52% as etheral sulfates (Fig. 5.3). In dogs given single oral doses of 300 mg/kg bw, 2.3% of the

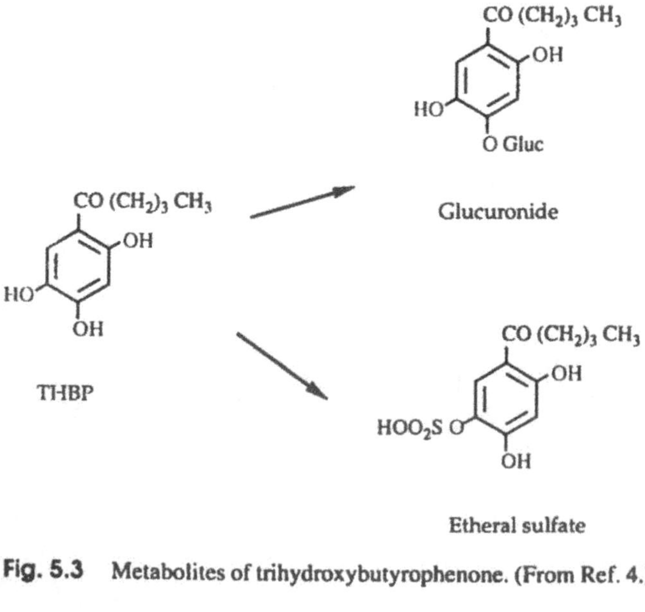

Fig. 5.3   Metabolites of trihydroxybutyrophenone. (From Ref. 4.)

dose was excreted unchanged in the urine and 5.3% in the feces, 30% as glucuronides and 45% as etheral sulfates (Table 5.4). The path of metabolism was largely by conjugation, and no evidence of reduction or oxidation was found. Etheral sulfate conjugation at the 5-hydroxyl group leads to 5-butyryl-2,4-dihydroxyphenly hydrogen sulfate isolated as the potassium salt. Glucuronic acid conjugation at the 4-hydroxyl group leads to 4-butyryl-2,5-dihydroxyphenyl glucosiduronic acid, which was not isolated.

Very little is known about the toxicity of THBP. The acute oral LD_{50} in rats was 3200–6400 mg/kg bw, and in mice, 800–1600 mg/kg bw (3). In a long-term study in rats fed 0.1, 0.3, 1, and 3% daily for 2 years and in dogs at levels of 0.1, 0.3, and 0.5 g/kg bw per day for 1 year, no adverse effects were observed. Dogs readily tolerated 0.5 g/kg and rats 1.5 g/kg with negligible storage in body fat and various organs including liver, brain, and kidneys (38).

Nordihydroguaiaretic Acid

Nordihydroguaiaretic acid [NDGA; 2,3-dimethyl-1,4-bis(3,4-dihydroxyphenyl) butane] is a naturally occurring polyhydroxy phenolic antioxidant prepared from an evergreen desert shrub, *Larrea divaricata*, and has also been commercially synthesized. NDGA is effective in oils, fats, and fat-containing foods. NDGA was removed from the list of GRAS compounds (39) because of its toxicity at high dose levels. It is no longer used as food antioxidant in many countries.

Table 5.4 Excretion of Trihydroxybutyrophenone and Metabolites After a Single Oral Dose

Animals	Dose (g)	Urinary THBP	Glucuronide	Etheral sulfate	Period of excretion (days)	Fecal THBP
Rat group						
1	0.13	7.8	21	44	3–4	1.8
2	0.13	6.3	23	66	3–4	1.5
3	0.15	8.0	15	50	4–5	2.2
4	0.15	3.1	32	47	3–4	1.3
Mean		6.3	23	52		1.7
Dog. no.						
13	2.85	2.2	36	30	4–5	4.9
16	4.34	1.3	41	31	4–5	6.2
14	4.2	2.2	19	65	3–4	4.8
15	3.64	3.6	24	54	4–5	5.5
Mean		2.3	30	45		5.3

Results are expressed as percentage of dose fed. Metabolites were detected chromatographically.
Source: Ref. 39.

TOXICOLOGICAL STUDIES

Absorption, Metabolism, and Excretion.   In rats, NDGA was readily converted to its metabolite orthoquinone (Fig. 5.4). Orthoquinone was found in all kidney extracts and urine of rats fed 0.5 and 1% NDGA. No free NDGA was observed. The orthoquinone was formed in the lower third of the ileum and cecum and was probably absorbed there. In rats, after a single administration of 250 mg of NDGA directly into the small intestine, 2760 µg of orthoquinone was observed in the ilium and 1620 µg in the cecum 7.5 h after dosing (40).

Effects on Enzymes.   Nordihydroguaiaretic acid was found to inhibit a number of enzyme systems. Tappel and Marr (41) observed specific inhibition of peroxidase, catalase, alcohol dehydrogenase, and nonspecific inhibition of ascorbic acid oxidase, D-amino acid oxidase, the cyclophorase system, and urease. Placer et al. (42) observed that NDGA inhibited the esterase activity in liver and serum, which may have an adverse effect on fat metabolism.

Acute Toxicity.   The acute oral LD_{50} in rats was 2000–5500 mg/kb bw and in guinea pigs 830 mg/kg bw. In mice the acute oral and intraperitoneal LD_{50} were 2000–5000 mg/kg bw and 550 mg/kg bw, respectively (6).

Short-Term Studies.   Short-term studies in mice and dogs have been reported. NDGA at dose levels of 0.25 and 0.5% for 6.5–7.5 months in mice or at levels of 0.1, 0.5, and 1% for 1 year in dogs had no deleterious effects on growth rate or food intake, and no pathological changes were observed (43,44). In guinea pigs, NDGA induced skin sensitivity (45).

Long-Term Studies.   Long-term studies have been conducted in rats fed 0.1, 0.5, and 1% NDGA for 2 years, Cranston et al. (43,44) observed a reduction in growth rate, cecal hemorrhage, and mesenteric cysts at higher dosage levels.

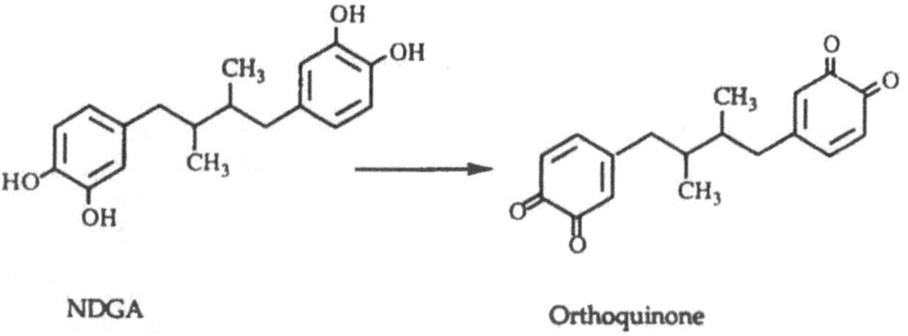

NDGA   Orthoquinone

Fig. 5.4   Metabolite of nordihydroguaiaretic acid. (From Ref. 4.)

Lehman et al. (6) confirmed the observations and reported that 0.5% was the lowest effective dose. Grice et al. (40) reported that at dose levels of 0.5 and 1% for 1.5 years, NDGA had a strong toxic effect in the rat and induced extensive cystic reticuloendotheliosis of the paracecal lymph nodes, an increase in kidney weight, and loss of tubular function, with distended tubular cells indicating impaired kidney function. NDGA did not affect absolute kidney weights but significantly influenced the kidney-to-body weight ratio (Table 5.5). Rats fed 2% NDGA for shorter periods showed similar pathological changes. The orthoquinone may induce loss of tubular function in the kidneys by affecting the permeability of the lysosomal membranes or by inhibiting the lysosomal enzymes (46).

5.2.2 "Hindered" Phenols

Butylated Hydroxyanisole

Butylated hydroxyanisole (BHA; *tert*-butyl-4-hydroxyanisole) is perhaps the most extensively used antioxidant in the food industry. BHA is used in fats and oils, fat-containing foods, confectioneries, essential oils, food-coating materials, and waxes. BHA is a mixture of two isomers, 2-*tert*-butyl-4-hydroxyanisole (2-BHA) and 3-*tert*-butyl-4-hydroxyanisole (3-BHA), with the commercial compound con-

Table 5.5 Average Terminal Body Weight and Absolute and Relative Kidney Weights of Rats Fed Nordihydroguaiaretic Acid at 0–1% of the Diet for 74 Weeks[a]

Dietary level	Number of rats	Terminal body weight (g)	Kidney weight	
			Absolute (g)	Relative (g/100 g bw)
Males				
0	8	395	2.82	0.71
0.5	8	335**	2.75	0.82**
1.0	7	322**	2.65	0.82**
Females				
0	6	270	2.31	0.86
0.5	9	238**	2.26	0.95*
1.0	9	230**	2.36	1.03*

[a]Values marked with asterisks differ significantly from those of controls: *$P = 0.07$; **$P < 0.01$.
Source: Ref. 40.

taining 90% of the 3-isomer (47). In a recent reevaluation, JECFA allocated an ADI of 0–0.5 mg/kg bw (48).

TOXICOLOGICAL STUDIES

Absorption, Metabolism, and Excretion.   The absorption and metabolism of BHA has been studied in rats, rabbits, dogs, monkeys, and humans. BHA was rapidly absorbed from the gastrointestinal tract in rats (49), rabbits (50), dogs, and humans (51), rapidly metabolized, and completely excreted. No evidence of tissue accumulation of BHA was observed in rats or dogs (49,52,53). The major metabolites of BHA were the glucuronides, ether sulfates, and free phenols (Fig. 5.5). The metabolites were excreted in the urine, and unchanged BHA was eliminated in the feces. The proportions of the different metabolites varied in different species and also for different dosage levels. In rabbits dosed orally with 1 g of BHA, 46% of glucuronides, 9% ether sulfates, and 6% free phenols were observed in the urine (50). In rats at lower doses, the metabolism was similar to that of rabbits. Urinary excretion was 86% by 24 h and 91% in 4 days (54). In dogs, nearly 60% of a 350 mg/kg dose was excreted unchanged in the feces. The remaining was excreted as ether sulfate, *tert*-butyl hydroquinone (TBHQ), an unidentified phenol, and some glucuronides. In humans, 22–72% of an oral dose

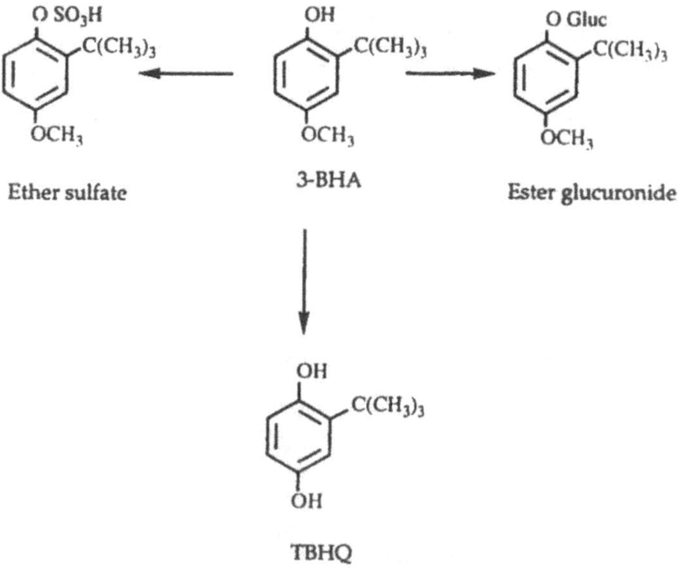

Fig. 5.5   Major metabolites of butylated hydroxyanisole.

of BHA at levels of 0.5–0.7 mg/kg bw was recovered as glucuronides in 24 h, and less than 1% free BHA and very little ether sulfates were observed (51). El-Rashidy and Niazi (55) reported substantial quantities of TBHQ glucuronide or sulfate as a metabolite of the 3-isomer. Later studies confirmed the formation of TBHQ in rats (56,57). Tissue retention of BHA was greater in humans than in rats (58). Much lower doses of BHA were required to produce a given plasma level in humans than in rats (56).

Acute Toxicity.   The acute oral LD_{50} was 2200–5000 mg/kb bw in rats and 2000 mg/kg bw in mice (3,6).

Short-Term Studies.   Short-term studies have been conducted in a number of species including rats, rabbits, and dogs. Rats administered 500–600 mg/kg bw for a period of 10 weeks showed decreased growth rate and reduced activity of the enzymes catalase, peroxidase, and cholinesterase (59). In rabbits, large doses of BHA (1 g/day) administered by stomach tube for 5–6 days caused a 10-fold increase in sodium excretion and a 20% increase in potassium excretion in the urine (60). No adverse effects were observed in dogs fed 0.3, 30, or 100 mg/kg bw BHA for 1 year (52).

In high doses (500 mg/kg bw per day), BHA induced an increase in the relative liver weight in rats and mice. In rats, the changes follow a complex course that depends on the mode of administration. When rats were given BHA by stomach tube, the relative liver weight increase followed a bimodal time course, with maxima on days 2 and 10 and a highly significant increase on day 7. On dietary administration, liver enlargement was not apparent until day 5, and a single maximum was observed by day 11 (61,62). A preliminary ultrastructural study by Allen and Engblom (63) did not reveal any nucleolar abnormalities in the liver. BHA has been reported to induce a number of hepatic enzymes in rats and mice such as epoxide hydrolase, glutathione-*S*-transferase, glucose-6-phosphate dehydrogenase, and biphenyl-4-hydroxylase (Table 5.6) (61,62,64–66). In dogs, BHA at levels of 1 and 1.3% induced liver enlargement, proliferation of the smooth endoplasmic reticulum, the formation of hepatic myelinoid bodies, and an increase in hepatic enzyme activity (67). Given in doses of 500 mg/kg bw to young rhesus monkeys for 28 days, BHA induced liver hypertrophy and the proliferation of the smooth endoplasmic reticulum. Some differences were observed between monkeys and rats. In monkeys, the activity of microsomal glucose-6-phosphatase was decreased and the nitroanisole demethylase activity was increased, whereas in rats no changes were observed at similar dose levels (63,65).

Long-Term and Carcinogenicity Studies.   In earlier long-term studies, BHA was found to be without any toxic effects in rats after 22 months (59,68,69) and in dogs after 15 months (53). However, in later studies, Ito et al. (70–72) reported that in F344 rats, administration of BHA at a 2% level resulted in a high incidence of

Table 5.6 Effects of Butylated Hydroxyanisole Treatment on Hepatic Microsomal and Cytoplasmic Enzyme Activities of Mice[a]

Enzymes	Control group	BHA-treated group
Number of animals	30	12
Microsomal enzymes		
NADPH-cytochrome c reductase (μmol cytochrome c reduced min^{-1} mg^{-1})	0.712 ± 0.028	1.31 ± 0.07
NADH-cytochrome c reductase (μmol cytochrome c reduced min^{-1} mg^{-1})	4.03 ± 0.24	8.22 ± 0.35
Cytochrome P-450 (nmol/mg)	1.53 ± 0.14	1.12 ± 0.10
Cytochrome b_5 (nmol/mg)	0.656 ± 0.061	1.51 ± 0.07
Glucose-6-phosphatase (μmol P_i min^{-1} mg^{-1})	0.509 ± 0.029	0.413 ± 0.066
Aminopyrine demethylase (nmol formaldehyde min^{-1} mg^{-1})	18.7 ± 1.46	13.9 ± 0.84
Aniline hydroxylase (nmol p-aminophenol min^{-1} mg^{-1})	2.25 ± 0.16	6.08 ± 0.59
Benzo[a]pyrene hydroxylase (nmol 3-hydroxybenzo[a]pyrene min^{-1} mg^{-1})	2.07 ± 0.16	1.46 ± 0.19
UDP-glucuronyltransferase (nmol UDP-p-aminophenol min^{-1} mg^{-1})	25.2 ± 3.15	117.3 ± 8.3
Cytoplasmic enzymes		
Glucose-6-phosphate dehydrogenase (nmol NADPH min^{-1} mg^{-1}	23.3 ± 5.1	88.2 ± 5
UDP-glucose dehydrogenase (nmol NADH min^{-1} mg^{-1})	9.79 ± 1.24	59.8 ± 10.4

[a]BHA was fed at 0.75% (w/w) in the diet for 10 days.
Source: Ref. 64.

papilloma in almost 100% of the treated animals and squamous cell carcinoma of the forestomach in about 30% of the treated animals (Table 5.7). At lower dose levels of 0.5%, no neoplastic lesions were observed, but forestomach hyperplasia was observed. Most of the changes were close to the limiting ridge between the forestomach and the glandular stomach. Ito et al. (72) observed that in addition to 3-BHA, two metabolites p-*tert*-butylphenol and 2-*tert*-butyl-4-methylphenol, also induced papillomas in the forestomach. Verhagen et al. (73) observed that in rats not only the forestomach but also the glandular stomach, small intestine, colorectal tissues, and possibly esophageal tissues were susceptible to the proliferative effects

Table 5.7 Proliferative and Neoplastic Lesions of the Forestomach Epithelium in F344 Rats Given Diet Containing Butylated Hydroxyanisole

BHA in diet (%)	Effective no. of rats[a]	No. of rats with changes in forestomach[b] (%)		
		Hyperplasia	Papilloma	Squamous-cell carcinoma
0	50	0 (0)	0 (0)	0 (0)
0.125	50	1 (2)	0 (0)	0 (0)
0.25	50	7 (14) *	0 (0)	0 (0)
0.5	50	16 (32) **	0 (0)	0 (0)
1	50	44 (88) **	10 (20)*	0 (0)
2	50	50 (100)**	50 (100)**	11 (22)**

[a]Number surviving at least to week 50.
[b]Asterisked values differ significantly from the control value: *$P < 0.01$; **$P < 0.001$.
Source: Ref. 72.

of BHA. Hamsters were found to be more susceptible to BHA than rats (74). In hamsters fed 1 or 2% BHA for 24 and 104 weeks, forestomach papillomas were observed in almost all treated animals and carcinomas in 7–10% in the 104-week group. A lower incidence of lesions was observed in mice fed 0.5 and 1% BHA (75).

In order to determine which isomer of BHA was carcinogenic or whether the isomers had a synergistic action, feeding studies were conducted with the pure isomers and crude BHA in hamsters for 1–4 weeks. Severe adverse effects were observed with crude BHA and the 3-isomer. The 2-isomer had no effect (76). In rats given 1 g/kg bw of the two isomers, 2-BHA was also active in the induction of forestomach papillomas (77). The forestomach hyperplasia was found to be reversible, but the time taken for recovery depended on the duration and level of treatment. In rats fed 0.1–2% BHA for 13 weeks, on cessation of treatment the forestomach reverted to normal after 9 weeks. In rats fed 2% BHA for 1, 2, or 4 weeks followed by a 4-week recovery period, the mild hyperplasia and epithelial changes observed in the 1-week group almost completely disappeared. The more severe changes observed in the 2- and 4-week groups regressed partially during the recovery period (77).

Because of the possible relevance of these observations to humans, studies were conducted in other species including monkeys and pigs, which, like humans, do not have a forestomach. In female cyanomalgus monkeys, BHA (125 or 500 mg/kg bw) given by gavage for 12 weeks failed to induce any histopathological changes in the stomach and esophagus. However, a 40% increase in the mitotic index was observed at the lower end of the esophagus (Table 5.8) (78). In dogs fed BHA at levels of 0.25, 0.5, 1, and 1.3% for 6 months, no histopathological changes were noticed in the stomach, esophagus, or duodenum. Liver weights were increased without any related histopathological changes (67,79). Administration of BHA at levels of 50, 200, and 400 mg/kg bw per day to pregnant pigs from mating to day 110 of the gestation period resulted in proliferation and parakeratotic changes in the esophageal epithelium in a few pigs in the two groups with the higher dose levels. But papillomas were not observed, and no changes were found in the glandular stomach (80). In Japanese house musk shrews (*Sancus murinus*), which have no forestomach, BHA was fed at levels of 0.5, 1, and 2% for 80 weeks. All the animals in the 2% group died of hemorrhage in the gastrointestinal tract. Adenomatous hyperplasia in the lungs was observed at the 0.5 and 1% levels at a significantly higher rate (81).

The mechanism by which 3-BHA induces carcinomas in the forestomach is not clear. Studies by De Stafney et al. (82) suggest that two factors may be of importance. One of these entails thiol depletion. The second is an attack by the reactive metabolites of 3-BHA or secondary products produced by these metabolites on cellular constituents. Studies by Williams (83) also indicated that BHA has an effect on membrane systems, blocking the exchange between the hepatocytes and the epithelial cells. The data strongly suggest that BHA is an epigenetic carcinogen that produces forestomach neoplasia through a promoting effect.

Butylated hyroxyanisole has a promoting or inhibitory effect on the carcinogenic effects of a number of chemical carcinogens. BHA enhanced forestomach carcinogenesis initiated by either N-methyl-N'-nitro-N-nitrosoguanidine or N-methylnitrosourea (MNU) in rats. BHA had a promoting effect on the urinary bladder carcinogenesis initiated by MNU or N-butyl-N-(4-hydroxybutyl)nitrosamine and thyroid carcinogenesis initiated by MNU in rats. It had an inhibitory effect on the liver carcinogenesis initiated by either diethylnitrosamine or N-ethyl-N-

Table 5.8 Mitotic Index of the Esophageal Epithelium from Cyanomalgus Monkeys given 0–250 mg BHA/kg per Day by Gavage on 5 days per week for 12 weeks

BHA treatment (mg kg^{-1} day^{-1})	No. of monkeys	No. of cells counted	No. of cells in mitosis	Mitotic cells (% of total)
0 (control)	8	4860 ± 223	43 ± 7	0.87 ± 0.11
125	7	4921 ± 154	38 ± 6	0.77 ± 0.11
250	7	5485 ± 468	91 ± 10	1.66 ± 0.17*

*$P \le 0.05$.
Source: Ref. 78.

hydroxyethylnitrosamine and on mammary carcinogenesis initiated by 7,12-dimethylbenz[a]anthracene (72).

Reproduction. Butylated hydroxyanisole has not been reported to have any adverse effects on reproduction data or in teratogenicity studies in mice, rats, hamsters, rabbits, pigs, and rhesus monkeys (80,84–88). It has been reported to induce some behavioral abnormalities in mice. Weanling mice exposed to BHA via their mothers during pregnancy and lactation (0.5% level) and then directly for up to 3 weeks showed a significant increase in exploratory activity, decreases in sleeping and in self-grooming, slower learning, and a decrease in the orientation reflex (89). In another study, Stokes et al. (90) observed a decrease in serotonin levels and cholinesterase activity and changed noradrenaline levels in the brain of newborn mice exposed to BHA in utero, and it is postulated that these changes may have an effect on the behavioral modifications observed.

Mutagenicity. Butylated hydroxyanisole was not mutagenic in a number of test systems. It was nonmutagenic in five tester strains in the Ames *Salmonella*/microsome test at concentrations of up to 10 mg/mL (83). In the hepatocyte primary culture/DNA repair test, BHA was negative (91). In mammalian cell mutagenesis assay using adult rat liver epithelial cells (92) and in V79 Chinese hamster lung cells (93), BHA was negative. BHA did not induce sister chromatid exchanges in Chinese hamster ovary cells (83). In tests for chromosomal aberrations, BHA was negative in Chinese hamster lung cells and in Chinese hamster DON cells (94,95). In vivo, BHA was negative in rat bone marrow cells and in the rat dominant lethal assay (96).

Butylated Hydroxytoluene

Butylated hydroxytoluene (BHT; 2,6-di-*tert*-butyl-*p*-cresol) is one of the antioxidants used extensively in the food industry. It is used in low-fat foods, fish products, packaging materials, paraffin, and mineral oils. BHT is also widely used in combination with other antioxidants such as BHA, propyl gallate, and citric acid for the stabilization of oils and high-fat foods. ADI values for BHT have changed over the years because of its toxicological effects in different species. The latest temporary value allocated by JECFA is 0–0.125 mg/kg bw (1).

TOXICOLOGICAL STUDIES

Absorption, Metabolism, and Excretion. The absorption, metabolism, and excretion of BHT have been studied in rats, rabbits, dogs, monkeys, and humans. In general, the oxidative metabolism of BHT was mediated by the microsomal monooxygenase system. In rats, rabbits, dogs, and monkeys, oxidation of the *p*-methyl group predominated, whereas in humans the *tert*-butyl groups were oxidized. Oxidation of both *p*-methyl and *tert*-butyl groups was observed in mice.

The metabolism of BHT is more complicated and slower than that of BHA. The relatively slow excretion of BHT has been attributed to the enterohepatic circulation (97–99).

In rats given 0.5 and 1% BHT for 5 weeks, the concentration of BHT increased rapidly in liver and body fat. Approximately 30 ppm was observed in the body fat in males and 45 ppm in females and 1–3 ppm in the liver. On cessation of treatment the concentration in the tissue decreased with a half-life of 7–10 days. In rats given single oral doses of ^{14}C-labeled BHT (1–100 mg/rat), nearly 80–90% was recovered in 4 days, with up to 40% in the urine. Approximately 3.8% was retained in the alimentary tract (98,100–103). The major urinary metabolites observed were BHT-acid (3,5-di-*tert*-butyl-4-hydroxybenzoic acid) (both free and as ester glucuronide) and BHT-mercapturic acid (di-*tert*-butylhydroxybenzyl acetyl cysteine) in addition to many other compounds including BHT alcohol (Ionox-100 or 3,5-di-*tert*-butyl-4-hydroxybenzyl alcohol), BHT aldehyde (3,5-di-*tert*-butyl-4-hydroxybenzaldehyde), and BHT dimer. Free BHT acid was the major metabolite in feces. About 10% of the dose was excreted unchanged (101,102, 104,105). Tye et al. (102) observed distinct sex differences in the mode of excretion. Female rats excreted about 40–60% of a single oral dose in feces and about 20–40% in the urine. Males excreted about 70–95% in the feces and 5–9% in the urine. Females showed more tissue retention, especially in the gonads. Significant biliary excretion of BHT and metabolites has also been observed. Four major metabolites have been identified: BHT acid, BHT alcohol, BHT aldehyde, and BHT quinone methide (2,6-di-*tert*-butyl-4-methylene-2,5-cyclohexadienone) (Fig. 5.6) (98,100,106).

The half-life of a single oral dose of BHT in mice was found to be 9–11 h in major tissues such as the stomach, intestine, liver, and kidney. The half-life was 5–10 days when daily doses were given for 10 days. The major metabolite in the urine was the glucuronide conjugate of the acid and free acid in the feces. Excretion was mainly in the feces (41–65%) and urine (26–50%). The formation of BHT quinone methide has been observed in vitro in liver microsomes and in vivo in mouse liver (101).

The major metabolites observed in rabbits were BHT alcohol, BHT acid, and BHT dimer. Excretion of all metabolites was essentially complete in 3–4 days (107). Urinary metabolites constituted 37.5% glucuronides, 16.7% etheral sulfates, and 6.8% free phenols. Unchanged BHT was observed only in the feces (108,109). Significant biliary excretion of BHT and metabolites has also been reported (110). In dogs, the metabolism was similar to that of rats, and significant biliary excretion was observed (111). In monkeys, the major metabolite was the ester glucuronide of BHT acid, and the rate of excretion was similar to that of humans (112). Limited studies in humans (single oral doses of approximately 0.5 mg/kg bw) have indicated that the major metabolite is in the form of an ether-insoluble glucuronide identified as 5-carboxy-7-(1-carboxy-1-methyl ethyl)-3,3-dimethyl-2-hydroxy-2,3-dihydro-

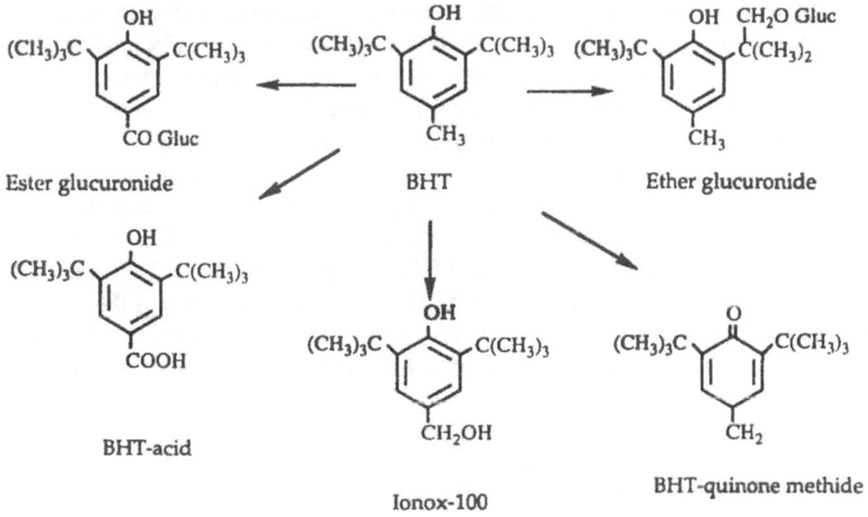

Fig. 5.6   Major metabolites of butylated hydroxytoluene.

benzofuran (99). Daniel et al. (58) studied the excretion of single oral doses of [^{14}C]BHT (40 mg/kg bw) in humans.

Approximately 50% was excreted in the urine in the first 24 h followed by slower excretion for the next 10 days. Tissue retention was found to be greater in humans than in rats. Studies by Wiebe et al. (99) suggest that biliary excretion may be an important route for the elimination of BHT and also that enterohepatic circulation occurs in humans. Verhagen et al. (104) reported differences in the metabolism in rats and humans, especially in terms of plasma kinetics and plasma concentrations, and concluded that the differences were too wide to allow a hazard estimation for BHT consumption by humans on the basis of its metabolism.

Acute Toxicity.   The acute oral LD$_{50}$ in mg/kg bw in rats was 1700–1970; in rabbits, 2100–3200; in guinea pigs, 10,700; in cats, 940–2100 (113), and in mice, 2000 (114).

Short-Term Studies.   In rats, BHT at the level of 0.3 or 0.5% caused an increase in the level of serum cholesterol and phospholipids within 5 weeks. Brown et al. (69) observed reduced growth rates and increases in liver weight in rats fed BHT at 0.5% in the diet. But at lower dose levels (0.1%) no adverse effects were observed (115). In rabbits at the 2% dose level, BHT caused an acute effect on electrolyte excretion similar to that of BHA, whereas lower levels were without any adverse effect (60). No symptoms of intoxication or histopathological changes were observed in dogs fed 0.17–0.94 mg BHT/kg bw 5 days a week for 12 months (113).

In high doses, BHT had a toxic effect on liver, lung, and kidney and also on the blood coagulation mechanism. Early studies in rats and mice revealed that BHT at 500 mg/kg per day induced liver enlargement in 2 days and stimulated microsomal drug-metabolizing enzyme activity. The effects were found to be reversible (116). Creaven et al. (65) observed that in rats, BHT at levels of 0.01–0.5% for 12 days resulted in increased liver weights and induction of liver microsomal biphenyl-4-hydroxylase activity. At levels of 500 mg/kg bw for 14 days, a reduction in the activity of glucose-6-phosphatase was observed, indicative of early liver damage (115,117). In rats, administration of BHT by gavage at levels of 25, 250, or 500 mg/kg bw for 21 days resulted in a dose-related hepatomegaly, and at the highest dose a progressive periportal hepatocyte necrosis (Table 5.9) (118). The periportal lesions were associated with a proliferation of the bile ducts, persistent fibrosis,

Table 5.9 Summary of Hepatic Histopathology in Rats Treated with Butylated Hydroxytoluene for 7 or 28 Days

	No. of rats with lesion[a] after treatment with BHT in doses (mg kg^{-1} day^{-1}) of:		
Observation	25	250	500
After 7 days			
Periportal region			
Hepatocyte necrosis	0	0	2
Fibrosis	0	0	3
Hepatocyte hypertrophy	0	0	3
Hepatocyte hyperplasia	0	0	4
Glycogen accumulation	0	4	4
After 28 days			
Periportal region			
Hepatocyte necrosis	0	0	6
Fibrosis	0	0	5
Bile-duct cell proliferation	0	0	4
Hepatocyte hypertrophy	0	0	2
Hepatocyte hyperplasia	0	0	3
Pigment-laden macrophages	0	0	3
Glycogen depletion	0	0	7
Glycogen accumulation	0	8	0
Midzonal glycogen accumulation	0	0	5

[a]Out of a total of five per group treated for 7 days and 10 per group treated for 28 days.
Source: Ref. 118.

and inflammatory cell reactions. At sublethal dose levels of 1000 and 1250 mg BHT/kg bw for up to 4 days, centrilobular necrosis was observed within 48 h. At a lower dose level (25 mg), no adverse effects were observed. In mice, BHT at levels of 0.75% for 12 months resulted in bile duct hyperplasia (119). The liver hypertrophy was accompanied by a proliferation of the smooth endoplasmic reticulum, an increase in the cytochrome P-450 level, and induction of a number of enzymes, including glutathione-S-transferase, glutathione reductase, thymidine kinase, nitroanisole demethylase, epoxide hydrolase, and aminopyrene demethylase. The changes were reversible after the cessation of the treatment (62,118,120, 121). At 500 mg/kg bw for 14 days, BHT caused slight hepatomegaly, moderate proliferation of the smooth endoplasmic reticulum, a reduction in glucose-6-phosphatase, and an increase in nitroanisole demethylase activity in monkeys. At the lower dose of 50 mg/kg bw, no adverse effects were observed (63).

In a more recent study, Takahashi (122) reported that BHT in very high doses (1.35–5% for 30 days) caused a dose-related toxic nephrosis with tubular lesions in mice. The lesions appeared as irregular patches or wedge-shaped proximal tubules, necrosis, and cyst formation. Renal toxicity has also been reported in rats (123,124).

Butylated hydroxytoluene was reported to cause extensive internal and external hemorrhages in rats due to a disruption of the blood coagulation mechanism, resulting in increased mortality (125,126). The minimum effective dose was found to be 7.5 mg/kg bw per day. The disruption observed in the blood coagulation was due to hypoprothrombinemia brought about by inhibition of phylloquinone epoxide reductase activity in the liver by BHT quinone methide, one of the reactive metabolites of BHT (127). Administration of vitamin K prevented the BHT-induced hemorrhage. Takahashi and Hirage (128) suggested that BHT may inhibit absorption of vitamin K in the intestines or uptake by the liver. An increased fecal excretion of vitamin K was observed in rats receiving 0.25% BHT for 2 weeks. BHT was also reported to affect platelet morphology, fatty acid composition of the platelets, and vascular permeability, which may play a role in the hemorrhagic effect (129).

In mice, Takahashi (122) observed that BHT at levels of 0.5, 1, or 2% for 21 days caused massive hemorrhages in the lungs and blood pooling in various organs but only a slight reduction in blood coagulating activity. It was suggested that the hemorrhages might be due to a severe lung injury and not to a coagulation defect as observed in rats. BHT did not cause significant hemorrhaging in guinea pigs at dietary levels of 0–2%. The prothrombin index was slightly reduced at the 1% level. BHT quinone methide was not detected in guinea pigs, whereas 7–40 mg/g liver was detected in rats. In rabbits, dogs, and Japanese quail fed BHT for 14–17 days at levels of 170 or 700 mg/kg bw, 173, 400, or 760 mg/kg bw, and 1%, respectively, no hemorrhages were observed (Table 5.10) (122,126).

A number of studies have shown that BHT causes acute pulmonary toxicity in

Table 5.10 Hemorrhagic Effects of BHT in Various Species

Species and strain	Sex	Dose of BHT	Mean intake of BHT[a] (mg kg⁻¹ day⁻¹)	% Total population population with hemorrhages		Mean prothrombin index[b] (%)	Hepatic level of quinone methide[c] (μg/g liver)
				Dead	Surviving		
Rat							
Sprague-Dawley	M	0	0 (31)	0	0	101	
		1.2%	693 (50)	44	50	18***	NC
	F	0	0 (5)	0	0	101	
		1.2%	1000 (10)	0	30	73***	NC
Wistar	M	0	0 (5)	0	0	100	
		1.2%	638 (10)	10	90	22***	38
	F	0	0 (5)	0	0	100	
		1.2%	854 (10)	0	100	38***	7
Donryu	M	0	0 (5)	0	0	100	
		1.2%	1120 (10)	10	70	13***	41
	F	0	0 (5)	0	0	100	
		1.2%	1000 (10)	0	0	92	27
Fischer	M	0	0 (5)	0	0	100	
		1.2%	821 (10)	30	70	5***	16
	F	0	0 (5)	0	0	100	
		1.2%	895 (10)	20	70	18***	11
Mouse							
ddY	M	0	0 (5)	0	0	100	
		1.2%	1701 (10)	0	30	79**	ND
ICR	M	0	0 (10)	0	0	100	
		1.2%	1344 (10)	0	0	96	ND
DBA/2	M	0	0 (10)	0	0	100	
		1.2%	847 (10)	0	0	138**	NC

BALB/cAN	M	0	0 (5)	0	0	100	NC
		1.2%	1730 (10)	0	0	84**	NC
C3H/He	M	0	0 (10)	0	0	100	NC
		1.2%	1858 (10)	0	0	115***	NC
C57BL/6	M	0	0 (5)	0	0	100	NC
		1.2%	1925 (10)	0	0	91*	NC
Hamster Syrian golden	M	0	0 (5)	0	0	101	
			380 (4)	0	0	101	
			760 (6)	0	0	87	ND^d
Guinea pig Hartley	M	0	0 (5)	0	0	100	
			190 (5)	0	0	78	
			380 (5)	0	0	73	ND^d
Japanese quail White egged	M	0	0 (5)	0	0	100	
		1%	1056 (5)	0	0	53*	ND

[a] The numbers in brackets are the numbers of animals in each group.
[b] The values marked with asterisks differ significantly from the corresponding control values: $*P < 0.05$; $**P < 0.01$; $***P < 0.001$. The times of treatment for rats, hamsters, guinea pigs, and quail were 3 weeks, 1 week, 3 days, 3 days, and 17 days, respectively.
[c] NC = Not calculated; ND = Not detected.
[d] Determined in separate experiments in hamsters or guinea pigs at 1.2 or 1% in the diet, respectively, for 3 days.
Source: Ref. 126.

mice at levels of 400–500 mg/kg bw. The effects include hypertrophy, hyperplasia, and a general thickening of the alveolar walls of the lungs. A substantial proliferation of the pulmonary cells accompanied by a dose-dependent increase in total DNA, RNA, and lipids in the lungs was also observed within 3–5 days of a single intraperitoneal (IP) injection of BHT (130–132). The effect was generally reversible in 6–10 days of cessation of the treatment. However, exposure to a second stress such as hyperbaric O_2 after administration of BHT impeded the repair process, resulting in pulmonary fibrosis (133). Some of the morphological and cytodynamic events include perivascular edema and cellular infiltration in type I epithelial cells followed by multifocal necrosis, destruction of the air-blood barrier, and fibrin exudation by day 2 after a single IP injection of 400 mg/kg bw BHT (134). Ultrastructural studies indicated that the type I cells were damaged by day 1 and cell destruction was complete within 2–3 days. Elongation of the type II cells with large nuclei and abundant cytoplasm was evident in 2–7 days (135). It has been postulated that BHT causes cell lysis and death as a result of interaction with the cell membrane (136). However, the mechanism of BHT toxicity still remains unclear.

Long-Term and Carinogenicity Studies. In early studies, Deichmann et al. (113) observed no adverse effects in rats fed 0.2, 0.5, 0.8, or 1% BHT for 2 years. In a 104-week long-term feeding study in rats, Hirose et al. (137) reported that BHT at the 0.25 and 1% levels was not carcinogenic. Treated rats of both sexes showed reduced body weight gain and increased liver weights. Only males showed increased γ-glutamyl transferase levels. Tumors were observed in various organs, but their incidence was not statistically significant compared to controls (Table 5.11). In a two-generation carcinogenicity study with in utero exposure in rats, BHT was fed at levels of 25, 100, or 500 mg/kg bw per day from 7 weeks of age to the weaning of the F_1 generation. The F_1 generation were given 25, 100, or 250 mg/kg bw per day from weaning to 144 weeks of age. At weaning, the BHT-treated F_1 rats, especially the males, had lower body weights than untreated controls. Dose-related increases in the numbers of hepatocellular adenomas and carcinomas were statistically significant in male F_1 rats tested for heterogeneity or analyzed for trend. In F_1 females the increases were statistically significant only for adenomas in the analysis of trend. However, all tumors were detected when the F_1 rats were more than 2 years old (138). Unlike BHA, BHT had no adverse effects on the forestomach of rats or hamsters at the 1% level (76,77).

Carcinogenicity studies have been conducted in various strains of mice. In a 2-year study in B6C3F1 mice, Shirai et al. (139) reported that BHT at the level of 0.02, 0.1, 0.5% was not carcinogenic. A reduction in body weight gain was noticed, the effect being more pronounced in males. Nonneoplastic lesions related to BHT treatment were lymphatic infiltration of the lung in females and of the urinary bladder in both sexes at the highest dose level. Tumors were observed in various

Table 5.11 Tumor Incidence in Rats Fed BHT

	Males			Females		
Treatment group: Site and type of tumor	Control	0.25%	1%	Control	0.25%	1%
No. of rats[b]:	26	43	38	32	46	51
Liver						
Hyperplastic nodule	2 (7.7)	2 (4.7)	1 (2.6)	0	3 (6.5)	3 (5.9)
Pancreas						
Carcinoma	0	0	1 (2.6)	0	1 (2.2)	4 (7.8)
Islet-cell adenoma	0	1 (2.3)	2 (5.3)	0	0	0
Mammary gland						
Fibroadenoma	—	—	—	6 (18.8)	8 (17.4)	8 (15.7)
Adenoma	—	—	—	1 (3.4)	1 (2.2)	1 (2.0)
Uterus						
Leiomyoma	—	—	—	1 (3.4)	1 (2.2)	0
Carcinoma	—	—	—	1 (3.1)	2 (4.3)	1 (2.0)
Pituitary gland						
Adenoma	2 (7.7)	3 (7.0)	1 (2.6)	0	6 (13.0)[c]	3 (11.8)
Carcinoma	0	2 (4.7)	5 (13.2)	3 (9.4)	3 (6.5)	7 (13.7)
Adrenal gland						
Adenoma	1 (3.8)	3 (7.0)	0	0	2 (4.3)	1 (2.0)
Carcinoma	0	0	0	0	0	1 (2.0)
Others	2 (7.7)	2 (4.7)	4 (10.5)	2 (6.3)	4 (8.7)	3 (11.8)
Total	6 (23.1)	13 (30.2)	10 (26.3)	11 (34.4)	25 (54.3)	25 (49.0)

[a] Percent of group given in parentheses.
[b] Animals that survived more than 69 weeks were included.
[c] Differs significantly (chi-square test) from the corresponding control value ($P < 0.05$).
Source: Ref. 137.

organs, with a high incidence in the lung, liver, and the lymph nodes. But the incidence was not statistically significant. In another study in the same strain at higher dose levels of 1 or 2% in the diet, a significant dose-dependent increase in hepatocellular adenomas and foci of alterations in the liver were observed in males but not in females (140). Clapp et al. (141) reported an increase in the incidence of lung tumors and hepatic cysts in BALB/c mice fed 0.75% BHT for 16 months. Brooks et al. (142) reported a dose-related increase in both benign and malignant tumors in the lung in both sexes of CF1 mice and benign ovarian tumors in females. In C3H mice, which are more likely to develop spontaneous liver tumors with age, BHT fed at levels of 0.05 or 0.5% for 10 months increased the incidence of liver tumors in males, but it was not dose-related. The incidence of lung tumors was increased in males at both dietary levels but in females only at the higher dose level (143).

A number of studies have been conducted on the modifying effects of BHT on chemical carcinogenesis. These effects depend on a number of factors including target organs, type of carcinogen, species and strain differences, type of diet used, and time of administration. In general, BHT inhibited the induction of neoplasms in the lung and forestomach in mice and neoplasms in the lung, liver, and forestomach in rats when given before or with the carcinogen. BHT had a promoting effect on urinary bladder, thyroid, and lung carcinogenesis (144).

Reproduction. In an earlier study, Brown et al. (69) reported that rats fed 0.1 or 0.5% BHT showed a 10% incidence of anophthalmia. These findings were not confirmed in any other laboratory. BHT had no adverse effects on reproduction data and was not teratogenic in single or multigeneration reproduction studies in rats, mice, hamsters, rabbits, and monkeys at lower doses, and the no-effect level was equivalent to 50 mg/kg bw (84,88,145–149). At higher dose levels, some of the significant effects observed in rats include a dose-related response in litter size, number of males per litter, and body weight gain during lactation, but the effects were significant only at 500 mg/kg bw per day. In rabbits given 3–320 mg/kg bw per day by gavage during embryogenesis, an increase in intrauterine deaths was observed at high doses. In mice at 500 mg/kg bw per day, prolonged time to birth of first litters and a reduction in pup numbers and pup weight were observed.

In a developmental neurobehavioral toxicity test, the offspring of rats fed 0.5% BHT before conception and throughout pregnancy and lactation showed delayed eyelid opening, surface righting development, and limb coordination in swimming in males and reduced female open-field ambulation. However, the results did not suggest any specific toxicity of BHT for the central nervous system (150). In another study weanling mice fed 0.5% BHT for 3 weeks, whose parents had been maintained at the same level during their entire mating, gestation, and preweaning period, decreased sleeping, increased social and isolation-induced aggression, and learning disabilities were observed under the experimental conditions employed (89).

Mutagenicity. Butylated hydroxytoluene was found to be negative in several strains of *Salmonella typhimurium* with or without metabolic activation (151,152). In in vitro tests using mammalian cells, BHT was weakly positive in the test for gene mutation in Chinese hamster V79 cells (153). BHT was positive in tests for chromosomal aberrations in human lymphocyte cultures (154) and in Chinese hamster ovary cells (155). In in vivo tests, BHT was negative in tests for chromosomal damage in bone marrrow cells and liver cells of rats (156,157). In mice, BHT was negative in three dominant lethal tests, but in rats at high doses BHT was positive in two dominant lethal tests (158,159). In general, the mutagenic effects were observed only at higher levels of BHT.

tert-*Butyl Hydroquinone*

tert-Butyl hydroquinone (TBHQ) was introduced in the 1970s and was approved as a food grade antioxidant in 1972. TBHQ is used for the stabilization of fats and oils, confectionery products, and fried foods and is regarded as the best antioxidant for the protection of frying oils and the fried product (160). TBHQ is currently being used in the United States and some other countries but is not permitted for food use in the EEC countries and Japan (161,162) due to lack of adequate toxicological data. In the 1987 reevaluation, JECFA allocated a temporary ADI of 0–0.2 mg/kg bw (1).

TOXICOLOGICAL STUDIES

Absorption, Metabolism, and Excretion. Absorption and metabolism studies have been conducted in rats, dogs, and humans. In all three species, more than 90% of an orally administered dose of TBHQ was rapidly absorbed and nearly 80% was excreted in the urine in the first 24 h. The excretion was essentially complete within 48 h. In rats given single oral doses of TBHQ (100 mg/kg), about 57–80% was excreted as the 4-*O*-sulfate, 4% as the 4-*O*-glucuronide conjugate, and about 4–12% of unchanged TBHQ was observed (Fig. 5.7). In long-term studies, an increase in the amount of glucuronide was observed. A similar pattern was observed in dogs, but the proportion of glucuronide was higher. No tissue accumulation of TBHQ was observed. In humans given single oral doses of 0.5–4 mg/kg in a high-fat vehicle, most of the dose was recovered within 40 h in the urine. The proportions of the major metabolites observed were 73–88% of 4-*O*-sulfate, 15–22% of 4-*O*-glucuronide, and less than 1% unchanged TBHQ. In a low-fat vehicle, absorption was much lower, and urinary elimination accounted for less than half the intake (Table 5.12) (163).

Acute Toxicity. The acute oral and intraperitoneal LD_{50} in mg/kg bw in rats was 700–1000 and 300–400, respectively. In mice, the acute oral LD_{50} was 1260 mg/kg bw, and in guinea pigs it was 790 mg/kg bw (163,164).

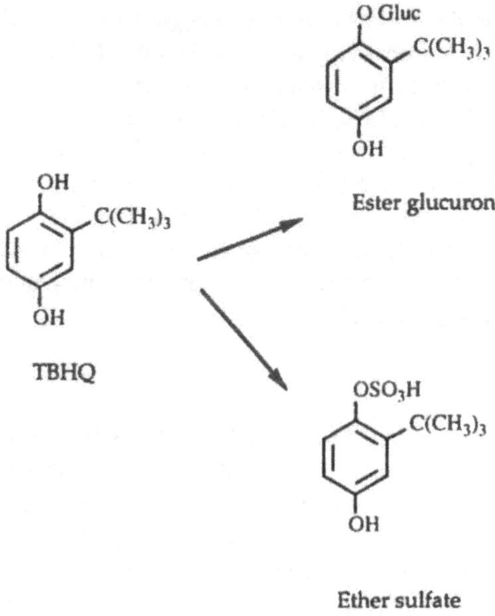

Fig. 5.7 Major metabolites in *tert*-butyl hydroquinone.

Table 5.12 Disposition of Single Oral Doses of TBHQ by Rats, Dogs, and Humans[a]

Species	Dose (g/kg)	No. of animals	% Dose eliminated in urine			Days	Total
			Unchanged	O-SO$_2$O	O-glu		
Rate	0.10	2 × 6	117	10	4	3	96
	0.20	2 × 6	0.6–0.8	60, 61	2, 8.5	3	70
	0.40	2 × 6	4	57	4	4	65
Dog	0.10	2 × 3	3	69, 85	31, 24	3	100, 112
Humans							
HF	0.001	2	<0.1	73, 88	15, 22	2–3	95, 103
LF	0.001–0.002	8	<0.1	18–51	0–6	2–3	18–51

[a]Values are percent of given dose as averages or ranges. Rats received oral intubations of a 10% corn oil solution of TBHQ. Dogs received TBHQ as a meat capsule. Humans received TBHQ incorporated in high-fat (HF) or low-fat (LF) food.
Source: Ref. 163.

Short-Term Studies. In rats given 100 or 200 mg/kg bw TBHQ intraperitoneally for 1 month or fed 1% TBHQ for 22 days, no adverse effects were observed on growth rate, mortality, or microscopic pathology (165). TBHQ had a slight inductive effect on some of the liver microsomal mixed function oxidases such as *p*-nitroanisole demethylase and aniline hydroxylase in a 21-day feeding study in rats at the 0.2% level (Table 5.13). However, the effects were slight compared to those of BHA and BHT. No adverse effects on the enzyme systems were observed in long-term studies (163).

Long-Term and Carcinogenicity Studies. In a 20-month study in rats at dose levels of 0.016, 0.05, 0.16, and 0.5%, TBHQ had no deleterious effects on food intake, growth rate, mortality, or organ weights. No changes were observed in hematological, biochemical, and extensive histological studies. In a 2-year dog study at levels of 0.05, 0.15, and 0.5%, no abnormalities were observed in histopathological studies, but at the highest dose level reductions in red blood cell counts, hemoglobin levels, and hematocrit values were observed (163).

tert-Butyl hydroquinone was also tested for its ability to induce cellular proliferation in the forestomach of rats and hamsters. In rats fed at a dose level of 0.25% for 9 days, TBHQ had no significant effect, but at the 1% level a slight but significant increase in cell proliferation was observed (166). At the 2% level, TBHQ caused mild hyperplasia of the forestomach with focally increased hyperplasia of the basal cells, but no differentiation of the basal hyperplasia was observed (77). In hamsters, a 0.5% dose level for 20 weeks resulted in a slight increase in cell proliferation (76). TBHQ has also been reported to have a weak promoting effect on urinary bladder carcinogenesis induced by *N*-butyl-*N*-(4-hydroxybutyl)nitrosamine at a dose level of 0.2% (167).

Table 5.13 Effects of Dietary Feeding of TBHQ on Rat Liver Microsomal Enzyme Activities[a]

		Specific activities		
Dietary intake	Feeding (days)	Glucose-6-phosphatase[b]	*p*-Nitroanisole demethylase[c]	Aniline hydroxylase[c]
No additive	21	1.92	106	30
0.2% BHA	21	1.31	219	51
0.2% TBHQ	21	1.91	164	47
0.05% TBHQ	21	1.44	94	34

[a]Average values from eight rats at each level.
[b]μM P$_i$/15 min per mg protein.
[c]Optical density × 10^5/min/per mg protein.
Source: Ref. 163.

Reproduction. Reproduction studies have been carried out at the highest dose level of 0.5% in rats. The studies include a three-generation study with two litters per generation, a one-generation study, and a separate teratological study during organogenesis. At the 0.5% level a reduction in food intake with a consequent decrease in the body weight of the pups was observed. TBHQ had no effect on reproduction data and was not teratogenic (163,168).

Mutagenicity. The mutagenicity of TBHQ has been studied in a number of test systems. TBHQ was nonmutagenic in five strains of *Salmonella typhimurium* with and without metabolic activation in the Ames test. TBHQ was negative in CHO/HGPRT forward mutation assay and V79 Chinese hamster cells. In a mouse lymphoma forward mutation assay, TBHQ was slightly positive with and without a metabolic activation system. In tests for chromosomal aberrations in mammalian cells in vivo, a rat dominant lethal test and a mouse bone marrow cytogenetic assay were negative (1). TBHQ was found to be slightly positive in tests for chromosomal aberrations and sister chromatid exchanges in V79 Chinese hamster lung cells and Chinese hamster fibroblast cell lines (169,170). In in vivo tests, Giri et al. (171) reported a clastogenic effect of TBHQ in mouse bone marrow cells. Two hundred milligrams of TBHQ per kilogram body weight administered intraperitoneally resulted in a significant increase in chromosomal abnormalities such as breaks, gaps, and centric fusions. A significant depression in mitotic index was also observed in treated animals. Mukherjee et al. (172) reported sister chromatid exchanges (SCEs) in mouse bone marrow metaphase cells (Table 5.14). Concentrations of 0.5, 2, 20, 50, 100, and 200 mg/kg of TBHQ in corn oil were injected intraperitoneally. A positive dose–response effect in the SCE frequency was observed using the Cochran Armitage trend test. Two milligrams of TBHQ per kilogram body weight was found to be the minimum effective dose. At higher concentrations of 50, 100, and 200 mg/kg, TBHQ also induced a significant delay in cell cycle. The results indicate that at higher levels TBHQ may be slightly mutagenic.

Tocopherols

Tocopherols are a group of chemically related compounds that occur naturally in plant tissues, especially in nuts, seeds, fruits, and vegetables, and show both antioxidant and vitamin E activity. Vegetable oils from nuts and seeds form a rich source of vitamin E. The tocopherols include α, β, γ, and δ homologs and the corresponding tocotrienols. α-Tocopherol and its acetate have also been chemically synthesized. Synthetic products, which are designated as *dl*-α-tocopherol and *dl*-α-tocopheryl acetate, are actually mixtures of four racemates. The tocopherols are generally considered the major lipid-soluble antioxidants. *d*-α-Tocopheryl polyethylene glycol 1000 succinate (TPGS) is a synthetic water-soluble form of vitamin E prepared by the esterification of *d*-α-tocopheryl acid succinate with

Table 5.14 Sister Chromatid Exchange (SCE) and the Proliferation Rate Indices (PRIs) in Bone Marrow Cells of Mice Following In Vivo Treatment with TBHQ[a]

Treatment (mg/kg)	SCE/cell[b]	PRI[c]
0 (corn oil)	2.24 ± 0.06	2.07 ± 0.07
0.5	2.32 ± 0.15	2.0 ± 0.04
2	2.81 ± 0.14[e]	1.9 ± 0.06
20	4.31 ± 0.17[e]	1.95 ± 0.03
50	5.5 ± 0.16[e]	1.84 ± 0.05[e]
100	7.08 ± 0.22[e]	1.8 ± 0.07[e]
200	7.65 ± 0.10[e]	1.77 ± 0.07[e]
20 CPA[d]	14.89 ± 0.38	1.84 ± 0.05

[a]Group mean ± standard error mean.
[b]Individual animal mean SCE frequency ± standard error mean for N cells.
[c]PRI based on 100 metaphase cells analyzed per animal.
[d]Positive control cyclophosphamide in corn oil.
[e]Significantly different from concurrent control at $\alpha = 0.008$ ($\alpha = 0.05$ Bonferroni-corrected for six pairwise comparisons).
Source: Ref. 172.

polyethylene glycol. TPGS provides 260 mg of d-α-tocopherol per gram and forms a clear solution in water that is stable under normal handling and storage conditions (173). The tocopherols are effective in stabilizing animal fats, fat-containing foods, fried products, carotenoids, and vitamin A (174,175). Both natural and synthetic forms of α-tocopherol are used commercially in addition to mixed tocopherol concentrates and esters. The tocopherols as a group are permitted for use as food additives, and JECFA has allocated an ADI of 0.15–2 mg/kg bw (as α-tocopherol) (1).

TOXICOLOGICAL STUDIES

Absorption, Metabolism, and Excretion. The mechanism of absorption and metabolic fate of α-tocopherol is not fully known. In mammals, the most functional absorption site is generally located in the medial small intestine. Tocopherols are partially hydrolyzed before absorption and enter the systemic circulation via the lymphatic system. In the lymph, tocopherols circulate bound to nonspecific lipoproteins (176). Pearson and MacBarnes (177) observed that in rats α- and γ-tocopherols are absorbed to a greater extent (32 and 30%, respectively) than the β- and δ-tocopherols (18 and 1.8%, respectively). In humans, the percent absorption of dl-α-tocopherol is inversely proportional to the dose. With a 10-mg dose, 96.9% absorption was observed; with a 100-mg dose, 81.5%; and with a 200-mg dose, 55.2% (178). Studies with ^{14}C-labeled dl-α-tocopheryl acetate in rats revealed that absorption occurs slowly, and maximum concentration was observed after several hours of administration. High tissue levels remained for a longer period. Maximum uptake was observed in the liver, adrenal glands, and ovaries. Intermediate uptake was observed in the heart, kidneys, adipose tissue, and skeletal muscle (179, 180). In liver cell particles, nearly 85% of the α-tocopherol was located in the structural part of the cell, especially in the mitochondria and microsomes. Bieri (181) observed two different rates of mobilization of vitamin E in rats, one with a half-life of about 1 week and the other with a half-life of 1–2 months. Most of the tocopherols remained unchanged in the tissue, and some metabolism was observed in the liver and kidney. The metabolites observed in urine include trace amounts of tocopherylquinone and dimers similar to the in vitro oxidation products of tocopherol (182,183). The urinary excretion of tocopherol accounted for only 1% of the dose, whereas fecal elimination ranged from 10 to 75% of the dose (179,183). In humans, tocopherol is possibly metabolized via tocoquinone to tocopheronolactone (Fig. 5.8) (184).

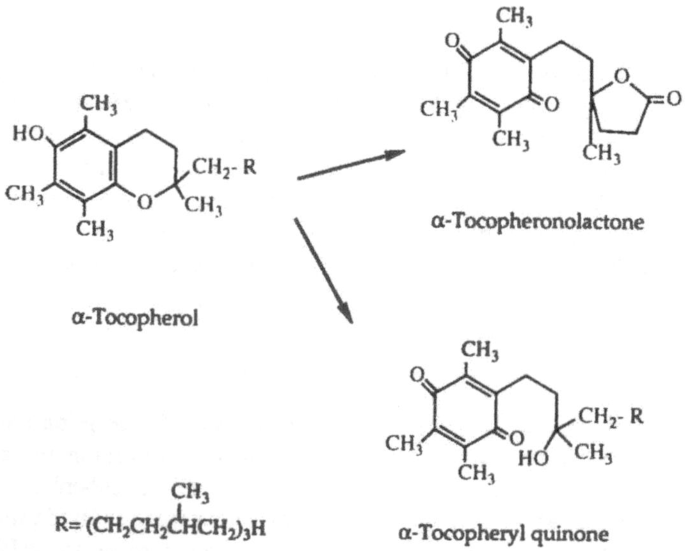

α-Tocopheronolactone

α-Tocopherol

$R = (CH_2CH_2CHCH_2)_3H$
       CH_3

α-Tocopheryl quinone

Fig. 5.8   Major metabolites of α-tocopherol.

Acute Toxicity. The acute oral LD$_{50}$ of α-tocopherol is not known. α-Tocopherol has been tolerated even at higher doses of 2000 mg/kg bw in rabbits, 5000 mg/kg bw in rats, and 50,000 mg/kg bw in mice (185). The acute oral LD$_{50}$ values of TPGS and *d*-α-tocopheryl succinate were >7000 mg/kg bw in rats (173).

Short-Term Studies. No adverse effects were observed in short-term studies in rats and mice fed α-tocopherol at dose levels of 0.4 and 5% daily for 2 months (186). Weissberger and Harris (187) reported that in weanling rats fed 10 or 100 mg of vitamin E for 19 weeks, phosphorus metabolism was stimulated at the higher dose level. In another feeding study in weanling rats given 0.0035% and 25, 50, 100, and 1000 times this concentration for 13 weeks, significant reductions in feed intake and protein efficiencies were observed after 8 weeks. No changes were observed in hemoglobin levels, serum cholesterol, or urinary creatine levels. Serum glutamate pyruvate transaminase activity increased with the highest dose (188). Administration of TPGS at levels of 0.002, 0.2, or 2% for 90 days had no effect on body weight gain, food consumption, hematology, organ weights, serum constituents, or histopathology (173).

dl-α-Tocopheryl acetate at 2.2 g/kg bw in the diet for 55 days induced a depression in growth, bone calcification, disturbed thyroidal function, prolongation of prothrombin time, and an increase in reticulocyte count in 23-day-old chicks (189). In a 13-week study in Fischer 344 rats administered *d*-α-tocopheryl acetate by gavage at levels of 125, 500, or 2000 mg/kg bw per day, no adverse effects were observed on food intake or body weight. An increase in liver weight without any histopathological changes was observed in females at the highest dose level. High levels caused prolongation of prothrombin and activated partial thromboplastin times, reticulocytosis, and a decrease in hematocrit and hemoglobin levels in males (Table 5.15). Vitamin E caused hemorrhages in the orbital cavity of the eye and in the meninges. Also in higher doses an antagonistic effect was observed between vitamin E and vitamin K. These symptoms were reversible with supplementation of vitamin K. Vitamin E at all levels caused interstitial inflammation and adenomatous hyperplasia of the lung. Male rats generally had more severe lesions and a higher incidence of lesions than females (Table 5.16) (190).

Takahashi et al. (191) reported the hemorrhagic toxicity of *d*-α-tocopherol in rats (Table 5.17). Dietary administration of *d*-α-tocopherol at levels of 0.63% and 1% for 7 days resulted in external hemorrhages, and severe internal hemorrhages in a number of organs including the stomach and around the inferior vena cava, cranial cavity, and spinal column were observed at both test levels. The toxicity was found to be dose-dependent. Significant prolongation of prothrombin time and kaolin-activated partial thromboplastin time were observed at both test levels. Intraperitoneal administration of *d*-α-tocopherol at levels of 1.14, 1.82, and 2.91 mmol/kg for 7 days was less toxic, and bleeding in the abdominal cavity was found in only one animal at the highest dose level. Dietary administration resulted in a

Table 5.15 Hemoglobin Changes in Male Fischer 344 Rats Dosed for 90 Days with *dl*-α-Tocopheryl Acetate

Determination	Values[a] for groups given *d*-α-tocopheryl acetate in doses (mg kg^{-1} day^{-1}) of:				
	0 (VC)[b]	0 (UC)[c]	125	500	2000
Hematocrit (%)	41.2 ± 0.4[d] (10)	41.4 ± 0.5 (10)	40.8 ± 0.5 (9)	39.4 ± 0.8 (10)	27.4 ± 6.6[e] (5)
Hemoglobuin (g/dL)	15.7 ± 0.1[d] (10)	15.9 ± 0.2 (10)	15.4 ± 0.1 (9)	14.8 ± 0.3 (10)	11.4 ± 1.9[e] (5)
RBC (10^6/mm^3)	8.69 ± 0.13[d] (10)	8.91 ± 0.13 (10)	8.59 ± 0.1 (9)	8.27 ± 0.21 (10)	5.68 ± 1.4[e] (5)
Reticulocytes (% RBCs)	1.3 ± 0.1[d] (10)	1.3 ± 0.1 (10)	1.3 ± 0.1 (9)	2.6 ± 0.7 (10)	12.6 ± 6.4[e] (5)
PT (s)	10.1 ± 0.4[d] (10)	10.6 ± 0.3 (10)	11.4 ± 0.6 (8)	13.4 ± 1.0 (9)	20 ± 5.9[e] (4)
APTT (s)	15.5 ± 1.2[d] (10)	18.4 ± 1.6 (10)	18.8 ± 1.5 (8)	28 ± 2.7* (9)	72.9 ± 5.9[e] (4)
Fibrinogen (mg/dL)	258.1 ± 15.8 (9)	265.4 ± 10.1 (10)	246.2 ± 8.5 (9)	233.8 ± 19.7 (10)	350.4 ± 30.7[e] (5)

[a]Values are means ± SD for the samples indicated in brackets.
[b]VC = Vehicle control.
[c]UC = Untreated control.
[d]Significant dose-related trend (*P* < 0.02).
[e]Differ significantly from VC (*P* < 0.05).
Source: Ref. 190.

Table 5.16 Incidence and Severity of Lung Lesions in Fischer 344 Rats Dosed with dl-α-Tocopheryl Acetate[a]

Lung Lesions	Dose (mg kg^{-1} day^{-1})				
	0 (VC)[b]	0 (UC)[c]	125	500	2000
Males					
Chronic interstitial inflammation	0	0	5/10 (1.8)	10/10 (2.5)	10/10 (2.8)
Adenomatous hyperplasia	0	0	2/10 (1.5)	10/10 (3.0)	10/10 (3.0)
Females					
Chronic interstitial inflammation	0	2/10	6/10 (1.7)	8/10 (2.5)	8/10 (2.8)
Adenomatous hyperplasia	0	0	1/10 (1)	7/10 (2.5)	4/10 (2.3)

[a]Incidence = (number with lesion)/(number of animals examined). Lesion severity (in parentheses) is calculated from: 0 = normal, 1 = minimal, 2 = mild, 3 = moderate, and 4 = severe.
[b]VC = Vehicle control.
[c]UC = Untreated control.
Source: Ref. 190.

five- to eightfold increase in the concentration of tocopherol in the plasma of rats whereas intraperitoneal administration resulted in only a two- to fourfold increase, indicating that the difference in toxicity may be due to differences in the rate of absorption of d-α-tocopherol. It has also been observed that α-tocopherolquinone, a metabolite of α-tocopherol, is the causative agent for hemorrhages (192,193).

In humans, adverse effects of hypervitaminosis E include nausea, gastrointestinal disturbances, weakness, fatigue, creatinuria, and impairment of blood coagulation (194–197). Allergic reactions in some individuals using topical creams and

Table 5.17 Hemorrhages in Surviving Rats; Prothrombin and Kaolin-Activated Partial Thromboplastin Time (KPTT) Indices Following Dietary Administration of d-α-Tocopherol (Toc)[a]

Dietary (%)	Chemical intake (mmol/kg)	Survivors with hemorrhage	Prothrombin index (%)	KPTT
Basal	—	0	100 ± 5	100 ± 4
0.63 Toc	1.44	5	22 ± 3[b]	23 ± 3[b]
1.0 Toc	2.31	5	13 ± 3[b]	16 ± 3[b]

[a]N = 6.
[b]P < 0.01 by Dunnett's two-sided test.
Source: Ref. 191.

sprays containing α-tocopherol were shown by patch tests to be due to α-tocopherol (198–200). In a toxicological position report on vitamin E, Kappus and Diplock (201) observed that intakes of up to 720 mg/day were tolerated without side effects. At levels above 720 mg, in particular, at doses of 1600–3000 mg/day, some adverse effects may be observed that subside rapidly on reduction of the dose or discontinuation of vitamin E.

Long-Term and Carcinogenicity Studies. In rats fed 0.025, 0.25, 2.5, 10, or 25 g/kg diet of *dl*-α-tocopheryl acetate for 8–16 months, a reduction in body weight, a decrease in bone ash content, an increase in heart and spleen weights, elevated hematocrit, a reducion in prothrombin time, and increases in plasma alkaline phosphatase levels and in lipid levels in the liver were observed (202). A 2-year study in rats fed *dl*-α-tocopheryl acetate at levels of 500, 1000, and 2000 mg/kg bw per day revealed excessive hemorrhagic mortalities only in males. A dose-related increase in alanine aminotransferase activity was also observed in males. In female rats, an increase in relative liver weight was accompanied by histopathological changes indicating liver damage. At higher doses, a trend toward mammary tumor incidence was observed (203). Moore et al. (204) reported that α-tocopherol at the 1% level induced forestomach lesions in a 10-month study in hamsters, although at a lower level than BHA.

Tocopherols show inhibitory action against some known carcinogens. α-Tocopherol inhibited methylcholanthrene-induced buccal pouch tumors in hamsters, and dimethylhydrazine induced colonic tumors in mice (205,206). The anticarcinogenic properties of the tocopherols have been summarized by Tomassi and Silano (207).

Reproduction. In a reproduction study in weanling rats fed a normal level of *dl*-α-tocopheryl acetate (35 mg/kg diet) and 25, 50, 100, and 1000 times the control amount for 8 weeks, the number of pups born alive was less at the highest dose level (188). In a two-generation study in rats fed very high levels of 2252 mg/kg bw during pregnancy and lactation, α-tocopherol induced eye abnormalities in the progeny of both the first and second generations. Lower levels of 90 mg/kg bw had no adverse effects. α-Tocopherol was not teratogenic in rats, mice, or hamsters at dose levels of 16, 74, 250, 345, and 1600 mg/kg bw per day (208). In rats fed TPGS at dose levels of 0.002, 0.2, and 2% for 90 days, no adverse effects were observed on mean gestation period, litter size, sex ratio, or mortality rate. TPGS given on days 6–16 of gestation was also not teratogenic (173). The tocopherols were found to have an adverse effect on the testes at high doses. α-Tocopheryl acetate at 75 mg for 8–30 days caused significantly lower testicular weights and a transient disruption in spermatogenesis in hamsters (209). In mice given 10 mg of α-tocopheryl acetate per day for 20 days, an enlargement of the testicular interstitial cells and smooth endoplasmic reticulum was observed, indicating an enhanced steroid production (210). Vitamin E has been reported to increase the urinary excretion of androgens and estrogens in humans (211,212).

Mutagenicity. dl-α-Tocopherol and its acetate have not been found to be mutagenic and are reported to reduce the genotoxic effect of other known mutagens. dl-α-Tocopherol reduced the mutagenic effects of dimethylbenz[a]anthracene in leukocyte cultures (213). In five strains of *S. typhimurium*, dl-α-tocopherol markedly reduced the mutagenic effects of malonaldehyde and β-propiolactone (214). dl-α-Tocopherol and its acetate were not mutagenic in human lymphocyte cultures in vitro or in a sex-linked lethal test in *Drosophila* (215,216).

Gum Guaiac

Gum guaiac is a naturally occurring antioxidant from the wood of *Guajacum officinale* L. or *G. sanctum* L., family Zygophyllaceae. Gum guaiac was used in the 1940s as an antioxidant for oils and fats and also as a stabilizer for pastry. Though gum guaiac is an effective antioxidant, it is no longer used, mainly because of availability problems and also because of its tendency to produce various undesirable colors in shortenings. In the 1974 evaluation, JECFA allocated an ADI of 0–2.5 mg/kg bw.

TOXICOLOGICAL STUDIES

Very little information is available on the absorption and toxicological aspects of gum guaiac. However, gum guaiac has been used in medicine for a long time without any adverse effects and is generally considered to be a harmless substance. Absorption studies have been carried out in rats, dogs, and cats. Gum guaiac was poorly absorbed, being mostly excreted unchanged in the feces, with the remainder destroyed in the colon (217).

The acute oral LD_{50} in rats was >5000 mg/kg bw, and in guinea pigs it was >1120 mg/kg bw. In mice both oral and IP LD_{50} were >2000 mg/kg bw (6). In short-term studies in rats (0–0.5% for 4 weeks), cats (500 or 1000 mg for 34–117 weeks), and dogs (500 or 1000 mg for 62–103 weeks), no adverse effects were observed on growth rate, general behavior, hemoglobin levels, or red and white blood cell counts. The lungs, kidneys, liver, and spleen were found to be normal upon histological examination. Long-term and lifetime studies in rats showed no changes in growth rate, mortality, reproduction, or pathological examination (6,217). No adverse effects were observed in humans given 50–100 mg of gum guaiac daily for 18–104 weeks (217).

Ionox Series

The Ionox series comprises Ionox-100 [2,6-di-*tert*-butyl-4-hydroxymethylphenol], Ionox-201 [di-(3,5-di-*tert*-butyl-4-hydroxybenzyl) ether], Ionox-220 [di-(3,5-di-*tert*-butyl-4-hydroxy phenyl)methane], Ionox-312 [2,4,6-tri-(3′,5′-di-*tert*-butyl-4-

hydroxybenzyl)phenol], and Ionox-330 [2,4,6-tri-(3′,5′-di-*tert*-butyl-4-hydroxy-benzyl)mesitylene]. All these compounds are derivatives of BHT. Ionox-100 is approved for use in food products, and Ionox-330 is approved for incorporation into food packaging materials.

TOXICOLOGICAL STUDIES

The absorption, metabolism, and excretion of the Ionox compounds was reviewed by Hathway (218). Ionox-330 and Ionox-312 were not absorbed in the gastrointestinal tract in dogs, rats, pigs, or humans. Unchanged Ionox-330 was quantitatively eliminated in the feces. No biliary excretion was observed in rats or pigs. Rats eliminated 75% of a single oral dose of ^{14}C-labeled Ionox-330 (285.7 mg/kg bw) in 24 h and the remainder in 48 h. Dogs eliminated the entire dose (90 mg/kg bw) within 24 h (219). A similar trend was observed with Ionox-312 (218).

Ionox-100 was completely absorbed, rapidly metabolized, and quantitatively eliminated. In rats, a single oral dose (250 mg/kg bw) of ^{14}C-labeled Ionox-100 was eliminated in 11 days; 15.6–70.8% was found in the urine and 27–72.2% in the feces. Dogs showed a similar pattern of elimination. The elimination of Ionox-100 and its metabolites was faster than that of BHT and its metabolites. No unchanged Ionox-100 was found. The major metabolites were BHT acid and its glucuronide, similar to the structurally related compound BHT (Fig. 5.9) (111). In rabbits, BHT aldehyde and unchanged Ionox-100 have been reported (220).

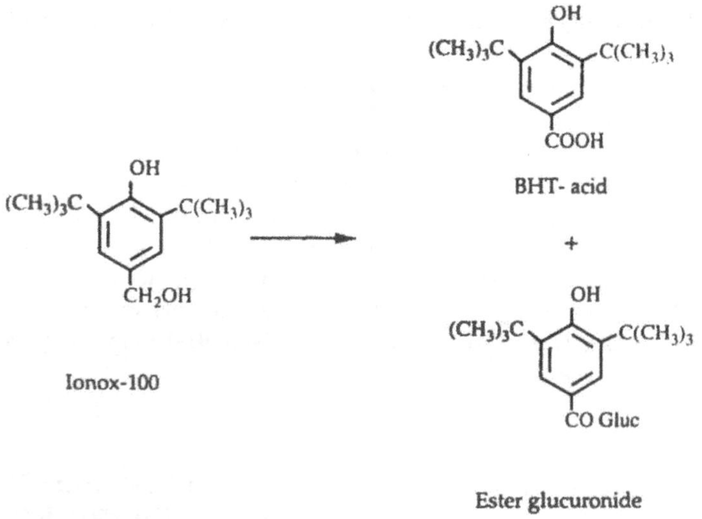

Fig. 5.9   Metabolites of Ionox-100.

Ionox-220 and -201 were found to have intermediate rates of absorption, metabolism, and elimination. In rats, 89.4–97.5% of a single oral dose (10 mg/kg bw) was eliminated in the feces in 20 days. Approximately 20% of a single oral dose of Ionox-220 was absorbed, and part of it was metabolized. Some of the metabolites identified were 3,5-di-*tert*-butyl-4-hydroxybenzoic acid and its glucuronide, quinone methide, and 2,6-di-*tert*-butyl-*p*-benzoquinone. Most of the dose was in the form of unchanged Ionox-220 (221). A similar pattern was observed with Ionox-201. Approximately 32% of a single oral dose was absorbed in the rats, and about 65% was eliminated unchanged. The major metabolites were BHT aldehyde, BHT acid and its glucuronide, and 3,3′,5,5′-di-*tert*-butyl-4-stilbenequinone (222).

The acute oral LD_{50} of Ionox-100 in rats and mice was found to be greater than 7 g/kg bw. In a 2-year study in rats at dose levels of 0.2 or 0.35%, no significant differences in survival rate, organ weights, histopathology, hematology, or serum enzyme levels were found between treated and control animals. Ionox-100 was without any adverse effects in a three-generation reproduction study in rats at the 0.35% level (223).

The acute oral LD_{50} of Ionox-330 in rats was >5000 mg/kg bw. In a 90-day study in rats with various dosage levels, Ionox-330 had no adverse effects except for a slight suppression in growth at the highest dosage level (3.1%) (224).

5.2.3 Miscellaneous Primary Antioxidants

Ethoxyquin

Ethoxyquin (EQ; 6-ethoxy-1,2-dihydro-2,2,4-trimethylquinoline) was first permitted for use in the United Kingdom for the prevention of scald in apples and pears. EQ is effective as an antioxidant in feeds, particularly for the stabilization of carotenoids in dehydrated alfalfa (225) and in fish meals, squalene, and fish oils (226–228). It readily undergoes oxidation to form a stable free radical ethoxyquin nitroxide that is more effective as an antioxidant than ethoxyquin (228,229). An ADI of 0–0.06 mg/kg bw has been allocated in the Joint FAO/WHO meetings on pesticide residues (230).

TOXICOLOGICAL STUDIES

Absorption, Metabolism, and Excretion. Ethoxyquin is readily absorbed and rapidly metabolized following the pattern of a nonphysiological substance. Studies on the metabolism and metabolites of EQ have been carried out by Skaare (231), Skaare and Solheim (232), Wilson et al. (233), and Wiss et al. (180). In rats given a single oral dose of ^{14}C-labeled EQ at the 100 mg/kg level, nearly 67–80% was recovered after 24 h in urine and feces. Nearly 50% of the radioactivity was recovered from the urine and 20% from the feces. Approximately 95% of the dose was excreted in 6 days. The major metabolites in urine were 6-hydroxy-2,2,4-trimethyl-1,2-dihydroquinoline and the oxidation product 2,2,4-trimethyl-6-quinoline (Fig. 5.10). Other metabolites observed were hydroxylated EQ and dihydroxy EQ (232). In another study in rats, Skaare (231) observed 28 and 36% of biliary excretion of a single oral dose of [^{14}C]EQ (100 mg/kg bw) in 12 and 24 h, respectively. However, EQ was not metabolized extensively before biliary excretion, and 75–85% of the ^{14}C excreted was identified as unchanged EQ. Some of the metabolites identified were 2,2,4-trimethyl-6-quinoline and hydroxylated and dihydroxylated EQ. The highest concentration of EQ was observed in the liver and kidneys in rats. Heart, skeletal muscles, and brain showed the least concentrations and eliminated the material rapidly. Spleen, blood, and abdominal fat had intermediate concentrations and eliminated the material at slower rates (Table 5.18). A small fraction of the injected EQ was observed in the milk, indicating an in utero transfer of the compound (233). Ethoxyquin was located primarily in the supernate rather than in the structural parts of the liver cells (181). In the cow, of the 155 mg of [^{14}C]EQ administered, 45.3 mg was recovered in the feces and 107.9 mg in the urine. The highest concentration of 0.036 ppm was found in the milk 33 h after ingestion (233).

Acute Toxicity. The acute oral LD$_{50}$ in rats and mice was found to be 178 mg/kg bw (234).

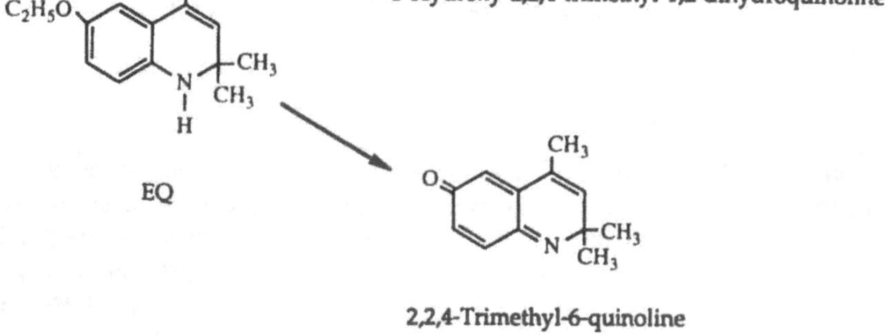

6-Hydroxy-2,2,4-trimethyl-1,2-dihydroquinoline

EQ

2,2,4-Trimethyl-6-quinoline

Fig. 5.10   Major metabolites of ethoxyquin.

Table 5.18 Concentration of ^{14}C-Labeled Ethoxyquin (EQ) in Rats After a Single Oral Dose[a]

Organ	Days after administration				
	1	2	7	14	28
	Concentration of EQ (ppm)				
Liver	2.04	1.32	0.34	0.11	0.038
Kidney	1.36	0.80	0.41	0.099	0.048
Fat	0.45	0.28	0.19	0.83	0.13
Spleen	0.30	0.31	0.32	0.045	0.063
Heart	0.23	0.072	0.075	0.015	0.014
Muscle	0.089	0.055	0.038	0	0.006
Brain	0.045	0.031	0.018	0	0.003
Blood			0.048	0.044	0.044

[a]Dosage = 6.7–9.2 mg EQ/kg bw.
Source: Ref. 233.

Short-Term Studies. Short-term studies have been conducted in rats and mice. No adverse effects were observed in rats fed EQ at levels of up to 0.4% for 20 days (234). In another study in rats fed EQ at the 0.5% level, reductions in food intake and growth rate were observed over a 60-day period. By the 14th day, a 50% increase in liver weight was observed. Concentrations of hepatic microsomal proteins, cytochrome P-450, and cytochrome b5 were markedly reduced. Induction of hepatic drug-metabolizing enzymes such as biphenyl-4-hydroxylase and ethylmorpholine-*N*-demethylase, and a reduction in the activity of glucose-6-phosphatase were observed. A 25% increase in total hepatic DNA was observed, suggesting that the liver enlargement may have resulted from both cellular hypertrophy and hyperplasia (235–237). However, the effects were found to be reversible on cessation of the treatment (Table 5.19). In rats given a single oral dose of 500 mg EQ/kg bw by stomach tube or 0.015% in drinking water (37.5 mg/kg bw per day) for 60 days, a significant increase in plasma glutamic oxaloacetic transaminase activity and hepatic lesions were observed (238). When EQ was given by gavage, ultrastructural changes observed in the liver were a proliferation of the smooth endoplasmic reticulum, dilatation of the perinuclear space, a disorganization of the mitochondrial membrane, desquamation, and fragmentation of the cells. In oral treatment, a slight proliferation of the endoplasmic reticulum was observed (239). In a more recent study, Kim (240) observed that in mice fed EQ at 0.125 and 0.5% for 14 weeks, relative liver weights and hepatic glutathione levels significantly increased. Nearly a twofold increase in the hepatic mito-

Table 5.19 Recovery from the Effects of Dietary Ethoxyquin of Some Hepatic Microsomal Enzymes and Other Parameters in Rats[a]

| | Period after withdrawal of EQ | | | |
| | 0 days | | 30 days | |
Parameters	Control	Test	Control	Test
1. Body wt (g)	98 ± 3	105 ± 3	293 ± 7	300 ± 14
2. Liver wt (g)	3.6 ± 0.1	7.0 ± 0.1***	12.3 ± 0.7	12.7 ± 0.9
3. DNA content				
a. mg/g liver	3.2 ± 0.1	2.1 ± 0.1***	2.3 ± 0.1	2.3 ± 0.1
b. mg/liver	11.5 ± 0.4	14.7 ± 1.3	28.3 ± 1.3	29.2 ± 2.4
4. Microsomal protein				
a. mg/g liver	28.4 ± 0.6	40.5 ± 1.9	29.4 ± 0.9	27.7 ± 1.6
b. mg/liver	102 ± 4	284 ± 4***	362 ± 16	352 ± 43
5. Cytochrome b_5				
a. nmol/g liver	14.4 ± 0.8	25.4 ± 0.8***	12.6 ± 0.6	13.6 ± 0.4
b. nmol/liver	52 ± 2	178 ± 2***	155 ± 6	173 ± 19
6. Cytochrome P-450				
a. nmol/g liver	21.9 ± 0.8	44 ± 3***	29.4 ± 0.9	28 ± 2.9
b. nmol/liver	79 ± 2	308.8 ± 8***	362 ± 26	356 ± 52
7. Biphenyl-4-hydroxylase				
a. μmol/h per g liver	5.1 ± 0.5	8 ± 0.9***	3 ± 0.1	3.1 ± 0.2
b. μmol/h per liver	18 ± 1	56 ± 5***	37 ± 1	39 ± 1
8. Ethylmorphine-N-demethylase				
a. μmol/h per g liver	38 ± 1	33 ± 1***	34 ± 3	33 ± 1
b. μmol/h per liver	137 ± 3	231 ± 13**	418 ± 52	419 ± 33
9. Glucose-6-phosphatase				
a. μmol/min per g liver	12.2 ± 0.1	5.5 ± 0.1***	11.2 ± 0.3	11 ± 0.3
b. μmol/min per liver	44 ± 1	39 ± 5	138 ± 10	140 ± 8

[a]EQ was administered at 0.5% during a 14-day pretreatment period. Values are means ± SEM. Each group had four animals. Significant effects are shown: *$P < 0.5$; **$P < 0.01$; ***$P < 0.001$.
Source: Ref. 236.

chondrial glutathione levels was observed only at the 0.5% level, which may be of significance in various cellular and subcellular activities.

Long-Term and Carcinogenicity Studies. In a chronic toxicity study in rats, EQ was administered at dose levels of 0.0062, 0.0125, 0.025, 0.05, 0.1, 0.2, and 0.4% for 200, 400, and 700 days. In the 200-day group, the kidneys became significantly heavier at the 0.2 and 0.4% levels. In the males, significantly heavier livers were observed at the 0.1% and higher levels. Heavier kidneys were noticed from 0.025%. In the 400-day group at 0.2 and 0.4% EQ, females showed no distinct lesions, but clear lesions were observed in the kidneys, livers, and thyroid glands in males. Kidneys showed tubular atrophy, fibrosis, focal tubular dilatation, and lymphocytic infiltration characteristic of chemical pyelonephrites. Renal calcification and necrosis was also observed. The effects were dose-dependent. A decrease in stored colloid and a diffuse increase in the height of follicular epithelium were observed in the thyroid glands, indicating mild hyperplasia. In males, of the 700-day group at the 0.2% level, the lesions were comparable to those of the 400-day group. But in females some patchy changes were observed in the kidneys. Occasional mammary, uterine, and adrenal tumors were also observed. At the lower dosage level of 0.0062%, there were mild changes in males and none in females (234). In a study in rats, Manson et al. (241) observed that the degree of nephrosis was dependent on age, sex, and duration of treatment. In another study on the nephrotoxicity of EQ in Fischer 344 rats, Hard and Neal (242) reported that EQ increased the cellular accumulation of lipofuscin-related pigments involving proximal tubules in female rats. They also observed that in female rats, papillary changes developed at a later stage than in males and the lesions never progressed beyond interstitial degeneration.

The modifying effects of EQ on various chemical carcinogens have been studied in rats. EQ administered by gavage or in drinking water significantly enhanced the hepatotoxic effect of N-nitrosodimethylamine (237). EQ significantly enhanced the colon carcinogenesis initiated by 1,2-dimethylhydrazine (243) and liver carcinogenesis initiated by diethylnitrosamine (144). Ethoxyquin enhanced the preneoplastic and neoplastic lesions in the kidneys and inhibited liver carcinogenesis initiated by N-ethyl-N-hydroxyethylnitrosamine (244). It enhanced the urinary bladder carcinogenesis induced by N-butyl-N-(4-hydroxybutyl)nitrosamine (245). Hasegawa et al. (246) observed that EQ inhibited the lung carcinogenesis and significantly enhanced the thyroid carcinogenesis initiated by N-bis(2-hydroxypropyl)nitrosamine in male rats.

Skin Toxicity. In rabbits and guinea pigs, repeated application of EQ produced a slight erythema followed by a papular eruption and in some instances scab formation. When the treatment was stopped, the lesions disappeared, leaving a normal appearing skin after a few weeks (234). EQ is also reported to cause allergic contact dermatitis in apple packers and animal feed mill workers (247–249).

Reproduction. Reproduction studies have been conducted in rats (234) and rabbits (250). In rats fed EQ at levels of 0.025–0.1% and in rabbits at 0.0025–0.01%, no reproductive or teratogenic effects were observed following administration of EQ before mating, throughout pregnancy, or during organogenesis. Ethoxyquin was also reported to prevent the occurrence of congenital malformities associated with vitamin E deficiency (21).

Anoxomer

Anoxomer (Poly AO 79) is a synthetic phenolic polymer developed by Weinshen-ker et al. (251). Anoxomer is a condensation product of divinylbenzene, hydroxyanisole, TBHQ, and *tert*-butyl phenol. Anoxomer is not a single molecular weight compound but is a distribution of molecular weights centered about a "peak" at 4000 (252). Anoxomer is highly effective in frying fats and oils and provides carry-through protection to the fried product. Anoxomer was accepted as a food grade antioxidant in 1982. JECFA has allocated an ADI of 0–8 mg/kg bw (253).

TOXICOLOGICAL STUDIES

Anoxomer is very poorly absorbed and metabolized because of its large molecular size. Absorption studies indicate that the compound is minimally absorbed (0.1–0.06% of the dose) in rats, mice, guinea pigs, rabbits, and humans (Table 5.20) (254–256). Anoxomer had no adverse effects on liver weight, hepatic cytochrome P-450 level, or hepatic mixed function oxygenase systems in rats at levels of 250 and 2500 mg/kg bw for 60 days (257).

Anoxomer was found to be nontoxic in short-term studies in rats, mice, and dogs at single doses of up to 10 g/kg bw and in subchronic feeding studies in rats at doses of up to 5% in the diet and in dogs (252,256). In in vitro mammalian and microbial mutagenicity studies, Anoxomer was negative (258). Anoxomer had no adverse effects on reproduction data in rats or rabbits and was not teratogenic (252).

Trolox-C

Trolox-C (6-hydroxy-2,5,7,8-tetramethyl chroman-2-carboxylic acid), a synthetic derivative of α-tocopherol, was developed by Scott et al. (259). In thin-layer tests

Table 5.20 Recovery of Radioactivity After Oral Administration of Poly AO-79 in Humans[a]

Time after dosing (h)	Urinary recovery (% of dose[b])	Fecal recovery (% of dose[b])
12	0.01	—
24	0.003	17.48
48	0.0	51.2
72	0.002	21.75
96	0.001	5.79
Total	0.016	93.66

[a]$N = 6$.
[b]Initial dose = 50 μCi of [^{14}C] Poly AO-79.

in vegetable oils and animals fats, Trolox C was found to have two to four times the antioxidant activity of BHA or BHT and was more active than tocopherols, propyl gallate, and ascorbyl palmitate. In vegetable oils, a combination of Trolox-C with certain amino acids was found to be highly effective (260,261). However, Trolox-C is not being used commercially at present.

TOXICOLOGICAL STUDIES

Trolox-C was found to have a low order of toxicity. In mice, the acute oral, intraperitoneal, and subcutaneous LD_{50} was 1630, 1700, and 1930 mg/kg bw, respectively. In adult rats the acute oral and IP LD_{50} was 4300 and 1800 mg/kg bw, respectively. In neonatal rats, acute oral LD_{50} was 1120 mg/kg bw, and in rabbits it was >2000 gm/kg bw. In a short-term study in dogs fed Trolox-C in doses pyramiding daily up to 320 mg/kg bw for over 14 days, no adverse effects were observed on hematocrit, hemoglobin levels, differential leukocyte counts, blood chemistry, plasma glucose levels, activities of serum glutamic oxaloacetic trans-aminase, glutamic pyruvic transaminase, or alkaline phosphatase (260).

5.3   TOXICOLOGICAL ASPECTS OF SYNERGISTIC OR SECONDARY ANTIOXIDANTS

5.3.1   Oxygen Scavengers

Sulfites

Sulfites represent a group of compounds comprising sulfur dioxide (SO_2), sodium sulfite (Na_2SO_3), sodium metabisulfite ($Na_2S_2O_5$), and sodium bisulfite ($NaHSO_3$). Sulfites are widely used as antimicrobials in wine making and corn wet milling and as antioxidants in fruits and vegetables to prevent enzymatic browning and preserve freshness. Sulfites are also added to dehydrated fruits, vegetables, soups, fruit juices, and beer. Sulfites have been GRAS substances since 1959. However, in 1986 the use of sulfiting agents in raw packaged or unpackaged fruits and vegetables was banned by the FDA because of some adverse reactions reported in sulfite-sensitive individuals. Most of the adverse reactions were associated with the use of sulfites in salad bars in restaurants. Sulfites precipitate attacks of asthma in some individuals, and up to 17 fatalities have been attributed to sulfite ingestion in sulfite-sensitive asthmatic individuals. The patients most likely to have sulfite sensitivity seem to be those severely ill with asthma and dependent on corticosteriod medication for control of the disease (262). An ADI of 0–0.7 mg/kg bw has been allocated to sulfites (as sulfur dioxide) by JECFA (13).

TOXICOLOGICAL STUDIES

Absorption, Metabolism, and Excretion. Sulfites are readily absorbed and quickly metabolized either by oxidation or by the formation of S-sulfonates (263,264). In rats given single oral doses of sodium metabisulfite in a 0.2% solution, 55% of the sulfur was eliminated as sulfates in the urine within the first 4 h (265). In rats, mice, and monkeys, following oral administration of 10 or 50 mg of SO_2 (as sodium bisulfite) per kilogram body weight, 70–95% was absorbed and eliminated in the urine within 24 h. Most of the remaining dose was eliminated in the feces, and only 2% or less remained in the body after 1 week (266).

In mammals, the primary route of sulfite metabolism was via enzymatic oxidation to sulfates by the mitochondrial enzyme sulfite: cytochrome c reductase (sulfite oxidase), which occurs predominantly in the liver and in lower concentrations in almost all other tissues of the body. Sulfite oxidase had a very high capacity for sulfite oxidation with first-order rate constant on the order of $0.7–1$ min^{-1}, equivalent to a half-life of approximately 1 min for sulfite (267). Gunnison et al. (267) reported that rats exhibit approximately three to five times as much sulfite oxidase activity as rabbits and monkeys. Small but significant amounts of sulfites were metabolized to S-sulfonate compounds in rats, monkeys, and rabbits (263,268). At higher concentrations, sulfites reacted with β-mercaptopyruvate, an intermediate in sulfur-amino acid catabolism, forming inorganic thiosulfate, which as been detected in the urine of normal humans and rats (269). Gunnison and Palmes (263,270) also detected free sulfites in the plasma of rats, rabbits, and monkeys both before and after administration of sulfites.

Reactions of Sulfites In Vitro. In vitro, sulfites react reversibly with open-chain aldehydes and ketones to form hydroxysulfonate compounds (271,272) and with pyridine and flavin nucleotides to form sulfonic acid compounds (273). Sodium bisulfite adds reversibly to uracil and cytosine and their derivatives (274–276) to form a sulfonate adduct. Sulfites induce sulfitolysis of thiamine, a irreversible reaction resulting in the cleavage of thiamine (277) and irreversible cleaving of disulfide bonds in free cystine, forming cysteine S-sulfonate (278). The aerobic oxidation of sulfites involves a free-radical mechanism resulting in the formation of free radicals that have potential biological significance (275). Significant cleavage of the glycosidic linkage of uridine and cytidine (279), oxidation of methionine and certain other diallyl sulfides (280), destruction of tryptophan (281), and oxidation of lipids (282) are some of the effects of the free radicals arising from sulfite autoxidation.

In vitro, sulfites inhibit a number of enzymes associated with NAD or flavin nucleotide cofactors. Ciaccio (283) observed a 50% inhibition of a number of NAD-dependent dehydrogenases (lactate dehydrogenase, malate dehydrogenase, alcohol dehydrogenase, and glutamate dehydrogenase) in the presence of 0.03–0.5 mM sulfites. Massey et al. (284) observed that a series of flavin-dependent enzymes such as D- and L-amino acid oxidases and lactate oxidase react reversibly with sulfites, forming flavin sulfite adducts that are unstable in the absence of sulfites. Sulfites also inhibit cytochrome oxidase (285), α-glucan phosphorylase (286), and most of the sulfatase enzymes (287).

Acute Toxicity. The acute intraperitoneal LD_{50} of sodium bisulfite was 498 mg/kg bw in rats, 300 in rabbits, 244 in dogs, and 675 in mice (288). Acute intravenous LD_{50} was 115 mg/kg bw in rats, 130 in mice, 95 in hamsters, and 65 in rabbits (289).

Short-Term and Long-Term Studies. Short-term studies have been conducted in rats and pigs. In rats fed fresh diet with 0.6% sodium metabisulfite for 6 weeks, a reduction in growth rate was observed that was attributed to lack of thiamine. After 75 days, rats fed the same diet showed signs of thiamine deficiency and additional toxic effects such as diarrhea and reduced growth rate that could not be completely corrected by the administration of thiamine (290). Thiamine deficiency was not observed when sulfite was administered in drinking water (291). In rats fed 0–8% sodium metabisulfite for 10–56 days in a thiamine-enriched diet, levels of 6% or above depressed food intake and growth (Table 5.21) accompanied by glandular hyperplasia, hemorrhage, ulceration, necrosis, and inflammation of the stomach. Anemia occurred in all animals receiving sodium metabisulfite at levels of 2% or above. Leukocytosis and splenic hematopoiesis were also observed. The effects were found to be reversible (292). Gunnison (269) observed that the anemia results from the interaction of sulfite with dietary constituents, possibly cyanocobalamine.

Table 5.21 Mean Values of Body Weight, Food Intake, and Food Efficiency of Male Rats Fed Dietary Levels of 0–6% Sulfite or 4% Sulfate for 8 Weeks

Dietary level (%)	Body weight (g) at end of week			Food consumption (g/rat per day), weeks 1–3	Food efficiency[a] weeks 1–3
	0	4	8		
0 (control)	124	237	295	15.5	0.32
$Na_2S_2O_5$					
0.5	125	227	285	14.9	0.30
1.0	125	238	298	16	0.31
2.0	125	219	275	14.9	0.27
6.0	125	149[b]	174[b]	11.2	0.08
Na_2SO_4					
4.0	124	231	274	15.9	0.29

[a]Food efficiency = [weight gain (g)]/[food intake (g)].
[b]$P < 0.001$.
Source: Ref. 292.

In long-term and multigeneration feeding studies, rats were fed 0.125, 0.25, 0.5, 1, or 2% sodium metabisulfite in a thiamine-enriched diet for 2 years. A significant reduction of thiamine levels in the urine and the liver was observed, and addition of thiamine prevented the symptoms at lower levels of sulfite. At the 1 and 2% levels, occult blood was present in the feces. At the 2% level, a slight growth retardation was observed in the F_1 and F_2 generations. Increase in the relative weight of the kidneys not accompanied by any functional or histological changes was observed in females of the F_2 generation. Pathological examination revealed hyperplastic changes in both the forestomach and glandular stomach at the 1 and 2% levels in all three generations. There was no indication that sulfites had any carcinogenic effect. Sulfites had no effect on reproduction data or early development (292).

In short-term and long-term feeding studies, pigs were fed sodium metabisulfite in thiamine-enriched diets at levels of 0.125, 0.25, 0.5, 1, or 2% for 15 weeks and up to 48 weeks. Thiamine levels were markedly reduced in the urine and liver in animals fed 0.16% or more, and thiamine supplementation prevented the deficiency with sulfite levels up to 0.83% (Table 5.22). No adverse effects were observed on health, mortality, or the blood picture. A reduction in body weight and food conversion was observed at 1.72%. An increase in the weight of liver, kidneys, heart, and spleen and mild inflammatory hyperplastic changes were observed in the gastric mucosa at the 0.83 and 1.72% levels. In paired feeding studies, esophageal intraepithelial microabscesses, epithelial hyperplasia, and accumulation of neutrophilic leukocytes in the papillae tips were observed (293).

Table 5.22 Thiamine Levels in Pigs Fed Sulfite at Dietary Levels of 0–1.72% for 48 Weeks[a]

Dietary level (%)	Thiamine in the urine (μg/100 mL)			Thiamine in the liver (μg/g) at week 48
	wk 4	wk 26	wk 47	
0[b]	27	1	30	1.8
0[c]	940	1153	1000	4
0.06	1360	1132	2000	4.2
0.16	890	594	540	3.6
0.35	310	648	280	2.8
0.83	72	220	25	2.25
1.72	30	4	5	1.15

[a]$N = 4$.
[b]Animals fed a basal diet not supplemented with thiamine.
[c]Animals supplemented with 50 mg thiamine/kg in their diet.
Source: Ref. 293.

Mutagenicity. The genotoxic effect of sulfites has been studied in a number of test systems and reviewed by Gunnison (269) and Shapiro (294). Sulfites were mutagenic in a number of in vitro systems such as *Escherichia coli* (295), yeasts (294), and Chinese hamster ovary cells (296) and enhanced the UV mutagenicity in Chinese hamster cell lines by the inhibition of excision repair of DNA (297). The mutagenic effects of sulfites have been attributed to their reaction with cytosine and uracil and to the free radicals formed during the autoxidation of sulfites. Sulfites were negative in a dominant lethal test in mice (298) and did not induce chromosomal aberrations in mouse oocytes in vitro following a single IV injection of up to 2 mmol/kg (299).

Ascorbic Acid and Sodium Ascorbate

L-Ascorbic acid (AA) or vitamin C (3-keto-L-glucofuranolactone) occurs widely in nature. Ascorbic acid and sodium ascorbate (SA) are used as oxygen scavengers and also as synergists in a wide variety of food products including canned or bottled products with a headspace, vegetable oils, beverages, fruits, vegetables, butter, cured meat, and fish products. In the presence of oxygen and metal ions in aqueous solutions, AA readily undergoes oxidation to form dehydroascorbic acid. AA and SA have been allocated an ADI of "not limited" by JECFA (300).

TOXICOLOGICAL STUDIES

Absorption, Metabolism, and Excretion. Absorption studies have been carried out in rats and humans. In rats after an IP injection of 1.5–5.9 mg of ^{14}C-labeled ascorbic acid, 19–29% was converted to CO_2 and only 0.4% was excreted as oxalic acid within 24 h (301). The average absorption of AA has been estimated to be 84% in humans (302). Kuebler and Gehler (303) observed that increasing oral intake from 1.5 to 12 g decreased the relative absorption of AA from about 50% to only 16%. Oxalic acid was the major urinary metabolite in humans. Other metabolites observed were dehydroascorbic acid, 2,3-dioxo-L-gulonic acid, and ascorbic acid sulfate (Fig. 5.11) (304). Ascorbic acid is excreted by glomerular filtration and active tubular reabsorption. High doses of AA (4 g or more) increased the urinary excretion of oxalate. About 40% of the urinary oxalate was derived from AA, but the mechanism of its formation is not clear (304). Briggs et al. (305) observed that at daily intakes of 4 g of ascorbic acid, the urinary oxalate level incrased 10-fold, which may lead to the formation of kidney stones. However, Hornig and Moser (306), in an evaluation of the safety of high vitamin C intakes, observed that even with large daily intakes the amount of oxalic acid formed was far too little to contribute significantly to oxalate formation.

Acute Toxicity. In mice and rats, the acute oral and IV LD$_{50}$ of ascorbic acid was >5000 and >1000 mg/kg bw, respectively, and in guinea pigs it was >5000 and >500 mg/kg bw, respectively (307).

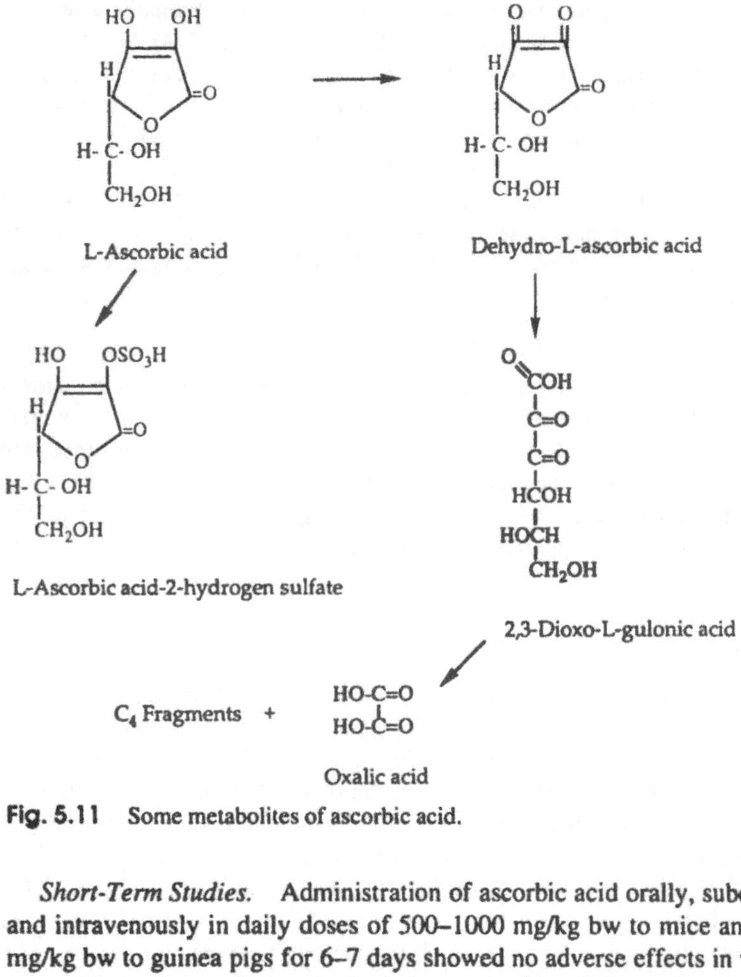

Fig. 5.11   Some metabolites of ascorbic acid.

Short-Term Studies.   Administration of ascorbic acid orally, subcutaneously, and intravenously in daily doses of 500–1000 mg/kg bw to mice and 400–2500 mg/kg bw to guinea pigs for 6–7 days showed no adverse effects in weight gain, general behavior, or histopathological examination of the various organs (307). Ohno and Myoga (308) observed that in guinea pigs, an intake of vitamin C at 10–100 times the normal daily requirement (600 mg/kg bw per day) daily for 112 days resulted in fat deposition and congestion in the liver leading in death.

Long-Term and Carcinogenicity Studies.   In rats given daily doses of 1000, 1500, or 2000 mg/kg bw of ascorbic acid for 2 years, no adverse effects were observed in hematological examination, urinalysis, liver, or renal function tests. Pathological examination revealed no toxic lesions attributable to AA (13). Ascorbic acid was not carcinogenic in rats and mice fed 2.5 or 5% in the diet for 2 years (309). In rats, sodium ascorbate at the 5% level in the diet has been shown to promote urinary bladder carcinogenesis initiated by *N*-butyl-*N*-(4-hydroxybutyl)-

nitrosamine or *N*-methylnitrosourea, forestomach carcinogenesis initiated by *N*-methylnitrosourea or *N*-methyl-*N'*-nitro-*N*-nitrosoguanidine, and colon cancer initiated by 1,2-dimethylhydrazine (76). However, AA was shown to have anticarcinogenic and antimutagenic properties, and it also prevented nitrosation reactions, thus inhibiting the formation of carcinogenic nitroso compounds (310). Ascorbic acid suppressed the development of UV-induced skin cancer (311) and inhibited the mutagenicity of a number of compounds (214,312,313).

Reproduction. Information on reproduction is conflicting. In some studies, it has been observed that large daily doses of AA increased abortion in guinea pigs and rats (314,315). In later studies, Alleva et al. (316) and Frohberg et al. (317) reported no adverse effects in rats, mice, hamsters, and guinea pigs at dosages comparable to those of the earlier studies (Table 5.23). The reason for the discrepancy is not known.

Mutagenicity. Ascorbic acid was reported to be mutagenic in a number of in vitro systems (318–321). However, in in vivo systems AA was negative in Chinese hamster bone marrow tests, rat dominant lethal assay, and guinea pig intrahepatic host mediated test (320,322,323). In in vitro systems, AA was mutagenic only in the presence of oxygen and metal ions in the incubation medium. Under these conditions AA completely oxidizes to form hydrogen peroxide, which is a known mutagen (306). These observations indicate that ascorbic acid is not mutagenic.

Table 5.23 Effects of Ascorbic Acid on Pregnancy in Rats and Hamsters[a]

| | | | Pups/litter | |
Group	Dose (mg/kg)	N	Dead[b]	Alive[b]
Rats				
1	0	14	0.36 ± 0.2	11.1 ± 0.6
2	50	11	0.4 ± 0.31	11.7 ± 0.8
3	150	11	0.6 ± 0.27	11 ± 0.5
4	450	11	1.2 ± 0.53	10.5 ± 0.9
Hamsters				
1	0	11	0.09 ± 0.09	10.1 ± 0.5
2	50	12	0.08 ± 0.08	9.9 ± 0.09
3	150	10	0.1 ± 0.1	10.5 ± 0.7
4	450	14	0.29 ± 0.16	11 ± 0.5

[a]Single oral dose of ascorbic acid given daily days 1 – 19 (rats) or on day 15 (hamsters) of pregnancy. Controls received water.
[b]Mean ± SE.
Source: Ref. 316.

Studies in Humans. Studies in humans indicate that ascorbic acid has a diuretic effect at 5 mg/kg bw, and glycosuria was observed with doses of 30–100 mg/kg bw. With very high doses of 6000 mg/day, adverse effects observed were nausea, vomiting, diarrhea, flushing of the face, headache, fatigue, and disturbed sleep. The main reaction observed in children was skin rashes (324).

Ascorbyl Palmitate

Ascorbyl palmitate, the ester of ascorbic acid and palmitic acid, is highly effective in frying fats, oils, and fried products. Alone, ascorbyl palmitate is better than BHA and BHT, and in combination with other antioxidants it improves the shelf life of all vegetable oils (176). In the 1940s, food grade ascorbyl palmitate normally contained a significant amount of ascorbyl stearate, and toxicological studies were done of samples containing 5–20% stearate and 80–90% palmitate. An ADI of 0–1.25 mg/kg bw was allocated by JECFA (13) for ascorbyl palmitate or ascorbyl stearate or the sum of both.

TOXICOLOGICAL STUDIES

No information is available on the metabolism of ascorbyl palmitate, but it is assumed that ascorbyl palmitate is hydrolyzed to ascorbic acid and palmitic acid (325). The acute oral LD_{50} in mice is 25 g/kg bw (13). In short-term studies in rats fed 2 or 5% ascorbyl palmitate for 9 months, at the 5% level a significant retardation in growth rate, bladder stones, and hyperplasia of the bladder epithelium were observed. At the 2% level only a slight retardation in growth was observed. In long-term studies in rats fed 1 or 5% ascorbyl palmitate for 2 years, no adverse effects were observed on growth rate or mortality, or in pathological studies (326).

Erythorbic Acid and Sodium Erythorbate

Erythorbic acid (3-keto-D-glucofuranolactone) is the D-isomer of ascorbic acid. Erythorbic acid has no vitamin C activity and does not occur naturally in food products. Erythorbic acid and its sodium salt are effective in the stabilization of nitrate- and nitrite-cured meat products, in dehydrated fruit and vegetable products, and as synergists for tocopherols in fats and oils. Erythorbic acid in combination with citric acid can be an effective alternative to the use of sulfites in frozen seafood, salad vegetables, and apples (327–329). In model solutions, erythorbic acid was more rapidly oxidized than ascorbic acid (330). Erythorbic acid is not permitted for food use in the European Economic Community countries. In a 1974 reevaluation, JECFA allocated an ADI of 0–5 mg/kg bw (13).

TOXICOLOGICAL STUDIES

Erythorbic acid is readily absorbed and rapidly metabolized. In guinea pigs, erythorbic acid decreased the tissue uptake of ascorbic acid, thereby lowering its level in various organs, including the spleen, adrenal glands, and kidneys (331–333). No information is available on the acute toxicity of erythorbic acid. In short-term studies in rats fed 1% erythorbic acid for 36 weeks, no adverse effects were observed on growth rate and mortality. Gross and histopathological studies of various organs revealed no changes (326). In long-term studies in rats fed 1% erythorbic acid, no adverse effects were observed on growth rate or mortality and no histopathological changes were observed (6). In a carcinogenicity test in F344 rats, sodium erythorbate given at levels of 1.25 or 2.5% in drinking water for 104 weeks was not carcinogenic (334).

Sodium erythorbate at a 5% level was found to have a promoting effect on the N-butyl-N-4-hydroxybutlynitrosamine-initiated urinary bladder carcinogenesis in rats. However, when tested alone it was without any effect (245). Erythorbic acid was weakly positive in the Ames S. typhimurium reverse mutation assays with and without metabolic activation. In an in vitro assay for chromosomal aberrations in Chinese hamster fibroblast cell lines, erythorbic acid was negative (335).

5.3.2 Chelating Agents

Phosphoric Acid and Phosphates

Phosphoric acid and its salts are widely used in the food industry as chelating agents, emulsion stabilizers, anticaking agents, and antimicrobial agents. Some of the food products in which phosphates are used are meat products, poultry, cheese, and soft drinks. Phosphoric acid is an essential biological constituent of bone and teeth and also plays an important role in carbohydrate, fat, and protein metabolism. Phosphates that are used as chelators include sodium and potassium orthophosphates (Na_2HPO_4, NaH_2PO_4, K_2HPO_4, KH_2PO_4), sodium pyrophosphates ($Na_4P_2O_7$, $Na_2P_2H_2O_7$), sodium tripolyphosphate ($Na_5P_3O_{10}$), and sodium polyphosphates ($NaPO_3)_{n=4-15}$. Phosphoric acid also functions as a synergist with other antioxidants in vegetable shortenings. JECFA has allocated an ADI of 0–70 mg/kg bw that applies to the sum of added phosphate and food phosphate (13).

TOXICOLOGICAL STUDIES

The orthophosphates are readily absorbed through the intestinal wall. The pyrophosphates and polyphosphates are converted to monophosphate by an inorganic diphosphatase before absorption. The hexametaphosphates are more slowly degraded than tripolyphosphates. The degree of hydrolysis and absorption decreases with an increase in molecular weight (336–338). The level of inorganic phosphate in the blood was found to be regulated by the action of the parathyroid hormone. This hormone also inhibits the tubular reabsorption of phosphates by the kidneys. Phosphates are mainly excreted in the feces as calcium phosphate. Hence, excessive intake of phosphate may result in a depletion of calcium and bone mass (13,339).

Phosphates in general have no adverse effects at low dosages. At high levels, adverse effects that have been observed are retarded growth, increase in kidney weight, hypertrophy of the parathyroid glands, and metastatic calcification in the soft tissues, especially the kidneys, stomach, and aorta. Tables 5.24 and 5.25 present the acute toxicity levels of various phosphates and the maximum tolerated dietary levels in various animals.

Phosphoric Acid.   In rats in a 44-day study, phosphoric acid at levels of 2.94% caused extensive kidney damage (340). At lower levels of 0.75%, phosphoric acid had no adverse effects (341). Feeding studies in dogs showed that phosphoric acid was well tolerated at up to 13 g/day before signs of enteritis appeared. In humans, phosphoric acid had no adverse effects at dosages of up to 26 g/day (342).

Monosodium Phosphate.   Monosodium phosphate at the 3.4 g/kg bw level caused kidney damage in rats in a 42-day study (343). In guinea pigs, the level of magnesium and potassium also had a significant effect on the physiological damage by phosphates. Diets containing 0.9% calcium, 1.7% phosphorus, 0.04% magnesium, and 0.41% potassium resulted in a severe reduction in weight gain, stiffness in the leg joints, and numerous deposits of calcium phosphates in the foot pads. The symptoms were significantly reduced or eliminated when the diet contained 0.35% magnesium and 1.5% of potassium with the same levels of phosphorus and calcium.

Disodium and Dipotassium Phosphates.   Hahn et al. (344) reported that in rats disodium phosphate at levels of 1.8, 3, and 5% for 6 months caused extensive physiological damage only at the 3 and 5% levels. Dipotassium phosphate at the 0.87 and 5% levels for 150 days had no adverse effects in rats (345). No adverse effects were observed on the levels of iron, calcium, and copper in the blood in either study. In a three-generation study in rats fed diets containing 0.5–5% of a mixture of mono- and disodium phosphates, renal damage and a reduction in growth rate and fertility were observed at levels of 1% or more. A reduction in the life span in the 5% group was also observed (346).

Pyrophosphates.   In rats fed 0.5–5% pyrophosphate for 6 months (347), 4 months (348), and in a multigeneration study (346), 1% or more of the pyrophosphate produced kidney damage and calcification.

Polyphosphates.   In rats, sodium tripolyphosphate or sodium hexametaphosphate at the 3 and 5% levels for 6 months caused severe kidney damage equivalent to that found when similar amounts of disodium phosphate were fed in the diet, indicating the hydrolysis of polyphosphates to orthophosphates before absorption

Table 5.24 Acute Toxicity Levels of Phosphates in Animals

Phosphate	Animal	Route[a]	LD_{50} (mg/kg bw)	Lethal dose (approx.) (mg/kg bw)
H_3PO_4	Rabbit	IV		1010
NaH_2PO_4	Mouse	O		>100
	Guinea pig	O		>2000
	Rat	IP		>36
	Rabbit	IV		>985
				≤1075
$NaH_2PO_4 + Na_2HPO_4$	Rat	IV	>500	
$NaH_2P_2O_7$	Mouse	O	2,650	
		SC	480	
		IV	59	
	Rat	O	>4,000	
$Na_4P_2O_7$	Rat	IP	233	
		IV	100–500	
	Mouse	IP		~40
		O	2,980	
		SC	400	
		IV	69	
	Rabbit	IV		~50
$Na_5P_3O_{10}$	Mouse	O		>100
	Mouse	O	3,210	
		SC	900	
	Rat	IP	134	
$Na_6P_4O_{13}$	Rat	O	3,920	
		SC	875	
$(KPO_3)n$ + pyro	Rat	O	4,000	
		IV	~18	
Hexametaphosphate[b]	Rabbit	IV		~140
	Mouse	O		>100
	Mouse	O	7,250	
		SC	1,300	
		IV	62	
$(NaPO_3)_{n=6}$	Rat	IP	192	
$(NaPO_3)_{n=1}$	Rat	IP	200	
$(NaPO_3)_{n=27}$	Rat	IP	326	
$(NaPO_3)_{n=47}$	Rat	IP	70	
$(NaPO_3)_{n=65}$	Rat	IP	40	
$(NaPO_3)_3$ cyclic	Mouse	O	10,300	
		SC	5,940	
		IV	1,165	

[a]IP = Intraperitoneal; IV = intravenous; O = oral—in diet, stomach tube, etc.; SC = subcutaneous.
[b]Average chain length not given. Original references in Ellinger (339).
Source: Ref. 339. (With permission from CRC Press, Boca Raton, Florida.)

Table 5.25 Dietary Levels of Phosphates Producing No Adverse Effects

Phosphate[a]	Test animal	Length of test[b]	Maximum level tolerated[c]	Effect of excess phosphate
Orthophosphate				
H_3PO_4 (36.4%)	Human	Variable	17–6 g/day	No adverse effects
	Rat	>12 mo	>0.75%	No adverse effects
Na and K	Rat	44 days	<2.94%	Kidney damage
orthophosphates	Rat	44 days	<3.93%	Kidney damage
MSP	Human	Variable	5–7 g/day	No adverse effects
	Rat	42 days	>3.4 g/kg/day	Kidney damage
	Guinea pig	200 day	>2.2%, <4.0%	Ca deposits reduced growth
		12–32 wk	4–8%	Ca deposits reduced growth
MSP + DSP	Rat	3 gen[b]	>0.5%, <1%	Kidney damage
DSP	Rat	6 mo	>1.8%, <3.0%	Kidney damage
DSP	Rat	1 mo	<5%	Kidney damage
DKP	Rat	150 days	>5.1%	No adverse effects
Pyrophosphates				
TSPP	Rat	6 mo	>1.8%, <3.0%	Kidney damage
	Rat	16 wk	<1%	Kidney damage
SAPP + TSPP + $(KPO_3)n$	Rat	3 gen[b]	>0.5%, <1%	Kidney damage

Tripolyphosphates				
STP	Rat	6 mo	>1.8%, <3.0%	Kidney damage
		1 mo	>0.2%, <2%	Kidney damage
		2 yr	>0.5%, <5%	Kidney damage
	Dog	1 mo	>0.1 g/kg per day	No adverse effects
		5 mo	<4.0 g/kg per day	Kidney, heart damage
Polyphosphates				
SHMP	Rat	150 days	>0.9%, <3.5%	Slight growth reduction
		1 mo	>0.2%, <2%	Kidney damage
		3 gen[b]	>0.5%, <5%	Kidney damage
	Dog	1 mo	>0.1 g/kg per day	No adverse effects
		5 mo	<2.5 g/kg per day	Kidney, heart damage
Graham's salt	Rat	6 mo	>1.8%, <3%	Kidney damage
$(KPO_3)n$ + SAPP + TSPP	Rat	3 gen[b]	>0.5%, <1%	Kidney damage
Cyclic phosphates				
$(NaPO_3)_3$	Rat	1 mo	>2%, <10%	Kidney damage
		2 yr	>0.1%, <1%	Retarded growth
		3 gen[b]	>0.05%	No adverse effects
	Dog	1 mo	>0.1 g/kg per day	No adverse effects
		5 mo	<4 g/kg per day	Kidney, heart damage
$(NaPO_3)_4$	Rat	1 mo	>2%, <10%	Kidney damage
	Dog	1 mo	>0.1 g/kg per day	No adverse effects
		5 mo	<4 g/kg per day	Kidney, heart damage

[a]MSP = Monosodium orthophosphate; DSP = disodium orthophosphate; DKP = dipotassium orthophosphate; TSPP = tetrasodium pyrophosphate; SAPP = sodium acid pyrophosphate; STP = sodium tripolyphosphate; SHMP = sodium hexametaphosphate.
[b]gen = Generations.
[c]% = Percent phosphate in the diet (often calculated from %P).
Source: Ref. 339. (With permission from CRC Press, Boca Raton, Florida.)

(344,347). Hodge (unpublished report cited in Ref. 339) found that in dogs 0.1 g/kg bw per day of sodium tripolyphosphate or sodium hexametaphosphate produced no adverse effects. On increasing the dosage to 4 g/kg bw per day over a 5-month feeding period, a reduction in body weight was observed.

Human Studies. Phosphoric acid at levels of 2000–4000 mg for 10 days or 3900 mg for 14 days had no adverse effects on urine composition (349). Long-term administration of 5000–7000 mg of monosodium phosphate also had no adverse effect (350). Phosphate-enriched diets (700 mg of calcium and 1 g of phosphorus per day for 4 weeks followed by 700 mg of calcium and 2.1 g of phosphorus per day for 4 weeks) resulted in mild diarrhea, soft stools, an increase in serum and urinary phosphorus, and a decrease in serum and urinary calcium. An increase in the excretion of hydroxyproline and cyclic AMP was also observed (351).

Ethylenediaminetetraacetic Acid

Ethylenediaminetetraacetic acid (EDTA) and its disodium salt (Na_2EDTA) and calcium disodium salt ($CaNa_2EDTA$) are used as chelating agents in the food industry. They are highly effective in a wide variety of products, including fats and oils, salad dressings, sauces, dairy products, meat, and processed fruits and vegetables and in fruit juices for the stabilization of vitamin C. EDTA forms stable water-soluble complexes with many di- or polyvalent metal ions. Maximum chelating efficiency occurs at higher pH values where the carboxyl groups are dissociated. The $CaNa_2$ salt is considered to have the lowest toxicity, and JECFA has allocated an ADI of 2.5 mg/kg bw (13).

TOXICOLOGICAL STUDIES

Absorption, Metabolism, and Excretion. Absorption studies were carried out in rats and humans. EDTA is poorly absorbed and rapidly excreted from the body. In rats fed ^{14}C-labeled $CaNa_2EDTA$ at doses of 50 mg/kg bw, only 2–4% of the dose was absorbed and 80–90% was excreted in the feces within 24 h (352). In rats given a single oral dose of 95 mg of Na_2EDTA, 93% was eliminated in 32 h. The amount of EDTA recovered was also found to be directly proportional to the dose level, suggesting that EDTA was absorbed from the gastrointestinal tract by passive diffusion (353). The elimination of EDTA from the kidneys was mainly by tubular excretion and glomerular filtration (352). In normal healthy humans given a dose of 1.5 mg of ^{14}C-labeled $CaNa_2EDTA$ in a gelatin capsule, only 5% absorption was observed (354). After an intravenous injection of 3000 mg of [^{14}C]$CaNa_2EDTA$, the compound was almost completely eliminated in 12–16 h, and only 2.5% of the dose was recovered in the urine (355).

Reactions of EDTA. Ethylenediaminetetracetic acid chelates with a number of physiologically important metal ions such as calcium, zinc, and iron and may induce essential mineral deficiency. CaNa2EDTA enhanced the excretion of Zn (356). EDTA was found to inhibit the activity of a number of heavy metal–containing enzymes because of its metal-chelating capacity. At concentrations of 10^{-3} M, EDTA inhibited aldehyde oxidase and homogentisinase (357). Na2EDTA at 5.5×10^{-6} M was a strong inhibitor of γ-amino levulinic acid dehydrogenase (358). At 1722 mg/kg bw, CaNa2EDTA inhibited alkaline phosphatase in the liver, prostate, and serum in rats (359).

Acute Toxicity. The acute oral LD50 of Na2EDTA in rats was 2000–2200 mg/kg bw (360), and in rabbits, 2300 mg/kg bw (361). The acute oral LD50 of CaNa2EDTA in rats was 10,000 mg/kg bw, in rabbits 7000 mg/kg bw, and in dogs 12,000 mg/kg bw (362).

Short-Term Studies. Short-term studies have been conducted in rats, rabbits, and dogs. In rats given 250 or 500 mg/kg bw of CaNa2EDTA intraperitoneally daily for 21 days, no deleterious changes were observed in weight gain or histology of the liver, lung, spleen, adrenals, small gut, or heart. A moderate renal hydropic change with focal subcapsular swelling and proliferation in glomerular loops at the 500 mg level was observed (363). In rats, intravenous infusion of 0.1 M CaNa2EDTA at the rate of 6 mmol/kg for 24 h produced vacuolization and increased lysosomal activity in the kidneys (364). On dietary administration of 0.5 and 1% CaNa2EDTA for 205 days in rats, no significant effects were observed in weight gain, mortality, or histopathology of the liver, kidney, or spleen. No change was observed in blood coagulation time, total bone ash, or blood calcium levels (353). In rats fed 0.5 or 1% Na2EDTA for 6 months with a low mineral diet, at the 1% level a reduced growth rate in males, lowered blood cell counts, a prolonged blood coagulation time, a slight but significant increase in blood calcium level, lower bone ash levels, considerable erosion of the molars, and diarrhea were observed. Rats fed the same dose of Na2EDTA in a normal mineral diet showed no dental erosion (353). In another study in rats given 250, 400, or 500 mg/kg bw of Na2EDTA intraperitoneally for 3–31 days, at the 400 and 500 mg levels all the rats became lethargic, and 100% mortality was observed in 9–14 days. Moderate dilatation of the bowel, subserosal hemorrhages, and swollen kidneys were observed. At the 250 mg level one rat showed hemorrhage of the thymus. All the groups showed hydropic necrosis of the renal proximal convoluted tubules with epithelial sloughing. The effects were found to be reversible on cessation of the treatment (363). The nephrotoxic effects of EDTA were caused mainly by the chelation of metal ions leading to an alteration of their distribution in the tissue.

In rabbits given 0.1, 1, 10, or 20 mg/kg bw Na2EDTA IV or orally at 50, 100, 500, or 1000 mg/kg bw for 1 month, all the animals at the highest oral level had severe diarrhea and died. Histopathological studies showed degenerative changes in the liver, kidney, parathyroid, and endocrine glands and edema in muscles, brain, and heart at all treatment levels (361). In dogs fed 50, 100, or 250 mg/kg bw of

CaNa$_2$EDTA for 12 months, no adverse effects were observed in survival rate, blood chemistry, or histology of the various organs (362).

Long-Term Studies. In rats fed 0.5, 1, or 5% Na$_2$EDTA for 2 years, no significant effects on weight gain, blood coagulation time, red blood cell counts, or bone ash were observed. At the 5% level, severe diarrhea was observed. No significant changes were observed in gross and microscopic examination of the various organs due to Na$_2$EDTA (360). In a four-generation study spanning a period of 2 years in rats fed 50, 125, or 250 mg/kg bw of CaNa$_2$EDTA, no adverse effects were observed in weight gain, feed efficiency, hematopoisis, organ weights, or histopathology of various organs. Fertility, lactation, and weaning were not affected (362).

Skin Toxicity. Ethylenediaminetetraacetic acid caused acute allergic conjunctivitis and periorbital dermatitis in some cases following the use of eyedrops containing EDTA at the 0.1% level. However, no adverse effects were observed with a 10% aqueous solution of CaNa$_2$EDTA (365).

Mutagenicity. The genotoxic effect of EDTA has been studied with various test systems, and the subject was reviewed by Heindorff et al. (366). EDTA was nonmutagenic in bacterial systems such as *E. coli* and *S. typhimurium* (367). It induced inhibition of DNA synthesis in kidney cortex cells (368), in regenerating rat liver cells (369), and in phytohemagglutinin-stimulated lymphocytes (370). EDTA also induced dominant lethal mutations in insects such as wasps (371) and *Drosophila* (372,373). EDTA was positive in tests for chromosomal aberrations in bone marrow cells and splenic cells in mice (374) and in human leukocytes (375). In a bone marrow micronucleus assay, Muralidhara and Narasimhamurthy (376) reported that short-term oral administration of Na$_2$EDTA at 5–20 mg/kg bw in mice induced a dose-dependent increase in the incidence of micronucleated polychromatic erythrocytes (Table 5.26). EDTA interferes with the DNA repair process that takes place after exposure to mutagens. In Chinese hamster cells and human cell lines, the fast repair component detectable after treatment with ionizing radiation or bleomycin was inhibited by EDTA (377–379). EDTA also had a modulating effect on the mutagenic response of a variety of chemical and physical mutagens in plant and insect systems (366).

Reproduction. In a reproduction study in rats fed 0.5, 1, and 5% Na$_2$EDTA for 12 weeks, rats at the 5% level failed to produce litters (360). In a teratogenic study, rats were injected intramuscularly with 20–40 mg EDTA/day during pregnancy on days 6–8, 10–15, and 16 to the end of pregnancy. With 40 mg injected during days 6–8 or 10–15, some dead or malformed fetuses with polydactyly, double tail, generalized edema, or circumscribed edema were observed (380). The adverse effects could be due to the binding of Zn by EDTA resulting in maternal Zn deficiency, which is known to cause gross congenital malformations in rats.

Table 5.26 Incidence fof Micronuclei (MN) in Bone Marrow Erythrocytes of Male Mice Administered Acute Oral Doses of EDTA

Dose (mg/kg bw)	PCE with MN (%)	PCE/NCE ratio
Control (water)	0.35 ± 0.014	1.04
EDTA		
5	0.543 ± 0.05*	1.08
10	0.65 ± 0.094*	0.98
15	0.85 ± 0.08**	0.97
20	1.433 ± 0.136**	0.95
CPA 20	2.25 ± 0.166	0.59

*$P < 0.05$; **$P < 0.01$.
PCE = Polychromatic erythrocytes; NCE = normochromatic erythrocytes
Values are means ± SEM of four animals per dose, analyzed by Student's t-test.
Source: Ref. 376.

Swernerton and Hurley (381) reported that when Na$_2$EDTA was administered with Zn supplementation, no teratogenic effects were observed in rats.

Tartaric Acid

L-Tartaric acid (2,3-dihydrosuccinic acid) occurs naturally in many fruits and is a by-product of wine making. Tartaric acid and its salts are used as synergists and acidulants in a number of products such as confectioneries, bakery products, and soft drinks. An ADI of 0–30 mg/kg bw (calculated as L-tartaric acid) has been allocated by JECFA (13).

TOXICOLOGICAL STUDIES

Absorption, Metabolism, and Excretion. L-Tartaric acid is almost completely absorbed in the gastrointestinal tract, and significant amounts are metabolized to CO_2 in body tissues. Studies in rats and humans have indicated a species difference in the metabolism. Also L-tartaric acid, the naturally occurring form of tartaric acid, was eliminated more rapidly than the synthetic racemate DL-tartaric acid (382). Following oral administration of L-tartaric acid, rats excreted 73% of the dose unchanged in the urine. In humans, 17% of an oral dose was recovered in the urine, whereas after parenteral administration, recovery was almost quantitative. The difference in excretion indicates the metabolism of tartaric acid in the intestine, presumably by bacterial action (383,384).

In rats administered single oral doses of ^{14}C-labeled monosodium L-tartrate (400 mg/kg bw), excretion in the urine, feces, and expired air within 48 h was 70, 13.6, and 15.6%, respectively. With intravenous administration, the corresponding figures were 81.8, 0.9, and 7.5%, respectively (385). In another study in rats, ^{14}C-labeled monosodium DL-tartrate was administered orally, intraperitoneally, and by direct injection to the cecum. With oral administration, 51% was excreted in the urine and 21% was observed in the expired air. Following intraperitoneal administration, 63% of the dose was excreted unchanged in the urine within 24 h and 9% was excreted as CO_2 in 6 h. However, after cecal injection, less than 2% was excreted in the urine and 67% in the expired air. In humans, the major portion of the oral dose was excreted as CO_2 (46%), and only 12% was found in the urine. With IV administration, 64% was excreted in the urine in 22 h and 18% was excreted as CO_2 in 8 h (386).

Acute Toxicity.   In mice the acute oral LD_{50} was 4360 mg/kg bw, and in rabbits approximately 5290 mg/kg bw (387).

Short-Term Studies.   In short-term studies in dogs fed daily oral doses of 900 mg/kg bw for 90–114 days, blood chemistry remained normal except in one dog, in which azotemia developed followed by death in 90 days. Weight changes were equivocal. In some cases casts appeared in the urine (388). No adverse effects were observed in rats fed 7.7% sodium L-tartrate (389). Renal damage was observed in rabbits and rats only after intravenous administration of L-tartaric acid in doses of 0.2–0.3 g (390,391).

Long-Term Studies.   In long-term studies in rats fed 0.1, 0.5, 0.8, or 1% L-tartaric acid for 2 years, no significant effects were observed in growth rate or mortality or in gross and pathological studies of the organs (392). In another long-term study, 21-day-old weanling rats were fed L-tartaric acid at levels of up to 1.2% without any adverse effects (389). L-Tartaric acid was not teratogenic in mice, rats, hamsters, or rabbits at levels of 274, 181, 225, and 215 mg/kg bw, respectively (393).

Citric Acid

Citric acid and its salts are widely used as chelators and acidulants in the food industry. Citric acid is used as a synergist with both primary antioxidants and oxygen scavengers at levels of 0.1–0.3%. In fats and oils, citric acid chelates metal ions at levels of 0.005–0.2% (394). Citric acid is widely distributed in plant and animal tissues and is an intermediate in the Krebs cycle. The potassium and sodium salts have been extensively used in medicine for many years without any adverse effects. JECFA has allocated an ADI of "not limited" for citric acid (13).

TOXICOLOGICAL STUDIES

Citric acid and its salts have been found to have a low order of toxicity. The acute IP LD_{50} in mice in 96 mg/kg bw, and in rats it is 884 mg/kg bw. The acute IV (rapid infusion) LD_{50} in mice is 42 mg/kg bw (395). In short-term studies in dogs given daily doses of 1380 mg/kg bw for 112–120 days, no adverse effects were observed (388). In a two-generation study in rats fed 1.2% citric acid for 90 weeks, no adverse effects in growth rate, blood picture, histopathological studies, or reproduction were observed. No adverse effects were observed in the blood or urine chemistry or in histological studies in rabbits fed 7.7% sodium citrate for 60 days (389). In humans, both potassium and sodium salts have been used in daily doses of up to 10 g as mild diuretics without any adverse effects (396).

Citric acid was found to have an adverse effect on calcium absorption at high levels. Cramer et al. (397) observed that in rats fed sodium citrate or citric acid in a vitamin D-free diet with low levels of phosphorus, citric acid completely prevented the absorption of calcium. Gomori and Gulyas (398) observed that in dogs fed 520–1200 mg/kg bw of sodium citrate, urinary calcium levels increased. Gruber and Halbeisen (395) suggested that the effects of high levels of citric acid resembled symptoms of calcium deficiency. In rats, a dose level of 50% resulted in lowered food intake and slight abnormalities in blood chemistry (399,400).

Citrate Esters

Esters of citric acid such as isopropyl citrate mixture (see below) and stearyl citrate are also used as synergists in food products. Citrate esters are readily soluble in fats and oils from propylene glycol solution concentrates. Formulations containing various mixtures of antioxidants and citrates in propylene glycol are commercially available.

Isopropyl Citrate Mixture. Isopropyl citrate mixture consists of monoisopropyl citrate 27%, diisopropyl citrate 9%, triisopropyl citrate 2%, and mono- and diglycerides 62%. In rats, isopropyl citrate mixture is readily absorbed when incorporated in the diet at the 10% level (401). The acute oral LD_{50} in rats was 2800–2250 mg/kg bw and in dogs, 2250 mg/kg bw (402). No deleterious effects were observed in short-term studies in rats (1500–2000 mg/rat per day for 6 weeks), rabbits (3600 mg/rabbit per day for 6 weeks), and dogs (0.06% for 6 weeks) or in long-term studies in rats at levels of 0.28, 0.56, and 2.8% for 2 years. Multigeneration studies at levels of 2.8% in the diet in rats also showed no adverse effects (402). However, because of the known toxicity of isopropyl alcohol, an ADI of 0–14 mg/kg bw (calculated as monoisopropyl citrate) has been allocated by JECFA (13).

Stearyl Citrate. Stearyl citrate is hydrolyzed readily to stearyl alcohol and citric acid in dogs and to a lesser extent in rats (401). In rats, the oral LD_{50} was

>5400 mg/kg bw and in dogs, >5600 mg/kg bw (402). In short-term studies in rats at levels of 1.3, 2.5, 5, and 10% for 10 weeks, in rabbits at 2 and 10% for 6 weeks, and in dogs at levels of 3% for 12 weeks, no adverse effects were observed on body weight, growth rate, or histological studies. In a five-generation study in rats at levels of 1.9 and 9.5%, no adverse effects were observed on litter size, fertility, lactation, or growth rate of progeny (402). Stearyl citrate at levels of 2.5 and 10% reduced the digestibility of fat to 77 and 71%, respectively, in rats. In dogs at the 3% level, fat digestibility was reduced to 80% (401). An ADI of 0–50 mg/kg bw has been allocated by JECFA (13).

Phytic Acid

Phytic acid (myoinositol hexaphosphoric acid) is a major component of all seeds, constituting 1–5% by weight of many cereals, nuts, oilseeds, and edible legumes, and occurs as a mixture of calcium, magnesium, and potassium salts. Some of the important physiological functions of phytic acid include storage of phosphorus, high-energy phosphate groups, and cations and functioning as a cell wall precursor. Phytic acid is a potential chelating agent. At low pH it precipitates Fe^{3+} quantitatively. At intermediate and high pH, it forms insoluble complexes with all other polyvalent cations. Phytic acid prevents browning and putrefaction of fruits and vegetables by inhibiting polyphenol oxidase activity. It is effective in preventing both autoxidation and hydrolysis of soybean oil and stabilizes a number of food products such as fish products, lipid-containing foods, and natural and artificial coloring agents. Phytic acid is also effective in removing iron from dilute molasses, wine, and other beverages (403–405). Phytic acid is used widely in medicine. In some countries, for example, Japan, it is also used as a food additive. In the United States, phytic acid has not been considered for inclusion in the food additive list because of lack of information on its use in food manufacture (406).

TOXICOLOGICAL STUDIES

Phytic acid undergoes hydrolysis in the gastrointestinal tract to form inositol and orthophosphate, mainly by the action of the enzyme phytase. Phytases are widely distributed in plants, animals, and fungi (407). Phytases have been isolated and characterized in rats, calves, and humans (408). In humans, the hydrolysis of phytates most probably occurs by the action of microbial phytases or by nonenzymatic cleavage, and nearly 85% of the ingested phytic acid was hydrolyzed (409,410). Nearly 71 and 39% of the ingested phytate was hydrolyzed in weanling and mature rats, respectively (411).

Phytic acid forms insoluble complexes with a number of essential minerals such as calcium, iron, and zinc, decreasing their bioavailability and resulting in mineral deficiency in humans (403). Phytic acid also precipitates most proteins at low pH in the absence of cations, and at high pH in the presence of cations it forms a ternary protein, metal, and phytate complex (412). Phytic acid inhibits trypsin (413) and binds to hemoglobin, thereby reducing its affinity for O_2 or CO (414,415). The nutritional implications of phytic acid have been exhaustively reviewed (416–418).

Toxicological studies have been carried out in recent years with just one long-term study and a subacute study in rats. The LD_{50} in rats calculated by the moving average method was 450–500 mg/kg bw for males and 480 mg/kg bw for females (419). In a subacute study in F344 rats, phytic acid administered at levels of 0.6, 1.25, 2.5, 5, or 10% in drinking water for 12 weeks resulted in high mortality at the 5 and 10% levels. In the 1.25 and 2.5% groups, the reduction in body weight was less than 10% compared to the control. In a 100–108-week study at dose levels of 1.25 and 2.5%, a 50% mortality was observed in females at the 2.5% level. The mean final body weights was significantly lower than the control at both dose levels, the effect being more marked in females. No significant differences in hematology or clinical chemistry values or relative organ weights were observed. Necrosis and calcification of the renal papillae were observed at both dose levels, the incidence being higher in females. In males, hyperplasia of the renal pelvis at higher doses was significantly higher than that in controls (Table 5.27) (420).

In a teratogenicity study in mice, pregnant mice were given oral doses of 1.6, 3.1, and 6.3% phytic acid in aqueous solution on days 7–15 of gestation. Although

Table 5.27 Histological Findings in the Renal Papillae and Pelvis of F344/DuCrj Rats Treated with Phytic Acid 'Daiichi' (PA)[a]

Treatment group (% PA in drinking water)	No. rats	Renal papillae		Renal pelvis				Nonrenal pelvic area calcification
				Hyperplasia[b]				
		Necrosis	Calcification	±	+	++	Papilloma	
Males								
2.5	57	1	3	6*	21**	0	3	3
1.25	59	1	0	2	9*	0	0	0
0 (control)	58	0	0	0	1	0	0	0
Females								
2.5	55	10**	17**	2	6	2	4	25**
1.25	58	6*	6*	2	5	0	3	26**
0 (control)	58	0	0	0	2	0	0	0

[a]Values marked with asterisks differ significantly (chi-square test) from that for the corresponding control group (*$P < 0.05$; **$P < 0.01$).
[b]Hyperplasia was graded as follows: ± = mild; + = moderate; ++ = severe.
Source: Ref. 420.

no clear teratogenic effects were observed with external or skeletal malformations, the incidence of late resorbed fetuses increased in the 6.3% group (421).

Lecithin

Lecithin (phosphatidylcholine) is a naturally occurring phospholipid and makes up 1–2% of many crude vegetable oils and animal fats. The commercial source of lecithin is predominantly soybeans. Commercial lecithin preparation is a mixture of phospholipids and contains phosphatidylethanolamine and phosphatidylinositol in addition to lecithin. Lecithin functions as a potent synergist in fats and oils with a range of primary antioxidants and oxygen scavengers at elevated temperatures above 80°C. Phosphatidylethanolamine is also highly effective as a synergist. Phosphatidylserine is less effective, and phosphatidylinositol has no synergistic effect (422–425). An ADI of "not limited" has been allocated by JECFA (13).

TOXICOLOGICAL STUDIES

No information on the acute toxicity of lecithin is available in the literature. In cats, rapid infusion of a 1.2% egg yolk phosphatide emulsion containing 5% glucose had no adverse effects. A rapid infusion of soybean phosphatides caused a fall in blood pressure with apnea (426). In humans, administration of large doses of 25–40 g/day for some months resulted in a lowering of serum cholesterol (427). Although no conventional toxicological studies have been carried out, lecithin is considered a nontoxic substance based on extensive nutritional and clinical experience in humans.

5.3.3 Secondary Antioxidants

Thiodipropionic Acid and Its Dilauryl and Distearyl Esters

Thiodipropionic acid (TDPA), dilauryl thiodipropionate (DLTDP), and distearyl thiodipropionate (DSTDP) are generally regarded as secondary antioxidants or synergists. Free TDPA has the ability to chelate metal ions (428) and also functions as a sulfide, decomposing alkyl hydroperoxides into more stable compounds. In model systems containing performic acid, nonanal, ethyl oleate, or cholesterol, Karahadian and Lindsay (429) observed that DLTDP was preferentially oxidized to sulfoxide and prevented the formation of nonanoic acid, 9-epoxy oleate, and 5,6-epoxy cholesterol, respectively. Though TDPA and its salts have been approved for food use, they are not being used as direct additives in food products. JECFA has allocated an ADI of 0–3 mg/kg bw. This evaluation is largely based on the toxicological studies of Lehman et al. (6).

TOXICOLOGICAL STUDIES

Absorption, Metabolism, and Excretion. In rats, single oral doses of ^{14}C-labeled TDPA in the range of 241–650 mg/kg bw were almost completely absorbed from the gastrointestinal tract and rapidly eliminated. During a period of 4 days, 87–95% of the dose was recovered—78–88% in the urine, 0.1–0.9% in the feces, and 3–8% as CO_2. TDPA was excreted in urine either largely unchanged or as an acid-labile conjugate that was not a glucuronide. Single oral doses of ^{14}C-labeled DLTDP (107–208 mg/kg bw) were completely absorbed and rapidly eliminated, mostly in the urine (85–88%), with just 1.8–3.5% in the feces and 3–4% as CO_2. TDPA and an acid-labile conjugate were the metabolites in the urine, indicating the hydrolysis of DLTDP (430).

Acute Toxicity. The acute oral, IP, and IV LD_{50} of TDPA in mice was 2000, 250, and 175 mg/kg bw, respectively, and in rats it was 3000, 500, and >300 mg/kg bw, respectively. The acute oral and IP LD_{50} of DLTDP in mice was >2000 mg/kg bw, and in rats the acute oral LD_{50} was >2500 mg/kg bw. The acute oral and IP LD_{50} of DSTDP in mice was >2000 mg/kg bw, and in rats the acute oral LD_{50} was >2500 mg/kg bw (6).

Short-Term and Long-Term Studies. In short-term studies in rats (3% level for 120 days), guinea pigs (0.5% level for 120 days), and dogs (0.1 and 3% for 100 days), no adverse effects were observed with TDPA and DLTDP. In long-term studies in rats fed TDPA or its esters at levels of 0.5, 1, or 3% for 2 years, no discernible adverse effects were observed as determined by growth rate, mortality, and pathological examination (6).

5.3.4 Miscellaneous Antioxidants

Nitrates and Nitrites

Sodium nitrate and sodium nitrite are widely used in meat curing and in fish products. Some of the important functions of these compounds include the formation of the stable color pigment nitrosylmyoglobin by reacting with myoglobin, texture improvement, cured meat flavor formation, and antimicrobial and antioxidant activities. It is postulated that nitrite functions as an antioxidant by converting heme proteins to their catalytically inactive and stable nitric oxide forms; by chelating metal ions, especially nonheme iron, copper, and cobalt, present in meat; by stabilizing lipids per se in the muscle against oxidation; and by the formation of nitroso compounds that possess antioxidative properties (431–434). Nitrates and nitrites occur in all common food products. Certain vegetables, for example, beets, celery, lettuce, radish, and spinach, tend to accumulate more nitrates than others. The use of nitrates and nitrites has become a controversial issue because of their involvement in the formation of nitrosamines in situ in the food or in the body.

Nitrosamines are powerful carcinogens and mutagens and also have embryopathic and teratogenic properties (435). JECFA has allocated an ADI of 0–5 mg/kg bw for sodium nitrate and a temporary ADI of 0–0.2 mg/kg bw for sodium nitrite (13).

TOXICOLOGICAL STUDIES

Absorption, Metabolism, and Excretion.   Sodium nitrate is readily absorbed and almost completely eliminated unchanged. A small amount of the nitrate is converted to nitrite by bacterial action in the gastrointestinal tract (436,437). In humans, approximately 20% of the salivary nitrate is reduced by bacterial action to nitrite (438,439). Sodium nitrate inhibits the uptake of iodine by the thyroid gland in rats and sheep, which results in an interference in vitamin A formation from β-carotene, vitamin A deficiency, and lower liver stores (440–442).

Sodium nitrite is also readily absorbed from the gastrointestinal tract. Nearly 30–40% of the absorbed nitrite was excreted unchanged in the urine, but the fate of 60–70% was not clear. Some of the important reactions of nitrite include conversion of myoglobin to nitrosomyoglobin and of hemoglobin to methemoglobin, degradation of β-carotene under acidic conditions, and formation of dialkyl and acylalkyl nitrosamines and nitrosoguanidines with secondary amines (442–445).

Acute Toxicity.   The acute oral LD$_{50}$ of sodium nitrate in rats is 3236 mg/kg bw (13). The acute oral LD$_{50}$ of sodium nitrite in mice is 220 mg/kg bw, and in rats, 85 mg/kg bw (446).

Short-Term Studies.   No adverse effects were observed in short-term studies in dogs fed 2% sodium nitrate for 105 and 125 days (443). Potassium nitrate at 1.5% in the forage caused all symptoms of nitrate poisoning in cattle (436). In rats given 100, 300, or 2000 mg/liter of sodium nitrite in drinking water for 60 days, abnormalities in EEG were observed at all dose levels (447). At levels of 170 and 340 mg/kg bw per day for 200 days, methemoglobinemia, raised hematocrit, increased spleen weights in females, increased heart weight in males, and increased kidney weights in both sexes were observed (448). A reduction in motor activity was observed in mice given sodium nitrite in drinking water at levels of 100, 1000, 1500, and 2000 mg/liter (447).

Long-Term and Carcinogenicity Studies.   In rats fed sodium nitrate at levels of 0.1, 1, 5, and 10% for 2 years, no adverse effects were observed except for a slight reduction in growth rate at the 5 and 10% levels (449). Sodium nitrite at low levels was without any adverse effects. In a long-term study using canned meat (40% of the diet) treated with 0.5 or 0.02% sodium nitrite and glucono-δ-lactone, no adverse effects were observed in hematology, histopathology, or clinical biochemistry in rats. No evidence of any preneoplastic change or tumor formation was observed (450). No adverse effects were observed in a three-generation study with 100 mg/kg bw of sodium nitrite in rats (443).

Exposure to high levels of nitrites has been linked to a higher incidence of cancer of the stomach, esophagus, and lymphocytes. However, some of the findings have not been confirmed (451). Shank and Newberne (452) reported that in rats and hamsters, sodium nitrite in combination with morpholine, a cyclic amine, induced a dose-related hepatocellular carcinoma and angiosarcoma. Morpholine is an anticorrosive agent and an unintentional additive to foods such as canned ham. Nitrite and morpholine, when combined at concentrations found in food products, could result in the formation of nitrosated morpholine at levels high enough to be significantly carcinogenic. Shank and Newberne (452) demonstrated a dose–response relationship for the induction of liver and lung tumors by dietary sodium nitrite and morpholine, these tumors being identical to those induced by N-nitrosomorpholine (NNM), a preformed nitrosamine. The nitrite concentration in the diet seemed to have a greater effect on the incidence of hepatocellular carcinoma and angiosarcoma in the rat than did the concentration of morpholine (Table 5.28). Incidence of lung adenomas caused by morpholine and high levels of sodium nitrite in mice has also been reported (453).

Reproduction. Sodium nitrite was found to have an adverse effect on reproduction. In a reproduction study in rats, sodium nitrite at levels of 2000 or 3000 mg/liter resulted in mortality of the newborn and also maternal anemia (447). In guinea pigs fed 0.03–1% sodium nitrite, no live births were observed at or above the 0.5% level. Maternal deaths, abortions, fetal resorptions, mummification, placental lesions, and inflammation of the uterus were observed (454). In cattle, sodium nitrite at levels to produce 40–50% methemoglobinemia caused no adverse effects (455). Sodium nitrate had no adverse effects on reproduction in guinea pigs (454).

Amino Acids

Amino acids are effective both as primary antioxidants and as synergists (456). At low concentrations, most of the amino acids have antioxidant properties, but at higher concentrations they function as prooxidants (457). Also, at low pH most of the amino acids are prooxidants whereas higher pH favors antioxidant activity.

Glycine, tryptophan, methionine, histidine, proline, and lysine have been found to be effective in fats and oils. Glycine has been listed as a GRAS substance for addition to fats and oils at up to 0.01%. Covalent attachment of tryptophan, cystine, methionine, and histidine significantly enhanced the antioxidant action of Trolox-C, a synthetic derivative of α-tocopherol (458).

TOXICOLOGICAL STUDIES

Excessive intake of amino acids has been reported to cause a reduction in growth rate, feed intake, and survival rate in rats. In a 3-week study in weanling rats at

Table 5.28 Percent Incidence of Hepatocellular Carcinoma and Angiosarcoma Among Rats Fed Nitrite and Morpholine

| Dietary level (ppm) | | | | Liver cell | Liver | Lung | Other | Metastasis from liver |
NaNO$_2$	Morpholine	NNM[a]	No. of rats	carcinoma	angiosarcoma	angiosarcoma	angiosarcoma	to lung
0	0	0	156	0	0	0	0	—
1000	0	0	96	1	0	0	1	0
0	1000	0	104	3	0	2	1	0
1000	1000	0	159	97	14	23	1	49
1000	50	0	117	59	5	6	0	17
1000	5	0	154	28	12	8	1	7
50	1000	0	109	3	2	1	0	0
5	1000	0	172	1	2	1	1	0
50	50	0	152	2	1	1	1	0
5	5	0	125	1	2	2	1	0
0	0	5	128	58	15	9	1	22
0	0	50	94	93	21	20	1	58

[a]NNM = N-nitrosomorpholine.
Source: Ref. 452.

dietary levels of 1–10% of 17 different amino acids, Daniel and Waisman (459) found methionine, cystine, phenylalanine, and tryptophan to be highly toxic; histidine, tyrosine, and lysine to be toxic; glycine, proline, valine, and arginine to be slightly toxic; and serine, aspartic acid, glutamic acid, isoleucine, leucine, and alanine to be nontoxic at all levels tested. Similar results were obtained when the amino acids were given intraperitoneally in three daily doses of 25–350 mg/55 g rat. In studies in rats fed 5% glycine, reductions in weight gain and the ability to use food, reduction in the activity of certain enzymes (e.g., transaminase activity), and increased plasma cholesterol level were observed (460).

Spices

Spices and herbs have antioxidant and antimicrobial properties. Various ground spices and spice extracts are effective in fats, meat products, and baked goods. Rosemary, sage, allspice, cloves, oregano, and thyme are some of the spices with significant antioxidant properties in model systems or food products (461,462). Vanillin, used extensively as a food flavoring agent, and curcumin, the major pigment in turmeric, also have antioxidant properties (463,464). The active compounds in spices are phenolic and function as primary antioxidants. Spice extracts find limited use as antioxidants only in products compatible with the specific spice flavor because of their color, odor, and taste.

Rosemary (*Rosmarinus officinalis* L., family Labiatae) is generally recognized as the most potent antioxidant herb. Industrial processes have been developed to produce a purified antioxidant that is bland, odorless, and tasteless (465,466). Rosemary antioxidant is available commercially as a fine powder soluble in fats and oils and insoluble in water. Rosemary antioxidant is recommended for use in amounts of 200–1000 mg/per kilogram of the food product to be stabilized. Approximately 90% of the antioxidant activity in rosemary extract has been attributed to carnosol, a phenolic diterpene. Other active molecules so far identified are rosmarinic acid, carnosic acid, and rosmaridiphenol (422,467). No toxicological or biochemical information is available on rosemary antioxidant. Rosemary extract at the 1% level was found to significantly inhibit the initiation and promotion of DMBA-induced mammary tumorigenesis in rats (468).

Flavonoids

Flavonoids form a group of naturally occurring phenolics widespread in common edible plants. The estimated daily intake of mixed flavonoids is nearly 1 g/day in western diets (469). Flavonoids function both as primary antioxidants and as chelators. Various flavonols, chalcones, and flavones were found to be very effective as antioxidants in fats and oils (470). At present, flavonoids are not being used commercially due to insufficient or in some instances adverse toxicological information.

TOXICOLOGICAL STUDIES

Flavonoids in general are poorly absorbed in the gastrointestinal tract in mammalian systems, efficiently metabolized, and eliminated (471,472). They also undergo extensive microbial degradation to form inactive derivatives in the large bowel (473,474). In a carcinogenicity bioassay in mice, rats, and hamsters, quercetin was negative even at high doses of 10% in the diet (475–478). However, in another study in rats, quercetin at the 0.1% level for 1 year caused intestinal and bladder cancer (479). Quercetin was found to be a potent mutagen in a number of in vitro systems and a weak mutagen in in vivo systems (480,481). Ahmad et al. (482) observed that several flavonoids, including quercetin and rutin, caused DNA strand scission in the presence of copper (Table 5.29).

Vitamin A

Vitamin A or all-*trans*-retinol has very limited use as an antioxidant because of its high susceptibility to oxidation on exposure to air and light, conditions under which it becomes prooxidant. However, vitamin A was found to be effective in fats and oils kept in the dark. Vitamin A also inhibited the formation of free acids in vegetable oils. Vitamin A occurs widely in all animal tissues, mainly in the liver,

Table 5.29 Comparison of S_1 Nuclease Hydrolysis Following Damage to DNA Induced by Flavonoids and Cu(II) and Stoichiometry of Cu(II) Reduction by Flavonoids[a]

Flavonoid	DNA hydrolyzed (%)[b]	mol Cu(II) reduced/mol flavonoid[c]
Denatured DNA control	100	—
Native DNA control	2.8	—
Quercetin	49.5	5
Myricetin	55	6
Epicatechin	41.3	4
Rutin	35.8	5
Galangin	24.8	3
Apigenin	35.8	4
Morin	NT[d]	3

[a]Flavonoids and Cu(II) were each 0.1 mM.
[b]DNA hydrolysis (%) refers to loss of acid-precipitable DNA following treatment with S_1 nuclease.
[c]Deduced from Job plots.
[d]NT = Not tested.
Source: Ref. 482.

eggs, and milk as the ester of acetate or palmitate. In the liver, palmitate is the primary storage form of vitamin A. The recommended daily intake is 750 mg/kg bw (483).

TOXICOLOGICAL STUDIES

In the intestinal mucosa, dietary retinol is esterified with long-chain fatty acids, incorporated into chylomicra, and transported via the lymphatic system to the liver (484,485). Vitamin A is mobilized from the liver and delivered to various organs in bound form via a specific vitamin A-binding protein. The major metabolite of vitamin A is retinoic acid. Retinoic acid is excreted in the urine either as free acid or as a glucuronide conjugate (Fig. 5.12). Other metabolites have not been characterized (486).

The preclinical and clinical toxicity of vitamin A was reviewed by Kamm et al. (487). The acute oral LD_{50} in mice is 1510–2570 mg/kg bw (488). In rodents, in general, signs of hypervitaminosis A has been associated with anorexia, weight loss, hair loss, emaciation, anemia, spontaneous fractures due to the destruction of bone matrix, and extensive hemorrhages. Effects on internal organs include fatty infiltration of the liver, testicular hypertrophy, and fatty changes in the heart and kidney (487). The bone toxicity of vitamin A has been reported in dogs (489), cats (490), calves (491), and hogs (492). Vitamin A was nonmutagenic in the Ames test (488) and also did not induce sister chromatid exchanges in Chinese hamster ovary cells (493).

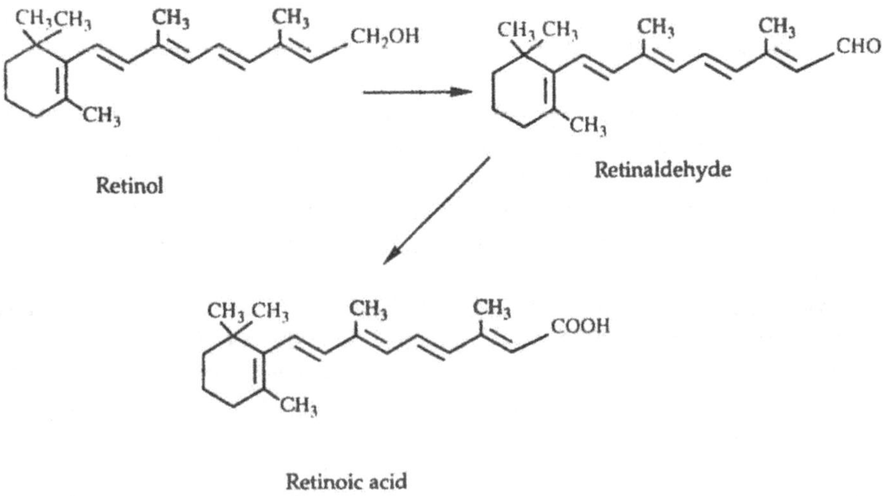

Fig. 5.12   Major metabolites of vitamin A.

High intakes of vitamin A have been associated with adverse effects on reproduction such as testicular changes and the inhibition of ovulating activity in rats (494,495). High doses of vitamin A have been reported to be teratogenic in a number of species, including mice, hamsters, dogs, pigs, monkeys, and humans. More than 70 types of abnormalities affecting almost every organ or tissue system have been reported. The teratogenic effects of vitamin A are mainly due to the major metabolite retinoic acid, a potent teratogen (496–498).

In humans, the symptoms of hypervitaminosis A include skin scaling, tenderness of bones, headache, anorexia, weight loss, general fatigue, edema, and hemorrhage. Most of the symptoms disappeared on discontinuation of vitamin A, except for the changes in bones, which took a longer time for reversal (499–501).

β-Carotene

β-Carotene (provitamin A) is mainly used as a food colorant. β-Carotene and related carotenoids are effective quenchers of singlet oxygen (502,503) and can act as antioxidants by preventing the formation of hydroperoxides (504). β-Carotene occurs widely as a lipoprotein complex in all green parts of plants. It is found in abundance in carrots, some tropical fruits such as papayas and mangoes, and all dark green leafy vegetables. β-Carotene and related carotenoids have also been chemically synthesized. β-Carotene and related carotenoids have also been chemically synthesized. β-Carotene has the highest provitamin A activity of all carotenoids. It is very sensitive to oxidative decomposition when exposed to air or light. An ADI of 0–5 mg/kg bw has been allocated by JECFA (165).

TOXICOLOGICAL STUDIES

In humans, 10–41% of an ingested dose of β-carotene was absorbed and 30–90% was excreted in the feces (505). In the presence of oxygen, β-carotene is cleaved by β-carotene-15,15′-oxygenase in the intestine and liver to two molecules of retinaldehyde (506). In the intestinal mucosa, the liver, and the eye, the retinaldehyde is reduced to retinol by retinaldehyde reductase (Fig. 5.13) (507–510).

β-Carotene is not toxic to rats, dogs, or humans. The acute oral LD_{50} in dogs was >8000 mg/kg bw (511), and in rats, >1000 mg/kg bw (512). Doses of 100 mg of β-carotene/kg bw administered 5 days a week for 13 weeks had no significant effect on growth rate, and no evidence of toxicity was observed in dogs. In a four-generation study in rats fed 0.1% β-carotene for 110 weeks, no adverse effects were observed (512,513). In humans administered 100,000 units of β-carotene per day for 3 months, no signs of vitamin A toxicity were observed (514).

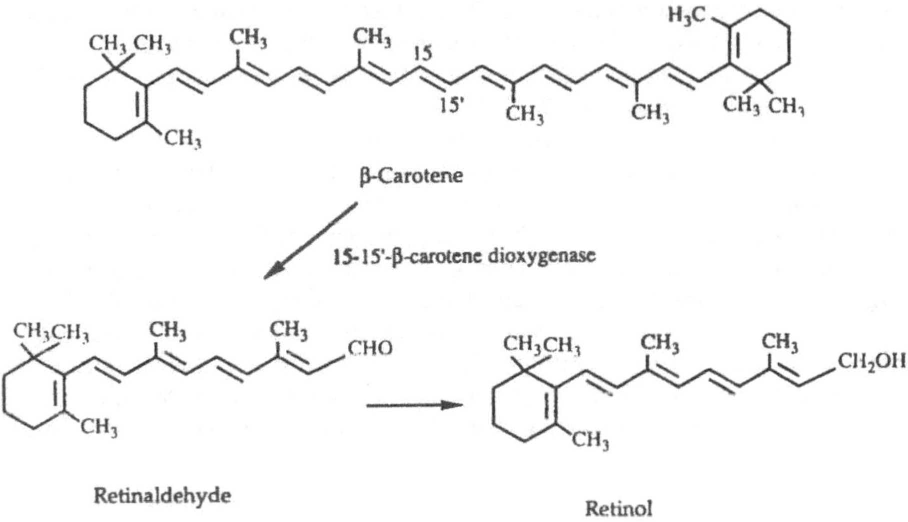

Fig. 5.13   Metabolites of β-carotene.

Tea Extracts

Tea extracts form a potential source of natural antioxidants for food products. Extracts of black tea and green tea were found to be highly effective in fats and oils (515,516). The polyphenols present in tea consist mainly of catechins and have antioxidant properties. The acute oral and intraperitoneal LD_{50} in mice were 10 g and 0.7 g, respectively. In subchronic tests in rats at levels as high as 1000 mg/kg bw, growth rate, feed efficiency, and protein efficiency ratio were not affected. No adverse effects were observed in hematological analysis, gross and microscopic examination of the organs, or litter size (517).

Zinc

Zinc, one of the essential nutrients, also functions as an antioxidant. Zinc strongly inhibits lipid peroxidation at the membrane level, possibly by altering or preventing iron binding (518). The major dietary sources of zinc are seafoods and meat. Other dietary sources include brewer's yeast, whole grains, wheat bran, and wheat germ. An RDA of 15 mg/day has been specified in the United States by the Food and Nutrition Board of the National Academy of Sciences (519).

TOXICOLOGICAL STUDIES

Zinc is absorbed mainly from the intestines and transported to the liver in the portal plasma bound to albumin. About 30–40% of the zinc is extracted by the liver and subsequently released into the blood. Zinc uptake by the pancreas, liver, kidney, and spleen was found to be very rapid compared to that of the central nervous system, bones, muscles, and red blood cells. Unabsorbed and endogenous zinc is excreted mainly in the feces (520–523).

Zinc was formerly considered relatively nontoxic to animals and humans. However, in a 1990 review, Fosmire (524) observed that in humans zinc supplementation even at moderate concentrations may have adverse effects under certain circumstances. At levels of 225–400 mg/day, the acute toxic symptoms included nausea, vomiting, epigastric pain, lethargy, and fatigue. At pharmacological dosages of 100–300 mg/day, zinc induced copper deficiency symptoms as well as impaired immune function and had adverse effects on the ratio of low-density lipoprotein to high-density lipoprotein. Even lower levels of zinc supplementation, closer in amount to the RDA, have been suggested to interfere with the utilization of copper and iron and to adversely affect high-density lipoprotein concentrations.

Selenium

Selenium is an essential nutrient with antioxidant properties and is necessary for the synthesis and activity of glutathione peroxidase, a primary cellular antioxidant enzyme. Dietary sources of selenium include cabbage, celery, radish, brewer's yeast, fish, whole grains, and meat. Selenium is present in organic form in these sources and is the major nutritional form in animals and humans. According to the National Research Council (519), levels of 50–100 mg/day are considered a safe and adequate daily dietary intake for adult humans.

TOXICOLOGICAL STUDIES

In animals, nearly 90% of dietary selenium is absorbed from the gastrointestinal tract, and in humans 55–70% is absorbed. Selenium is transported in the plasma bound to plasma proteins, but the mechanism of transport is not clear (525). In the tissues, the selenate is reduced to selenide, which is incorporated into selenocysteine, the form at the active site of glutathione peroxidase (526). Selenomethionine, the organic form of selenium, can be incorporated directly into tissue proteins (527). Selenium is excreted mainly in the urine in humans, rats, and other monogastric species (525).

The acute oral and intravenous LD_{50} of selenium as sodium selenite or selenate in rats, rabbits, and cats is 1.5–3 mg/kg. Selenium was found to be toxic at higher doses. In rats and dogs, acute or subacute selenium poisoning in general resulted in anorexia, anemia, and severe pathological changes in the liver (528,529). High dietary selenium was found to be carcinogenic and also increased the carcinogenic effects of certain chemicals in rats and hamsters (530,531). In humans, overexposure to selenium has resulted in loss of hair and nails, skin lesions, abnormalities of the nervous system, disturbances of the digestive tract, and possibly tooth decay. However, hepatotoxic effects were not observed (532).

5.4 TOXICOLOGICAL IMPLICATIONS ASSOCIATED WITH LEACHING OF ANTIOXIDANTS FROM PLASTICS USED IN FOOD PACKAGING

In recent years the possible health hazards associated with the migration of additives from food packaging materials have attracted a lot of attention. Efforts are being made to assess the significance of migration of various additives and their decomposition products into the food system. Plastics are used extensively as films, containers, and coating materials for packaging food products. The commonly used plastics include high-density and low-density polyethylenes, polypropylenes, polyvinyl chloride, polystyrenes, and copolymers of ethylene and vinyl acetate. Apart from high molecular weight polymers, plastics also contain a number of low molecular weight additives such as antioxidants, heat stabilizers, UV absorbers, and plasticizers. The additives are added either during the processing and fabrication stage as processing aids or after molding as end product use additives. Most of the additives have a high mobility in the polymer material and are readily leached from it into the food product.

Polymers undergo oxidative degradation at all stages of fabrication in the presence of oxygen and also during long-term storage. The process of oxidation involves the formation of polymer hydroperoxides, which decompose to form reactive hydroxy radicals, thus initiating a chain reaction in the polymer. Antioxidants are used to remove the hydroperoxides that are formed or to retard their decomposition (preventive antioxidants) and also to remove the reactive radicals that are formed (chain-breaking antioxidants). Preventive antioxidants include phosphite esters, sulfides, and dithiophosphates. Chain-breaking antioxidants include aromatic amines, hindered phenols, quinones, and nitro compounds. Antioxidants are usually incorporated in the range of 0.1–0.5% by weight of the polymer (533). Phenolic antioxidants like BHA and BHT are also commonly added by spray or roll coat application to films or dissolved in paraffin wax used for coating containers. High-fat candies, bakery products, and dry breakfast cereals are some of the edible products for which antioxidant-treated packaging material is used.

Al-Malaika (533) reviewed the effect of some of the parameters on the leaching of antioxidants in the human environment. Migration of antioxidants from packaging material to food product is determined by the interaction of a number of physical and chemical parameters, including the nature of the polymer and the food product; the molecular weight, volatility, and solubility of the antioxidant; temperature; light; irradiation; and duration of contact with the food product. Migration studies have been conducted in food systems and in food-simulating model systems. In a study on the migration of Irganox-1010, a complex high molecular weight antioxidant, Schwope et al. (534) observed that the antioxidant migrated into *n*-heptane, 100% ethanol, and corn oil more rapidly from ethylene vinyl acetate films than from low-density polyethylene films. In aqueous media, little migration of the antioxidant was observed from ethylene vinyl acetate films, whereas the migration from low-density polyethylene films was relatively high. Chang et al. (535) reported similar findings with BHT, a low molecular weight volatile compound compared to Irganox-1010. In another study comparing the migration of BHT and Irganox-1010 from low-density polyethylene films, Schwope et al. (536) observed that BHT migrated more rapidly than Irganox-1010. Figge (537) observed that higher temperatures increased the migration of Ionox-330 from high-density polyethylene, polyvinyl chloride, or polystyrene films. Gamma irradiation resulted in a 30% loss of Irganox-1010 and Irgafos-168, an arylphosphite antioxidant, in a range of polymers including polyvinyl chloride, polyethylenes, and polypropylenes. Little of Irgafos-168 remained after a dose of 10 kGy, but its triarylphosphate oxidation product was found in the extracts of the irradiated polymer. With increasing irradiation, Irganox-1010 showed no increase in migration into water-based food simulants and a decrease in migration into a fatty food simulant (538). Antioxidants used in plastics are extensively tested for toxicity like other food additives before being accepted for incorporation into food packaging materials. However, very little attention has been paid to the assessment of health risks presented by the chemical transformation products formed during processing (533). Further work is required to elucidate the biotransformation, metabolism, and toxicity of these indirect food additives.

REFERENCES

1. FAO/WHO, WHO Tech. Rep. Ser. No. 21, World Health Organization, Geneva, Switzerland, 1987.
2. G. J. Van Esch, *Voeding*, *16*: 683 (1955).
3. J. C. Dacre, *J. N.Z. Inst. Chem.*, *24*: 161 (1960).
4. F. C. Johnson, *CRC Crit. Rev. Food Technol.*, 2: 267 (1971).
5. J. M. Orten, A. C. Kuyper, and A. H. Smith, *Food Technol.*, 2: 308 (1948).
6. A. J. Lehman, O. G. Fitzhugh, A. A. Nelson, and G. Woodard, *Adv. Food Res.*, *3*: 197 (1951).
7. K. J. H. Sluis, *Food Manuf.*, *26*: 99 (1951).
8. A. R. Johnson and F. R. Hewgill, *Aust. J. Exp. Biol. Med. Sci.*, *39*: 353 (1961).
9. C. A. Van Der Heijden, P. J. C. M. Janssen, and J. J. T. W. A. Strik, *Food Chem. Toxicol.*, *24*: 1067 (1986).
10. J. C. Dacre, *Food Cosmet. Toxicol.*, *12*: 125 (1974).
11. F. D. Tollenar, *Proc. Pacific Sci. Congr.*, *5*: 92 (1957).
12. S. C. Allen and F. D. DeEds, *J. Amer. Oil Chem. Soc.*, *28*: 304 (1951).
13. FAO/WHO, WHO Food Additive Ser. No. 5, World Health Organization, Geneva, Switzerland, 1974.

14. G. J. Van Esch and H. Van Genderen, Netherland Inst. Public Health Rep. No. 481, The Netherlands, 1954.
15. K. M. Abdo, J. E. Huff, J. K. Haseman, and C. J. Alden, *Food Chem. Toxicol.*, *24*: 1091 (1986).
16. M. M. King and P. B. McCay, *Proc. Amer. Assoc. Cancer Res.*, *21*: 113 (1980) (Abstract).
17. M. M. King and P. B. McCay, *Cancer Res.*, *43*: 2485S (1983).
18. Z. Mikhailova, R. Vachkova, L. Vasileva, T. Stavreva, N. Donchev, I. Goranov, and E. Tyagunenko, *Voprosy-Pitaniya*, 2: 49 (1985).
19. S. Tanaka, K. Kawashima, S. Nakaura, S. Nagao, and Y. Omori, *Shokuhin Eisegaku Zasshi*, *20*: 378 (1979).
20. FDA, NTIS PB-223816, Maspeth, NY, 1973.
21. D. W. King, *J. Nutr.*, *83*: 123 (1964).
22. J. M. DeSesso, *Teratology*, *24*: 19 (1981).
23. R. Brun, *Berufsderm*, *12*: 281 (1964).
24. R. Brun, *Dermatologica*, *140*: 390 (1970).
25. W. Burkhardt and U. Fierz, *Dermatologica*, *129*: 431 (1964).
26. W. G. Van Ketel, *Contact Dermatitis*, *4*: 60 (1978).
27. B. M. Hausen and W. Beyer, *Contact Dermatitis*, *26*: 253 (1992).
28. M. P. Rosin and H. F. Stich, *J. Environ. Pathol. Toxicol.*, *4*: 159 (1980).
29. Litton Bionetics Inc., NTIS PB-245 441, Kensington, MD, 1974.
30. A. S. Raj and M. Katz, *Mutat. Res.*, *136*: 247 (1984).
31. T. Kawachi, T. Yahagi, T. Kada, Y. Tazyma, M. Ishidate, M. Sasaki, and T. Sugiyama, in *Molecular and Cellular Aspects of Carcinogen Screening Tests* (R. Montesano, H. Bartsch, and L. Tomatis, Eds.), International Agency for Research on Cancer, IARC Sci. Publ. No. 27, Lyon, France, 1980.
32. Y. Fukuhara, D. Yoshida, and F. Goto, *Agric. Biol Chem.*, *45*: 1061 (1981).
33. L. W. Lo and H. F. Stich, *Mutat. Res.*, *57*: 57 (1978).
34. G. A. Garton and R. T. Williams, *Biochem. J.*, *44*: 234 (1949).
35. G. Woodard, E. C. Hagan, and J. L. Radomski, *Fed. Proc.*, *8*: 348 (1949).
36. A. J. Carlson and N. R. Brewer, *Proc. Soc. Exp. Biol.*, *84*: 684 (1953).
37. B. N. Stuckey and W. M. Gearhart, *Food Technol.*, *11*: 676 (1957).
38. B. D. Astill, D. W. Fassett, and R. L. Roudabush, *Biochem. J.*, *72*: 451 (1959).
39. *Federal Register*, 39 (185, Sept. 23) 34172-34173 (1974).
40. H. C. Grice, G. Becking, and T. Goodman, *Food Cosmet. Toxicol.*, *6*: 155 (1968).
41. A. L. Tappel and A. G. Marr, *J. Agric. Food Chem.*, *2*: 554 (1954).
42. Z. Placer, Z. Veselkova, and R. Petrasek, *Nahrung*, *8*: 707 (1964).
43. E. M. Cranston et al., 1947. Unpublished report cited in Ref. 13.
44. E. M. Cranston, M. J. Jensen, A. Moren, T. Brey, E. T. Bell, and R. N. Bieter, *Fed. Proc.*, *6*: 318 (1947).
45. F. Griepentrog, *Arzneim. Forsch.*, *11*: 920 (1961).
46. T. Goodman, H. Grice, G. C. Becking, and F. A. Salem, *Lab. Invest.*, *23*: 93 (1970).
47. D. F. Buck, *Manuf. Confect.*, *65*: 45 (1985).
48. FAO/WHO, WHO Tech Rep. Ser. No. 776, World Health Organization, Geneva, Switzerland, 1989.
49. B. D. Astill, D. W. Fassett, and R. L. Roudabush, *Biochem. J.*, *75*: 543 (1960).

50. J. C. Dacre, F. A. Denz, and T. H. Kennedy, *Biochem. J.*, *64*: 777 (1956).
51. B. D. Astill, J. Mills, D. W. Fasset, R. L. Roudabush, and C. J. Terhaar, *J. Agric. Food Chem.*, *10*: 315 (1962).
52. H. C. Hodge, D. W. Fassett, E. A. Maynard, W. L. Downs, and R. D. Coye, Jr., *Toxicol. Appl. Pharmacol.*, *6*: 512 (1964).
53. O. H. M. Wilder, P. C. Ostby, and B. R. Gregory, *J. Agric Food Chem.*, *8*: 504 (1960).
54. W. S. Golder, A. J. Ryan, and S. E. Wright, *J. Pharm. Pharmacol.*, *14*: 268 (1962).
55. R. El-Rashidy and S. Niazi, *Biopharm. Drug Dispos.*, *4*: 389 (1983).
56. H. Verhagen, H. H. W. Thijssen, F. ten Hoor, and J. C. S. Kleinjans, *Food Chem. Toxicol.*, *27*: 151 (1989).
57. K. E. Armstrong and L. W. Wattenberg, *Cancer Res.*, *45*: 1507 (1985).
58. J. W. Daniel, J. C. Gage, D. I. Jones, and M. A. Stevens, *Food Cosmet. Toxicol.*, *5*: 475 (1967).
59. I. A. Karplyuk, *Trudy 2-oi (vtor) Nauchn. Konf. Po Vopr. Probl. Zhira. V Pitanii Leningrad*, 1962, p. 318.
60. F. A. Denz and J. G. Llaurado, *Brit. J. Nutr.*, *38*: 515 (1957).
61. A. D. Martin and D. Gilbert, *Biochem. J.*, *106*: 22p (1968) (Abstract).
62. Y. N. Cha and H. S. Heine, *Cancer Res.*, *42*: 2609 (1982).
63. J. R. Allen and J. F. Engblom, *Food Cosmet. Toxicol.*, *10*: 769 (1972).
64. Y. N. Cha and E. Beuding, *Biochem. Pharm.*, *28*: 1917 (1979).
65. P. J. Creaven, W. H. Davies, and R. T. Williams, *J. Pharm. Pharmacol.*, *18*: 485 (1966).
66. A. M. Benson, R. P. Batzinger, S.-Y. L. Ou, E. Beuding, Y.-N. Cha, and P. Talalay, *Cancer Res.*, *38*: 4486 (1978).
67. G. J. Ikeda, J. E. Stewart, P. P. Sapienza, J. O. Peggins, T. C. Michel, V. Olivito, H. Z. Alam, and M. W. O'Donnell, Jr., *Food Chem. Toxicol.*, *24*: 1201 (1986).
68. O. H. M. Wilder and H. R. Kraybill, Amer. Meat Inst. Foundation, Univ. Chicago, Chicago, IL, 1948.
69. W. D. Brown, A. R. Johnson, and M. W. Halloran, *Aust. J. Exp. Biol. Med.*, *37*: 533 (1959).
70. N. Ito, A. Hagiwara, M. Shibata, T. Ogiso, and S. Fukushima, *Gann*, *73*: 332 (1982).
71. N. Ito, S. Fukushima, A. Hagiwara, M. Shibata, and T. Ogiso, *J. Natl. Cancer Inst.*, *70*: 343 (1983).
72. N. Ito, M. Hirose, S. Fukushima, H. Tsuda, T. Shirai, and M. Tatematsu, *Food Chem. Toxicol.*, *24*: 1071 (1986).
73. H. Verhagen, C. Furhee, B. Schutte, F. T. Bosman, G. H. Blijham, P. Th. Henderson, F. ten Hoor, and J. C. S. Kleinjans, *Carcinogenesis*, *11*: 1461 (1990).
74. N. Ito, S. Fukushima, K. Imaida, T. Sakata, and T. Masui, *Gann*, *74*: 459 (1983).
75. T. Masui, M. Hirose, K. Imaida, S. Fukushima, S. Tamano, and N. Ito, *Gann*, *77*: 1083 (1986).
76. N. Ito, M. Hirose, Y. M. Kurata, E. Ikawa, Y. Nera, and S. Fukushima, *Gann*, *75*: 471 (1984).
77. H. J. Altmann, W. Grunov, U. Mohr, H. B. Richter-Reichhelm, and P. W. Wester, *Food Chem. Toxicol.*, *24*: 1183 (1986).
78. F. Iverson, J. Truelove, E. Nera, E. Lok, D. B. Clayson, and J. Wong, *Food Chem. Toxicol.*, *24*: 1197 (1986).

79. M. Tobe, T. Furuya, Y. Kawasaki, K. Naito, K. Sekita, K. Matsumoto, T. Ochai, A. Usui, T. Kokubo, J. Kanno, and Y. Hayashi, *Food Chem. Toxicol.*, *24*: 1223 (1986).
80. G. Wurtzen and P. Olsen, *Food Chem. Toxicol.*, *24*: 1229 (1986).
81. H. Amo, H. Kubota, J. Lu, and M. Matsuyuma, *Carcinogenesis*, *11*: 151 (1990).
82. C. M. DeStafney, U. D. G. Prabhu, V. L. Sparnius, and L. W. Wattenberg, *Food Chem Toxicol.*, *24*: 1149 (1986).
83. G. M. Williams, *Food Chem. Toxicol.*, *24*: 1163 (1986).
84. D. J. Clegg, *Food Cosmet. Toxicol.*, *3*: 387 (1965).
85. E. Hansen and O. Meyer, *Toxicology*, *10*: 195 (1978).
86. FDA, NTIS PB-221783, Maspeth, NY, 1972.
87. FDA, NTIS PB- 267200, Maspeth, NY, 1974.
88. J. R. Allen, *Arch. Environ. Health*, *31*: 47 (1976).
89. J. D. Stokes and C. R. Scudder, *Dev. Psychobiol.*, *7*: 343 (1974).
90. J. D. Stokes, C. R. Scudder, and A. G. Karczmar, *Fed. Proc.*, *31*: 596 (1972) (Abstract).
91. G. M. Williams, *Cancer Res.*, *37*: 1845 (1977).
92. C. Tong and G. M. Williams, *Mutat. Res.*, *74*: 1 (1980).
93. C. G. Rogers, B. N. Nayak, and C. Heroux-Metcalf, *Cancer Lett.*, *27*: 61 (1985).
94. M. Ishidate, Jr. and S. Odashima, *Mutat. Res.*, *48*: 337 (1977).
95. S. Abe and M. Sasaki, *J. Natl. Cancer Inst.*, *58*: 1635 (1977).
96. Litton Bionetics, Inc., NTIS PB-245 460, Kensington, MD, 1974.
97. A. J. Ryan and S. E. Wright, *Food Technol. Aust.*, *16*: 626 (1964).
98. L. G. Ladomery, A. J. Ryan, and S. E. Wright, *J. Pharm. Pharmacol.*, *19*: 383 (1967).
99. L. I. Wiebe, J. R. Mercer, and A. J. Ryan, *Drug Metab. Dispos.*, *6*: 296 (1978).
100. J. W. Daniel and J. C. Gage, *Food Cosmet. Toxicol.*, *3*: 405 (1965).
101. M. Matsuo, K. Mihara, M. Okuno, H. Ohkawa, and J. Miyamoto, *Food Chem Toxicol.*, *22*: 345 (1984).
102. R. Tye, R., J. D. Engel, I. Rapien, and J. Moore, *Food Cosmet. Toxicol.*, *3*: 547 (1965).
103. J. W. Daniel, J. I. Gage, and D. I. Jones, *Biochem. J.*, *106*: 783 (1968).
104. H. Verhagen, H. H. G. Beckers, P. A. W. V. Comuth, L. M. Mass, F. ten Hoor, P. Th. Henderso, and J. C. S. Kleinjans, *Food Chem. Toxicol.*, *27*: 765 (1989).
105. K. Yamamoto, K. Tajima, and T. Mitzutani, *J. Pharm. Dyn.*, *2*: 164 (1979).
106. K. Tajima, K. Yamamoto, and T. Mitzutani, *Chem. Pharm. Bull.*, *29*: 3738 (1981).
107. J. C. Dacre, *Biochem. J.*, *78*: 758 (1961).
108. M. Akagi and I. Aoki, *Chem. Pharm. Bull.*, *10*: 101 (1962).
109. I. Aoki, *Chem. Pharm. Bull.*, *10*: 105 (1962).
110. R. El-Rashidy and S. Niazi, *J. Pharm. Sci.*, *69*: 1455 (1980).
111. A. S. Wright, D. A. A. Akintonwa, R. S. Crowne, and D. E. Hathway, *Biochem. J.*, *97*: 303 (1965).
112. A. L. Branen, *J. Amer. Oil Chem. Soc.*, *52*: 59 (1975).
113. W. B. Deichmann, J. J. Clemmer, R. Prakoczy, and J. Biachine, *AMA Arch. Ind. Health*, *11*: 93 (1955).
114. I. A. Karplyuk, *Voprosy Pitaniya*, *18*: 24 (1959).
115. I. F. Gaunt, G. Feuer, F. A. Fairweather, and D. Gilbert, *Food Cosmet. Toxicol.*, *3*: 433 (1965).
116. D. Gilbert and L. Goldberg, *Food Cosmet. Toxicol.*, *3*: 417 (1965).

117. G. Feurer, I. F. Gaunt, L. Goldberg, and F. A. Fairweather, *Food Cosmet. Toxicol.*, *3*: 457 (1965).

118. C. J. Powell, S. M. Connelly, S. M. Jones, P. Grasso, and J. W. Bridges, *Food Chem. Toxicol.*, *24*: 1131 (1986).

119. N. K. Clapp, R. L. Tyndall, and R. B. Cumming, *Food Cosmet. Toxicol.*, *11*: 847 (1973).

120. C. M. Botham, D. M. Conning, J. Hayes, M. H. Litchfield, and T. F. McElligott, *Food Cosmet. Toxicol.*, *8*: 1 (1970).

121. S. Kawano, T. Nakao, and K. Hiraga, *Jpn. J. Pharmacol.*, *30*: 861 (1980).

122. O. Takahashi, *Food Chem. Toxicol.*, *30*: 89 (1992).

123. Y. Nakagawa and K. Tayama, *Arch. Toxicol.*, *61*: 359 (1988).

124. O. Meyer, L. Blom, and P. Olsen, *Arch. Toxicol. Suppl. 1*, 355 (1978).

125. O. Takahashi and K. Hiraga, *Toxicol. Appl. Pharmacol.*, *43*: 399 (1978).

126. O. Takahashi, S. Hayashida, and K. Hiraga, *Food Chem. Toxicol.*, *18*: 229 (1980).

127. O. Takahashi, *Biochem. Pharm.*, *37*: 2857 (1988).

128. O. Takahashi and K. Hiraga, *J. Nutr.*, *109*: 453 (1979).

129. O. Takahashi and K. Hiraga, *Food Chem. Toxicol.*, *22*: 97 (1984).

130. A. A. Marino and J. T. Mitchell, *Proc. Soc. Exp. Biol. Med.*, *140*: 122 (1972).

131. H. P. Witschi and W. Saheb, *Proc. Soc. Exp. Biol. Med.*, *147*: 690 (1974).

132. W. Saheb and H. P. Witschi, *Toxicol. Appl. Pharmacol.*, *33*: 309 (1975).

133. H. P. Witschi, W. M. Hascheck, A. J. P. Liein-Szanto, and P. J. Hakkinen, *Amer. Rev. Respir. Dis.*, *123*: 98 (1981).

134. I. Y. R. Adamson, D. H. Bowden, M. G. Cote, and H. P. Witschi, *Lab. Invest.*, *36*: 26 (1977).

135. K. I. Hirai, H. P. Witschi, and M. G. Cote, *Exp. Mol. Pathol.*, *27*: 295 (1977).

136. D. Williamson, P. Esterez, and H. P. Witschi, *Toxicol. Appl. Pharmacol.*, *43*: 577 (1978).

137. M. Hirose, M. Shibata, A. Hagiwara, K. Imaida, and N. Ito, *Food Chem. Toxicol.*, *19*: 147 (1981).

138. G. Wurtzen and P. Olsen, *Food Chem. Toxicol.*, *24*: 1121 (1986).

139. T. Shirai, A. Hagiwara, Y. Kurata, M. Shibata, S. Fukushima, and N. Ito, *Food Chem. Toxicol.*, *20*: 861 (1982).

140. K. Inai, T. Kobuke, S. Nambu, T. Takemoto, E. Kou, H. Nishina, M. Fujihara, S. Yonehara, S. Suehiro, T. Tsuya, K. Horiuchi, and S. Tokuoka, *Gann*, *79*: 49 (1988).

141. N. K. Clapp, R. L. Tyndall, R. B. Cumming, and J. A. Otten, *Food Chem. Toxicol.*, *12*: 367 (1974).

142. T. M. Brooks, P. F. Hunt, E. Thorpe, and A. T. Walker, Unpublished data cited in *Fed. Reg.*, *42*: 27603 (1974).

143. R. C. Lindenschmidt, A. F. Tryka, M. E. Goad, and H. P. Witschi, *Toxicology*, *38*: 151 (1986).

144. N. Ito, S. Fukushima, and M. Hirose, in *Toxicological Aspects of Food* (K. Miller, Ed.), Elsevier Applied Sci., New York, 1987, p. 253.

145. J. P. Frawley, F. E. Kohn, J. H. Kay, and J. C. Calendra, *Food Chem. Toxicol.*, *3*: 377 (1965).

146. A. R. Johnson, *Food. Cosmet. Toxicol.*, *3*: 371 (1965).

147. P. Olsen, O. Meyer, N. Bille, and G. Wurtzen, *Food Chem. Toxicol.*, *24*: 1 (1986).

148. FDA, NTIS PB-221782, Maspeth, NY, 1972.
149. FDA, NTIS PB-267201, Maspeth, NY, 1974.
150. C. V. Vorhees, R. E. Butcher, R. L. Brunner, and T. J. Sobotka, *Food Chem. Toxicol.*, *19*: 153 (1981).
151. G. J. Hageman, H. Verhagen, and J. C. S. Kleinjans, *Mutat. Res.*, *208*: 207 (1988).
152. L. A. Shelef and B. Chin, *Appl. Environ. Microbiol.*, *40*: 1039 (1980).
153. Yu. V. Paschin and L. M. Bahitova, *Mutat. Res.*, *137*: 57 (1984).
154. L. J. Sciorra, B. N. Kaufman, and R. Maier, *Food Cosmet. Toxicol.*, *12*: 33 (1974).
155. R. M. Patterson, L. A. Keith, and J. Stewart, *Toxicol. In Vitro*, *1*: 55 (1987).
156. D. Harman, H. J. Curtis, and J. Tilley, *J. Gerontol.*, *25*: 17 (1970).
157. W. R. Bruce and J. A. Heddle, *Can. J. Genet. Cytol.*, *21*: 319 (1979).
158. Stanford Research Institute, NTIS PB- 221 827, Menlo Park, CA, 1972.
159. Stanford Research Institute, NTIS PB- 278 026, Menlo Park, CA, 1977.
160. D. F. Buck, *Cereal Foods World*, *29*(5): 301 (1984).
161. E. R. Sherwin, *J. Amer. Oil Chem. Soc.*, *53*: 430 (1976).
162. EEC Commission Documents, EEC Commission Document III/26/82, Brussells, 1982.
163. B. D. Astill, C. J. Terhaar, W. J. Krasavage, G. L. Wolf, R. L. Roudabush, D. W. Fassett, and K. Morgareidge, *J. Amer. Oil Chem. Soc.*, *52*: 53 (1975).
164. FAO/WHO, WHO Food Additive Ser. No. 6, World Health Organization, Geneva, Switzerland, 1975.
165. D. W. Fassett, C. J. Terhaar, and B. D. Astill (1968), unpublished report cited in Ref. 164.
166. E. A. Nera, E. Lok, F. Iverson, E. Ormsby, K. F. Karpinski, and D. B. Clayson, *Toxicology*, *32*: 197 (1984).
167. S. Tamano, S. Fukushima, T. Shirai, M. Hirose, and N. Ito, *Cancer Lett.*, *35*: 39 (1987).
168. W. J. Krasavage, *Teratology*, *16*: 31 (1977).
169. K. Matsuoka, M. Matsui, N. Miyata, T. Sofuni, and M. Ishidate, Jr., *Mutat. Res.*, *241*: 125 (1990).
170. C. G. Rogers, B. G. Boyes, T. I. Matula, and R. Stapley, *Mutat. Res.*, *280*: 17 (1992).
171. A. K. Giri, S. Sen, G. Talukder, and A. Sharma, *Food Chem. Toxicol.*, *22*: 459 (1984).
172. A. Mikherjee, G. Talukder, and A. Sharma, *Environ. Mol. Mutagen*, *13*: 234 (1989).
173. W. J. Krasavage and C. J. Terhaar, *J. Agric. Food Chem.*, *25*: 273 (1977).
174. H. Klaui, *Flavors*, July/August 1976, p. 165.
175. W. M. Cort, *J. Amer. Oil Chem. Soc.*, *51*: 321 (1974).
176. H. E. Gallo-Torres, in *Vitamin E: A Comprehensive Treatise*, Vol. 1 (L. J. Machlin, Ed.), Marcel Dekker, New York, 1980, p. 170.
177. C. K. Pearson and M. MacBarnes, *Int. Z. Vitaminforsch.*, *40*: 19 (1970).
178. H. Schmandke, C. Sima, and R. Maune, *Int. Z. Vitaminforsch.*, *39*: 296 (1969).
179. H. E. Gallo-Torres, in *Vitamin E: A Comprehensive Treatise*, Vol. 1 (L. J. Machlin, Ed.), Marcel Dekker, New York, 1980, p. 193.
180. O. Wiss, H. R. Bunnel, and U. Gloor, *Vit. Horm.*, *20*: 441 (1962).
181. J. G. Bieri, *Ann. N.Y. Acad. Sci.*, *203*: 181 (1972).
182. A. S. Csallany, H. H. Draper, and S. N. Shah, *Arch. Biochim. Biophys.*, *98*: 142 (1962).
183. E. J. Simon, C. S. Gross, and A. T. Milhorat, *J. Biol. Chem.*, *221*: 797 (1956).
184. H. Schmandke and G. Schmidt, *Int. Z. Vitaminforsch.*, *38*: 75 (1968).

185. A. Hanck, *Arzneimitteltherapie Heute*, Band 42, Aesopus-Verlag, FRG, 1986, p. 36.
186. V. Demole, *Int. Z. Vitaminforsch.*, *8*: 338 (1939).
187. L. H. Weissburger and P. L. Harris, *J. Biol. Chem.*, *151*: 543 (1943).
188. H. A. Dysmsza and J. Park, *Fed. Amer. Soc. Exp. Biol.*, *34*: 912 (1975) (Abstract).
189. B. E. March, E. Wong, L. Seier, J. Sim, and J. Biely, *J. Nutr.*, *103*: 371 (1973).
190. K. M. Abdo, G. Rao, C. A. Montgomery, M. Dinowitz, and K. Kanagalingam, *Food Chem. Toxicol.*, *24*: 1043 (1986).
191. O. Takahashi, H. Ichikawa, and M. Sasaki, *Toxicology*, *63*: 157 (1990).
192. R. E. Olson and J. P. Jones, *Fed. Proc.*, *38*: 710 (1979) (Abstract).
193. W. J. Bettger and R. E. Olson, *Fed. Proc.*, *41*: 344 (1982) (Abstract).
194. M. Briggs, *N. Engl. J. Med.*, *290*: 579 (1974).
195. S. Dahl, *Lancet*, *1*: 465 (1974).
196. A. C. Tsai, J. J. Kelley, B. Peng, and N. Cook, *Amer. J. Clin. Nutr.*, *31*: 831 (1978).
197. R. W. Hillman, *Amer. J. Clin. Nutr.*, *5*: 597 (1957).
198. R. H. Brodkin and J. Bleiberg, *Arch. Dermatol.*, *92*: 76 (1965).
199. W. Minkin, H. J. Cohen, and S. B. Frank, *Arch. Dermatol.*, *107*: 774 (1973).
200. J. L. Aeling, P. J. Panagotacos, and R. J. Andreozzi, *Arch. Dermatol.*, *108*: 579 (1973).
201. H. Kappus and A. T. Diplock, *Free Rad. Biol. Med.*, *13*: 55 (1992).
202. N. Y. Yang and I. D. Desai, *J. Nutr.* *107*: 1410 (1977).
203. G. H. Wheldon, A. Bhatt, P. Keller, and H. Hummler, *Int. J. Vit. Nutr. Res.*, *53*: 287 (1983).
204. M. A. Moore, H. Tsuda, W. Thamvit, T. Masui, and N. Ito, *J. Natl. Cancer Inst.*, *78*: 289 (1987).
205. G. Shklar, *J. Natl. Cancer Inst.*, *68*: 791 (1982).
206. M. G. Cook and P. McNamara, *Cancer Res.*, *40*: 1329 (1980).
207. G. Tomassi and V. Silano, *Food Chem. Toxicol.*, *24*: 1051 (1986).
208. FDA, NTIS PB-233809, Maspeth, NY, 1973.
209. J. C. Czyba, *Comples Rend. Soc. Biol.*, *160*: 765 (1966).
210. I. Ichihara, *Okajimas Folia Anat.*, *43*: 203 (1967).
211. D. Solomon, D. Strummer, and P. P. Nair, *Ann. N.Y. Acad. Sci.*, *203*: 103 (1972).
212. H. Winkler, *Zentr. Gynakol.*, *67*: 32 (1943).
213. R. J. Shamberger, F. F. Baugham, S. L. Kalchert, C. E. Willis, and G. C. Hoffman, *Proc. Natl. Acad. Sci. USA*, *70*: 1461 (1973).
214. R. J. Shamberger, C. L. Corlett, K. D. Beaman, and B. L. Kasten, *Mutat. Res.*, *66*: 349 (1979).
215. E. Gebhart, H. Wagner, K. Grziwok, and H. Behnsen, *Mutat. Res.*, *149*: 83 (1985).
216. C. Beckman, R. M. Roy, and A. Sproule, *Mutat. Res.*, *105*: 73 (1982).
217. V. Johnson, A. J. Carlson, N. Kleitman, and P. Bergstrom, *Food Res.*, *3*: 555 (1938).
218. D. E. Hathway, *Adv. Food Res.*, *15*: 1 (1966).
219. A. S. Wright, R. S. Crowne, and D. E. Hathway, *Biochem. J.*, *95*: 98 (1965).
220. M. Akagi and I. Aoki, *Chem. Pharm. Bull.*, *10*: 200 (1962).
221. A. S. Wright, R. S. Crowne, and D. E. Hathway, *Biochem. J.*, *99*: 146 (1966).
222. A. S. Wright, R. S. Crowne, and D. E. Hathway, *Biochem. J.*, *102*: 351 (1967).
223. J. C. Dacre, *Toxicol. Appl. Pharmacol.*, *17*: 669 (1970).
224. D. E. Stevenson, P. L. Chambers, and C. G. Hunter, *Food Chem. Toxicol.*, *3*: 281 (1965).

225. E. M. Bickoff, A. L. Livingston, J. Guggolz, and C. R. Thompson, *J. Agric. Food Chem.*, *2*: 1229 (1954).
226. A. Atkinson, R. P. Van der Merwe, and L. G. Swart, *Agroanimalia*, *4*: 63 (1972).
227. H. S. Olcott, *J. Amer. Oil Chem. Soc.*, *35*: 597 (1958).
228. J. T. Weil, J. Van der Veen, and H. S. Olcott, *Nature*, *219*: 168 (1968).
229. J. S. Lin and H. S. Olcott, *J. Agric. Food Chem.*, *23*: 798 (1975).
230. FAO/WHO, WHO Tech. Rep. Ser. No. 458, FAO Agricultural Studies No. 84, World Health Organization, Geneva, Switzerland, 1969.
231. J. U. Skaare, *Xenobiotica*, *9*: 659 (1979).
232. J. U. Skaare and E. Solheim, *Xenobiotica*, *9*: 649 (1979).
233. R. H. Wilson, J. O. Thomas, R. A. Thompson, H. F. Launer, and G. O. Kohler, *J. Agric. Food Chem.*, *3*: 206 (1959).
234. R. H. Wilson and F. DeEds, *J. Agric. Food Chem.*, *3*: 203 (1959).
235. D. V. Parke, A. Rahim, and R. Walker, *Biochem. J.*, *130*: 84p (1972) (Abstract).
236. D. V. Parke, A. Rahim, and R. Walker, *Biochem. Pharmacol.*, *23*: 1871 (1974).
237. M. A. Cawthorne, J. Bunyan, M. V. Sennit, J. Green, and P. Grasso, *Brit. J. Nutr.*, *24*: 357 (1970).
238. J. U. Skaare, I. Nafstad, and H. K. Dahle, *Toxicol. Appl. Pharmacol.*, *42*: 19 (1977).
239. I. Nafstad and J. U. Skaare, *Toxicol. Lett.*, *1*: 295 (1978).
240. H. L. Kim, *J. Toxicol. Environ. Health*, *33*: 229 (1991).
241. M. M. Manson, J. A. Green, B. J. Wright, and P. Carthrew, *Arch. Toxicol.*, *66*: 51 (1992).
242. G. C. Hard and G. E. Neal, *Fundam. Appl. Toxicol.*, *18*: 278 (1992).
243. T. Shirai, E. Ikawa, M. Hirose, W. Thamvit, and N. Ito, *Carcinogenesis*, *6*: 637 (1985).
244. H. Tsuda, T. Sakata, T. Masui, K. Imaida, and N. Ito, *Carcinogenesis*, *5*: 525 (1984).
245. S. Fukushima, Y. Kurata, M. Shibata, E. Ikawa, and N. Ito, *Cancer Lett.*, *23*: 29 (1984).
246. R. Hasegawa, F. Furukawa, K. Toyoda, Y. M. Takahashi, M. Hirose, and N. Ito, *Jpn. J. Cancer Res.*, *81*: 871 (1990).
247. C. Savini, R. Morelli, E. Piancastelli, and S. Restani, *Contact Dermatitis*, *21*: 342 (1989).
248. D. Burrows, *Brit. J. Dermatol.*, *92*: 167 (1975).
249. E. Van Hecke, *Contact Dermatitis*, *3*: 341 (1977).
250. R. S. Isensein, *Am. J. Vet. Res.*, *31*: 907 (1970).
251. N. M. Weinshenker, L. A. Bunes, and R. Davis, U.S. Patent 3,996,199 (1976).
252. N. M. Weinshenker, *Food Technol.*, *34*: 40 (1980).
253. FAO/WHO, WHO Tech. Rep. Ser. No. 710, World Health Organization, Geneva, Switzerland, 1984.
254. T. M. Parkinson, S. C. Halladay, and F. E. Enderlin, *J. Amer. Oil Chem. Soc.*, *55*: 242A (1978).
255. T. M. Parkinson, T. Honohan, F. E. Enderlin, S. C. Halladay, R. L. Hale, S. A. de Keczer, P. L. Dubin, B. A. Ryerson, and A. R. Read, *Food Cosmet. Toxicol.*, *16*: 321 (1978).
256. P. D. Walson, D. E. Carter, B. A. Iverson, and S. C. Halladay, *Food Cosmet. Toxicol.*, *17*: 201 (1979).
257. S. C. Hallady, B. A. Ryerson, C. R. Smith, J. P. Brown, and T. M. Parkinson, *Food Cosmet. Toxicol.*, *18*: 569 (1980).

258. J. P. Brown, R. J. Brown, and G. W. Roehm, in *Progress in Genetic Toxicology* (D. Scott, B. A. Bridges, and F. H. Sobels, Eds.), Elsevier/North-Holland Biomedical Press, Amsterdam, Netherlands, 1977.
259. J. W. Scott, W. M. Cort, J. H. Harley, D. R. Parrish, and G. Saucy, *J. Amer. Oil Chem. Soc.*, *51*: 200 (1974).
260. W. M. Cort, J. W. Scott, M. Araujo, W. J. Mergens, M. A. Cannalinga, M. Osadca, H. Harley, D. R. Parrish, and W. R. Pool, *J. Amer. Oil Chem. Soc.*, *52*: 174 (1975).
261. W. M. Cort, J. W. Scott, and J. H. Harley, *Food Technol.*, *29*(11): 46 (1975).
262. R. Bush, *Food Technol.*, *40*: 49 (1986).
263. A. F. Gunnison and E. D. Palmes, *Chem.-Biol. Interact.*, *21*: 315 (1978).
264. N. Oshino and B. Chance, *Arch. Biochim. Biophys.*, *170*: 514 (1975).
265. B. Bhagat and M. F. Lockett, *J. Pharm. Pharmacol.*, *12*: 690 (1960).
266. W. B. Gibson and F. M. Strong, *Food Cosmet. Toxicol.*, *11*: 185 (1973).
267. A. F. Gunnison, C. A. Bresnahan, and E. D. Palmes, *Toxicol. Appl. Pharmacol.*, *42*: 99 (1977).
268. J. Wever, *Food Chem. Toxicol.*, *23*: 895 (1985).
269. A. F. Gunnison, *Food Chem. Toxicol.*, *19*: 667 (1981).
270. A. F. Gunnison and E. D. Palmes, *Toxicol. Appl. Pharmacol.*, *24*: 266 (1973).
271. D. H. Petering and N. T. Shih, *Environ. Res.*, *9*: 55 (1975).
272. L. C. Schroeter, *Sulfur Dioxide: Applications in Foods, Beverages, and Pharmaceuticals*, Pergamon Press, Oxford, England, 1966.
273. N. T. Shih and D. H. Petering, *Biochem. Biophys. Res. Commun.*, *55*: 1319 (1973).
274. H. Hayatsu, Y. Wataya, K. Kai, and S. Iida, *Biochemistry*, *9: 2858 (1970)*.
275. H. Hayatsu, *Prog. Nucleic Acid Res. Mol. Biol.*, *16*: 75 (1976).
276. R. Shapiro, B. Braverman, J. B. Louis, and R. E. Servis, *J. Biol. Chem.*, *248*: 4060 (1973).
277. R. R. Williams, R. E. Waterman, J. C. Keresztesy, and E. R. Buchman, *J. Amer. Chem. Soc.*, *57*: 536 (1935).
278. R. Cecil, in *The Proteins*, Vol. 1 (H. Neurath, Ed.), Academic Press, New York, 1963, p. 379.
279. N. Kitamura and H. Hayatsu, *Nucleic Acids Res.*, *1*: 75 (1974).
280. S. F. Yang, *Biochemistry*, *9*: 5008 (1970).
281. S. F. Yang, *Environ. Res.*, *6*: 395 (1973).
282. D. Kaplan, C. McJilton, and D. Luchtel, *Arch. Environ. Health*, *30*: 507 (1975).
283. E. I. Ciaccio, *J. Biol. Chem.*, *241*: 1581 (1966).
284. V. Massey, F. Muller, R. Feldberg, M. Schuman, P. Sullivan, L. G. Howell, S. G. Mayhew, R. G. Matthews, and G. P. Foust, *J. Biol. Chem.*, *244*: 3999 (1969).
285. S. J. Cooperstein, *J. Biol. Chem.*, *238*: 3606 (1963).
286. A. Kamogawa and T. Fukui, *Biochim. Biophys. Acta*, *302*: 158 (1973).
287. A. B. Roy, *Adv. Enzymol.*, *22*: 205 (1960).
288. J. W. Wilkins, Jr., J. A. Greene, Jr., and J. M. Weller, *Clin. Pharmacol. Ther.*, *9*: 328 (1968).
289. J. O. Hoppe and F. C. Goble, *J. Pharmacol. Exp. Ther.*, *101*: 101 (1951).
290. B. Bhagat and M. F. Lockett, *Food Cosmet. Toxicol.*, *2*: 1 (1964).
291. M. Lhuissier, D. Hugot, J. Leclerc, and J. Causeret, *Cah. Nutr. Diet.*, *2*: 23 (1967).
292. H. P. Til, V. J. Feron, and A. P. De Groot, *Food Cosmet. Toxicol.*, *10*: 291 (1972).

293. H. P. Til, V. J. Feron, and A. P. De Groot, *Food Cosmet. Toxicol.*, *10*: 463 (1972).
294. R. Shapiro, *Mutat. Res.*, *39*: 149 (1976).
295. F. Mukai, I. Hawryluk, and R. Shapiro, *Biochem. Biophys. Res. Commun.*, *39*: 983 (1970).
296. W. D. MacRae and H. F. Stich, *Toxicology*, *13*: 167 (1979).
297. R. G. Mallon and T. B. Rossman, *Mutat. Res.*, *88*: 125 (1981).
298. W. M. Generoso, S. W. Huff, and K. T. Cain, *Mutat. Res.*, *56*: 363 (1978).
299. G. M. Jagiello, J. S. Lin, and M. B. Ducayen, *Environ. Res.*, *9*: 84 (1975).
300. FAO/WHO, WHO Tech. Rep. Ser. No. 669, World Health Organization, Geneva, Switzerland, 1981.
301. C.O'H. Curtin and C. G. King, *J. Biol. Chem.*, *216*: 539 (1955).
302. A. B. Kallner, D. Hartman, and D. H. Hornig, *Amer. J. Clin. Nutr.*, *34*: 1347 (1981).
303. W. Kuebler and J. Gehler, *Int. J. Vit. Res.*, *40*: 442 (1970).
304. C. W. M. Wilson, in *Vitamin C: Recent Aspects of Its Physiological and Technological Importance* (G. G. Birch and K. H. Parker, Eds.), Wiley, New York, 1974, p. 203.
305. M. H. Briggs, P. Garcia-Webb, and P. Davies, *Lancet*, *2*: 201 (1973).
306. D. H. Hornig and U. Moser, in *Vitamin C (Ascorbic Acid)* (J. N. Counsell and D. H. Hornig, Eds.), Applied Science, London, England, 1981, p. 225.
307. V. Demole, *Biochem. J.*, *28*: 770 (1934).
308. T. Ohno and K. Myoga, *Nutr. Rep. Int.*, *24*: 291 (1981).
309. National Institutes of Health, NIH Publ. No. 83-2503, Research Triangle Park, NC, 1983.
310. S. S. Mirvish, *Ann. N.Y. Acad. Sci.*, *258*: 175 (1975).
311. H. S. Black and J. T. Chan, *J. Invest. Dermatol.*, *65*: 412 (1975).
312. J. B. Guttenplan, *Nature*, *268*: 368 (1977).
313. J. B. Guttenplan, *Cancer Res.*, *38*: 2018 (1978).
314. E. P. Samborskaya, *Byull. Eksp. Biol. Med.*, *57*: 105 (1964).
315. E. P. Samborskaya and T. D. Ferdman, *Byull. Eksp. Biol. Med.*, *62*: 96 (1966).
316. F. R. Alleva, J. J. Alleva, and T. Balazs, *Toxicol. Appl. Pharmacol.*, *35*: 393 (1976).
317. H. Frohberg, J. Gleich, and H. Kieser, *Arzneim. Forsch.*, *23*: 1081 (1973).
318. H. F. Stich, J. Karim, J. Koropatnick, and L. Lo, *Nature*, *260*: 722 (1976).
319. H. Omura, K. Shinohara, H. Maeda, M. Nonaka, and H. Murakami, *J. Nutr. Sci. Vitaminol.*, *24*: 185 (1978).
320. G. Speit, M. Wolf, and W. Vogel, *Mutat. Res.*, *78*: 273 (1980).
321. M. P. Rosin, R. H. C. San, and H. F. Stich, *Cancer Lett.*, *8*: 299 (1980).
322. P. S. Chauhan, P. S., M. Aravindakshan, and K. Sundaram, *Mutat. Res.*, *53*: 166 (1978).
323. E. P. Norkus, W. Kuenzig, and A. H. Conney, *Mutat. Res.*, *117*: 183 (1983).
324. F. Widenbauer, *Klin. Wochschr.*, *15*: 1158 (1936).
325. Y. Y. Gwo, J. R. Flick, and H. P. Dupuy, *J. Amer. Oil Chem. Soc.*, *62*: 1666 (1985).
326. O. G. Fitzhugh and A. A. Nelson, *Proc. Soc. Exp. Biol.*, *61*: 195 (1946).
327. K. N. Movaghar, U.S. Patent 4,948,609 (1990).
328. H. Kanematsu, M. Aoyama, T. Maruyama, I. Niiya, M. Tsukamoto, S. Tokairin, and T. Matsumoto, *Yukagaku*, *33*: 361 (1984).
329. Y. Nakao, S. Takagi, and H. Nakatani, U.S. Patent 3,666,488 (1972).
330. F. J. Yourga, W. B. Esselen, and C. R. Fellers, *Food Res.*, *9*: 188 (1944).

331. D. Hornig, F. Weber, and O. Wiss, *Experientia, 30*: 173 (1974).
332. D. Hornig and H. Weiser, *Int. J. Vit. Nutr. Res.*, *46*: 40 (1976).
333. N. Arakawa, E. Suzuki, T. Kurata, M. Otsuka, and C. Inagaki, *J. Nutr. Sci. Vit.*, *32*: 171 (1986).
334. I. Abe, S. Saito, K. Hori, M. Suzuki, and H. Sato, *Exp. Mol. Pathol.*, *41*: 35 (1984).
335. M. Ishidate, Jr., T. Sofumi, K. Yoshikawa, M. Hayashi, T. Nohmi, M. Sawada, and A. Matsuoka, *Food Chem. Toxicol.*, *22*: 623 (1984).
336. D. Grossman and K. Lang, *Biochem. Z.*, *336*: 351 (1962).
337. R. E. Gosselin, A. Rothstein, G. J. Miller, and H. L. Berke, *J. Pharmacol. Exp. Ther.*, *106*: 180 (1952).
338. J. P. Ebel, *Ann. Nutr. Alim.*, *12*: 57 (1958).
339. R. H. Ellinger, in *CRC Handbook of Food Additives* Vol. 1 (T. E. Furia, Ed.), CRC Press, Boca Raton, FL, 1972, p. 617.
340. E. M. McKay and J. Oliver, *J. Exp. Med.*, *61*: 319 (1935).
341. S. L. Bonting, Ph.D. Thesis, Univ. Amsterdam, Amsterdam, Netherlands, 1952.
342. G. Nazario, *Rev. Inst. Adolfo Lutz*, 2: 141 (1952).
343. D. McFarlane, *J. Pathol. Bacteriol.*, *52*: 17 (1941).
344. F. Hahn, H. Jacobi, and E. Seifen, *Arzneim. Forsch.*, *8*: 286 (1958).
345. H. A. Dysmsza, G. Reussner, and R. Thiessen, *J. Nutr.*, *69*: 419 (1959).
346. G. J. Van Esch, H. H. Vink, S. J. Wit, and H. Van Genderen, *Arzneim. Forsch.*, *7*: 172 (1957).
347. F. Hahn and E. Seifen, *Arzneim. Forsch.*, *9*: 501 (1959).
348. P. K. Datta, A. C. Frazer, M. Sharrat, and H. G. Sammons, *J. Sci. Food Agric.*, *13*: 556 (1962).
349. F. Lauersen, *Z. Lebensm. Untersuch.*, *96*: 418 (1953).
350. K. Lang, *Z. Lebensm. Untersuch.*, *110*: 450 (1959).
351. R. Raines Bell, H. H. Draper, D. Y. M. Tzeng, H. K. Shin, and G. R. Schmidt, *J. Nutr.*, *107*: 42 (1977).
352. H. Foreman, M. Vier, and M. Magee, *J. Biol. Chem.*, *203*: 1045 (1953).
353. M. S. Chan, Ph.D. Thesis, Univ. Massachusetts, Amherst, MA, 1956.
354. H. Foreman and T. T. Trujillo, *J. Lab. Clin. Med.*, *43*: 566 (1954).
355. J. Srbova and J. Teisinger, *Pracovni Lekarstui.*, *9*: 385 (1957).
356. H. M. Perry, Jr. and E. F. Perry, *J. Clin. Invest.*, *38*: 1452 (1959).
357. W. W. Westerfield, *Fed. Proc.*, *20*: 158 (1961).
358. K. D. Gibson, A. Neuberger, and J. C. Scott, *Biochem. J.*, *61*: 618 (1955).
359. V. Nigrovic, *Arch. Exp. Pathol. Pharmacol.*, *249*: 206 (1964).
360. S. S. Yang, Ph.D. Thesis, Univ. Massachusetts, Amherst, MA, 1952.
361. S. Shibata, *Nippon Yakurigaku Zasshi*, *52*: 113 (1956).
362. B. L. Oser, M. Oser, and H. C. Spencer, *Toxicol. Appl. Pharmacol.*, *5*: 142 (1963).
363. M. D. Reuber and G. C. Schmieller, *Arch. Environ. Health*, *5*: 430 (1962).
364. V. B. Braide, *Res. Vet. Sci.*, *20*: 295 (1976).
365. J. Z. Raymond and P. R. Gross, *Arch. Dermatol.*, *100*: 436 (1969).
366. K. Heindorff, O. Aurich, A. Michaelis, and R. Rieger, *Mutat. Res.*, *115*: 149 (1983).
367. J. E. McCann, E. Chio, E. Yamasaki, and B. N. Ames, *Proc. Natl. Acad. Sci. USA*, 72: 5135 (1975).
368. J. Lieberman and P. Ove, *J. Biol. Chem.*, *237*: 1634 (1962).

369. M. Fujioka and J. Lieberman, *J. Biol. Chem.*, *239*: 1164 (1964).
370. R. H. Alford, *J. Immunol.*, *104*: 698 (1970).
371. L. E. LaChance, *Rad. Res.*, *11*: 218 (1959).
372. M. Ondrej, *Induction Mutot. Mutat. Process Proc. Symp.*, Prague, Czechoslovakia, 1965, p. 26.
373. A. P. Baranauskajte, O. I. Vasiljauskajte, and V. P. Ranchyalis, *Lietuvos. TSR Mokslu. Akad. Darb. Ser. B.* 2: 107 (1972).
374. R. K. Das and G. K. Manna, *Proc. Indian Sci. Congr.*, *59*: 413 (1972).
375. V. R. Basrur and D. G. Baker, *Lancet*, *1*: 1106 (1963).
376. Muralidhara and K. Narasimhamurthy, *Food Cosmet. Toxicol.*, *29*: 845 (1991).
377. G. A. Leontjiva, Y. A. Mantzighin, and A. I. Gaziev, *Int. J. Radiat. Biol.*, *30*: 577 (1976).
378. M. A. Sognier and W. N. Hittelman, *Mutat. Res.*, *62*: 517 (1979).
379. W. J. Kleijer, J. L. Hoeksema, M. L. Sluyter, and D. Bootsma, *Mutat. Res.*, *17*: 385 (1973).
380. H. Tuchmann-Duplessis and L. Mercier-Parot, *Compt. Rend.*, *243*: 1064 (1956).
381. H. Swernerton and L. S. Hurley, *Science*, *173*: 62 (1971).
382. J. D. Lewis, *Acta Pharmacol. Toxicol.*, *41*: 144 (1977).
383. J. Gry and J. C. Larsen, *Arch. Toxicol.*, *36*: 351 (1978).
384. P. Finkle, *J. Biol. Chem.*, *100*: 349 (1933).
385. L. F. Chasseaud, W. H. Down, and D. Kirkpatric, *Experientia*, *33*: 998 (1977).
386. V. S. Chadwick, A. Vince, M. Killingley, and O. M. Wrong, *Clin. Sci. Mol. Med.*, *54*: 273 (1978).
387. A. Locke, R. B. Locke, H. Schlesinger, and H. Carr, *J. Amer. Pharm. Assoc.*, *31*: 12 (1942).
388. S. Krop and H. Gold, *J. Amer. Pharm. Assoc.*, *34*: 86 (1945).
389. E. W. Packman, D. D. Abbot, and J. W. E. Harrison, *Toxicol. Appl. Pharmacol.*, *5*: 163 (1963).
390. O. Bodansky, H. Gold, W. Zahm, H. Civin, and C. Salzman, *J. Amer. Pharm. Assoc.*, *31*: 1 (1942).
391. H. Gold and W. Zahm, *J. Amer. Pharm. Assoc.*, *32*: 173 (1943).
392. O. G. Fitzhugh and A. A. Nelson, *J. Amer. Pharm. Assoc.*, *36*: 217 (1947).
393. FDA, NAS/NRC Questionnaire on Tartaric Acid, Contract 71-75, Food and Drug Administration, Rockville, MD, 1973.
394. J. D. Dziezak, *Food Technol.*, *40*(9): 94 (1986).
395. C. M. Gruber and W. A. Halbeisen, *J. Pharmacol. Exp. Ther.*, *94*: 65 (1948).
396. *Martindale's Extra Pharmacopoeia*, 26th ed., The Pharmaceutical Press, London, England, 1972.
397. J. W. Cramer, E. I. Porrata-Doria, and H. Steenbock, *Arch. Biochim. Biophys.*, *60*: 58 (1956).
398. G. Gomori and E. Gulyas, *Proc. Soc. Exp. Biol. Med.*, *56*: 226 (1944).
399. H. Yokotani, T. Usui, T. Nakaguchi, T. Kanabayashi, M. Tanda, and Y. Aramaki, *J. Takeda Res. Lab.*, *30*: 25 (1971).
400. H. J. Horn, E. G. Holland, and L. W. Hazelton, *J. Agric. Food Chem.*, *5*: 759 (1957).
401. C. E. Calbert, S. M. Greenberg, G. Kryder, and H. J. Deuel, Jr., *Food Res.*, *16*: 294 (1951).

402. H. J. Deuel, Jr., S. M. Greenberg, C. E. Calbert, R. Baker, and H. R. Fisher, *Food Res.*, *16*: 258 (1951).
403. E. Graf, *J. Amer. Oil Chem. Soc.*, *60*: 1861 (1983).
404. E. Graf, J. R. Mahoney, R. G. Bryant, and J. W. Eaton, *J. Biol. Chem.*, *259*: 3620 (1984).
405. E. Graf, K. L. Empson, and J. W. Eaton, *J. Biol. Chem.*, *262*: 11647 (1987).
406. *Federal Register*, *47*: 27806 (1982).
407. D. J. Cosgrove, *Rev. Pure Appl. Chem.*, *16*: 209 (1966).
408. D. J. Bitar and J. G. Reinhold, *Biochim. Biophys. Acta*, *268*: 442 (1972).
409. V. Subrahmanyan, M. Narayana Rao, G. Rama Rao, and M. Swaminathan, *Bull. Central Food Technol. Res. Inst.*, *4*: 87 (1955).
410. D. M. Hegsted, M. F. Trulson, and F. J. Stare, *Physiol. Rev.*, *34*: 221 (1954).
411. T. S. Nelson and L. K. Kirby, *Nutr. Rep. Int.*, *20*: 729 (1979).
412. K. Okubo, D. Myers, and G. A. Iacobucci, *Cereal Chem.*, *53*: 513 (1976).
413. M. Singh and A. D. Krikorian, *J. Agric. Food Chem.*, *30*: 799 (1982).
414. R. E. Benesch and R. Benesch, *Adv. Prot. Chem.*, *28*: 211 (1974).
415. R. E. Isaacks, D. R. Harkness, P. H. Goldman, J. L. Adler, and C. Y. Kim, *Hemoglobin*, *1*: 577 (1977).
416. N. R. Reddy, S. K. Sathe, and D. K. Salunkhe, *Adv. Food Res.*, *28*: 1 (1982).
417. M. Cheryan, *CRC Crit. Rev. Food Sci. Nutr.*, *13*: 297 (1980).
418. J. W. Erdman, *J. Amer. Oil Chem. Soc.*, *56*: 736 (1979).
419. H. Ichikawa, S. Ohishi, O. Takahashi, H. Kobayashi, K. Yuzawa, N. Hoshikawa, and T. Hashimoto, *Annu. Rep. Tokyo Metrop. Res. Lab. Public Health*, *38*: 371 (1987).
420. Y. Hiasa, Y. Kitahori, J. Morimoto, N. Konishi, S. Nakakoa, and H. Nishioka, *Food Chem. Toxicol.*, *30*: 117 (1992).
421. A. Ogata, H. Ando, Y. Kubo, M. Sasaki, and N. Hosokawa, *Annu. Rep. Tokyo Metrop. Res. Lab. Public Health*, *38*: 377 (1987).
422. J. Loliger, in *Free Radicals and Food Additives* (O. I. Arouma and B. Halliwell, Eds.), Taylor and Francis, London, England, 1991, p. 121.
423. B. J. F. Hudson and M. Ghavami, *Lebensm. Wiss. Technol.*, *17*: 191 (1984).
424. A. L. Tappel, *Geriatrics*, *23*(10): 97 (1968).
425. S. Z. Dziedzic and B. J. F. Hudson, *J. Amer. Oil Chem. Soc.*, *61*: 1042 (1984).
426. O. Schuberth and A. Wretlind, *Acta Chir. Scand. Suppl.*, *278*: 1 (1961).
427. J. M. Merrill, *J. Amer. Med. Assoc.*, *170*: 2202 (1959).
428. A. W. Schwab, H. A. Moser, R. S. Curley, and C. D. Evans, *J. Amer. Oil Chem. Soc.*, *30*: 413 (1953).
429. C. Karahadian and R. C. Lindsay, *J. Amer. Oil Chem. Soc.*, *65*: 1159 (1988).
430. R. C. Reynolds, B. D. Astill, and D. W. Fassett, *Toxicol. Appl. Pharmacol.*, *28*: 133 (1974).
431. W. A. Waters, *J. Amer. Oil Chem. Soc.*, *48*: 427 (1971).
432. D. S. MacDonald, J. I. Gray, and L. N. Gibbins, *J. Food Sci.*, *45*: 893 (1980).
433. P. A. Morrissey and J. Z. Tichivangana, *Meat Sci.*, *14*: 175 (1985).
434. J. Kanner, I. Ben-Gera, and S. Berman, *Lipids*, *15*: 944 (1980).
435. P. N. Magee, *Food Cosmet. Toxicol.*, *9*: 207 (1971).
436. W. B. Bradley, H. T. Eppson, and O. A. Beath, *J. Amer. Vet. Med. Assoc.*, *47*: 541 (1939).

437. W. Kuebler, Z. *Kinderheilk.*, *81*: 405 (1958).
438. S. S. Mirvish, *J. Natl. Cancer. Inst.*, *71*: 630 (1983).
439. B. Spiegelhalder, G. Eisenbrand, and R. Preussmann, *Food Cosmet. Toxicol.*, *14*: 545 (1976).
440. J. B. Wyngaarden, J. B. Stanbury, and B. Rapp, *Endocrinology*, *52*: 568 (1953).
441. R. L. Reid, G. A. Jung, R. Weiss, A. J. Post, F. P. Horn, E. B. Kahle, and C. E. Carlson, *J. Animal Sci.*, *29*: 181 (1969).
442. R. J. Emerick and O. E. Olson, *J. Nutr.*, *78*: 73 (1962).
443. A. J. Lehman, *Quart. Bull. Assoc. Food Drug Off.*, *22*: 136 (1958).
444. J. Sander and F. Seif, *Arzneim. Forsch.*, *19*: 1091 (1969).
445. N. P. Sen, D. C. Smith, and L. Schwinghamer, *Food Cosmet. Toxicol.*, *7*: 301 (1969).
446. M. Greenberg, W. B. Birnkrant, and J. J. Schiftner, *J. Amer. Public Health*, *35*: 1217 (1945).
447. N. Gruener and H. I. Shuval, *Environ. Qual. Safety*, *2*: 219 (1973).
448. J. Musil, *Acta Biol. Med. Ger.*, *16*: 380 (1966).
449. A. B. Rosenfield and R. Huston, *Minn. Med.*, *33*: 787 (1950).
450. M. J. Van Logten, E. M. den Tonkelaar, R. Kroes, J. M. Berkvens, and G. J. Van Esch, *Food Cosmet. Toxicol.*, *10*: 475 (1972).
451. FDA, Department of Health and Human Services, Washington, DC, 1980.
452. R. C. Shank and P. M. Newberne, *Food Chem. Toxicol.*, *14*: 1 (1976).
453. M. Greenblatt, S. Mirvish, and B. T. So, *J. Natl. Cancer Inst.*, *46*: 1029 (1971).
454. S. D. Sleight and O. A. Atallah, *Toxicol. Appl. Pharmacol.*, *12*: 179 (1968).
455. A. J. Winter and J. F. Hokanson, *Amer. J. Vet. Res.*, *25*: 353 (1964).
456. S. J. Bishov and A. S. Henick, *J. Food Sci.*, *40*: 345 (1975).
457. R. J. Marcuse, *J. Amer. Oil Chem. Soc.*, *39*: 97 (1962).
458. M. J. Taylor, T. Richardson, and R. D. Jasensky, *J. Amer. Oil Chem. Soc.*, *58*: 622 (1981).
459. R. G. Daniel and H. A. Waisman, *Growth*, *32*: 255 (1968).
460. Anonymous, *Chem. Eng. News*, *48*(22): 27 (1971).
461. J. S. Pruthi, *Spices and Condiments: Chemistry, Microbiology, and Technology*, Academic Press, London, England, 1980.
462. J. R. Chipault, G. R. Mizuna, J. M. Hawkins, and W. O. Lundberg, *Food Res.*, *17*: 46 (1952).
463. J. Burri, M. Graf, P. Lambelet, and J. Loliger, *J. Sci. Food Agric.*, *48*: 49 (1989).
464. W. M. Cort, *Food Technol.*, *28*(10): 60 (1974).
465. S. S. Chang, B. Ostric-Matijaseric, O. A. L. Hsieh, and C. Huang, *J. Food Sci.*, *42*: 1102 (1977).
466. U. Bracco, J. Loliger, and J. L. Viret, *J. Amer. Oil Chem. Soc.*, *58*: 686 (1981).
467. M. C. Houlihan, C. H. Ho, and S. S. Chang, *J. Amer. Oil Chem. Soc.*, *62*: 96 (1985).
468. K. W. Singletary and J. M. Nelshoppen, *Cancer Lett.*, *60*: 169 (1991).
469. J. Kuhnau, *World Rev. Nutr. Diet.*, *24*: 117 (1976).
470. D. E. Pratt and B. J. F. Hudson, in *Food Antioxidants* (B. J. F. Hudson, Ed.), Elsevier Applied Science, London, 1990, p. 171.
471. R. Gugler, M. Leschik, and H. J. Dengler, *Eur. J. Clin. Pharmacol.*, *9*: 229 (1975).
472. I. Ueno, N. Nakano, and I. Hirono, *Jpn. J. Exp. Med.*, *53*: 41 (1983).
473. L. A. Griffiths and G. E. Smith, *Biochem. J.*, *128*: 901 (1972).

474. F. J. Simpson, J. A. Jones, and E. A. Wolin, *Can. J. Microbiol.*, *15*: 972 (1969).
475. I. Hirono, I. Ueno, S. Hosaka, H. Takanashi, T. Matsushima, T. Sugimura, and S. Natori, *Cancer Lett.*, *13*: 15 (1981).
476. K. Morino, N. Matsukura, T. Kawachi, H. Ohgaki, T. Sugimura, and I. Hirono, *Carcinogenesis*, *3*: 93 (1982).
477. D. Saito, A. Shirai, T. Matsushima, T. Sugimura, and I. Hirono, *Teratogen Carcinogen Mutagen*, *1*: 213 (1980).
478. S. Hosaka and I. Hirono, *Gann*, *72*: 327 (1981).
479. A. M. Pamukçu, S. Yalciner, J. F. Hatcher, and G. T. Bryan, *Cancer Res.*, *40*: 3468 (1980).
480. R. K. Sahu, R. Basu, and A. Sharma, *Mutat. Res.*, *89*: 69 (1981).
481. J. T. MacGregor, C. M. Wehr, G. D. Manners, L. Jurd, J. K. Minkler, and A. V. Carrano, *Mutat. Res.*, *124*: 255 (1983).
482. S. M. Ahmad, F. Fazal, A. Rahman, S. M. Hadi, and J. H. Parish, *Carcinogenesis*, *13*: 605 (1992).
483. FAO/WHO, WHO Tech. Rep. Ser. No. 362, World Health Organization, Geneva, Switzerland, 1967.
484. D. S. Goodman, H. S. Huang, and T. Shiratori, *J. Lipid Res.*, *6*: 390 (1965).
485. D. S. Goodman, R. Blomstrand, B. Werner, H. S. Huang, and T. Shiratori, *J. Clin. Invest.*, *45*: 1615 (1966).
486. C. A. Frolik, in *The Retinoids*, Vol. 2 (B. M. Sporn, A. B. Roberts, and D. S. Goodman, Eds.), Academic Press, London, England, 1984, p. 177.
487. J. J. Kamm, K. O. Ashenfelter, and C. W. Ehmann, in *The Retinoids*, Vol. 2 (B. M. Sporn, A. B. Roberts, and D. S. Goodman, Eds.), Academic Press, London, England, 1984, p. 287.
488. J. J. Kamm, *J. Amer. Acad. Dermatol.*, *6*: 652 (1982).
489. C. L. Maddock, S. B. Wolbach, and S. Maddock, *J. Nutr.*, *39*: 117 (1949).
490. A. A. Seawright, P. B. English, and R. J. W. Gartner, *J. Comp. Pathol.*, *77*: 29 (1967).
491. R. M. Grey, S. W. Nielsen, J. E. Rousseau, Jr., M. C. Calhoun, and H. D. Eaton, *Pathol. Vet.*, *2*: 446 (1965).
492. R. E. Wolke, S. W. Nielsen, and J. E. Rousseau, Jr., *Amer. J. Vet. Res.*, *29*: 1009 (1968).
493. C. C. Huang, J. L. Huseh, H. H. Chen, and T. R. Batt, *Carcinogenesis*, *3*: 1 (1982).
494. R. J. Gellert, *J. Reprod. Fertil.*, *50*: 223 (1977).
495. T. L. Lamano-Carvalho, R. A. Lopes, R. Azoubel, and A. L. Ferreira, *Int. J. Vit. Nutr. Res.*, *48*: 307 (1978).
496. R. E. Shenefelt, *Amer. J. Pathol.*, *66*: 589 (1972).
497. F. W. Rosa, A. L. Wilk, and F. O. Kelsey, *Teratology*, *33*: 355 (1986).
498. J. A. Geelen, *Crit. Rev. Toxicol.*, *6*: 351 (1979).
499. J. A. Toomey and R. A. Morissette, *Amer. J. Dis. Child*, *73*: 473 (1947).
500. G. M. Pease, *J. Amer. Med. Assoc.*, *182*: 980 (1962).
501. W. F. Koerner and J. Voellm, *Int. J. Vit. Nutr. Res.*, *45*: 363 (1975).
502. C. S. Foote and R. W. Denny, *J. Amer. Oil Chem. Soc.*, *90*: 6233 (1968).
503. J. Terao, *Lipids*, *24*: 659 (1989).
504. M. H. Gordon, in *Food Antioxidants* (B. J. F. Hudson, Ed.), Elsevier Applied Science, London, England, 1990, p. 1.
505. W. Kuebler, *Wiss. Veroeffentl. Deut. Ges. Ernahrung*, *9*: 222 (1963).

506. D. S. Goodman and H. S. Huang, *Science, 149*: 879 (1965).
507. J. A. Olson, *J. Biol. Chem.*, *236*: 349 (1961).
508. H. S. Huang and D. S. Goodman, *J. Biol. Chem.*, *240*: 2839 (1965).
509. R. D. Zachman and J. A. Olson, *J. Biol. Chem.*, *236*: 2309 (1961).
510. W. S. Blaner and J. E. Churchich, *Biochem. Biophys. Res. Commun.*, *94*: 820 (1980).
511. C. Nieman and H. J. Klein Obbink, *Vit. Horm.*, *12*: 69 (1954).
512. J. Zbinden and A. Studer, *Z. Lebensm. Unters. Forsch.*, *108*: 113 (1958).
513. R. E. Bagdon, G. Zbinden, and A. Studor, *Toxicol. Appl. Pharmacol.*, *2*: 225 (1960).
514. R. Greenberg, T. Cornbleet, and A. I. Jeffny, *J. Invest. Dermatol.*, *32*: 599 (1959).
515. M. H. Lee and R. L. Sher, *J. Chin. Agric. Chem. Soc.*, *22*: 226 (1984).
516. J. Mai, L. J. Chambers, and R. E. McDonald, U. K. Patent GB 2151123A (1985).
517. M. H. Lee, R. L. Sher, C. H. Sheu, and Y. C. Tsai, *J. Chin. Agric. Chem. Soc.*, *22*: 128 (1984).
518. A. W. Girotti, J. P. Thomas, and J. E. Jordon, *J. Free Radiat. Biol. Med.*, *1*: 395 (1985).
519. National Research Council, *Recommended Dietary Allowances*, 9th ed., National Academy of Sciences, Washington, DC.
520. J. R. McKenney, R. O. McClennan, and L. K. Bustad, *Health Physiol.*, *8*: 411 (1962).
521. K. T. Smith, M. L. Failla, and R. J. Cousins, *Biochem. J.*, *184*: 627 (1979).
522. R. L. Aamodt, W. F. Rumble, G. S. Johnston, D. Foster, and R. I. Henkin, *Amer. J. Clin. Nutr.*, *32*: 559 (1979).
523. K. Michael Hambidge, C. E. Casey, and N. F. Krebs, in *Trace Elements in Human and Animal Nutrition* (W. Mertz, Ed.), Academic Press, Orlando, Florida, 1986, p. 1.
524. G. J. Fosmire, *Amer. J. Clin. Nutr.*, *51*: 225 (1990).
525. O. A. Levander, in *Trace Elements in Human and Animal Nutrition* (W. Mertz, Ed.), Academic Press, Orlando, Florida, 1986, p. 208.
526. R. A. Sunde, *J. Amer. Oil Chem. Soc.*, *61*: 1891 (1984).
527. A. Ochoa-Solano and C. Gitler, *J. Nutr.*, *94*: 243 (1968).
528. M. I. Smith, E. F. Stohlman, and R. D. Lille, *J. Pharmacol. Exp. Therap.*, *60*: 449 (1937).
529. A. L. Moxon and M. Rhian, *Physiol. Rev.*, *23*: 305 (1943).
530. D. F. Birt, A. D. Julins, and P. M. Pour, *Proc. Amer. Assoc. Cancer Res.*, *25*: 133
531. WHO, *Environmental Health Criteria 58: Selenium*, World Health Organization, Geneva, Switzerland, 1987, p. 306.
532. J. Parizek, *Food Chem. Toxicol.*, *28*: 763 (1990).
533. S. Al-Malaika, in *Free Radicals and Food Additives* (O. I. Arouma and B. Halliwell, Eds.), Taylor and Francis, London, England, 1991, p. 151.
534. A. D. Schwope, D. E. Till, D. J. Ehntholt, K. R. Sidman, R. H. Whelan, P. S. Schwartz, and R. C. Reid, *Food Chem. Toxicol.*, *25*: 327 (1987).
535. S. S. Chang, G. A. Senich, and L. E. Smith, Final Report, National Bureau of Standards, NBSIR 82-2472, 1982.
536. A. D. Schwope, D. E. Till, D. J. Ehntholt, K. R. Sidman, R. H. Whelan, P. S. Schwartz, and R. C. Reid, *Food Chem. Toxicol.*, *25*: 317 (1987).
537. K. Figge, *Food Cosmet. Toxicol.*, *10*: 815 (1972).
538. D. W. Allen, D. A. Leathard, C. Smith, and J. D. McGuinness, *Food Addit. Contam.*, *5*: 433 (1988).

6

Nutritional and Health Aspects of Food Antioxidants

S. S. Deshpande
Idetek, Inc., Sunnyvale, California

U. S. Deshpande
Cor Therapeutics, South San Francisco, California

D. K. Salunkhe
Utah State University, Logan, Utah

6.1 INTRODUCTION

During the past two decades, global epidemiological studies and multidisciplinary research approaches have been targeted at uncovering the causes and modifying factors associated with the etiology of several important diseases afflicting humankind. These surveys have clearly demonstrated that many of the chronic diseases that affect humans have an uneven geographic distribution (1,2). Although the general perception that several diseases, especially the various types of cancers, often result from an exposure to toxic environmental and agricultural chemicals, pesticides, herbicides, fungicides, or even some food additives is at least partially true, many dietary surveys attribute the incidence of such diseases to locally prevailing nutritional traditions (Table 6.1). Thus, the high incidence of coronary heart disease and cancers of the breast, prostate, pancreas, colon, ovary, and endometrium in the western world is often correlated with high-fat, high-cholesterol, low-fiber diets and the consumption of fried foods, whereas increased risks of hypertension and stroke and cancers of the stomach and esophagus in the Far East are linked to the consumption of salted, pickled, and smoked foods. Similarly, smoking is a worldwide recognized cause of heart disease and of cancers of the lung, pancreas, kidney, urinary bladder, and cervix (1–6). Such uneven geographical distributions of these chronic diseases, therefore, indicate a strong association between prevailing lifestyles, local dietary patterns and traditions, and habits.

361

Table 6.1 Geographic Pathology of Cancer and Nutrition Traditions

Organ	Lower risk population	Lower risk factors	Higher risk population	Higher risk factors
Colon	Japan	Low-fat diet	North America, western Europe, New Zealand, Australia, southern Scandinavia	High-fat, high-cholesterol, low-fiber diets; fried foods
	Mormons	High-fiber diet		
	7th day Adventists	Low or no fried food, high-fiber diet		
	Finland	High-fiber diet, lower intake of fried food		
Breast	Japan	Low-fat diet	North America, western Europe. New Zealand, Australia	High-fat, low-fiber diet
Endometrium	Japan	Low-fat diet	USA (California), Europe	High-fat diet, obesity, estrogen use
Prostate	Japan	Low-fat diet	USA, Scandinavia, western Europe	High-fat diet
Pancreas	Bombay, India		USA (California, blacks)	Early smoking habit; Western-style high-fat, high-cholesterol diet
Stomach	USA	Low pickled; salted foods, high fruits, salads, vitamins C, E	Japan, China, Chile, Colombia, eastern Europe	Salted, pickled foods, geochemical nitrate, and low intake of vitamins A, C, and E
Esophagus	Utah (USA), rural Norway	Low alcohol and smoking habits	France (Calvados, Normandy) USA (lower socioeconomic groups)	Alcohol and smoking
			India	Tobacco chewing
			Eastern Iran, southern Russia, central China	Low vitamins A, C, and E; salted, pickled foods?

Source: Ref. 1.

Epidemiological studies linking the prevalence of certain diseases to dietary patterns often tend to show an inverse correlation between the consumption of foods, particularly leafy green and yellow vegetables and fruits and certain diseases. Nevertheless, such studies on the role of specific nutrients in these foods and their direct association in the etiology of certain diseases are hampered by at least three possible sources of error (7): (i) error in the classification of individuals with respect to their nutrient intake, (ii) errors of interpretation arising from the fact that nutrients are correlated both negatively and positively with other nutrients, and (iii) uncertainty about the ability of studies within populations that are quite homogeneous with respect to intake of a nutrient to detect an effect of high or low intake. To overcome these sources of error, several studies by independent groups are required to establish a linkage between the intake of a particular nutrient and a beneficial health outcome. Only when a majority of the studies support the hypothesis of a health effect from a nutrient can the hypothesis be taken seriously. Furthermore, in complex biological systems, nutrients seldom act in isolation. Both synergistic and antagonist reactions are not uncommon in cellular metabolism. A more definitive proof, therefore, must await the elucidation of the protective mechanisms by which a given nutrient prevents the deleterious changes in the biochemical pathways implicated in the etiology of various diseases.

Epidemiological dietary surveys have nonetheless enabled us to establish a sound relationship between dietary nutrients and certain diseases to the point that preliminary recommendations to lower the risk for each of these diseases can be made. For example, a decrease in the calories derived from fat intake lowers the risk associated with several types of cancers including postmenopausal breast, distal colon, pancreas, prostate, ovary, and endometrium cancers. Similarly, micronutrients such as carotenoids and folic acid that are present in yellow and green vegetables appear to contribute to a lower risk of several types of diseases. Thus current research on the prevention of chronic diseases through nutrition suggests specific adjustments in the traditional intake of macronutrients such as fat, protein, complex carbohydrates, and dietary fiber. The recent trends in this regard, as defined by Weisburger (2), are summarized in Table 6.2.

Parallel to these developments in the field of nutritional epidemiology are the increasing efforts being made to understand the importance of preventive medicine and its role in human well-being. Such efforts may be directly attributed to the rising health care costs worldwide. These developments have been further accelerated by our realization of the importance and significance of biological oxidation reactions in the clinical field. In fact, free oxygen radical pathology offers a revolutionary example of this approach, because it is increasingly evident that several disease conditions are caused by or related to this biochemical phenomenon (Table 6.3).

The primary biological role of antioxidants is in preventing the damage that reactive free radicals can cause to cells and cellular components. In fact, almost all

Table 6.2 New Approach to Improved, Balanced
Nutrition to Promote Health and Prevent Disease

Old concept
  Recommended dietary allowance (RDA)
  Avoidance of deficiency diseases
New concept
  Optimal nutrition
  Avoidance of chronic diseases and protection against
    environmental toxicants
  Optimal levels of macronutrients
  Optimal levels of micronutrients

Source: Ref. 2.

Table 6.3 Examples of Some Disorders and Diseases
Associated With Free-Radical Pathology

Cancers
Coronary heart disease/atherosclerosis
Diabetes
Cataract
Adverse drug reactions
Toxic liver injuries
  CCl_4 and other halogenoalkanes
  Bromobenzene
  Allyl alcohol
  Iron overload
  Paracetamol
  Alcohol
Redox cycling mechanisms
  Quinones
  Nitroimidazoles
Arthritis
Immune hypersensitivity
Inflammatory disorders
Reperfusion injuries
  Thrombosis
  Organ storage
  Transplantation
Neurological degeneration
Aging
Traumatic inflammation

Source: Compiled from Refs. 8–10.

the nutrient constituents identified in the dietary surveys thus far as having a protective effect against specific diseases seem to have some kind of antioxidant properties. In this chapter, recent developments in the nutritional and health aspects of food antioxidants are reviewed. The information is presented under three broad subject categories. The first section deals with free-radical chemistry and its importance in cellular metabolism. This is followed by information on the natural defense mechanisms against free-radical-induced toxicity as well as the biological properties and significance of antioxidant nutrients. Finally, the preventive role of various antioxidants in the etiology of several important diseases is described.

6.2   FREE-RADICAL CHEMISTRY

To appreciate the importance of antioxidant defenses, it is essential to understand how free radicals are formed and subsequently damage the cellular components in our body. This process occurs routinely in the cellular metabolism.

The term *free radical* (commonly, just *radical*) is generically used in the field of chemistry to refer to various groups of atoms that behave as a unit, for example, the carbonate radical (CO_3^{2-}), nitrate radical (NO_3^-), and methyl radical (CH_3^-). More specifically, a free radical is "any species capable of independent existence that contains one or more unpaired electrons" (11). An unpaired electron is one that occupies an atomic or molecular orbital by itself. The presence of one or more unpaired electrons causes the species to be paramagnetic and sometimes makes it highly reactive.

Because electrons tend to occur in pairs such that their spins are antiparallel to each other, the pair per se has no net spin. This is energetically most favorable, and therefore almost all chemical bonds tend to have two electrons. Free radicals can be generated by the loss of a single electron from a nonradical or by the gain of a single electron by a nonradical. The two-electron bonds can break in two ways: symmetrically by homolytic fission or asymmetrically by heterolytic fission. In homolytic fission, a covalent bond is broken such that one electron (indicated by a dot in the following equations) from each member of the shared pair remains with each atom as shown below:

$$R-H \rightarrow R^{\bullet} + H^{\bullet}$$

In contrast, when bonds are broken heterolytically, only one atom receives both electrons:

$$R-H \rightarrow R:^- + H^+$$

Only chemical species formed by the homolytic processes are called *free radicals*, because they have an unpaired electron. Because electrons are more stable when paired together in orbitals, radicals are in general more reactive than nonradicals. Their reactivity, however, varies greatly depending on the chemical species.

Free radicals can react with other molecules in a number of ways (12,13). Thus two free radicals can share their unpaired electrons and join to form a covalent bond:

$$R^\bullet + R^\bullet \rightarrow R\text{---}R$$

Similarly, a radical might donate its unpaired electron to another molecule (a reducing radical), it might take an electron from another molecule to form a pair (an oxidizing radical), or it might add on to a nonradical species. However, if a radical gives one electron to, takes one electron from, or adds on to a nonradical, that nonradical becomes a free radical. Thus, a general feature of the reactions of free radicals is that they tend to proceed as chain reactions (13).

6.2.1 Causes for the Formation of Free Radicals

In biological systems, free radicals are generated through the xenobiotic metabolism of drugs and chemicals, irradiation including ionizing radiation and light, transition metal one-electron oxidations catalyzed by minerals such as iron and copper, and the secretion of oxidants from inflammatory leukocytes and other cell-mediated immunodefensive processes (10,14,15). Uncatalyzed and both non-enzymatic and enzymatic catalysis of bond homolysis also generates free radicals in biological systems.

Ionizing radiation is well known to produce free radicals in living systems. Similarly, light (especially the short-wavelength UV rays) can either excite molecules that transfer their energy to oxygen or directly excite oxygen to produce singlet oxygen. According to Pryor (15), light-induced free-radical reactions are especially deleterious to the human eye.

Environmental pollutants containing free radical themselves or toxins that generate free radicals also play an important role. In fact, radical production by such xenobiotics appears to be the most important mechanism in vivo that ultimately causes pathological damage. Xenobiotics can produce free radicals by three distinct mechanisms (15): (i) Some are themselves free radicals and consequently react directly to form biopolymer radicals; (ii) some, although not free radicals themselves, are so reactive that they cause free radicals to form in the cell; and (iii) some require enzymatic activation to produce free radicals.

Nitrogen dioxide (NO_2) and nitrogen oxide (NO) in polluted air, cigarette smoke, smog and soot in urban air, and automobile exhaust are important examples of toxins that are themselves free radicals. Nitrogen dioxide in smoggy air is known to react with polyunsaturated fatty acids (PUFAs) both in vitro and in vivo to cause the production of free radicals (16,17). In addition, toxic wastes, pesticides, and herbicides can all generate free radicals in biological systems.

Ozone is yet another example of a highly reactive environmental pollutant. Although not a free radical itself, ozone can readily react with biological materials to produce free radicals. In contrast, some xenobiotics require enzymatic activation to produce free radicals. Typical of this group is carbon tetrachloride, a toxin that is metabolized at a cytochrome P-450 site that ultimately leads to the formation of the trichloromethyl free radical (18). Some of the quinone drugs used in cancer therapy also are reduced enzymatically to produce free radicals (15). For quinoid species such as paraquat and adriamycin, a redox cycle exists in which superoxide is produced as follows (Q represents a generalized quinone):

$$Q + e^- \rightarrow Q^{\bullet -}$$
$$Q^{\bullet -} + O_2 \rightarrow Q + O_2^{\bullet -}$$

If oxygen is not available, a more complex cycle occurs in which the semiquinone radical-ion itself reacts with hydrogen peroxide (H_2O_2) or an iron-catalyzed cycle occurs:

$$Q^{\bullet -} + H_2O_2 \rightarrow Q + HO^- + HO^{\bullet}$$
$$Q^{\bullet -} + Fe^{3+} \rightarrow Q + Fe^{2+}$$
$$Fe^{2+} + H_2O_2 \rightarrow Fe^{3+} + HO^- + HO^{\bullet}$$

Uncatalyzed bond homolysis reactions leading to the formation of free radicals are not very common in biological systems. This is especially true for the spontaneous but rather slow unimolecular homolysis of lipid hydroperoxides. The bond dissociation energy of RO—OH is 42 kcal/mol and is independent of the nature of the R group (19). This bond is so strong that uncatalyzed homolysis would not reach appreciable rates until temperatures near 110°C (20). In contrast, the enzymatically catalyzed production of superoxide and its subsequent reactions are an important source of free radicals in vivo (15). Lipid peroxidation reactions, as described in Chapter 2, belong to this class.

The human body also produces free radicals during the course of its normal metabolism. They are also required for several normal biochemical processes. Although oxygen is essential in cellular energy-generating processes, it may also give rise to harmful free radicals, including superoxide and hydroxyl radicals. Even the phagocyte cells involved in the body's natural immune defenses generate free radicals in the process of destroying bacteria, fungi, viruses, and other antigenic substances (21).

Free radicals may also be produced as normal by-products of the healthy body's response to various stress elements. Aging, trauma, certain medications, and some diseases can all lead to an increase in the formation of free radicals. Some examples of biochemical systems that may produce $O_2^{\bullet -}$ in vivo are summarized in Table 6.4.

6.2.2 Mechanisms of Free-Radical Formation

Humans, being oxygen-dependent organisms, are always at risk of free-radical damage. The body naturally produces free radicals as it metabolizes oxygen. In biological systems, with the exception of xenobiotic metabolism, oxygen and its

Table 6.4 Biochemical Systems that Produce $O_2^{\bullet-}$ Radical In Vivo

Source of $O_2^{\bullet-}$	Examples
Small molecules	Hydroquinones, leukoflavins, adrenalin, thiols (by oxidation)
Enzymes and proteins	Xanthine oxidase (degradation of purines to nitrogenous excretory products), aldehyde oxidase, peroxidase (oxidative reaction), hemoglobin, myoglobin, iron-sulfur proteins
Cell organelles	Chloroplasts, mitochondria, peroxisomes, glyoxysomes
Whole cells	Polymorphonuclear leukocytes (during the oxygen burst that leads to the destruction of invading bacteria and viruses)

Source: Refs. 63 and 72.

derivatives, transition metals, and the oxidation products of PUFAs primarily contribute to various types of free radicals (11,13,22). The various mechanisms by which free radicals are formed are briefly described below.

Oxygen and Its Derivatives

The oxygen molecule as it occurs naturally is a free radical, as it contains two unpaired electrons located in different π^* antibonding orbitals and having the same parallel spins. If a single electron is added to the ground-state oxygen molecule, it must then enter one of the π^* antibonding orbitals, thus generating the superoxide radical $O_2^{\bullet-}$. Addition of one more electron to the superoxide radical then yields O_2^{2-}, the peroxide anion. Because the atoms in ground-state oxygen are effectively bonded with two covalent bonds, with one and a half in the superoxide radical and with one bond only in the peroxide anion the single oxygen–oxygen bond in O_2^{2-} is quite weak. The addition of two more electrons to the peroxide ion would eliminate the bond entirely, thus yielding 2 O^{2-} species. In biological systems, the two-electron reduction product of oxygen is hydrogen peroxide (H_2O_2) and the four-electron product is water.

Summarizing these reactions below, the reduction of molecular oxygen to water proceeds through the formation of superoxide radical to peroxide anion, the electrons being acquired from either the electron transport chain in the mitochondria or from cytochrome P-450 in the microsomal system.

$$O_2 \xrightarrow{\ e^- \text{ reduction}\ } O_2^{\bullet-} \text{ (superoxide radical)}$$

$$O_2 \xrightarrow{\ 2\,e^- \text{ reduction}\ } O_2^{2-} \text{ (peroxide anion)}$$

$$O_2^{2-} + 2\,H^+ \xrightarrow{\hspace{2cm}} H_2O_2 \text{ (protonated form of peroxide)}$$

$$O_2 + 4\,H^+ \xrightarrow{\ 4\,e^- \text{ reduction}\ } 2\,H_2O \text{ (protonated form of } O_2^{\bullet-}\text{)}$$

To understand the potential detrimental effect of oxygen radicals in biological systems, one must appreciate the need to remove both $O_2^{\bullet-}$ and H_2O_2 from the system as soon as they are formed. This is because divalent cations such as iron and copper may cause catalyzed interactions of these two metabolites, with the formation of a further free radical, the hydroxyl radical ($HO^{\bullet}$). These are known as the Fenton reactions. For example, ·

$$O_2^{\bullet-} + Fe^{3+} \rightarrow O_2 + Fe^{2+}$$
$$H_2O_2 + Fe^{2+} \rightarrow OH^- + HO^{\bullet} + Fe^{3+}$$

The hydroxyl radical is the most reactive radical known, attacking and damaging almost every molecule found in living cells (11,13,22). The reactions of the hydroxyl radical can be classified into three main types: hydrogen abstraction (e.g., reactions with alcohols and phospholipids), addition (e.g., reactions with aromatic ring structures such as with the purine and pyrimidine bases present in DNA and RNA), and electron transfer (e.g., reactions involving organic and inorganic compounds). Radicals produced by reactions with $HO^{\bullet}$ are generally less reactive.

The high reactivity of the hydroxyl radical is seen in its having a very short half-life in biological systems. Generally, a hydroxyl radical does not persist for even a microsecond before combining with a molecule in its immediate vicinity. However, the free-radical chain reactions it triggers might cause extensive biological damage far from the site of their production.

Singlet oxygen is yet another reactive metabolite of oxygen. It is formed as an excited state of oxygen by energy capture.

$$O_2 \xrightarrow{\;h\upsilon\;} {}^1O_2$$

Although not a free radical, singlet oxygen contains a peripheral electron in the structure that is excited to an orbital above that which it normally occupies. According to Diplock (22), singlet oxygen may be of significance in certain biochemical processes such as the microbicidal activity of polymorphs as well as in causing tissue damage.

Transition Metals

With the sole exception of zinc, all the metals in the first row of the d block in the Periodic Table contain unpaired electrons and can thus qualify as free radicals (11). Transition metal ions can stimulate lipid peroxidation by two mechanisms: by participating in the generation of initiating species such as hydroxyl radicals and by accelerating peroxidation by decomposing lipid hydroperoxides into peroxyl and alkoxyl radicals, which can themselves abstract hydrogen and thus perpetuate the chain reaction of lipid peroxidation. The lipid oxidation reactions were described in detail in Chapter 2. It must be mentioned that many transition elements are of great biological importance and are essential micronutrients in the human diet.

Miscellaneous

Reactions involving sulfur-, carbon-, and nitrogen-centered radicals have been reviewed by Halliwell and Gutteridge (11). Thiol (R—SH) compounds oxidize in the presence of transition metal ions to form, among other products, thiyl radicals, RS˙:

$$R—SH + Cu^{2+} \rightarrow RS˙ + Cu^+ + H^+$$

These sulfur-centered radicals have considerable reactivity and can easily combine with oxygen, NADH, and ascorbic acid, for example,

$$RS˙ + O_2 \rightarrow RSO_2˙$$

and

$$RS˙ + NADH \rightarrow RS^- + NAD˙ + H^+$$

Thiyl radicals can also be formed by the homolytic fission of the disulfide bonds in proteins:

$$Cys-S—S-Cys \rightarrow Cys-S˙ + S˙-Cys$$

The toxic actions of sporidesmin and diphenyl disulfide are believed to involve thiyl radicals.

Carbon-centered radicals are also formed in many biological systems such as during the metabolism of carbon tetrachloride (CCl_4) by liver microsomes, yielding a trichloromethyl radical:

$$CCl_4 \xrightarrow{\text{Cytochrome P-450 system}} ˙CCl_3 + Cl^-$$

The trichloromethyl radical often reacts rapidly with oxygen to give the corresponding peroxyl radicals:

$$˙CCl_3 + O_2 \rightarrow ˙O_2CCl_3$$

Nitrogen-centered radicals, such as the phenyldiazine radical ($C_6H_5N{=}N˙$), is formed during oxidation of phenylhydrazine by erythrocytes.

6.2.3 Cellular Sites of Free-Radical Generation

As mentioned earlier, free radicals can be formed in cells during normal metabolic processes following exposure to ionizing radiation and after exposure to drugs or xenobiotics. The latter two can be metabolized to free radicals in situ. Because of the ready availability of molecular oxygen and its ability to accept electrons, oxygen-centered free radicals are often the primary or secondary metabolites of cellular free-radical reactions (23). In this section, the various cellular sites of free-radical generation are briefly described.

Plasma Membrane

Free-radical production by microsomal and plasma membrane–associated enzymes such as lipoxygenase and cyclooxygenase is derived from the arachidonic acid metabolism. The enzymatic oxidation of arachidonic acid by membrane-bound cyclooxygenase, leading to biologically potent products such as prostaglandins, thromboxanes, leukotrienes, and slow-reacting substances of anaphylaxis, involves intermediates containing both carbon- and oxygen-centered free radicals. Similarly, phagocytic cells such as leukocytes and monocytes generate partially reduced oxygen species as a defense against invading microorganisms. It is now a widely recognized fact that the biosynthesis of prostaglandins and thromboxanes results in hemoprotein-, oxygen-, and carbon-centered free radicals capable of reacting with the biosynthetic enzymes themselves and other cell components (23).

Soluble Components of the Cytosol

Many cell components such as thiols, hydroquinones, catecholamines, flavins, and tetrahydropterins are capable of undergoing oxidation–reduction reactions and thus are quantitatively important contributors to intracellular free-radical production (23). In all these cases, superoxide radical, leading eventually to the formation of the peroxide anion and thus H_2O_2, is the primary product of the reactions.

In the cytosol, several enzymes also generate free radicals during their catalytic cycling. These enzymes include xanthine oxidase, aldehyde oxidase, dihydroorotate dehydrogenase, flavoprotein dehydrogenases, and tryptophan dehydrogenase.

Studies of the enzyme sources of cellular free-radical production have shown that modulation of enzyme activities, cofactor availability, substrate concentration, and oxygen tension can all contribute to affect the rates of intracellular radical production.

Mitochondria

Although the electron transport chain involving a four-electron transfer reduction of molecular oxygen to water by the enzyme cytochrome c oxidase takes place in mitochondria, the enzyme per se is not a source of mitochondrial oxygen production (24). Superoxide radical generation by mitochondria is influenced by the reduced states of the respiratory chain carriers located on the inner mitochondrial membrane. These endogenous factors, including the availability of NAD-linked substrates, succinate, ADP, and oxygen, influence both mitochondrial radical production and the regulation of cellular respiration.

Studies of mitochondria isolated from a variety of mammalian cells and tissues have shown that the ubiquinone–cytochrome b region is a major site of superoxide radical formation (25–27). The generation of superoxide radicals in this region is primarily due to the autoxidation of ubiquinone. However, because of the efficient endogenous superoxide dismutase system, little if any superoxide enters the cytoplasm from mitochondria. Most of it is converted to H_2O_2 before being released into the cytoplasm (28).

Endoplasmic Reticulum and Nuclear Membrane

Both the endoplasmic reticulum and nuclear membranes contain the cytochromes P-450 and b5, which can oxidize unsaturated fatty acids (29) and xenobiotics (30) and reduce oxygen (31). In addition, flavoprotein-linked cytochrome reductases that provide electrons for the cytochrome P-450- and b5-mediated reactions are also capable of autoxidizing to produce both superoxide radical and H_2O_2 (32). Microsomal and nuclear membrane cytochromes can directly form the superoxide radical by one-electron transfer or will form H_2O_2 by dissociation of peroxy–cytochrome complexes.

Flavin-containing oxidases are also implicated in the production of microsomal oxygen radicals. Hydroxyl radicals are generated by liver microsomes in both the absence and presence of cytochrome P-450 substrates and require the presence of NAD(P)H (23). It should be noted that unlike the intramitochondrially derived free radicals, which must escape the organelle's antioxidant defenses to initiate cytosol damage, those produced by the endoplasmic reticulum and nuclear membranes can easily undergo both intraorganelle and cytosolic reactions. The nuclear membrane-derived free radicals, because of their proximity, are particularly harsh on the DNA (23).

Peroxisomes

The high intraorganelle concentrations of oxidases make peroxisomes some of the most potent sources of cellular H_2O_2, although none has been shown to generate the superoxide radical as an immediate precursor to H_2O_2 (23,33). The peroxisomal H_2O_2-generating enzymes include, among others, D-amino acid oxidase, urate oxidase, L-α-hydroxy acid oxidase, and fatty acyl CoA oxidase. Because of the efficiency of the enzyme peroxisomal catalase, which normally metabolizes most of the H_2O_2 generated by the peroxisomal oxidases, little of it can diffuse out of peroxisomes into the cytoplasm. Nevertheless, because of the relatively long half-life of the peroxide anion compared to that of the hydroxyl radical, H_2O_2 does seem to have the ability to diffuse across at least two membrane barriers and through the cytoplasm to cause tissue damage (34).

The various cellular sites of free-radical generation in the biological systems are shown schematically in Fig. 6.1. Structural components, subcellular organelles,

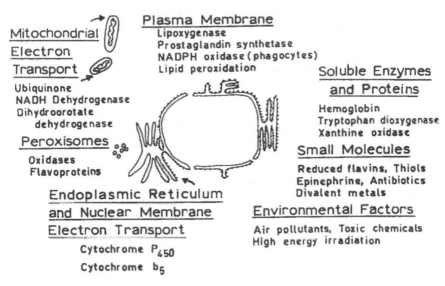

Fig. 6.1 Cellular sources of free radicals. Structural components, subcellular organelles, cytoplasmic contents, and xenobiotics all contribute to biological production of free-radical species. (From Ref. 23.)

cytoplasmic contents, and xenobiotics all contribute to the biological production of free-radical species.

6.2.4 Biological Effects of Free Radicals

Because of their very high chemical reactivity, free radicals have reasonably short lifetimes in biological systems. Extensive studies with model systems and with biological materials in vitro, however, have clearly shown that reactive free radicals are able to produce metabolic disturbances and to damage membrane structures in a variety of ways. Free radicals seek out unsaturated sites in biomolecules for rapid attack. Such random attack may produce undesirable chemical modifications of and damage to organic macromolecules such as proteins, carbohydrates, lipids, and nucleotides. Thus, if free radicals are produced during the normal cellular metabolism in sufficient amounts to overcome the normally efficient protective mechanisms, metabolic and cellular disturbances will occur.

Major disturbances induced by the free-radical reactions are shown in Fig. 6.2. The reactive free radicals may cause cellular damage in a variety of ways (12). These include (i) covalent binding of the free radical to membrane enzymes and/or receptors, thereby modifying the activities of membrane components; (ii) covalent binding to membrane components, thereby changing cellular structure and affecting membrane function and/or antigenic character; (iii) disturbance of transport

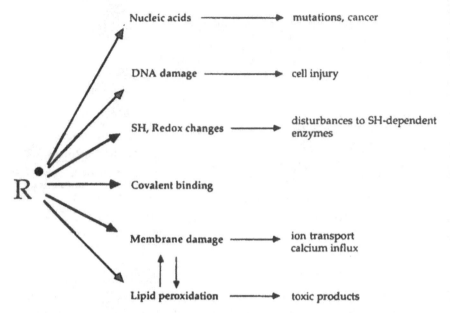

Fig. 6.2   Free radicals and cellular injury. Major routes are shown in which a free radical (R*) can interact with neighboring components in cells to disturb their metabolic function(s). (From Ref. 12.)

processes through covalent binding, thiol group oxidation, or change in PUFA/protein ratios; and (iv) initiation of lipid peroxidation of PUFAs with direct effects on membrane structure and associated influences of the products of peroxidation on membrane fluidity, cross-linking, structure, and function. The importance of the free-radical chain reactions and their cascading effects on a spreading network of disturbances are further illustrated in Fig. 6.3.

Among the most deleterious effects of free radicals are damage to DNA and phospholipids. In vitro studies have clearly shown that the presence of a free-radical-generating system, such as phagocytosing lymphocytes or xanthine oxidase, in close proximity to DNA will result in extensive damage to the DNA structure, leading in turn to harmful mutations and cytotoxic effects (12,22,35). The potential for such damage in the etiology of cancers and other diseases is therefore very large. It should, however, be remembered that the highly reactive free radicals are essentially trapped in the immediate vicinity of their site of formation as a consequence of their rapid interaction with neighboring molecules (36). Thus, in cellular terms, their radius of diffusion is often very small. The reactive free radicals formed in the endoplasmic reticulum are therefore unlikely to diffuse far enough to react with DNA in the nucleus. These restrictions on the diffusivity of the highly reactive free radicals, especially the HO* radical, necessitate the generation of an

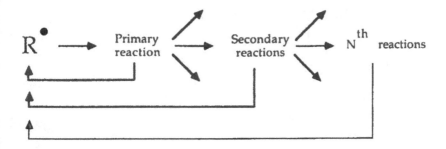

Fig. 6.3 A representation of a free-radical-mediated disturbance producing a diverging set of metabolic perturbations, with some or all of these repaired or prevented by effective cell defenses involving free-radical scavengers and antioxidants. (From Ref. 9.)

intermediate chemical reactivity in order to interact directly with DNA (37,38). Thus, during "normal" cellular metabolism, highly reactive free radicals cannot diffuse far enough, whereas free radicals of very low reactivity, although able to diffuse farther, will not be reactive enough to produce significantly important covalent adducts responsible for the mutations and cytotoxic effects (36).

That the chemical reactivity of different free radicals effectively controls their radii of diffusion away from their loci of formation in biological systems is illustrated in Fig. 6.4. For example, the very short half-life of only a few microseconds of $^{\cdot}CCl_3$ or $HO^{\cdot}$ radicals in biological systems results in a diffusion of the species to less than 100 nm from their site of formation (8,39). These considerations suggest that free-radical intermediates of xenobiotics, produced by the metabolic activation in the endoplasmic reticulum, will have moderate but not extreme chemical reactivity if the free radical is the active carcinogenic species (8). Normally, the most harmful and deleterious effects of free radicals on DNA structure will occur only if they are produced in close proximity to DNA. Such is the case upon exposure to ionizing radiation, which is well known to induce DNA mutations and other cytotoxic effects through the generation of free radicals.

Free-radical-induced chain reactions of lipid peroxidation are another major route for cellular disturbances. The trans oxidative products of the PUFAs, upon incorporation into cell membranes, induce changes in membrane structure and impaired function of phospholipid-dependent enzymes contained in cell membranes (35,40–42). The damage to phospholipids in cell membranes by free-radical peroxidation, however, has far-reaching consequences. Both arachidonic and

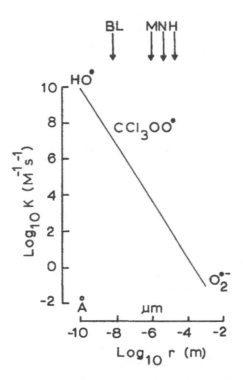

Fig. 6.4 A representation of a relationship between the chemical reactivity of free radicals and their radius of diffusion (r) in a biological environment. The ordinate shows the logarithm of the second-order rate constants (k) of the free radicals with a polyunsaturated fatty acid such as arachidonate in a biomembrane. The abscissa shows the logarithm of the diffusion radius of the free radical in the membrane. The vertical arrows indicate approximate sizes of (1) the thickness of a phospholipid (BL) bilayer, (2) the average length of a mitochondrion (M), (3) the diameter of a nucleus (N) in a liver cell, and (4) the average diameter of a liver hepatocyte (H). (From Ref. 9.)

linoleic acids, which are the precursors of prostaglandin synthesis, are lost in this process (41). This in turn leads to a state of imbalance in these biochemical components vital to cell regulation. Alterations in membrane characteristics further result in increased permeability, thereby interfering with the normally balanced transport of sodium, potassium, calcium, and magnesium and other ion-transport processes (40,41).

Although free radicals can cause harmful cellular injuries if produced in amounts that exceed the body's capacity to neutralize them, they nonetheless have several useful purposes in biological systems. For example, free radicals such as $O_2^{\bullet-}$ and $HO^{\bullet}$ as well as H_2O_2 are produced by leukocytes and macrophages to ward off attack by invading pathogens (43,44). Similarly, the detoxification of foreign substances such as certain food toxicants and additives, petrochemicals, inhaled fumes, and drugs is carried out in the endoplasmic reticulum of liver cells or other organs by reactions involving cytochrome P-450 and other enzymes. Such reactions do produce peroxides and hydroxyl radicals (40,41,45–48). Hydroxyl radicals are also involved in ethanol oxidation in vivo by the microsomal ethanol-oxidizing system (11). Some of the essential cellular reactions mediated via formation of free radicals in biological systems are as follows.

Reduction of ribonucleosides by ribonucleoside diphosphate reductase via tyrosine radical formation.

Cleavage of the indole rings of tryptophan, tryptamine, and serotonin via superoxide radical production catalyzed by dioxygenases.

Generation by pyruvate-metabolizing enzymes of carbon-centered radicals (from pyruvate) and sulfur-centered radicals (from coenzyme A).

Hydroxylases requiring ferryl species at the active site of cytochrome P-450 and peroxidases. However, no clear evidence exists for the participation of superoxide and hydroxyl radicals.

Carboxylation of glutamic acid by vitamin K-dependent enzymes generating superoxide radicals.

Phagocytosis involving polymorphonuclear leukocytes (e.g., polymorphs such as neutrophils, eosinophils, and basophils), monocytes, and lymphocytes that kill bacterial pathogens via superoxide and hydroxyl radicals and H_2O_2 production.

Synthesis of eicosanoids (prostaglandins and leukotrienes) from PUFAs by insertion of molecular oxygen in enzyme-catalyzed reactions by stereospecific free-radical mechanisms at the active sites.

6.2.5 Protective Mechanisms Against Free-Radical-Induced Pathology

To circumvent the damaging effects of free radicals, humans have developed a three-tier defense strategy. The foremost defense is to prevent the generation of the reactive forms of partially reduced oxygen. This is accomplished by the cytochrome oxidase system of the electron transport chain operating in the mitochondria, which catalyzes the tetravalent reduction of oxygen without the release of a significant amount of any of these reactive intermediates.

The second line of defense is provided by enzymes that catalytically scavenge the intermediates of oxygen reduction. The intracellular enzymes superoxide dismutase (SOD) and catalases are extremely important in defusing free radical oxidants before they react with critical cellular materials. These enzymes are probably present intracellularly because cells contain iron and probably copper ions in forms that can accelerate damaging free-radical reactions. Hence, rapid removal of both $O_2^{\cdot -}$ and H_2O_2 is essential before these species can come into contact with the intracellular metal ions. The enzymes glutathione peroxidase and glutathione reductase also play important protective roles in this regard.

The $O_2^{\bullet -}$ radical is eliminate by SOD, which dismutases $2O_2^{\bullet -}$ to H_2O_2 and O_2. The H_2O_2 is then reduced to water by the action of catalases. The removal of both $O_2^{\bullet -}$ and H_2O_2 prevents the formation of $HO^{\bullet}$ radicals. This is of extreme importance because the enzymatic scavenging of $HO^{\bullet}$ is impossible due to its extreme reactivity.

In contrast, the extracellular fluids contain little SOD or H_2O_2-scavenging enzymes, because at physiological generation rates, the $O_2^{\bullet -}$ and H_2O_2 are useful and thus should not be rapidly removed (13,49). However, it then becomes essential to prevent them from interacting to form $HO^{\bullet}$ and other highly toxic species. This is achieved by making transition metal ions unavailable to catalyze reactions in extracellular fluids. Hence, there is three times as much transferrin iron-binding capacity in human plasma as iron needing to be transported, so there are essentially no free iron ions in plasma (50). Iron ions bound to transferrin cannot stimulate lipid peroxidation or the generation of free $HO^{\bullet}$ radicals, and the same is true of copper ions bound to the plasma proteins ceruloplasmin or albumin (51).

Finally, these two levels of enzymatic defenses are supplemented by a third level of biochemical defenses that include vitamin E (tocopherols), vitamin C (ascorbic acid), and other natural and diet-derived synthetic antioxidants or free-radical scavengers. Although the term *antioxidant* is implicitly restricted to chain-breaking compounds, Halliwell and Gutteridge (11), taking a much broader view, define an antioxidant as "any substance that, when present at low concentrations compared to those of an oxidizable substrate, significantly delays or inhibits oxidation of that substrate."

To be effective against free-radical-mediated cell disturbances, the antioxidants or free-radical scavengers must have several important characteristics. These criteria, as described by Slater (9), are as follows.

The scavenger (antioxidant) must get to the right site within the cell of the relevant tissue in a concentration that is sufficient to allow effective competition with neighboring biomolecules.

The scavenger (antioxidant) must get to the right site at the right time in order to interact with transient damaging free-radical species as they are formed.

The scavenger (antioxidant) must be able to interact with the toxic species sufficiently rapidly to ensure successful competition with biologically sensitive loci in the immediate vicinity of free-radical production.

The scavenger (antioxidant) must have acceptable biological properties, that is, its inherent toxicity must be low.

Finally, in summary, the scavenger (antioxidant) must get to the right site at the right time and in the right concentration; moreover, it must have acceptable low intrinsic toxicity for use under conditions in vivo.

Furthermore, their individual properties cannot be judged in isolation. Antioxidants are members of a rather large and actively cooperating family of chemical substances. For example, α-tocopherol activity in humans is regenerated by ascorbic acid (52).

Antioxidants can act at different levels in the oxidative sequence (11) as follows.

1. Decreasing localized oxygen concentrations.
2. Preventing chain initiation by scavenging initiating radicals such as $HO^\bullet$.
3. Binding metal ions in forms that will not generate such initiating species as $HO^\bullet$, ferryl, or $Fe^{2+}/Fe^{3+}/O_2$ and/or will not decompose lipid peroxides to peroxy and alkoxy radicals.
4. Decomposing peroxides by converting them to non-radical products, such as alcohols.
5. Chain-breaking, that is, scavenging intermediate radicals such as peroxy and alkoxy radicals to prevent continued hydrogen abstraction. Chain-breaking antioxidants are often phenols or aromatic amines.

Antioxidants acting by the first three mechanisms are often known as preventive antioxidants (11). Those acting by mechanism 3 are not usually consumed during the course of the reactions. Antioxidants of the fourth type are also preventive antioxidants. However, depending on their chemical behavior, they may or may not be consumed during the reaction. For example, glutathione peroxidase acts by this mechanism and is not consumed. Similar to antioxidants of type 2 above, chain-breaking antioxidants combine with intermediate radicals and therefore are not consumed.

In the following two sections, the biological functions and significance of free-radical-scavenging natural enzyme systems and of various antioxidant nutrients are described.

6.3 BIOLOGICAL DEFENSE SYSTEMS

The reactive free radicals formed during normal cellular metabolism are capable of inducing a wide range of severely damaging reactions. To avoid such deleterious effects, biological systems have developed very efficient and widely distributed protective mechanisms of various kinds during the course of evolution. The deleterious effects of free radicals, therefore, are manifested only if these natural protective mechanisms are overcome or broken down. In this section, enzyme systems involved in eliminating and/or detoxifying the free radicals from cellular systems as well as the role of various natural and diet-derived antioxidant nutrients are described.

6.3.1 Enzyme Systems

Superoxide Dismutase (SOD)

The efficient removal of $O_2^{\bullet-}$ and H_2O_2 is essential to minimize the formation of $HO^\bullet$ in biological systems. It is widely recognized now that the most toxic free radical is not $O_2^{\bullet-}$ as such but the $HO^{\bullet}$ radical it may help to create (because of the extremely high reactivity of the latter in abstracting electrons from virtually any of the organic molecules in the immediate vicinity). SOD intervenes in this free-radical chain reaction by catalyzing the reduction of $O_2^{\bullet-}$ radical to H_2O_2. If the $O_2^{\bullet-}$ is properly quenched by SOD, it reduces the potential for $HO^{\bullet}$ radical generation. The SOD enzyme system consisting of metalloproteins, therefore, constitutes the first line of defense against free-radical-mediated cellular injury. All other chemical antioxidants and radical-scavenging nutrients play a secondary role in neutralizing $HO^{\bullet}$ radicals if they escape this first line of defense in a biological system.

Occurrence and Distribution   Superoxide dismutase was first discovered by Mann and Keilin (53) as a copper-containing blue-green protein, cuprein or hemocuprein, from bovine blood. Its biological function then was thought to be copper storage. Similar proteins were later isolated from horse liver (hepatocuprein) and brain (cerebrocuprein) (11). Their catalytic function, however, was not discovered until the late 1960s when McCord and Fridovich (54) demonstrated that the protein previously known as cuprein catalyzed the dismutation of superoxide radicals as follows:

$$2\,O_2^{\bullet-} + 2\,H^+ \rightarrow H_2O_2 + O_2$$

Superoxide dismutase is not known to react catalytically with any substrate other than the $O_2^{\bullet-}$ radical. With increasing realization of the importance of free-radical pathology in the etiology of several diseases, more papers were published on SOD than on any other type of enzyme in the 1980s (11).

The SOD system is virtually ubiquitous in distribution (55). It is found in virtually all oxygen-consuming organisms (56), in some aerotolerant anaerobes (57), and in some obligate anaerobes (58). It is also the fifth most common protein in the human body.

Chemical and Physical Properties   All SODs are metalloproteins containing either a copper, iron, or manganese ion at their active sites. They are therefore classified as Cu/Zn SOD, Fe SOD, or Mn SOD. The latter two are characteristic of prokaryotes and show extensive sequence homologies (59,60), whereas Cu/Zn SOD is characteristic of eukaryotic cytosols and shows no sequence homology with the other two (59,61). The latter probably represents an independent line of evolution. The physicochemical characteristics of the various SODs are briefly described below.

Copper/Zinc SOD   Copper/zinc SODs are found in virtually all eukaryotic cells such as yeasts, plants, and animals (Table 6.5), but not generally in prokaryotic cells such as bacteria or the blue-green algae. They have relative molecular masses around 32,000 and consist of two protein subunits, each bearing an active site containing one copper and one zinc ion.

X-ray crystallographic studies of bovine Cu/Zn SOD have shown that each

Table 6.5 Systems from Which Cu/Zn SOD Has Been Purified[a]

Mammalian tissues	Plant tissues
Bovine erythrocytes and retina	*Neurospora crassa*
Human erythrocytes	*Fusarium oxysporum*
Spermatozoa	Green peas
Rat liver	Maize seeds
Bovine liver	Wheat germ
Horse liver	Spinach chloroplasts
Bovine milk	Yeast (*Saccharomyces cerevisiae*)
Human milk	Pea seedlings*
Pig liver	Corn seedlings*
Fish	Tomatoes
Shark (*Prionace glauca*)*	Cucumber*
Cuttlefish (*Sepia officinalis*)*	Green peppers*
Ponyfish	Lentil (*Lens esculenta*)
Snapper	Other organisms
Sea bass	Fruit fly (*Drosophila melanogaster*)
Croaker	*Photobacterium leiognathi*
Merlin	Chicken liver
Trout	*Caulobacter crescentus*
Swordfish	*Trichinella spiralis*
	Housefly (*Musca domestica*)

[a]Unless indicated by an asterik, the enzyme was purified to homogeneity and shown to contain two subunits. All purified enzymes have one ion each of copper and zinc at each active site.
Source: Ref. 11.

subunit is composed primarily of eight antiparallel strands of β-pleated sheet structure (containing 46% of the approximately 153 amino acid residues) that form a flattened cylinder and three external loops that contain a further 48% of the amino acid residues (62). The copper ion is held at the active site by four histidine residues (numbers 44, 46, 61, and 118 in the amino acid sequence), whereas the zinc ion is bridged to the copper by interaction with the imidazole of His-61. The zinc ion also interacts with His-69 and His-78 as well as the carboxyl group of Asp-81. Histidine 61, which is involved in the positioning of both metals, is believed to be involved in the supply of protons needed for the dismutation reaction (11,63).

The zinc ion is completely buried within the protein structure, does not function in the catalytic cycle, and therefore appears to play only a structural role. In contrast, the copper ion functions in the dismutation reaction by undergoing alternate oxidation and reduction as follows:

$$SOD\text{–}Cu^{2+} + O_2^{\cdot-} \quad \rightarrow SOD\text{–}Cu^+ + O_2$$
$$SOD\text{–}Cu^+ + O_2^{\cdot-} + 2\,H^+ \rightarrow SOD\text{–}Cu^{2+} + H_2O_2$$

Net reaction: $O_2^{\cdot-} + O_2^{\cdot-} + 2\,H^+ \quad \rightarrow H_2O_2 + O_2$

Halliwell and Gutteridge (11) suggest the possibility of at least one other mechanism of action in which the first $O_2^{\cdot-}$ does not reduce the copper ion but forms a complex with it.

Mammalian Cu/Zn SODs are highly thermostable, resistant to protease attack, and tolerant to organic solvents, including chloroform and ethanol (63). Such great stability results from the close-packing in the hydrophobic interfaces between the dimeric subunits and between the two halves of the cylindrical structure. These enzymes are also resistant to denaturation by solvents such as SDS and urea (64,65), although inhibited by the cyanide and H_2O_2, the latter being a product of the dismutation reaction itself.

Manganese SOD   First isolated from *E. coli* as a pink rather than blue-green protein of about 40,000 relative molecular mass, Mn SODs have since been purified from a number of sources (Table 6.6). MnSODs purified from higher organisms consist of four protein subunits with 0.5 or 1.0 ion of Mn per subunit, whereas most bacterial Mn SODs are dimeric proteins (11). The subunit molecular weights are approximately 20,000, while those of the dimeric and tetrameric proteins vary between 39,500 and 94,000 (63).

Manganese SODs have extensive areas of α-helix in the tertiary structure and very little β-sheet, and the folded proteins are roughly rectangular. The metal-binding site is close to or even part of the subunit interface (66). All the Mn SODs studied to date are similar in their amino acid sequences and circular dichroism spectra (11).

The removal of Mn from the active site causes the enzyme to lose its catalytic activity. Generally, the activity cannot be replaced by any other transition metal ion, including iron, to yield a functional enzyme (11,63).

The Mn SODs catalyze exactly the same reaction as do the Cu/Zn SODs but are less active at higher pH. However, unlike the latter, Mn SODs are more susceptible to denaturation by heat or chemicals such as detergents (11,63). They are not, however, affected by H_2O_2.

Iron SOD   Iron SODs are generally dimeric proteins with each subunit having a molecular mass of about 22,000 and containing a functional iron ion, although at least two tetrameric enzymes exist (Table 6.7). The Fe SODs exhibit a very high degree of sequence homology with the Mn SODs. X-ray crystallographic studies have shown Fe SODs to be helical proteins. The monomer consists of six major helical segments and three strands of antiparallel β-sheet. The iron is ligated to four protein side chains, and the active center is exposed to the solvent (63,67). Halliwell

Table 6.6 Examples of Organisms from Which Mn-SOD Has Been Purified

Organism	Subunit structure	Moles Mn per mole enzyme
Higher organisms		
Maize	Tetramer	2
Bovine adrenal cells	Tetramer	2
Luminous fungus	Tetramer	2
(*Pleurotus olearius*)		
Pea	Tetramer	1
Chicken liver	Tetramer	2
Rat liver	Tetramer	4
Human liver	Tetramer	4
Yeast (*Saccharomyces cerevisiae*)	Tetramer	4
Bullfrog (*Rana catesbeiana*)	Tetramer	4
Bacteria		
Halobacterium halobium	Dimer	1–2
Rhodopseudomonas spheroides	Dimer	1
Escherichia coli	Dimer	1
Bacillus stearothermophilus	Dimer	1
Mycobacterium phlei	Tetramer	2
Mycobacterium lepraemurium	Dimer	1
Thermus thermophilus	Tetramer	2
Paracoccus denitrificans	Dimer	1–2
Streptococcus faecalis	Dimer	1
Streptococcus mutans	Dimer	1–2
Propionibacterium shermanii	Dimer	3
Bacillus subtilis	Dimer	1
Serratia marcescens	Dimer	1–2
Gluconobacter cerinus	Dimer	1
Acholeplasma laidlawii	Dimer	1

Source: Ref. 11.

and Gutteridge (11) speculate that the iron in the resting state is Fe^{3+} and that it probably oscillates between Fe^{3+} and Fe^{2+} states during the catalytic cycle.

Similar to Cu/Zn SOD, H_2O_2 inhibits Fe SOD activity, although, in common with Mn SOD, it is not affected by cyanide. It also exhibits decreased catalytic activity at higher pH. The physicochemical properties of the three metalloforms of SOD are summarized in Table 6.8. Data presented in Table 6.9 on the SOD activities in various animal tissues indicate an especially high concentration in the liver.

Table 6.7 Examples of Organisms from Which Fe SOD Has Been Purified

Organism	Subunit structure	Moles Fe per mole enzyme
Bacteria		
Streptococcus mutans	Dimer	1–2
Escherichia coli	Dimer	1–1.8
Desulphovibria desulphuricans	Dimer	1–2
Thiobacillus denitrificans	Dimer	1
Chromatium vinosum	Dimer	2
Photobacterium leiognathi	Dimer	1
Pseudomonas ovalis	Dimer	1–2
Methanobacterium bryantii	Tetramer	2–3
Azotobacter vinelandii	Dimer	2
Bacillus megaterium	Dimer	1
Mycobacterium tuberculosis	Tetramer	4
Propionibacterium shermanii	Dimer	2
Other organisms		
Tomato	Dimer	1–2
Mustard	Dimer	1–2
Water lily	Dimer	1
Red alga (*Porphyridium cruentum*)	Dimer	1
Blue-green alga (*Spirulina platensis*)	Dimer	1
Blue-green alga (*Anacystis nidulans*)	Dimer	1
Alga (*Euglena gracilis*)	Not reported	1
Trypanosome (*Crithidia fasciculata*)	Dimer	2–3
Ginkgo biloba	Dimer	1

Source: Ref. 11.

Mechanism of Dismutation The mechanism of $O_2^{\cdot-}$ dismutation by all three forms of SOD is the same; the active-site metal undergoes a cycle of reduction and reoxidation (55). In the first step, the metal is reduced by $O_2^{\cdot-}$ and one molecule of oxygen is released. In the second step, another molecule of $O_2^{\cdot-}$ oxidizes the reduced metal at the active site, which results in the formation of H_2O_2 (68). A general mechanism for catalysis is presented in the following two equations where Enz is SOD and M is the metal in the active site:

$$\text{Enz} \cdot M^n + O_2^{\cdot-} \rightarrow \text{Enz} \cdot M^{n-1} + O_2$$
$$\text{Enz} \cdot M^{n-1} + O_2^{\cdot-} + 2\,H^+ \rightarrow \text{Enz} \cdot M^n + H_2O_2$$

The dismutation is most rapid at acidic pH values (pH 4.8, the pK_a value of the hydroperoxyl radical) and becomes slower with increasing pH values. The loss of

Table 6.8 Physicochemical Properties of the Three Metalloforms of SOD

Property	Cu/Zn SOD	Mn SOD	Fe SOD
Molecular form	Dimer, identical subunits	Prokaryotic dimer, eukaryotic tetramer, identical subunits	Dimer, occasionally tetramer, identical subunits
Metal ion (prosthetic group)	One Cu^{2+}, one Zn^{2+} per subunit	One or two Mn^{2+} per dimer	One Fe^{3+} per subunit
Molecular weight	32,000	42,000–46,000 (dimer)	40,000–46,000 (dimer)
Inhibition by cyanide	At 1 mM CN^-	No effect	No effect
Inhibition by H_2O_2	At 0.5 mM H_2O_2	No effect	At 0.5 M H_2O_2
Inhibition by chloroform/ethanol	No effect	Inhibited	Inhibited
Effect of 10 mM azide	10% inhibition	30% inhibition	70% inhibition
Effect of SDS	Stable at 4% SDS solutions	Unstable to 1% SDS solutions	Unstable to 1% SDS solutions
Effect of pH on enzyme activity	No effect, pH 5–10	Activity decreases at pH > 8	Activity decreases at pH > 8

Source: Refs. 63 and 81.

Table 6.9 Representative Activities of SOD in Animal Tissues

Animal and tissue	SOD activity (units/mg protein)	Ref.
Mice		81
Pancreatic islets	331	
Liver	660	
Kidney	582	
Erythrocytes	52	
Heart	390	
Brain	408	
Skeletal muscle	282	
Rat		82
Liver	22	
Adrenal	20	
Kidney	13	
Erythrocytes	4	
Spleen	5	
Heart	9	
Pancreas (whole)	1.5	
Brain	3	
Lung	3	
Stomach	7	
Intestine	3	
Ovary	2	
Thymus	1	
Rat		83
Adipose tissue	11	

activity above pH 10 is attributed to a loss of positive charges in lysyl side chains. The positively charged lysyl side chains at lower pH, in combination with the electrostatic repulsion by negatively charged areas on the enzyme surface, serve to actively guide $O_2^{\bullet-}$ anions to the active-site channel (63).

Physiological Regulation   The biosynthesis of SOD seems to be under rigorous control in most biological systems. Exposure to high concentrations of $O_2^{\bullet-}$ increases SOD in most organisms (55). The inducer for SOD is not molecular oxygen itself but rather a product of its metabolism, since the level of SOD correlates with the intracellular level of $O_2^{\bullet-}$ generated in the presence of oxygen (69). Several redox-active compounds that increase the intracellular flux of $O_2^{\bullet-}$ also increase the levels and activity of SODs (70).

Physiological Functions   A large body of data indicates that SODs scavenge $O_2^{\bullet-}$ in vivo as they do in vitro and thus protect against many ill effects of the reduction products of oxygen. Although the role of SODs in the dismutation of $O_2^{\bullet-}$ in biological systems is well recognized, Gartner and Weser (67) raised several questions in this regard, some of which follow.

1. Why did nature not adapt a simpler mechanism for the removal of $O_2^{\bullet-}$ by its direct oxidation to molecular oxygen, thereby obviating the production of H_2O_2? This would have prevented the harmful reactions of H_2O_2 with transition metal ions in biological systems.
2. Is the diffusion of $O_2^{\bullet-}$ to SOD fast enough to prevent any side reactions?
3. Given the fact that $O_2^{\bullet-}$ is predominantly found in membranes, the presence of relatively low concentrations of SOD in lipophilic environments compared to its high levels in cytosol is quite puzzling.
4. The superoxide theory of oxygen toxicity does not adequately explain the presence of SOD in anaerobic microorganisms or the ability of some aerobes to survive without SOD in an oxygen-rich environment.

The fact that Cu/Zn SOD reacts with peroxyl radicals with rate constants comparable to its value for dismutating $O_2^{\bullet-}$ led Bors (71) to suggest that the function of SOD could be expanded to include the scavenging of radicals other than $O_2^{\bullet-}$ alone. Indeed, the high dismutase activity of Cu^{2+} and its complexes in vitro prompted Fee (72) to propose that the biological role of SOD is not dismutation of $O_2^{\bullet-}$ but some activity not yet determined and that the dismutase activity of the enzyme exists because of the fact that it contains a transition metal. However, using a similar rationale, other copper-containing proteins such as cytochrome c oxidase and ceruloplasmin should also perform similar functions although none do so catalytically. Moreover, Fe SOD and Mn SOD are much more effective catalysts than either free Fe or Mn ions at similar concentrations.

The vast body of literature including (i) evidence for the electrostatic facilitation of $O_2^{\bullet-}$ binding to the active sites of SOD; (ii) the amino acid sequence conservation both at the active site and in the electrostatic channel loop, coupled with molecular genetic studies (73); (iii) the positive correlation between aerotolerance and the concentration of SOD in anaerobic microorganisms (57); (iv) oxygen intolerance resulting from a loss of SOD activity (74); (v) high levels of SOD in rat lungs, induced by exposure to 85% oxygen, being positively correlated with their tolerance to 100% oxygen (75); and finally (vi) the protection offered by SOD against radiation damage (76–78) all indicate that the true biological function of SOD is almost certainly dismutation and that SOD is the only efficient catalyst of this reaction in vivo (79).

Catalase

Catalases help in preventing the accumulation of H_2O_2 within cells. They catalyze the reaction:

$$2 H_2O_2 \rightarrow 2 H_2O + O_2$$

Occurrence and Distribution   With few exceptions such as *Bacillus popilliae*, *Mycoplasma pneumoniae*, the green alga *Euglena*, several parasitic helminths, and the blue-green alga *Gloeocapsa*, catalases are widely distributed in aerobic cells, whereas most anaerobic organisms do not contain the enzyme activity. In animals, catalase is present in all major body organs, being especially concentrated in the liver and erythrocytes (11). The brain, heart, and skeletal muscle contain only low amounts, although the activity does vary between muscles and even among different regions of the same muscle (Table 6.10).

Catalases are located in animal and plant tissues in subcellular organelles bounded by a single membrane and known as *peroxisomes* (11,85). These organelles also contain some of the cellular H_2O_2-generating enzymes such as glycollate oxidase, urate oxidase, and flavoprotein dehydrogenases involved in the beta oxidation of fatty acids. The general hypothesis is that all reactions involving glyoxylate biosynthesis are carried out in microbodies to prevent the formation of large amounts of formic acid in cells.

Physicochemical Properties   Catalases are heme proteins. The enzyme consists of four identical subunits, each containing a heme [Fe(III)-protoporphyrin] group bound to its active site. It has a molecular weight of about 240,000 and contains 4 mol of protoheme. Each enzyme subunit also contains one molecule of

Table 6.10   Distribution of Catalase Activity (mg^{-1} Protein) in Normal Human Tissues

Tissue	Catalase activity
Liver	1300–1500
Erythrocytes	990–1300
Kidney cortex	110–430
Adrenal gland	300
Kidney medulla	220–700
Spleen	56
Lymph node	120
Pancreas	100–120
Lung	180–210
Heart	54
Skeletal muscle	25–36
Brain, gray matter	3–11
Brain, white matter	20
Adipose tissue	270–560

Source: Refs. 11 and 84.

NADPH, which helps to stabilize the enzyme. Dissociation of protein subunits results in a loss of enzyme activity (11).

Enzyme Mechanism   The mechanism of dismutation of H_2O_2 into oxygen and water by catalase was reviewed by Halliwell and Gutteridge (11). The reaction proceeds in two steps:

$$\text{Catalase–Fe(III)} + H_2O_2 \xrightarrow{k_1} \text{compound I}$$

$$\text{Compound I} + H_2O_2 \xrightarrow{k_2} \text{catalase–Fe(III)} + H_2O + O_2$$

The second-order rate constants k_1 and k_2 for rat liver catalase have values of 1.7×10^7 $M^{-1}s^{-1}$ and 2.6×10^7 $M^{-1}s^{-1}$, respectively. Although the exact structure of compound I is unknown, it is believed to be intermediate in structure between a ferric peroxide [Fe(III)—HOOH] and Fe(V)=O. Because of its enormous maximal velocity (V_{max}) for the destruction of H_2O_2, the enzyme is rarely saturated with its substrate.

Catalases also carry out certain peroxidase-type reactions in the presence of a steady supply of H_2O_2 that allow formation of compound I. The latter can oxidize the alcohols methanol and ethanol to their corresponding aldehydes, formaldehyde and acetaldehyde. Catalase activity is inhibited by azide and cyanide. Manganese-containing catalases that are insensitive to inhibition by azide or cyanide have been purified from several microorganisms (11).

Glutathione Peroxidase (GSH-Px)

Occurrence and Distribution   Glutathione peroxidase was first discovered in animal tissues by Mills (86) in 1957 as an enzyme that protected erythrocytes against hemoglobin oxidation and hemolysis. It is not generally present in higher plants or bacteria, although it has been reported in some algae and fungi (11). Its substrate is the low molecular weight thiol compound glutathione. Glutathione is found in animals, plants, and some bacteria, often in millimolar concentrations. It is a simple tripeptide (Glu-Cys-Gly) and in its reduced form is usually abbreviated to GSH. In the oxidized form, GSSG, two GSH molecules join together as the sulfhydryl groups of cysteine are oxidized to form a disulfide (—S—S—) bridge. Most free glutathione in vivo is present as GSH rather than GSSG, although up to one-third of the total cellular pool of GSH may be present as "mixed" disulfides with other compounds that also contain thiol groups. These compounds include the amino acid cysteine, coenzyme A, and the sulfhydryls of cysteine residues of several proteins (11). The enzyme GSH-Px catalyzes the oxidation of GSH to GSSH at the expense of H_2O_2 as follows:

$$H_2O_2 + 2\,GSH \xrightarrow{\text{GSH–Px}} GSSG + 2\,H_2O$$

Physicochemical Properties A protein of about 80,000 molecular mass, GSH-Px is made up of four apparently identical subunits, each of which contains one atom of selenium at its active site. Selenium is in group VI of the Periodic Table and has properties intermediate between those of a metal and a nonmetal. Sedimentation equilibrium analyses have shown that the molecular weight of GSH-Px differs from species to species, for example, 76,000 ± 1000 for rat liver GSH-Px (87), 83,800 ± 1200 for bovine erythrocyte GSH-Px (88), and 95,000 ± 3000 for human erythrocyte GSH-Px (89). Its molecular weight also can vary from tissue to tissue in the same species (90,91).

Rotruck and coworkers (92) were the first to report at least 2 g-atom of selenium per mole of GSH-Px. Later it was independently demonstrated that GSH-Px from bovine and human bone erythrocytes contained 4 g-atom of selenium per mole of GSH-Px (93,94). Similar values were confirmed for rat liver and human erythrocyte GSH-Px (87,89). In contrast to other peroxidases, GSH-Px contains no heme or flavin (85), and neutron activation analysis of crystalline enzyme has indicated that GSH-Px contains no metals other than selenium (94).

Selenium in the active site of the reduced GSH-Px is present as selenocysteine (95,96), the amino acid cysteine in which the normal sulfur atom has been replaced by a selenium atom (R-SeH instead of R-SH, where R is —CH_2—$CH(N^+H_3)$—COO^-). The GSH apparently reduces the selenium, and the reduced form of the enzyme then reacts with H_2O_2 with an approximate rate constant of 5×10^7 $M^{-1}s^{-1}$ (11). Trace amounts of selenium are therefore required in the human and animal diets, a deficiency of which can be explained by the resulting lack of GSH-Px. Selenium, however, is toxic to humans and animals in excess amounts.

Enzyme Mechanism The kinetic mechanism of GSH-Px is described as a series of three biomolecular steps (97,98). The first step involves the oxidation of the enzyme by the peroxide substrate and release of the corresponding alcohol (or water in the case of H_2O_2). This is followed by two successive additions of GSH and the release of GSSG. This mechanism takes into account the inability to saturate the enzyme with either substrate and the similar V_{max} values of most peroxide substrates.

Ganther et al. (99) proposed that during catalysis the selenium in GSH-Px cycles between a selenol (—SeOH) and a selenic acid (—SeH) or between a selenic acid and a selenious acid (—SeOOH), as shown in triangles A and B of Fig. 6.5. The relatively high values of GSH present in tissues would presumably keep the enzyme in the selenol-selenic acid cycle (triangle A) in vivo. GSH-Px activities normally found in various human tissues are presented in Table 6.11.

Physiological Functions The discovery that organic hydroperoxides as well as H_2O_2 are substrates for GSH-Px (100) provided an important clue to the biochemical function of GSH-Px and thus of selenium. Since a number of animal tissues contain both catalase and GSH-Px, these two enzymes must cooperate in

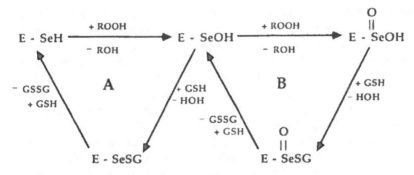

Fig. 6.5 Reaction mechanism of GSH-Px. (From Ref. 96.)

the removal of H_2O_2, especially since the rate of H_2O_2 reduction per heme or per selenium, respectively, is nearly identical for the two enzymes (97). For example, mammalian erythrocytes contain no subcellular organelles, and both these enzymes are present in the cytosol. Although, because of its greater concentration, catalase would seem to be far more important than GSH-Px for H_2O_2 destruction (101), the normally low rate of H_2O_2 production in these cells via SOD reactions seems to be mainly due to GSH-Px (11). Furthermore, GSH-Px-deficient erythrocytes are

Table 6.11 Distribution of Glutathione
Peroxidase (GSH-Px) Activity (mg^{-1} Protein) in
Normal Human Tissues

Tissue	GSH-Px activity
Liver	120–190
Erythrocytes	19
Kidney cortex	87–140
Adrenal gland	120
Kidney medulla	73–90
Spleen	50
Lymph node	160
Pancreas	43–110
Lung	53
Heart	69
Skeletal muscle	22–38
Brain, gray matter	66–71
Brain, white matter	76
Adipose tissue	77–89

Source: Refs. 11 and 84.

susceptible to hemolysis when exposed to oxidizing agents. In contrast, humans suffering from an inborn defect in the catalase gene, which produces an unstable mutant enzyme and thus decreases erythrocyte catalase activities, show no life-threatening harmful effects (11). In the erythrocytes, catalase becomes more important only if the concentration of H_2O_2 is raised by supplying these cells with drugs that tend to increase intracellular H_2O_2 generation.

Among the human cells, brain and spermatozoa contain little catalase but more GSH-Px activity, whereas liver contains high concentrations of both enzymes (Tables 6.10 and 6.11). Whereas catalase is largely or entirely in the peroxisomes, GSH-Px is found mainly in the cytosol and in the matrix of mitochondria (11). The distribution of GSH is similar. This suggests that the H_2O_2 produced by glycollate oxidase and urate oxidase in the peroxisomes is largely removed by catalase, whereas that produced in the mitochondria or the endoplasmic reticulum or produced by soluble cytosolic enzymes such as SOD is acted upon by GSH-Px.

It has become increasingly clear in recent years that a high degree of cooperativity exists between these different enzymes involved in the metabolism of $O_2^{\cdot -}$, H_2O_2, hydroperoxides, various other free radicals including $HO^{\cdot}$, and possibly singlet oxygen arising from normal cellular reactions as well as the products of the metabolism of toxic substances (Fig. 6.6). The effectiveness of an enzyme (or other antioxidant nutrient) in containing and ultimately destroying these reactive species depends on both the specificity of the enzyme and its subcellular location. Thus in a typical animal cell, lipid-soluble α-tocopherol (vitamin E) scavenges free radicals and possibly quenches singlet oxygen in the membranes, GSH-Px and SOD react with peroxides and superoxide, respectively, in the cytosol and mitochondrial matrix space, and catalase destroys H_2O_2 in the peroxisomes (11,101). The relative concentration and the importance of these protective species vary from species to species and result in the variety of selenium and/or vitamin E deficiency signs observed in various species.

Glutathione Reductase

The ratios of GSH/GSSG in normal cells generally are kept high (Table 6.12). To achieve this, the GSSG must be reduced back to GSH, a reaction catalyzed by glutathione reductase:

$$GSSG + NADPH + H^+ \xrightarrow{\text{Glutathione reductase}} 2\,GSH + NADP^+$$

Glutathione reductase can also catalyze the reduction of certain "mixed" disulfides such as that between GSH and conenzyme A (11). The NADPH required for these reactions is provided in animal tissues by the oxidative pentose phosphate pathway. As glutathione reductase operates to lower the NADPH/NADP$^+$ ratio, the pentose phosphate pathway speeds up to replace the NADPH.

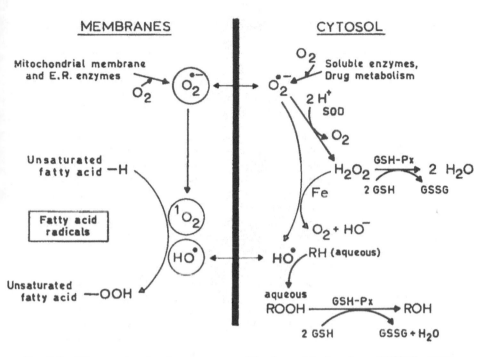

Fig. 6.6 Diagram showing the compartmentalization and the functions of GSH-Px, SOD, and α-tocopherol in a liver cell. (From Ref. 101.)

Glutathione reductases contain two protein subunits, each with the flavin FAD at its active site. Apparently, during the catalytic cycle the NADPH reduces the FAD, which then passes its electrons on to a disulfide (—S—S—) between two cysteine residues in the protein. The two -SH groups so formed then interact with GSSG and reduce it to 2 GSH, thereby re-forming the protein disulfide.

Glutathione-S-Transferase

Glutathione-S-transferase is involved in the metabolism of "foreign" compounds in the liver to yield mercapturic acids, which are then excreted from the body. The first step in this detoxifying mechanism involves the conjugation of these compounds to GSH. Several enzymes of this type are present in the liver and also in many other animal tissues (10,11,35). Glutathione conjugates are usually excreted into the bile by a transport mechanism similar to that by which the liver ejects GSSG when subjected to oxidative stress. Compounds that are metabolized by the mercapturic acid pathway include chloroform, bromobenzene, naphthalene, and paracetamole (11). This pathway, however, depletes the hepatic GSH pool, which in turn reduces the ability of the liver to cope with H_2O_2 and other oxygen radicals.

Table 6.12   Presence of Glutathione Species and the Enzymes Using It in Selected Organisms[a]

System	[GSH][b]	GSH/GSSH	GSH-Px activity	GSH-reductase activity
Spinach chloroplasts	3.0 mM	>10/1	Absent	High
Rat tissues				
Liver	7–8 mM	>10/1	High	High
Erythrocyte	2 mM	>10/1	Moderate	Moderate
Heart	2 mM	>10/1	Moderate	Moderate
Lung	2 mM	>10/1	Moderate	Moderate
Lens	6–10 mM	>10/1	Moderate	Moderate
Spleen	4–5 mM	>10/1	—	—
Kidney	4 mM	>10/1	Moderate	Moderate
Brain	2 mM	>10/1	Moderate	Moderate
Skeletal muscle	1 mM	>10/1	Low	Low
Blood plasma	0.02–0.03 mM	≈5/1	Low	Low
Adipose tissue	3.1 μg/10^6 cells	>100/1	Low	Low
Human tissue				
Liver	4 μmol/g wet wt	>10/1	High	High
Lens	6–10 mM	>10/1	Moderate	Moderate
Erythrocytes	240 μg/mL blood	>10/1	Moderate	Moderate
Neurospora crassa	20 μmol/g dry wt	150/1	Absent	Moderate
E. coli				
Aerobically grown	27 μmol/g	>10/1	Absent	—
Anaerobically grown	7 μmol/g			

[a]GSH = glutathione; GSSG = glutathione disulfide; GSH-Px = glutathione peroxidase enzyme.
[b]These GSH values (a) decrease with age in animals, (b) are different at different times of day in animals and at different points of the growth cycle in bacteria and fungi, (c) in liver, fall on starvation, (d) decrease with consumption of alcohol, and (e) will vary in the different cell types present in animal organs. *Source*: Ref. 11.

6.3.2   Glutathione (γ-Glutamyl-Cysteinyl-Glycine)

Glutathione (Glu-Cys-Gly) or GSH is the only significant electron donor substrate for the metabolic reactions involving GSH-Px, although the latter can use a variety of hydroperoxide acceptor substrates (102). In addition, GSH is also a scavenger of HO˙ radicals and singlet oxygen. Since GSH is present at high concentrations in many cells (Table 6.12), it may help to protect against these radical species. GSH can reactivate some enzymes inhibited by exposure to high oxygen concentration, presumably by reduction of the disulfide bonds of the enzyme. However, GSH is not essential for aerobic life; several aerobic bacteria do not contain it (11). As mentioned earlier, a lack of GSH in animal cells causes hemolysis.

Glutathione is a cofactor for several enzymes in widely different metabolic pathways including glyoxylase, maleylacetoacetate isomerase, prostaglandin endoperoxide isomerase, and DDT dehydrochlorinase (10,11,102,103). It also plays an important role in the synthesis of thyroid hormones, in the degradation of insulin in animals, and in the metabolism of herbicides, pesticides, and toxic "foreign" compounds, generally in both animal and plant tissues (11).

In contrast, the disulfide GSSG can inactivate a number of enzymes including adenylate cyclase, chicken liver fatty acid synthetase, rabbit muscle phosphofructokinase, and phosphorylase phosphatase, probably by forming mixed disulfides with them. Such a situation occurs when a large flux of H_2O_2 and/or $HO^\bullet$ radicals lowers the GSH/GSSG ratio in the cells. GSSG also inhibits protein synthesis in animal cells (11). Because of this, the cells presumably maintain a high GSH/GSSG ratio under normal conditions.

Halliwell and Gutteridge (11) reviewed other deleterious aspects of GSH biochemistry. Its rapid reaction with $HO^\bullet$ ($k_2 > 109$ M^{-1} s^{-1}) and oxidation by peroxides or by oxygen in the presence of transition metal ions such as Cu^{2+} or Fe^{2+} yield thiyl radicals. The reactions of GSH with $O_2^{\bullet-}$ can also lead to the formation of singlet oxygen ($^1O_2$) as follows:

$$GSH + O_2^{\bullet-} + H^+ \rightarrow GS^\bullet + H_2O_2$$
$$\text{or } GSH + HO_2^\bullet \rightarrow GS^\bullet + H_2O_2$$
$$GS^\bullet + O_2 \rightarrow GSO_2^\bullet$$
$$GSO_2^\bullet + O_2^{\bullet-} + H^+ \rightarrow {}^1O_2 + GSOOH$$
$$GSO_2^\bullet + GSO_2^\bullet \rightarrow GSSG + 2\,{}^1O_2$$
$$GSO_2^\bullet + GS^\bullet \rightarrow GSSG + {}^1O_2$$

A schematic representation of the GSH cycle in biological systems is shown in Fig. 6.7.

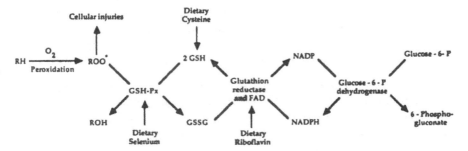

Fig. 6.7   Schematic representation of the GSH cycle in biological systems.

6.3.3 Vitamins and Provitamins

At the biochemical level, the primary defense mechanisms against the free-radical-mediated cell damage involving the major enzyme systems and their substrates described above are supported by secondary defenses. These include several vitamins that have antioxidant properties as well as those derived from the diet in which antioxidants are intentionally added to prevent or retard lipid peroxidation as well as to maintain the freshness and wholesomeness of processed food products. The physiological roles and properties of these compounds are described below.

Vitamin E

Vitamin E is the principal component of the secondary defense mechanisms against free-radical-mediated cellular injuries. In fact, it is the only natural physiological lipid-soluble antioxidant that can inhibit lipid peroxidation in cell membranes (104).

Vitamin E was first discovered in the early 1920s by Evans and Bishop as a fat-soluble nutritional factor that was found to be essential for normal reproduction in rats (105). It was later isolated as a closely related family of compounds, designated as *tocopherols*. Soon afterwards the detection of peroxides in the adipose tissue of animals fed diets deficient in vitamin E and the demonstration that several synthetic antioxidants prevent vitamin E deficiency signs in animals resulted in an "antioxidant" theory of vitamin E function (106–108).

A lack of vitamin E causes a wide variety of symptoms including sterility in male rats, dogs, cocks, rabbits, and monkeys; hemolysis in rats and chickens; muscular degeneration in rabbits, guinea pigs, monkeys, ducks, mice, and mink; "white-muscle" disease in lambs and calves; cerebellar degeneration and exudative diathesis in chicks; liver necrosis and enamel depigmentation in rats; anemia in monkeys; and steatitis or "yellow-fat" disease in a number of animals (105,109, 110). Although a short-term deficiency of vitamin E in the human diet does not cause any specific deficiency disease, an increasing body of scientific literature during the past two decades has shown vitamin E to be an essential vitamin in both animals and humans. In biological systems, vitamin E plays an important role against free-radical-mediated cellular injuries and in the enhancement of immune response as well as a preventive role against several cancers, cardiovascular diseases, cataract, and Parkinson's disease (11,111–119).

Chemistry, Nomenclature, and Properties   At least eight compounds exhibiting vitamin E activity and having a 6-chromanol ring structure and a side chain have been isolated from plant sources (Fig. 6.8). Based on the structure of the side chain, these compounds can be further classified as tocopherols (or tocols) and tocotrienols, the former containing a phytol side chain and the latter with a similar structure but containing double bonds at the 3'-, 7'-, and 11'-positions of the side chain.

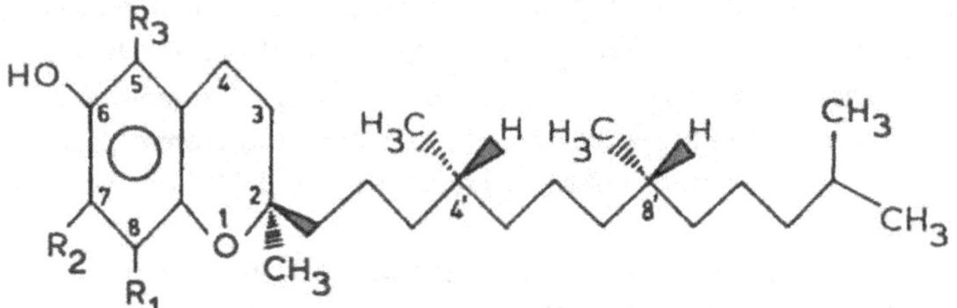

Tocopherols

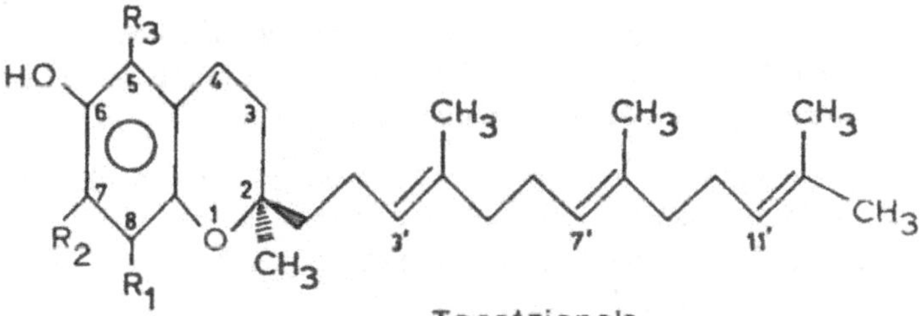

Tocotrienols

Fig. 6.8 Chemical structures of tocopherol and tocotrienol series of compounds having vitamin E activity.

Substitution	Tocopherol	Tocotrienol
$R_1, R_2, R_3 = CH_3$	α-Tocopherol (α-T)	α-Tocotrienol (α-T-3)
$R_1, R_3 = CH_3; R_2 = H$	β-Tocopherol (βT)	β-Tocotrienol (β-T-3)
$R_1 R_2 = CH_3; R_3 = H$	γ-Tocopherol (γ-T)	γ-Tocotrienol (γ-T-3)
$R_1 = CH_3; R_2, R_3 = H$	δ-Tocopherol (δ-T)	δ-Tocotrienol (δ-T-3)

Depending on the substitution pattern at the R_1, R_2, and R_3 positions of the chromanol ring, these compounds are further designated as α-, β-, γ-, and δ-tocopherols and tocotrienols, respectively (Fig. 6.8). The tocopherols contain three asymmetric carbons, at the C-2 position in the ring and in the C-4′ and C-8′ positions of the side chain, thus making a total of eight possible optical isomers. α-Tocopherol is the most active form of vitamin E, their biopotencies, defined as the absorption in the gut, being in the order α-T > β-T > γ-T > δ-T, and α-T-3 (which has approximately 25% of the biopotency of α-T) has about 5 times the potency of β-T-3.

Although the tocopherols do have some antioxidant activity, the terms α-*tocopherol* and *vitamin E* are now used in the literature almost interchangeably (11). To avoid misinterpretation, the American Institute of Nutrition (AIN), following the recommendations of the Commission of Biochemical Nomenclature (CBN) of the International Union of Pure and Applied Chemistry (IUPAC) (120), proposed the following system for the nomenclature of tocopherol-related compounds (109):

The term "vitamin E" should be used as the generic description for all tocol and tocotrienol derivatives qualitatively exhibiting the biological activity of α-tocopherol. Thus, phrases such as "vitamin E activity," "vitamin E deficiency," and "vitamin E in the form of . . ." represent the preferred usage.

The term "tocol" is the trivial designation of 2-methyl-2-(4′,8′,12′-trimethyltridecyl)chroman-6-ol.

The term "tocopherols" should be used as the generic description for all mono-, di-, and trimethyl tocols irrespective of biological activity. The term "tocopherols" is not synonymous with the term "vitamin E."

The only naturally occurring stereoisomers of α-tocopherol, formerly known as *d*-α-tocopherol or α-tocopherol, should be designated as *RRR*-α-tocopherol. The totally synthetic α-tocopherol, formerly known as *dl*-α-tocopherol, should be designated all-*rac*-α-tocopherol. Esters of tocopherols should be designated as tocopheryl esters (e.g., α-tocopheryl acetate).

Vitamin E is practically insoluble in water but is completely soluble in oils, fats, acetone, alcohol, chloroform, ether, benzene, and other nonpolar solvents (109,110, 117,121). It is stable to heat and alkali in the absence of oxygen and is unaffected by acids at temperatures up to 100°C. It is, however, slowly oxidized by oxygen to tocopheroxide, tocopherylquinone, and tocopheryl hydroquinone as well as to dimers and trimers. The oxidation rate is greatly enhanced in the presence of iron and copper. The esters of vitamin E are stable to oxidation; however, they cannot function as antioxidants.

Any significant structural changes in α-tocopherol, including the addition of a double bond to the side chain, shortening or lengthening of the side chain, masking of the phenolic hydroxyl group as an ether, and loss of any methyl group from or the oxidation of the chromanol ring, result in major loss of its biological activity (121). However, the substitution of an amine or *N*-methylamine for the phenolic hydroxyl or that of a dihydrobenzofuran for the chromanol ring appears to enhance the activity of tocopherols (122).

The α-tocopherol content of various human tissues is shown in Table 6.13. The adrenal and pituitary glands, testes, and platelets have the highest concentrations of the vitamin. It is most concentrated in cell fractions rich in membranes such as the mitochondria and microsomes.

Dietary Sources   Both tocopherol and tocotrienol compounds are widely distributed in nature (Table 6.14). α-Tocopherol accounts for almost all the vitamin activity in foods of animal origin, whereas in vegetable seed oils other isomers contribute substantially to the total vitamin E activity (109,123). For example, α-tocopherol represents only 8–10% of the total tocopherols in soybean oil but still contributes a major portion of the biological activity.

Vitamin E content of foods is influenced by a large number of factors, causing a wide variation in vitamin content of any given food product. For many foods,

Table 6.13 α-Tocopherol Content of Normal Human Tissues

Tissue	α-Tocopherol content	
	μg/g	mg/g lipid
Plasma	9.5	1.4
Erythrocytes	2.3	0.5
Platelets	30.0	1.3
Adipose tissue	150.0	0.2
Kidney	7.0	0.3
Liver	13.0	0.3
Muscle	19.0	0.4
Ovary	11.0	0.6
Uterus	9.0	0.7
Heart	20.0	0.7
Adrenal glands	132.0	0.7
Testis	40.0	1.0
Pituitary	40.0	1.2

Source: Ref. 109.

there is a three- to tenfold range in reported α-tocopherol values (123). For example, seasonal differences may cause a fivefold variation in the α-tocopherol content of milk. Similarly, the natural tocopherols are not very stable. Therefore, significant losses may occur during processing and storage of foods. Losses of natural tocopherols during storage of vegetable oils are usually minimal, but during cooking it can be appreciable (109). Likewise, cold-pressed oils tend to contain higher levels of α-tocopherol than those extracted by processing that involves heat treatment. Tocopheryl esters added to foods, however, are very stable.

Physiological Functions

In Vivo Antioxidant. As mentioned before, many early observations that the damaging effects of a deficiency of vitamin E in animal diets could be partially or completely alleviated by synthetic antioxidants led to the hypothesis that vitamin E functions in vivo as a protector against lipid peroxidation (106–108,125). The evidence included such in vitro observations as its direct reactions with and quenching of superoxide and peroxyl radicals and singlet oxygen as well as its ability to prevent lipid peroxidation.

Most in vivo observations in this respect have come from studies on animals fed vitamin E-deficient diets that showed increased levels of peroxides and aldehydes in many tissues and increased exhalation of ethane and pentane. Similarly, the

Table 6.14 α-Tocopherol Content of Food Products

Food category	α-Tocopherol content (μg/g)	Food category	α-Tocopherol content (μg/g)
Seeds and grains		Animal foods	
Corn	10	Beef	6
Milo	12	Pork	5
Oatmeal	17	Chicken	3
Poppyseed	18	Fish and shellfish	
Rice, brown	7	Halibut	9
Rice, white	1	Cod	2
Rye	10	Shrimp	9
Wheat grain	11	Milk	1
Wheat germ	117	Butter	24
Wheat bread, white	1	Lard	12
Wheat bread, whole	5	Eggs	11
Fruits		Vegetable oils	
Apple	3	Coconut	11
Banana	2	Corn	159
Grapefruit	3	Cottonseed	440
Orange	2	Olive	100
Peach	13	Peanut	189
Pear	5	Rapeseed	236
Strawberry	12	Safflower	396
		Soybean	79
Nuts		Sunflower	487
Almond	270	Wheat germ	1194
Brazil	65	Palm	211
Filbert	210	Margarine	
Peanut	72	Soft	139
Pecan	11	Hard	108
Walnut	5		

Source: Refs. 109, 123, and 124.

association of PUFAs with aggravated vitamin E-deficiency symptoms was also recognized (108). It was initially believed that dietary vitamin E was being destroyed when PUFAs were undergoing peroxidation in the diet. However, even when the oxidation was prevented or tocopherol was given separately from dietary PUFA, the deficiency symptoms were still observed. Furthermore, synthetic anti-oxidants such as ethoxyquin and diphenyl-p-phenylenediamine (DPPA), whose structures are quite unrelated to those of the tocopherols, were able to prevent several symptoms of vitamin E deficiency (106,126). This effect was not a result of the sparing of vitamin E in tissues, since rats that did not have sufficient tissue reserves of vitamin E to maintain normal reproduction responded to the administration of DPDD (127).

Based on these studies, Tappel (128) first proposed that vitamin E functions as an in vivo antioxidant that protects tissue lipids from free-radical damage. It is now widely recognized that tocopherol is located primarily in the membrane portion of the cell and is a part of the cell's defense against oxygen-centered radicals. Nevertheless, some controversy still exists regarding the antioxidant role of vitamin E in biological systems. Green and Bunyan (129) argue that a clear lack of evidence for the existence of peroxides in tissues such as the liver, testes, and muscle that have pathology resulting from a vitamin E deficiency contradicts its antioxidant role. However, the natural defense mechanism operating in biological systems in which peroxidized fatty acids are preferentially hydrolyzed by phospholipase A_2 and then rapidly destroyed by GSH-Px makes it difficult for the tissue to accumulate fatty acid peroxides (109). Similarly, many of the biochemical and pathological effects of vitamin E deficiency may be adequately explained by free-radical-mediated alteration of membrane structures. Likewise, almost all the enzymes that are affected by vitamin E status either are membrane-bound or are concerned with the GSH-Px system (130). Vitamin E, therefore, is unique in its more specific localization in membranes and the tenacity with which it remains in most tissues.

Vitamin E acts as a chain-breaking antioxidant in membranes. The hydrogen atom of the phenolic hydroxyl group on the chromanol ring is very easy to remove. It therefore reacts readily with lipid peroxy and alkoxy radicals by donating the labile hydrogen atom to them, thereby terminating the chain reaction of peroxidation by scavenging chain-propagating radicals.

The autoxidation and antioxidant reactions involving vitamin E are summarized by Machlin (109) as follows.

1. Initiation (formation of a free radical)

$$LH \xrightarrow{\text{Initiators}} L^\bullet$$

2. Reaction of radical with oxygen

$$L^\bullet + O_2 \rightarrow LO_2^\bullet$$

3. Propagation

$$LO_2^\bullet + LH \rightarrow L^\bullet + LOOH$$

4. Antioxidant reaction

$$LO_2^\bullet + E \rightarrow E^\bullet + LOOH$$

5. Regeneration

$E^{\cdot} + C \rightarrow E + C^{\cdot}$

$C^{\cdot} + NADPH \xrightarrow{\text{Semidehydro ascorbate reductase}} C + NADP$

$E^{\cdot} + 2\,GSH \rightarrow E + GSSG$

$GSSG + NADPH \xrightarrow{\text{GSH reductase}} 2\,GSH + NADP$

6. Termination

$E^{\cdot} + E^{\cdot} \rightarrow E{-}E$ (dimer)

$E^{\cdot} + LO_2^{\cdot} \rightarrow EOOL$ (?)

The abbreviations used in the above equations are $L^{\cdot}$ = fatty acid radical, $LO_2^{\cdot}$ = peroxy radical, LH = fatty acid, E = tocopherol, LOOH = hydroperoxide, C = ascorbic acid, $C^{\cdot}$ = ascorbyl radical, GSH = reduced glutathione, and GSSG = glutathione disulfide (oxidized form of GSH).

The tocopheroxy radical $(E^{\cdot})$ generated in the above reactions is poorly reactive and is therefore unable to attack adjacent fatty acid side chains. Its inability to abstract hydrogen from membrane lipids results from the fact that the unpaired electron on the oxygen atom can be delocalized into the aromatic ring structure. This increases the stability of the tocopheroxy radical. During peroxidation of cell membranes in vitro, some α-tocopherol is also converted to α-tocopherylquinone and other related products (11).

Platelet Function and Prostaglandin Metabolism. Platelets constitute one of the four main components of the hemostatic system that protects the body from the effects of trauma and preserves the integrity of blood vessels. The remaining three components of this system are the blood vessel wall and the coagulation and fibrinolytic systems (131). Platelets do not have a nucleus and thus lack the ability to divide. However, they are extremely complex cellular structures whose metabolic products influence the activity of the cell wall and are in turn affected by products released from the vascular endothelium. Platelets are also important for optimal operation of the coagulation system (132).

The facts that platelet activation is accompanied by a marked consumption of oxygen and that platelets contain a large amount of PUFA, primarily in the form of 5,8,11,14-eicosatetraenoic acid (arachidonic acid), provide a natural setting for the development of lipid peroxidation. The synthesis of eicosanoids (prostaglandins) from arachidonic acid metabolism via lipoxygenase and cycloxygenase pathways by the platelets involves free-radical-mediated reactions (131–133). Because of its well-known antioxidant effect, vitamin E is believed to be a potential inhibitor of platelet activation (134). Vitamin E interferes with some of the reactions leading to the formation of hydroxyeicosatetraenoic acid (HETE), thromboxane (TXA2), and prostaglandins (135–138). Similarly, in vitamin E-deficient animals, enhanced prostaglandin synthesis is observed in platelets, spleen, and bursa (137,139). Vitamin E increases the threshold concentration for arachidonic acid–induced platelet aggregation (140). For these reasons, high levels of vitamin E in diets that are marginally low in vitamin K can effectively inhibit the latter's ability to coagulate blood. Since at present there are very few agents that have platelet aggregation–inhibiting properties and none that constitute a normal part of the diet, the potential significance of vitamin E in preventing thrombotic diseases and inflammation needs further exploration.

Interrelationship of Vitamin E, Selenium, Vitamin C, and GSH in Membrane Protection   Although Tappel's (128) theory regarding the in vivo function of vitamin E as an antioxidant continues to constitute the cornerstone of most explanations of the biological activity of α-tocopherol in mammalian tissues, the inconsistencies in the hypothesis as pointed out by Green and Bunyan (129) led Diplock and Lucy (140) to propose an alternative hypothesis for the action of vitamin E. It was founded primarily on the physicochemical interaction between the phytol side chain of vitamin E and certain PUFAs. With the discovery of the selenium-containing GSH-Px, a lipid peroxide–destroying enzyme, the biochemical rationale for the close metabolic relationship between selenium and vitamin E was clearly demonstrated. Moreover, in biological systems, vitamin E is regenerated from the tocopheroxy radical by vitamin C and GSH (109,141,142). The interdependency and cooperativity of these various nutrients is schematically shown in Fig. 6.9. Vitamin C (ascorbic acid) reduces the tocopheroxy radical with concurrent formation of an ascorbate radical, which in turn can be enzymatically reduced back to ascorbate by an NADH-dependent system. Presumably, these reactions with ascorbate radical occur at the lipid/water interface of the membranes, because vitamin C is found primarily in the cytosol whereas vitamin E is located in the cell membranes. GSH can also reduce tocopheroxy radical, perhaps by an enzymatically mediated process (143).

Although the physiological functions are described above primarily from the viewpoint of preventing free-radical-mediated cell injuries, vitamin E also plays an important role in nucleic acid, protein, and lipid metabolism (109). It may also protect the sulfhydryl groups of dehydrogenases in the electron transport chain in mitochondria from oxidation or from reaction with metal ions. These and several other functions of vitamin E have been reviewed elsewhere (109,115–117).

Vitamin C (Ascorbic Acid)

Vitamin C is chemically the simplest of vitamins and therefore was among the first to be isolated, characterized, and purified and to have its structure determined. Its ene-diol structure, however, provides it with a highly complex chemistry. Thus it has a very complicated redox chemistry involving comparatively stable radical intermediates and heavily modified by the acidic properties of the molecule.

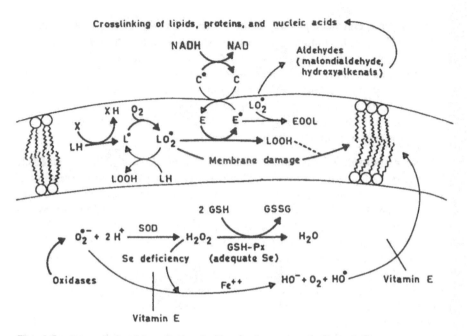

Fig. 6.9 Interrelationships of vitamin E, selenium, vitamin C, and GSH in protecting membranes. $O_2^{\bullet-}$ = superoxide radical, $HO^\bullet$ = hydroxyl radical, LH = fatty acid, $L^\bullet$ = fatty acid radical, $LO_2^\bullet$ = peroxy radical, E = tocopherol, $E^\bullet$ = tocopheroxy radical, LOOH = hydroperoxide, C = ascorbic acid, $C^\bullet$ = ascorbyl radical, GSH = reduced glutathione, GSSG = oxidized glutathione, GSH-Px = glutathione peroxidase, SOD = superoxide dismutase. (From Ref. 109.)

The early history of vitamin C is associated with the etiology, treatment, and prevention of scurvy. Because of its antiscorbutic properties, it was later named ascorbic acid. Vitamin C was first isolated in crystalline form by the Hungarian scientist Albert Szent-Gyorgyi as a reducing factor from the adrenal glands with the empirical formula $C_6H_8O_6$, which he called "hexuronic acid" (144). Its relationship to the antiscorbutic factor was simultaneously demonstrated by Svirbely and Szent-Gyorgyi (145) and by King and Waugh (146). Soon thereafter, Haworth and Hirst (147) announced the structure of vitamin C and, along with Szent-Gyorgyi, suggested that its name be changed to L-ascorbic acid to convey its antiscorbutic properties. Its first synthesis in DL form was reported by Reichstein et al. (148), the synthetic product having the same biological activity as the substance isolated from natural tissues.

Nearly all species of animals synthesize vitamin C and do not require it in their diets. However, humans, other primates, guinea pigs, fruit bats, some birds, including the red-vented bulbul and related Passeriformes species, and fish such as

Coho salmon, rainbow trout, and carp cannot synthesize the vitamin because of the loss of a liver enzyme, L-gulono-γ-lactone oxidase, during the course of evolution (149,150). According to Moser and Bendich (151), nature perhaps gave up ascorbic acid synthesis in these animals in order to conserve glucose, the precursor of vitamin C. This would mean that a significant amount of glucose, a major source of energy, would be conserved in species that cannot synthesize ascorbic acid.

Chemistry, Nomenclature, and Properties L-Ascorbic acid is an α-keto lactone with an almost planar five-membered ring. It has a double bond between the C-2 (or α) and C-3 (or β) carbons, with the two chiral centers at positions 4 and 5 providing four stereoisomers (Fig. 6.10). The configuration at C-5 is of the L-series and was confirmed by synthesis from L-xylose (148). Later on, L-ascorbic acid was also synthesized from D-glucose, whose chiral centers at C-2 and C-3 are in the proper configuration to become C-5 and C-4, respectively, of L-ascorbic acid (152). Among its isomers (Fig. 6.10), only erythorbic acid shows about 2.5–5% as much vitamin activity as that of L-ascorbic acid (151). The acidic nature of vitamin C in aqueous solution derives from the ionization of the enolic hydroxyl on C-3 ($pK_a = 4.25$), the resulting ascorbate anion being delocalized.

CBN-IUPAC has recognized ascorbic acid, L-ascorbic acid, L-xyloascorbic acid, 2-oxo-L-*threo*-hexono-1,4-lactone-2,3-ene-diol, and L-*threo*-hex-2-enoic

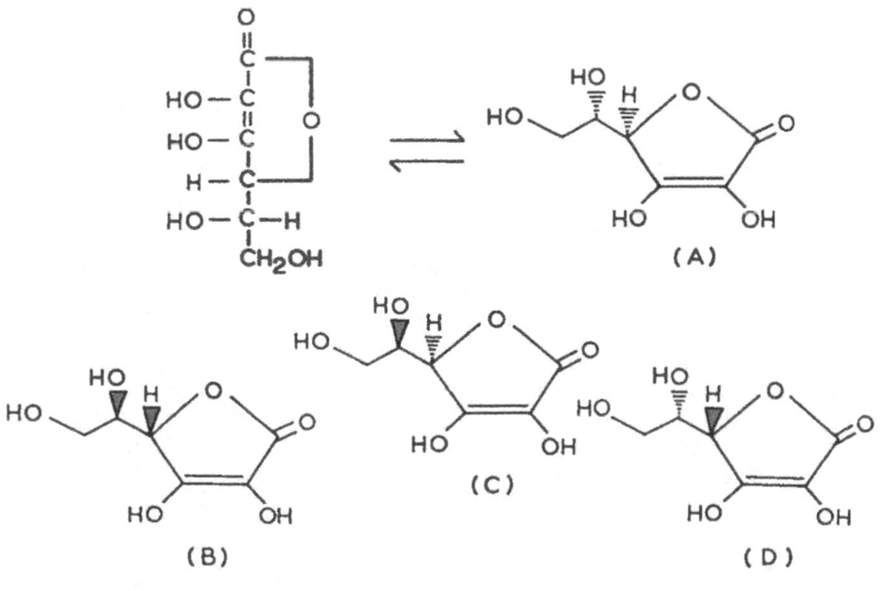

Fig. 6.10 Stereoisomers of vitamin C. (A) L-ascorbic acid; (B) D-ascorbic acid; (C) D-araboascorbic acid (erythorbic acid); (D) L -araboascorbic acid.

acid γ-lactone as chemical names for vitamin C (153). The name L-ascorbic acid was specifically designated to convey the vitamin's antiscorbutic properties.

The reversible oxidation–reduction with dehydro-L-ascorbic acid is L-ascorbic acid's most important chemical property and the basis for its known physiological activities and stabilities (151,154). Dehydroascorbic acid, the oxidized form of ascorbic acid that retains vitamin C activity, has a side chain that forms a hydrated hemiketal. The oxidation–reduction path of ascorbic acid and the reactions of its subsequent derivatives are shown in Fig. 6.11. Degradation reactions of L-ascorbic acid in aqueous solutions depend on several factors, such as pH (the range with highest stability being 4–6), temperature, and the presence of oxygen or metals such as copper (151). The loss of an electron by L-ascorbic acid gives the semidehydroascorbate radical, which can be further oxidized to give dehydroascorbate. The semidehydroascorbate radical is not particularly reactive and mainly undergoes a disproportionation reaction:

2 Semidehydroascorbate → ascorbate + dehydroascorbate

Dehydroascorbate is highly unstable and breaks down rapidly in a very complex way, eventually oxidizing to oxalic and L-threonic acids (Fig. 6.11). In acidic solutions, the degradation proceeds in forming L-(+)-tartaric acid, furfural and other furan derivatives, and some condensation products (155). Alkali-catalyzed degradation results in over 50 compounds, mainly mono-, di-, and tricarboxylic acids (156).

Aqueous solutions of L-ascorbic acid are stable unless transition metal ions are present (Fig. 6.12), which catalyze their rapid oxidation at the expense of molecular oxygen. Copper(II) is the most potent transition metal ion, producing hydrogen peroxide and hydroxyl radicals. The ability of vitamin C to degrade DNA and damage various animal cells in culture, including cancer cells, can probably be attributed to the formation of these species in the presence of traces of copper ions in the solution (11). Ascorbic acid–copper(II) complexes also inactivate many proteins, probably by formation of hydroxyl radicals and/or copper(III) species.

The distribution of vitamin C in rat and human tissue is summarized in Table 6.15. The adrenal and pituitary glands have the highest concentrations of ascorbic acid on a weight basis of any tissue in the body of animals that synthesize ascorbic acid (rats) as well as in humans (161).

Dietary Sources Vitamin C is ubiquitous. It is found throughout the plant and animal kingdom, where its roles are often not known or are poorly understood. The synthetic vitamin is also very widely used as a food additive, primarily because of its antioxidant and nutritional properties. It is classified as a "generally recognized as safe" (GRAS) compound. D-Erythorbic acid (D-araboascorbic acid) is also used in the food industry as a meat preservative even though it has little antiscorbutic activity.

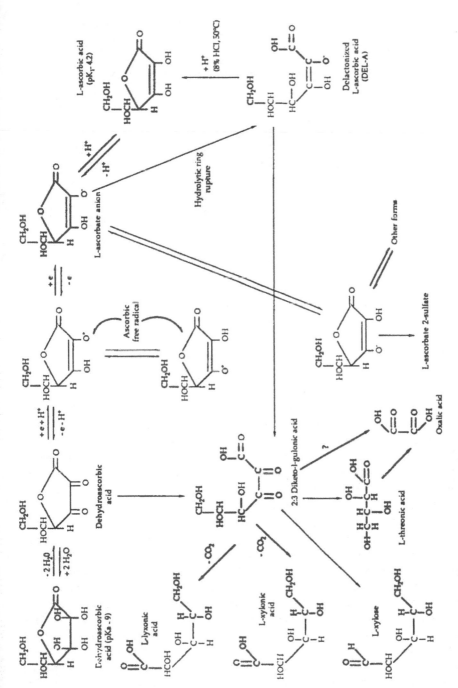

Fig. 6.11 L-Ascorbic acid and its degradation products.

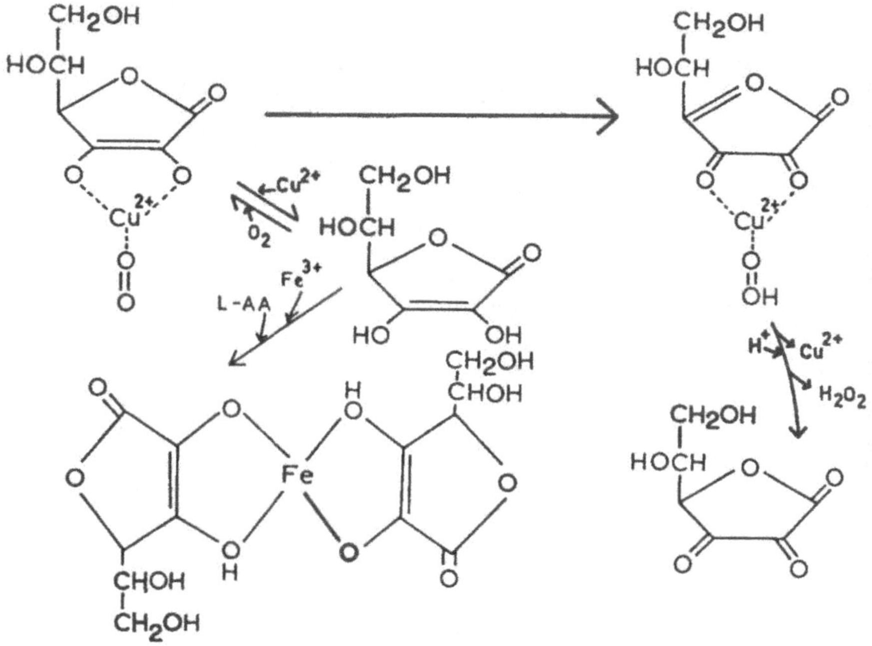

Fig. 6.12 Interactions of L-ascorbic acid with transition metal ions in aqueous systems.

The vitamin C contents of representative foods are summarized in Table 6.16. For humans, milk is the only animal product that provides a significant amount of the vitamin (1–5 mg/100 g), and, although there is some in liver and kidney, the best sources are fresh fruits (particularly citrus fruits, tomatoes, and green peppers), potatoes, and leafy vegetables. Some fruits such as guavas (300 mg/100 g) and black currants (150–230 mg/100 g) are particularly rich in the vitamin, but they contribute little to normal western dietary intake. For practical purposes, raw citrus fruits are good daily sources of L-ascorbic acid.

Although fruits and vegetables initially contain large amounts of ascorbic acid, most of it is lost due to oxidation during processing and storage (162). Plant material rich in ascorbic acid also contains ascorbic acid oxidase, which may normally be inactive or contained within vesicles. Fine chopping of such material is known to increase the enzyme's activity. The enzyme phenolase that produces enzymatic browning of fruits such as apples also causes loss of ascorbic acid. The enzyme functions with oxygen and ascorbic acid to reduce *ortho*-quinones back to *ortho*-diphenols. This results in the formation of dehydroascorbic acid, which is then rapidly converted to 2,3-diketogulonic acid. The process is catalyzed by copper(II) and other transition metal ions. The vitamin is also destroyed when foods contain-

Table 6.15 Tissue Concentrations (mg/100 g) of Ascorbic Acid

Tissue	Rat		Human	
Adrenal glands	280	−400	30	−40
Pituitary gland	100	−130	40	−50
Liver	25	−40	10	−16
Spleen	40	−50	10	−15
Lungs	20	−40	7	
Kidneys	15	−20	5	−15
Testes	25	−30	3	
Thyroid	22		2	
Thymus	40			
Brain	35	−50	13	−15
Pancreas			10	−16
Eye lens	8	−10	25	−31
Skeletal muscle	5		3	−4
Heart muscle	5	−10	5	−15
Bone marrow	12			
Plasma	1.6		0.4	−1.0
Saliva			0.07−	0.09

Source: Refs. 151 and 161.

ing it are kept warm for long periods (154). Although the vitamin C in frozen foods is stable and completely retained in storage, a significant amount is lost during thawing and cooking if not protected from oxidation (151).

Physiological Functions   Most physiological functions of ascorbic acid are related to its ability to act as an electron donor. In addition to its antiscorbutic properties, it can exert a significant influence on biological activities directly via its characteristic physicochemical properties of ionized state, oxidation/reduction, its capacity to lower interfacial tension, and its involvement in hydrogen bonding. Indirectly, it influences the concentration levels of the cyclic nucleotides cAMP and cGMP and affects the synthesis and levels of activity of several enzymes. Ascorbic acid is also involved in the production of hormones such as adrenalin, noradrenalin, and seratonin. It is also involved in the detoxification of toxins of both endogenous and exogenous origin. Vitamin C also has some antihistamine property and is directly involved in the collagen synthesis required for the healing of damaged tissues. Some of its major physiological functions related to free-radical pathology are described below.

Oxidation and Hydroxylation   Ascorbic acid is an essential factor in several hydroxylation reactions of the type $RH + O \rightarrow ROH$, primarily because of its ability

Table 6.16 Vitamin C Content of Some Common Foods

Category	Vit C (mg/100 g)	Category	Vit C (mg/100 g)
Fruits		Vegetables	
Acerola	1300	Peppers	125–200
Rose hips	1000	Kale	120–180
Berries	160–800	Parsley	170
Guava	300	Turnipgreen	139
Black currants	150–230	Horseradish	120
Lemons	50–80	Collard greens	100–150
Strawberries	40–90	Brussels sprouts	90–150
Oranges	40–60	Broccoli	70–160
Grapefruit	35–45	Spinach	50–90
Red currants	40	Watercress	79
Tangerines	30	Cauliflower	60–80
Pineapples	20–40	Kohlrabi	66
Raspberries	18–25	Cabbage	30–60
Melons	13–33	Turnip	15–40
Apples	10–30	Asparagus	15–30
Cherries	10	Leek	15–30
Peaches	7–14	Potato	10–30
Bananas	5–10	Bean	10–30
Plums	3	Pea	10–30
Milk		Onion	10–30
Human	3–6	Tomatoes	10–30
Cow	1–2	Squash	8–25
Meat (beef, pork)	0–2	Sweet corn	12
Fish	0–3	Rhubarb	10
Liver, kidney	10–40	Celery	7–10
Cereals		Carrot	5–10
Rice, oats, rye, wheat	0	Swiss chard	30

Source: Refs. 151, 154, 157–160.

to act as a redox couple, ascorbic acid/dehydroascorbic acid (H_2A/A), which undergoes cycling similar to that of cytochromes. It is also involved in the metabolism of several amino acids, leading to the formation of hydroxyproline, hydroxylysine, noradrenalin (norepinephrine), serotonin, homogentisic acid, and carnitine (163–166). Vitamin C also appears to be involved in the biotransformation of xenobiotics, such as drugs, poisons, and abnormal metabolites, that is carried out by the cytochrome P-450 enzyme system in the liver (166). Ascorbic acid also influences the activity of several oxidizing and hydroxylating enzymes (Table 6.17).

Antioxidant Properties   In recent years, the antioxidant properties of vitamin C have received considerable attention. Unlike the oxidation–reduction reactions in which ascorbate donates two electrons, the antioxidant reactions use its ability to donate a single electron to free-radical species. The products of such reactions are the quenched reactive species and the less reactive ascorbyl free radical. The ascorbyl radical then can be either reduced back to ascorbic acid or oxidized to form dehydroascorbic acid (141).

Ascorbate reacts rapidly with both superoxide and peroxyl radicals and even more rapidly with hydroxyl radicals (11,167,168). It also scavenges singlet oxygen (169), reduces thiyl radicals, and combines quickly with hypochlorous acid, a powerful oxidant generated at sites of inflammation (11). These free-radical-scavenging reactions are especially important in the eye and in the extracellular fluid of the lung, where they provide protection against other oxidizing agents such as ozone.

Ascorbic acid also plays a vital role in maintaining the balance between oxidative products and the various cellular antioxidant defense mechanisms. The interdependency of such reactions involving ascorbic acid, vitamin E, selenium, catalase, and GSH was described earlier.

Table 6.17   Enzymes Dependent on Ascorbic Acid for Maximal Activity

Enzyme	Function
Proline hydroxylase	*trans*-4-Hydroxylation of proline in procollagen biosynthesis
Procollagen-proline 2-oxoglutarate 3-dioxygenase	*trans*-3-Hydroxylation of proline in procollagen biosynthesis
Lysine hydroxylase	5-Hydroxylation of lysine in precollagen biosynthesis
γ-Butyrobetaine 2-oxoglutarate 4-dioxygenase	Hydroxylation of a carnitine precursor
Trimethyllysine-2-oxoglutarate dioxygenase	Hydroxylation of a carnitine precursor
Dopamine β-monooxygenase	Dopamine β-hydroxylation in norepinephrine biosynthesis
Peptidyl glycine α-amidating monooxygenase activity	Carboxy terminal α-amidation of glycine-extended peptides in peptide hormone processing
4-Hydroxyphenylpyruvate dioxygenase	Hydroxylation and decarboxylation of a tyrosine metabolite

Source: Ref. 151.

In addition to its role as an antioxidant, ascorbic acid is involved in several immunological and antibacterial reactions (170–172). It increases the mobility of white blood cells, stimulates the hexose monophosphate shunt within the cell, and protects the leukocyte membranes from oxidative damage. Vitamin C is also involved in the production of antibodies and white blood cells such as B-cells and T-cells (21). It also combats infections and promotes the production of interferon and interleukin-1, two lymphokines that contribute to the body's immune response (173,174).

The properties that help ascorbic acid to exert its protective beneficial effects in biological systems also have some drawbacks. For example, it can reduce Fe(III) to Fe(II), and in the presence of H_2O_2 it can stimulate hydroxyl formation by the Fenton reaction. Under certain clinical conditions, ascorbic acid can also stimulate iron-dependent peroxidation of membrane lipids and may cause severe reactions in cases of iron overload by stimulating hydroxyl radical formation in vivo (11).

Vitamin A

The association of a deficiency of vitamin A with xerophthalmia (night blindness) is well documented in the literature. Vitamin A deficiency is still a major nutritional problem throughout the developing world. The fat-soluble vitamin plays an important role in vision processes, although it does not enter directly into free-radical-mediated processes. It, however, does appear to have weak antioxidant properties (175). Retinoid compounds, including retinyl palmitate, retinyl acetate, 13-cis-retinoic acid, ethyl retinamide, 2-hydroxyethylretinamide, retinyl methyl ether, N-(4-hydroxyphenyl)retinamide and several other synthetic retinoids, are implicated as "blocking agents" for the prevention of several cancers (176). Experimentally, these compounds have been shown to suppress the expression of neoplasia in cells previously exposed to doses of carcinogenic agents.

The term *vitamin A* refers to all derivatives of β-ionone (other than carotenoids) that possess the biological activity of all-*trans*-retinol or are closely related to it structurally (177). Major compounds belonging to the vitamin A group are diagrammed in Fig. 6.13.

The richest sources of preformed vitamin A (i.e., retinol and its esters) include liver oils of the shark, marine fish such as halibut, and marine mammals such as polar bear. The common dietary sources include various dairy products, eggs, liver, and other internal organs such as the kidney and heart (177). Provitamin A activity in the form of carotenoids, however, is found in large amounts in carrots and green leafy vegetables such as spinach and amaranth. Although vitamin A is prone to oxidation, it is generally present together with natural antioxidants in an oil-rich state and is therefore fairly stable.

The primary biochemical function of vitamin A is in the vision process and mechanisms. The primary photochemical event in the process is the very rapid

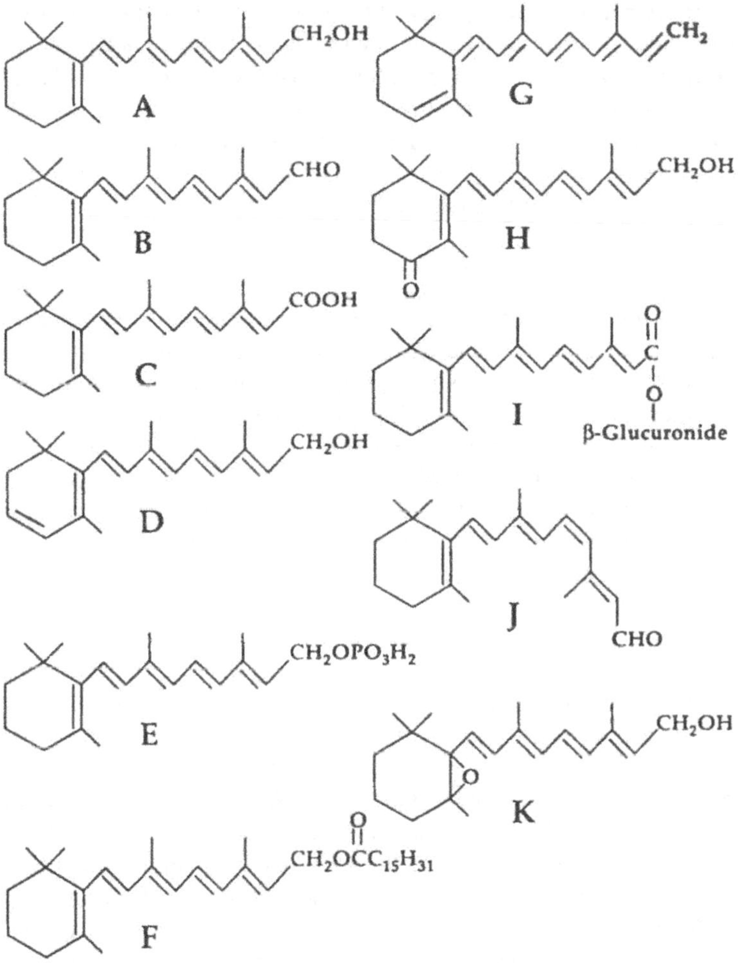

Fig. 6.13 Major compounds belonging to the vitamin A group. (A) Retinol; (B) retinal, retinaldehyde; (C) retinoic acid; (D) 3-dehydroretinol; (E) retinyl phosphate; (F) retinyl palmitate; (G) anhydroretinol; (H) 4-ketoretinol; (I) retinoyl β-glucuronide; (J) 11-*cis*-retinaldehyde; (K) 5,6-epoxyretinol.

isomerization of 11-*cis*-retinal, which is present in a protonated Schiff's base of a specific lysine residue of the visual pigment rhodopsin (177). A complex series of biochemical reactions in the visual cycle eventually lead to the formation of all-*trans*-retinal, which is then transported by the retinol-binding protein in plasma to retinal pigment epithelium via a receptor-mediated process. The all-*trans*-retinol can then be isomerized to 11-*cis*-retinol by all-*trans*:11-*cis*-retinol isomerase (178).

In addition to its role in the vision cycle, vitamin A is essential for normal growth and development and tissue differentiation in animals (177). Adequate levels of vitamin A in the human diet are also essential for normal maintenance of epithelial tissues and proper functioning of the immune system, including tumor surveillance (179). In addition, vitamin A and retinoids (i.e., vitamin A derivatives) may directly influence gene expression (180). High doses of vitamin A, however, are toxic and teratogenic in many animals, including humans (181).

Carotenoids

Carotenoids are a group of fat-soluble pigments that contribute to the yellow, orange, and/or red coloration of fruits and vegetables. They are also found in insects, birds, and other plant-eating animal species, including humans. Although close to 600 carotenoids have been identified and characterized in nature, only about 50 possess the biological activity of vitamin A (177,182). This biological activity is defined in terms of the carotenoids containing at least one β-ionone ring that is not hydroxylated in order to show the vitamin A activity. Such carotenoids are therefore sometimes referred to as "provitamin A" compounds. β-Carotene is the most abundant of the carotenoids found in human foods, and it also contains the highest vitamin A activity.

The recent renewed interest in carotenoids can be partly attributed to the role of β-carotene in reducing the risks of certain cancers through a mechanism that does not require its conversion to vitamin A (183). Several epidemiological studies have also suggested the possible association of lung cancer with a low intake of vegetables and fruits, the primary sources of carotenoids in the human diet. In 1982, the Committee on Diet, Nutrition, and Cancer of the National Research Council and the National Academy of Sciences in the United States concluded that the epidemiological evidence was sufficient to justify a higher intake of carotenoids as a possible means of lowering cancer risk (184). Recent studies demonstrating the antioxidant functions of β-carotene and carotenoids in general as well as their effective quenching of singlet oxygen have provided further stimulus to this research field. The chemistry and biological functions of various carotenoids in preventing free-radical-mediated cellular injuries are briefly discussed below.

Chemistry, Nomenclature, and Properties   Carotenoids represent a very large group of isoprenoid substances with various structural characteristics and biological activities. They are synthesized in plants and microorganisms from acetyl coenzyme A by a series of well-defined condensation reactions. Most carotenoids are 40-carbon compounds, less than 10% being hydrocarbons. Most have oxygen functions; some are conjugated to sugars and other molecules (185).

The first C_{40} carotenoid synthesized by the condensation reactions is phytoene (Fig. 6.14). This parent compound can then be dehydrogenated to other acyclic carotenoids and ultimately cyclized to carotenes (185). Cyclic carotenes may then

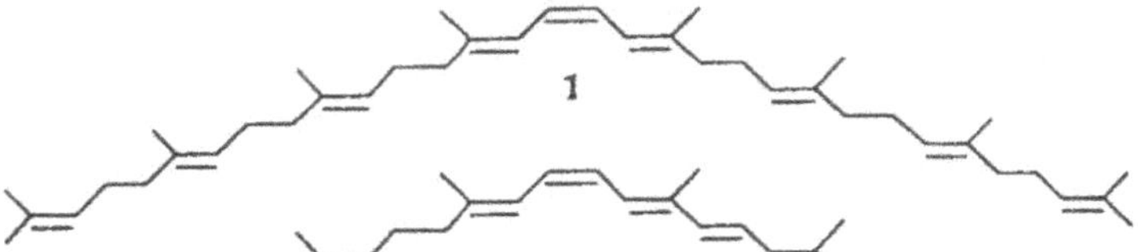

Fig. 6.14 Carotenoids commonly found in the human diet. (1) Phytoene; (2) phytofluene; (3) lycopene; (4) α-carotene; (5) β-carotene; (6) β-cryptoxanthin; (7) zeaxanthin; (8) lutein; (9) canthaxanthin; (10) violaxanthin; (11) neoxanthin; (12) astaxanthin.

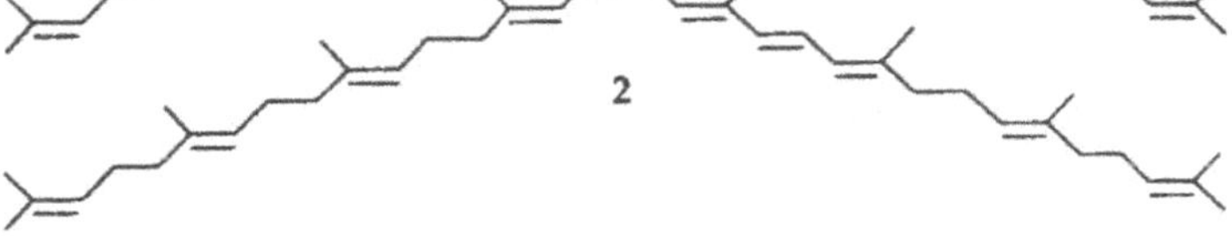

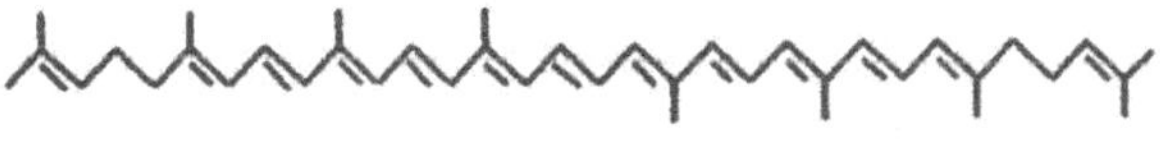

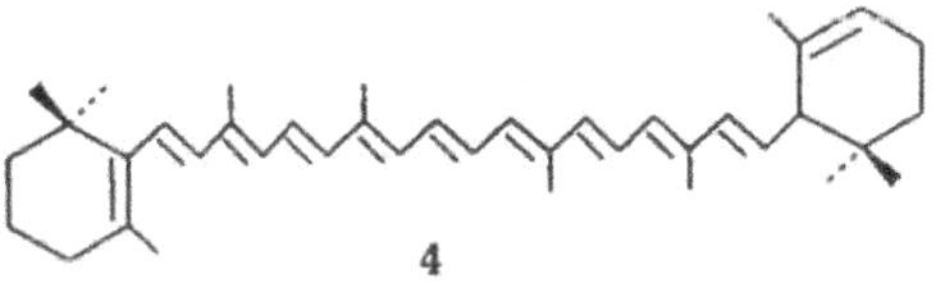

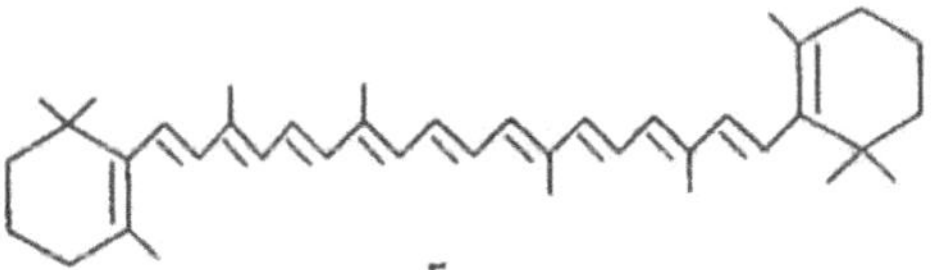

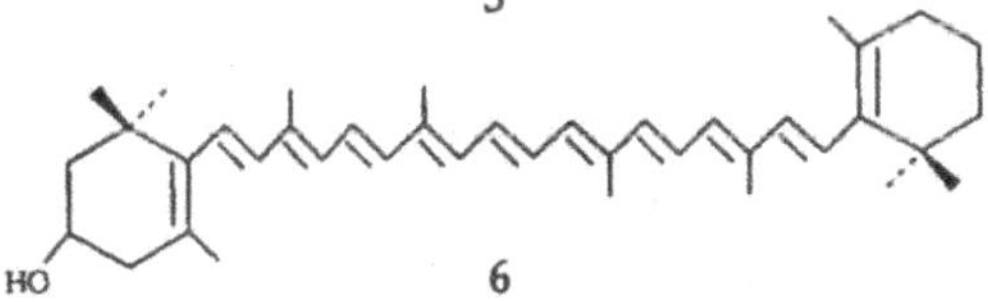

be oxidized to hydroxylated and epoxide derivatives, converted to allenic and less saturated derivatives, or cleaved oxidatively to shorter products. In addition, cis-trans isomers of a given carotenoid can generate a number of derivatives with different biological activities. Carotenoids that are normal constituents of the human diet are shown in Fig. 6.14.

Carotenoids are sensitive to alkali and very sensitive to air and light, especially at high temperatures. They are insoluble in water, ethanol, glycerol, and propylene glycol but soluble in edible oils at room temperature.

The conversion of biologically active carotenoids that contain at least one β-ionone ring that is not hydroxylated to vitamin A occurs by two primary oxidative reactions (187). No other route to vitamin A is known except via carotenoid

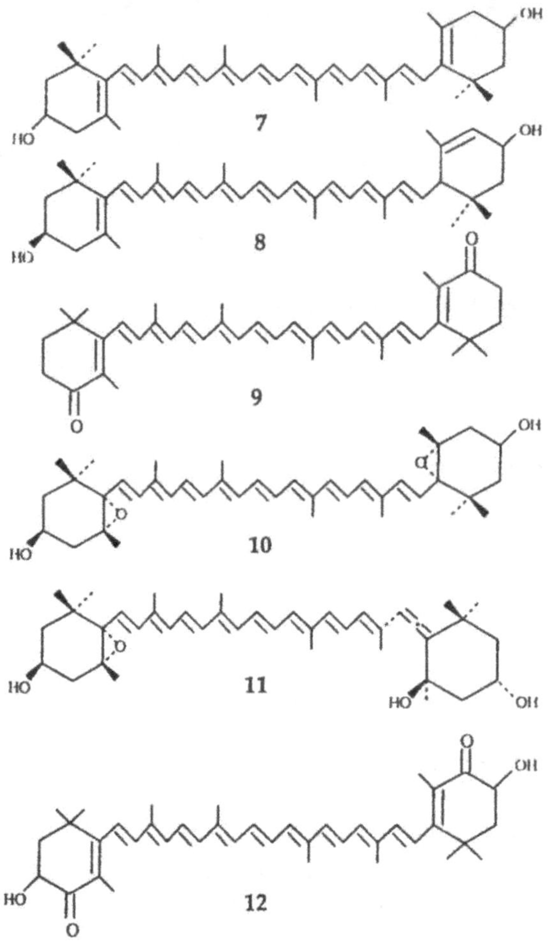

Fig. 6.14 (*Continued*)

cleavage. The two possible routes include a "central cleavage" at the central 15–15′ double bond and the other "excentric cleavage" at one or more of the other double bonds. The first route would thus yield two molecules of retinol by the central cleavage of β-carotene, whereas the second pathway yields one long and one short β-apocarotenol molecule, depending on which double bond is cleaved. The central cleavage is catalyzed by the enzyme β-carotenoid 15,15′-dioxygenase, which is found in mammalian tissues such as the intestine and liver (188,189). The enzyme requires molecular oxygen and is inhibited by sulfhydral- and iron-binding reagents. The dioxygenase is capable of cleaving many carotenoids, including several β-apocarotenols. The maximal activity of the intestinal form of this enzyme is sufficient to meet the nutritional needs of vitamin A but is not rapid enough to induce hypervitaminosis A (182). Among the carotenoids, even though β-carotene contains the highest provitamin A activity, its inefficient conversion to retinal, susceptibility to oxidative reactions leading to the formation first of retinoic acid and then to inactive products, and poor absorption compared to that of preformed vitamin A in the diet result only in about one-sixth the overall utilization of retinal. Therefore, in calculating the vitamin A activity, six units of β-carotene by weight is considered to be equivalent to one unit of retinal (190).

The excentric cleavage of carotenoids leading to the formation of retinal, although known to occur in plants and some microorganisms, appears not to play an important role in mammalian systems (182). Carotenoid dioxygenase activity, however, has been identified in mammals, yielding β-apocarotenals from β-carotene both in vivo and in vitro. Because the β-apocarotenals can be converted to retinal by the central cleavage enzyme, their nutritional activity may be readily explained by this known mechanism. The failure to detect β-apocarotenyl esters in human serum following administration of large doses of β-carotene nevertheless suggests that this pathway may not play a significant role in generating vitamin A activity. The possible transformations of β-carotene in mammalian systems are shown in Fig. 6.15.

Because the above two pathways lead to their conversion to nutritionally active forms and use singlet oxygen quenching as their selective biological function, Olson (177,191) recently classified carotenoids into four distinct types as follows:

Type 1. Biologically and nutritionally active (e.g., β-carotene)
Type 2. Biologically active but nutritionally inactive, at least in mammals (e.g., canthaxanthin, lycopene, violaxanthin)
Type 3. Biologically inactive but nutritionally active (e.g., β-apo-14'-carotenol)
Type 4. Biologically and nutritionally inactive (e.g., phytoene)

Since most carotenoids in the human diet, such as lycopene, lutein, zeaxanthin, neoxanthin, canthaxanthin, and violaxanthin, are not nutritionally active but are nevertheless normal constituents of human blood and human tissues, their biological activities, independent of their provitamin A activity, are receiving a great deal of attention from the scientific community. Thus the traditional view of the importance of carotenoids in mammals as solely the precursors of vitamin A is no longer considered valid.

Dietary Sources Animals, including humans, are not capable of de novo synthesis of β-carotene. This micronutrient must therefore be supplied in the diet. Fresh fruits and vegetables are excellent sources of carotenoids in the human diet (Table 6.18). Plants differ greatly in their carotenoid content as well as type. For

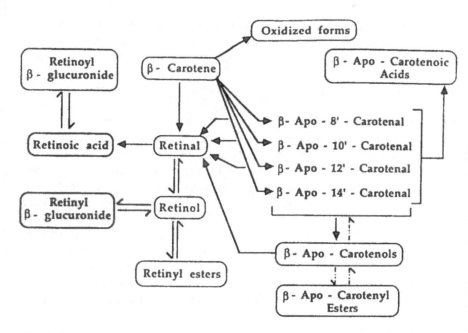

Fig. 6.15 Possible transformations of β-carotene in mammals. The solid arrows denote the known biological enzymatic reactions, the hyphenated arrows indicate probable but undemonstrated biological reactions, and the dash-dot arrows show nonbiological or possibly biological reactions. (Data sources Refs. 182 and 192).

example, carrots contain virtually all β-carotene, and sweet potatoes and spinach also contain high levels. In contrast, tomatoes are almost devoid of β-carotene; their predominant carotenoid is the nutritionally inactive lycopene. Other nutritionally inactive carotenoids such as xanthophylls and lutein occur at high levels in oranges and broccoli.

Dietary carotenoids are absorbed directly through the intestine; some are incorporated into the chylomicra, appearing in the blood via the lymph (191). The

Table 6.18 Dietary Sources of Carotenoids

Milk products: Milk, butter, cheese, ice cream
Seafood: Fish, bivalves, crustaceans
Fruits: Orange, pineapple, apricot, cranberry, grape, fig, mango, nectarine, pear, strawberry, watermelon, prune, peach
Vegetables: Carrot, squash, sweet potato, pumpkin, tomato, spinach, fenugreek, lettuce, asparagus, broccoli, pepper, cauliflower, corn, pea, bean
Miscellaneous: Palm oil, egg yolk, pasta products

fat-soluble carotenoids are transported in the blood primarily by low-density lipoproteins and, to a lesser extent, by high- and very low density lipoproteins (192). Most of the carotenoids distributed in the human tissue are localized in the adipose tissue (80–85%), followed by liver (8–12%) and muscle (2–3%). High concentrations are also found in the corpus luteum and adrenal glands. Of the total body pool of carotenoids, serum contains approximately 1% (192). Among the carotenoids, β-carotene, α-carotene, cryptoxanthin, lycopene, and lutein are the major components of human serum, with lycopene occurring in the highest concentration (193).

Physiological Functions   With the exception of their provitamin A activity, very little attention was paid in the past to the digestion, absorption, tissue localization, and metabolic fate of dietary carotenoids. However, the fact that many nutritionally inactive carotenoids are found in various human tissues and serum, the discovery that carotenoids can act as antioxidants in biological systems, and the growing epidemiological evidence of an inverse relationship between the consumption of foods rich in carotenoids and the risks of several cancers have stimulated a great deal of interest in the physiological functions of carotenoids and their possible beneficial role in the prevention of cancer. The important physiological functions of carotenoids related to free-radical pathology are described here.

Bendich and Olson (192) classified the biological activities of carotenoids into three distinct categories: functions, actions, and associations (Table 6.19). The biological functions may be defined as a preventive role of carotenoids in whose absence physiological capability may be impaired. The actions are defined in terms of physiological or pharmacological responses to the administration of carotenoids. The response itself may be either beneficial or adverse and may not be essential for physiological well-being. In contrast, the biological associations involve the possibility of a definitive relationship between carotenoids and some physiological or medical events such as the risk of a certain type of cancer (191). In terms of free-radical pathology, the most important biological functions of carotenoids appear to be their antioxidant nature, their ability to quench singlet oxygen, and possible roles in enhancing the immune responses and the inhibition of mutagenesis.

Antioxidant Functions   Carotenoids are known to scavenge and deactivate free radicals both in vitro and in vivo. β-Carotene especially has been shown to protect isolated lipid membranes from peroxidation, LDL-containing lipids from oxidation, and liver lipids from oxidation induced by trichloromethyl radical, although theoretically all carotenoids with a conjugated double bond system should act similarly (192,194).

Burton and Ingold (194) investigated the mechanisms by which β-carotene acts as an antioxidant. Unlike antioxidants that prevent the initiation of lipid peroxidation, β-carotene stops the chain reactions by trapping free radicals. It therefore

Table 6.19 Biological Activities of Carotenoids in Animals[a]

Functions
  A few carotenoids form vitamin A
Actions
  Serve as antioxidants (low PO_2)
  Serve as prooxidants (high PO_2) (+ / −)
  Enhance the immune response
  May enhance fertility (?)
  Reduced photoinduced neoplasm
  Inhibit mutagenesis
  Inhibit cell transformation in vitro
  Inhibit tumor development in vivo (+ / −)
  Prevent sister chromatid exchange (SCE)
  Reduce micronuclei induced by betel nut or tobacco chewing
Associations
  Intake of carotenoid-rich foods is associated with
    Lower lung cancer risk
    A lower risk of other cancers or of total cancers (+ / −)
  Serum β-carotene (carotenoid) levels are inversely associated with
    Smoking behavior
    Lung cancer risk (+ / −)
    Cervical dysplasia

[a]Question mark (?) indicates lack of agreement concerning the relationship;
(+ / −) indicates that some studies find significant effect or relationship and
others do not.
Source: Ref. 191.

appears to act as a chain-breaking antioxidant, although it does not have the characteristic structural features associated with this class of antioxidants. β-Carotene, in fact, has an extensive system of conjugated double bonds that impart a prooxidant character to the molecule. It is therefore very susceptible to attack by peroxyl radicals. Burton and Ingold (195) found that in in vitro systems, β-carotene interacts irreversibly with peroxyl radicals to form a carbon-centered carotenoid radical:

$$\beta\text{-Carotene} + ROO^\bullet \rightarrow \beta\text{-carotene}^\bullet$$

The carotenoid radical is resonance-stabilized to such an extent that its subsequent reaction with molecular oxygen to form a peroxy-β-carotene radical is reversible:

$$\beta\text{-Carotene}^\bullet + O_2 \rightarrow \beta\text{-carotene-OO}^\bullet$$

Under conditions of low oxygen pressure, the shifting of the equilibrium to the left greatly reduces the concentration of the highly reactive peroxy-β-carotene radicals. According to Burton and Ingold (194), the reactivity of β-carotene toward peroxyl radicals and the stability of the resulting β-carotene radical give the molecule its antioxidant capability. The first feature results in its competing with other lipids for the peroxyl radicals, whereas the second characteristic leads to the formation of the stable carbon-centered β-carotene radical, especially at low oxygen partial pressure. Furthermore, the β-carotene radical can also undergo termination by reaction with another peroxyl radical:

β-Carotene$^{\bullet}$ + ROO$^{\bullet}$ → inactive products

The rate of this reaction is also much faster than that of peroxyl–peroxyl self-termination reactions. Thus the expendible β-carotene can divert particularly damaging chain reaction products into much less deleterious side reactions.

β-Carotene, however, does not behave as an antioxidant at normal oxygen concentrations (196). The antioxidant function of β-carotene therefore complements the action of other antioxidant molecules such as catalase, GSH-Px, vitamin C, and vitamin E, which are very effective at normal oxygen concentrations. Vitamin A, in contrast, is a very weak antioxidant (175).

Singlet Oxygen Quencher   Carotenoids are very effective quenchers of singlet oxygen (197). Due to its instability and high energy level, singlet oxygen can potentially transfer this energy to other molecules, generating free radicals in the process. Singlet oxygen is therefore involved in oxidative reactions that may impair or destroy important cellular components such as membrane lipids, enzymes, and nucleic acids.

Foote and coworkers (198) first demonstrated that the ability of carotenoids to quench singlet oxygen is related to the number of double bonds they contain. It was later confirmed that carotenoids with 9, 10, and 11 conjugated double bonds are better quenchers of singlet oxygen than those with eight or fewer conjugated double bonds (199). Thus, β-carotene, isozeaxanthin, and lutein, which contain 11, 11, and 10 conjugated double bonds, respectively, all show quenching ability.

Carotenoids with nine or more conjugated double bonds can absorb the energy from the singlet oxygen, which is then distributed over all the single and double bonds in the molecule. The "energized" carotenoid then releases the absorbed energy in the form of heat, thereby restoring it to its normal energy level. Carotenoids are therefore not destroyed during the process and can repeat the process with additional singlet oxygen molecules. One molecule of β-carotene, for example, is estimated to quench up to 1000 molecules of singlet oxygen (197). Using a similar mechanism, carotenoids can also prevent damaging skin reactions due to photooxidation in humans suffering from the genetically inherited light-sensitive skin disease erythropoietic protoporphyria (200).

Immune Functions   Early observations on the anti-infective effects of carotenoids against ear, bladder, kidney, and gut infections were attributed to their provitamin A activity (201,202). Most such effects of carotenoids on immune functions now appear to be related to their antioxidant properties. The initial responses of leukocytes to infection often involve the production of free radicals and reactive oxygen molecules. However, overproduction of these chemical species often results in injury to the leukocytes themselves as well as to neighboring cells and tissues. Carotenoids containing nine or more conjugated double bonds can effectively quench the reactive oxygen species. Both in vitro and in vivo studies have shown that carotenoids in general, and β-carotene in particular, protect phagocytic cells from oxidative damage, enhance T- and B-lymphocyte proliferative responses, stimulate effector T-cell functions, and enhance the tumoricidal capacities of macrophages, cytotoxic T cells, and natural killer cells (203–206). These effects were seen with all carotenoids devoid of provitamin A activity but containing the antioxidant and singlet oxygen quenching capacities of β-carotene. The antitumor and antioxidant effects of carotenoids, coupled with the fact that vitamin A is a relatively poor antioxidant and cannot quench singlet oxygen, may very well explain the epidemiological data linking lower carotenoid status with higher incidences of certain cancers (203).

The antioxidant, antimutagenic, chemopreventive, and immunoenhancing functions of carotenoids are summarized in Fig. 6.16. More definitive studies detailing these in vivo functions of carotenoids are being carried out, especially considering the fact that β-carotene, unlike vitamin A, has proven to be nontoxic to humans when given in high dosages (208).

Vitamin B Complex

With the exception of a few, B-complex vitamins are not directly involved in the defense mechanisms against free-radical pathology. Nevertheless, several vitamins of this group, such as riboflavin, niacin, pyridoxine, and pantothenic acid, are cofactors of enzymes that may ultimately be involved directly or indirectly in these processes. Vitamins belonging to the B-complex group are water-soluble compounds. The biochemical functions of these vitamins are summarized in Table 6.20.

6.3.4 Antioxidant Minerals

The role of copper, zinc, iron, and selenium as cofactors for enzymes such as SOD, catalase, and GSH-Px was described earlier. Copper and iron, however, are also involved in free-radical chain propagation, especially in the membrane lipids. Zinc plays an important role in cellular immunity. It stimulates the disease-fighting T cells, helps the B cells release antibodies, and promotes the production of "activated" natural killer cells that help to destroy cancerous cells (209,210). Iron is required for the optimal activity of such enzymes as ribonucleotidyl reductase and myeloperoxidase, which play essential roles in the functions of lymphocytes and neutrophils, respectively. Iodine is an essential component of the thyroid hormone.

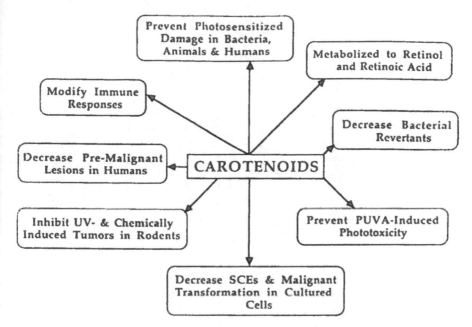

Fig. 6.16 Schematic representation of the biological functions of carotenoids. (From Ref. 207.)

Among these minerals, the antioxidant role of copper has only recently been recognized. It has both prooxidant and antioxidant properties in biological systems, and therefore its optimal range of intake may ultimately prove to be narrower than that of other minerals (211). It should be noted, however, that in spite of its high reactivity in biological systems, it is rarely available in sufficient quantities in the body to participate in free-radical reactions (212). Under normal dietary conditions, the absorption and excretion of copper are regulated to prevent excess copper from accumulating. Moreover, humans are more likely to consume too little rather than too much copper.

The actual dietary intake of copper was estimated to be 1.2 and 0.9 mg/day, respectively, in the male and female population groups in the United States. These values are considerably lower than the recommended intake of 1.5–3.0 mg of copper per day (213). Furthermore, copper utilization may be impaired by high intakes of dietary constituents such as zinc, ascorbic acid, iron, fructose, and sucrose (214). The antioxidant potential in people with low intakes of copper or high intakes of copper antagonists, therefore, may be of greater practical significance than its prooxidant functions.

Recently, Johnson et al. (212) proposed a theoretical model of how copper deficiency might impair antioxidant status (Fig. 6.17). Copper deficiency creates

Table 6.20 Biochemical Functions and Deficiency Symptoms of B-Complex Vitamins

Vitamin	Biochemical functions	Deficiency symptoms
B₁ (thiamine)	Oxidative decarboxylation of α-keto acids in the TCA cycle, involved in transketolase reactions and the pentose pathway, neurotransmission	Anorexia, heart involvement, and neurological symptoms
B₂ (riboflavin)	Coenzyme forms of FMN and FAD are involved in biological oxidation–reduction reactions as hydrogen acceptors; essential in the metabolism of carbohydrates, fats, and lipids; involved in conversion of pyridoxine and folic acid into their coenzyme forms	Seborrheic dermatitis around the nose and mouth, cheilosis, stomatitis, presenile cataract, anemia
Nicotinic acid (nicotinamide, niacin)	Integral part of the coenzymes NAD and NADP involved in hydrogen-donor reactions	Pellagra with typical dermal symptoms, insomnia, loss of appetite, weight loss, indigestion, diarrhea, abdominal pain, vertigo, nausea and vomiting
B₆ (pyridoxine)	Active form, pyridoxal-5-phosphate (PLP) serves as a coenzyme in several metabolic reactions including those of aminotransferases, deamidases, and nonoxidative decarboxylases; PLP-dependent enzymes also involved in the synthesis of epinephrine, norepinephrine, and porphyrins; involved in the conversion of trytophan to niacin	Hypochromic microcytic anemia, weight loss, abdominal distress, vomiting, hyperirritability, depression and convulsions; adversely affects both cellular and antibody functions of the immune system
Biotin	Serves as a prosthetic group of several enzymes including carboxylases, transcarboxylases, and decarboxylases, where it acts as a mobile carboxyl carrier	Deficiency in humans rarely seen due to its production by intestinal microflora
Pantothenic acid	Part of coenzyme A involved in acetyl transfers and acts as either a hydrogen donor or acceptor	Deficiency not usually seen except in case of severe malnutrition; symptoms include abnormal skin sensations of the feet and lower legs, fatigue, insomnia, vomiting, muscular weakness
Folic acid	Cofactor in amino acid and nucleotide metabolism as acceptors and donors of one-carbon units	Anemia
B₁₂ (cyanocobalamin)	Cofactor for cobamide enzymes	Addisonian pernicious anemia

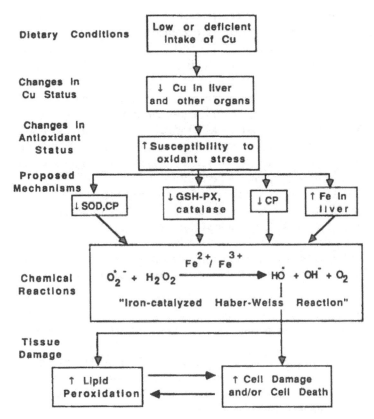

Fig. 6.17 Theoretical model of copper as an antioxidant nutrient. CP = ceruloplasmin, Cu = copper, Fe = iron, GSH-PX = glutathione peroxidase, H_2O_2 = hydrogen peroxide, $O_2^{\cdot-}$ = superoxide radical, $HO^{\cdot}$ = hydroxyl radical, SOD = superoxide dismutase. (From Ref. 212.)

several disturbances in the antioxidant protection system. These include increased liver iron concentrations and decreased activities of GSH-Px and catalase in numerous tissues, as well as those of Cu/Zn SOD and ceruloplasmin.

Animal studies with copper-deficient rats and mice have also shown enhanced sensitivity to factors that increase oxidant stress, such as carbon tetrachloride, adriamycin, ozone, and hyperoxia (212). This sensitivity to oxidant stress is generally manifested as increased mortality, convulsions, expiration of pentane or ethane, or decreased body weight. Cells, microsomes, or mitochondria isolated from copper-deficient animals also show increased lipid peroxidation in the presence of ascorbic acid, ferrous iron, ADP, and/or NADPH.

Selenium is also emerging as a dietary factor that may prove to be of major significance as a prophylactic agent against cancer. Selenium at high levels of intake can prevent cancer in animals (22). The protective effects of selenium against free-radical injury are primarily attributed to the selenoenzyme GSH-Px. Two other selenoenzymes may also protect against oxidant stress: phospholipid hydroperoxide GSH-Px, which metabolizes the fatty acid hydroperoxides, and selenoprotein P, which is made in the liver and secreted in the plasma (215,216). Selenoprotein P may also protect against oxidant stress or serve as a selenium transport protein.

The protective level of selenium, which is generally on the order of 5 ppm in the diet, is much higher than the physiological level of intake (i.e., 0.01–0.1 ppm) and indeed may approach the level of intake of selenium at which toxicity would be observed. Several mechanisms for the anticarcinogenic activity of selenium have been proposed, although none has been discovered that could apply to all types of cancer in which selenium has been found to be protective. Moreover, optimal anticarcinogenic activity requires selenium intake well beyond the saturation point of GSH-Px, suggesting that additional mechanisms may be involved. Nonetheless, a growing body of strong epidemiological evidence shows correlations between low blood levels of selenium (or low dietary intake) and an increased risk of some kinds of cancer (217). Such evidence falls into three categories:

1. Ecological evidence, which frequently, but not consistently, shows a correlation between high cancer incidence or death rate and a low apparent intake of selenium in the population from which the subjects originated.
2. Case-control evidence, which shows a correlation between low blood selenium and high incidence of cancer in subjects actually diagnosed as having overt cancer or a precancerous state. Such evidence is, however, flawed because it can never be clear whether the low blood selenium level was a cause or a consequence of the disease.
3. Prospective studies, in which the subsequent medical history of a large cohort of subjects is monitored for many years after sampling blood and measuring the selenium level. Several such studies have now shown that there is, for a number of cancer sites, an excellent correlation between low blood selenium level at a time when the disease may have been developing and a higher incidence of cancer some years later. However, these studies are ongoing, and only preliminary data are available.

More definitive studies on these minerals, especially selenium, are certainly required before any firm conclusions are drawn regarding their protective roles in the free-radical-mediated cellular diseases.

6.3.5 Uric Acid

Uric acid is a by-product of purine metabolism and has usually been thought to be a waste product with no biological function (218–221). Human tissues do not contain the enzyme urate oxidase. It therefore accumulates in the body. Its normal blood plasma concentration in humans ranges from 0.25 to 0.45 mM. Uric acid (Fig. 6.18) is produced as a by-product of xanthine dehydrogenase, an enzyme that oxidizes hypoxanthine to xanthine and then to uric acid while reducing NAD^+ to NADH (222).

Uric acid is a powerful scavenger of singlet oxygen and peroxyl and hydroxyl radicals and may function as an antioxidant in vivo (223). The loss of urate oxidase during human evolution, therefore, might have been beneficial. Uric acid also interferes with radical reactions by binding iron and copper ions in forms that do not participate in radical reactions. For example, it is a very powerful inhibitor of copper-dependent formation of $HO^{\cdot}$ from H_2O_2, apparently acting by binding the copper ions (222). Because of this ability to tightly bind iron and copper, uric acid also has the ability to inhibit lipid peroxidation. Moreover, it is a powerful scavenger of ozone and of hypochlorous acid, an oxidant produced by the enzyme myeloperoxidase.

The reactions of $HO^{\cdot}$ with uric acid produces a range of carbon-centered radicals that mostly react with oxygen to give urate peroxyl radicals (11):

$$R - \overset{|}{\underset{/}{C}} - H + HO^{\cdot} \rightarrow R - \overset{|}{\underset{/}{C^{\cdot}}} + H_2O$$

$$R - \overset{|}{\underset{/}{C^{\cdot}}} + O_2 \quad \rightarrow R - \overset{|}{\underset{/}{C}}OO^{\cdot}$$

Although much less reactive than $HO^{\cdot}$, urate-derived radicals are not completely harmless. They can inactivate the enzyme alcohol dehydrogenase from yeast and the human α-antiproteinase (220,222). Thus, like GSH, uric acid is not always a "perfect" antioxidant.

6.3.6 Antioxidant Additives

Antioxidants of both synthetic and natural origin are added intentionally to processed foods to prevent the peroxidation of lipids and thus flavor degradation

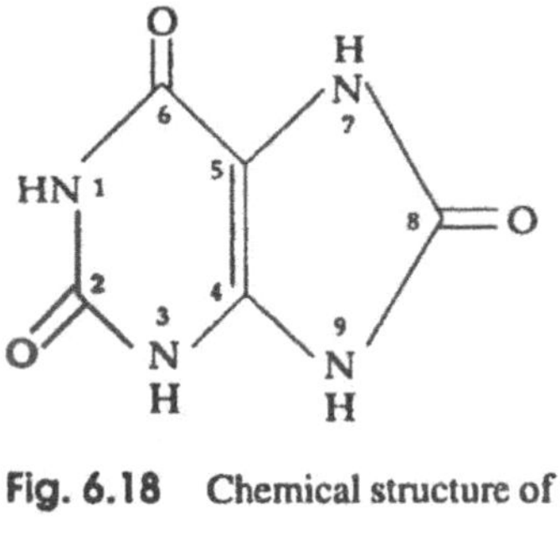

Fig. 6.18 Chemical structure of uric acid (8-hypoxanthine).

during storage. The oxidative degradation of PUFA in food products during processing and storage causes several undesirable changes such as limited shelf life; objectionable flavor, taste, or influence on rheological properties; loss of nutritive value; and potential health risks. To retard and/or prevent such deleterious changes, antioxidants are widely used as intentional additives in the food industry.

Although the primary function of these additives is to prevent food spoilage by oxidative reactions, they may indirectly also be responsible for preventing free radical-medicated processes in biological systems. Synthetic antioxidants, such as BHA, BHT and PG, have both inhibitory and promoting effects against various chemical carcinogenesis in experimental animals. Their promoting effects are described in Chapter 5. Therefore, the levels at which these compounds are added to various foods are strictly governed by the Food and Drug Administration's (FEA) guidelines. Because of such low levels of usage in foods, their inhibitory effects on chemical carcinogenesis, are likley to be minimal as compared to some of the vitamins and minerals described earlier.

The various oxidative degradation processes and the mechanisms by which different antioxidants exert their influence are shown in Fig. 6.19. The autoxidation process involves two interrelated oxidative cycles: a radical chain reaction that

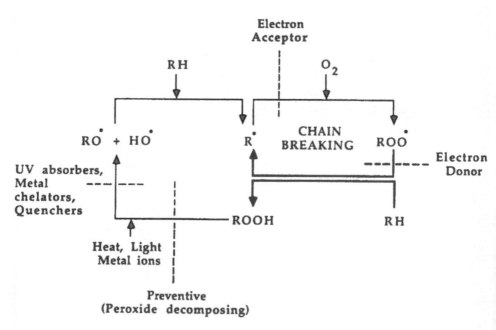

Fig. 6.19 Oxidative degradation processes and the mechanisms of antioxidant actions. RH = lipid, $R^\bullet$ = alkyl radical, $ROO^\bullet$ = alkylperoxyl radical, $HO^\bullet$ = hydroxyl radical, ROOH = hydroperoxide.

leads to alkyl ($R^{\bullet}$) and alkylperoxyl ($ROO^{\bullet}$) radicals and the formation of radical generators (224). Under normal conditions, the alkyl radicals react rapidly with oxygen to give alkylperoxyl radicals followed by hydrogen abstraction from the polymer substrate to give polymer hydroperoxides (ROOH). The hydroperoxides are the most important products of the first cycle. They are the sources of further initiating radicals in the second cycle, giving rise to the highly reactive hydroxyl radicals. The latter then react with the substrate to give alkyl radicals that feed back into the main autoxidation cycle.

Antioxidants interrupt the main oxidative cycle by removing the primary chain-propagating radicals, $R^{\bullet}$ and $ROO^{\bullet}$, either by oxidizing $R^{\bullet}$ (as chain-breaking electron acceptors) or by reducing $ROO^{\bullet}$ (as chain-breaking electron donors). Some antioxidants act by preventing the introduction of the chain-initiating radicals into the polymer system. The preventive antioxidants, therefore, either remove hydroperoxides or retard their breakdown to new initiating radicals (Fig. 6.19). In addition to preventing or retarding lipid peroxidation, the antioxidant radical generated in the process must have such low reactivity that no further reactions with lipid can occur.

Commercially available natural food grade antioxidants belonging to the chain-breaking hydrogen (electron) donor class include α-tocopherol, and Vitamin C. Food additives such as eugenol and vanillin, also have chain-breaking antioxidant properties. The extremely potent synthetic antioxidants belonging to this group (Fig. 6.20)—butylated hydroxyanisole (BHA), butylated hydroxytoluene (BHT), butylated hydroxyquinone (TBHQ), and various esters of gallic acid (e.g., propyl gallate)—are used in a wide range of manufactured foods. These synthetic antioxidants have been very thoroughly tested for their toxicological behavior.

The chain-breaking hydrogen donor antioxidants are primarily aromatic amines and hindered phenols. Their reactions with alkylperoxyl radicals generate a stable, nonpropagating phenoxyl radical in the case of hindered phenols or a transient aminyl radical. The latter produces further oxidation products that are themselves antioxidants (225). The exceptional stability of these antioxidant radicals is primarily due to the delocalization of the solitary electron over the aromatic ring structure.

The chain-breaking acceptor antioxidants are less well known but essentially involve similar radical-trapping chemistry as do the polymerization inhibitors in the absence of oxygen in trapping the alkyl radicals (224). Quinones and nitro compounds are examples of this class of antioxidants and are used widely as additives in packaging plastics. Similarly, UV absorbers, metal chelators and deactivators, and peroxide decomposers, which are either stoichiometric reducing agents for hydroperoxides or catalytic peroxidolytic agents, are also used as antioxidants to enhance the stability of packaging materials. This class of antioxidants, however, are not directly involved in the defense mechanisms against free-radical pathology in humans.

In addition to the antioxidants approved for food uses, several other compounds

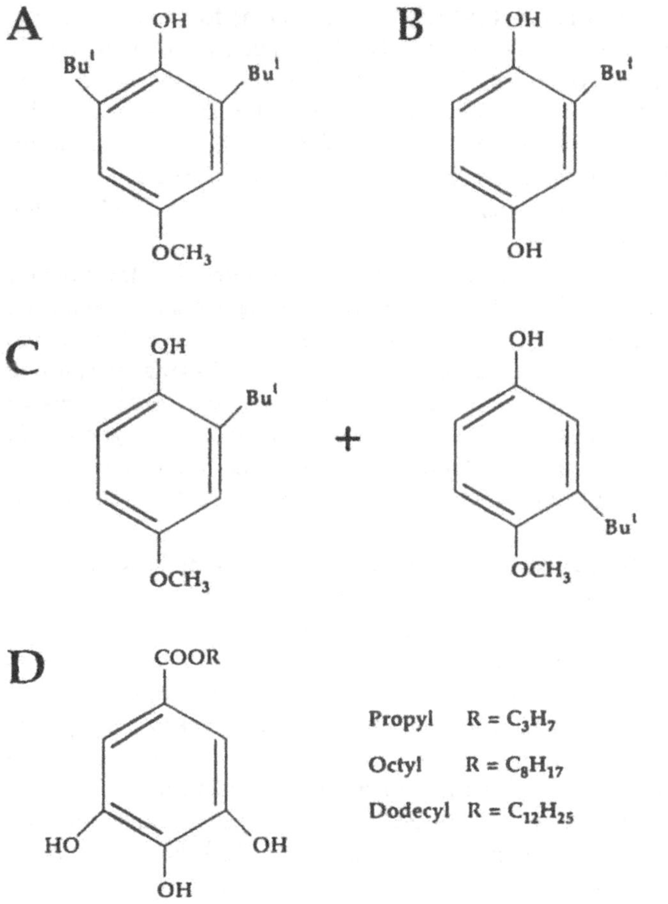

Fig. 6.20 Chemical structures of the most commonly used synthetic food antioxidants. (A) BHT (2,6-di-*tert*-butyl-4-methylphenol); (B) TBHQ (*tert*-butyl hydroquinone); (C) BHA (mixture of isomers of *tert*-butyl-*p*-hydroxyanisole); (D) gallate esters.

from natural food sources may also find their way into the human diet (Table 6.21). Natural antioxidants such as rosmarinic acid, phytic acid, and flavonoids have anticarcinogenic properties. Flavonoids are also known for their anti-inflammatory and cardioprotective properties.

6.4   ROLE IN DISEASE PREVENTION

Free-radical intermediates are involved in an important manner in several metabolic disturbances that result in cell and tissue injury. In fact, it is quite likely that

Table 6.21   Dietary Antioxidants from Natural Sources

Plant extracts: Cocoa shells, leaf lipids, oats, tea, olive, garlic, red onion skin, wheat gliadin, apple cuticle, korum rind, licorice, nutmeg, *Silybum marianum* seed oil, *Myristica fragrans*, mustard leaf seed, chia seed, peanut seed coat, rice hull, birch bark, carob pod

Spices: Rosemary, pepper, cloves, oregano

Animal/plant tissues: Vitamin E, vitamin C, carotenoids, uric acid, NDGA, phenolic acids, flavonoids, L-ascorbic acid esters, phytic acid, amino acids, tannins, sesame seed oil, flavones, saponin, vanillin

Fermentation products: *Penicillium commune*, *Penicillium herquei*, tempeh oil, micro-organisms

Antioxidants formed during food processing or food component interactions: Peptide + sugar; rosmarinic acid and phospholipids; ascorbyl palmitate, lecithin and tocopherols; phospholipids; amino acids; Maillard reaction products; soy protein hydrolysates

---

practically all human diseases involve some form of oxidation at the subcellular level, whether as causal or accompanying phenomena.

Several epidemiological studies suggest that groups of populations are at risk from living under predominantly prooxidative conditions. These situations may arise as a consequence of unbalanced dietary habits; consumption of excess fat of animal origin, smoked food, or alcohol; lack of vegetables; or environmental pollution (226). Others at risk include patients submitted to long-term prooxidative or radiative therapy and individuals suffering from congenital or acquired diseases with peroxidation mechanisms and subsequent increased consumption of antioxidants.

Clinical conditions in which the involvement of free radicals has been suggested are summarized in Table 6.22. There are several diseases in which the evidence for free-radical involvement in the disease pathology is stronger than average. Some specific examples are briefly described below.

6.4.1 Cancer

Cancer is a major public health problem. Apparently about 80–90% of all incident cancers are determined by potentially controllable external factors (227). Thus food as well as lifestyle may supply many carcinogenic substances. Although, in many cases, there is still a lack of definitive evidence about which dietary characteristics most influence cancer risk, substances such as coffee, alcohol, pyrolysis products, nitrites, amines, fat, smoke, pesticides, and most kinds of mutagens have been implicated in the etiology of cancer.

A cancerous tumor may be defined as an abnormal lump or mass of tissue. Its growth exceeds, and is not coordinated with, that of the normal tissue, continuing

Table 6.22 Clinical Conditions in Which the Involvement of Oxygen Radicals Has Been Suggested

Inflammatory—immune injury
  Glomerulonephritis
    (idiopathic, membranous)
  Vasculitis (hepatitis B virus, drugs)
Ischemia—reflow states
  Stroke, myocardial infarction,
    arrythmias
  Organ transplantation
  Inflamed rheumatoid joint
Iron overload
  Idiopathic hemochromatosis
  Dietary iron overload (Bantu)

Red blood cells
  Phenylhydrazine
  Primaquine, related drugs
  Lead poisoning
  Protoporphyrin photooxidation
  Malaria
Lung
  Cigarette smoke effects
  Emphysema
  Hyperoxia
  Bronchopulmonary dysplasia
  Oxidant pollutants (O_3, NO_2)
  Adult respiratory distress syndrome
    (some forms)
Gastrointestinal tract
  Endotoxic liver injury
  Halogenated hydrocarbon liver injury
    (e.g., bromobenzene, CCl_4,
    halothane)
  Diabetogenic action of alloxan
Brain, nervous system, neuromuscular
  disorders
  Hyperbaric oxygen
  Vitamin E deficiency
  Neurotoxins
  Parkinson's disease
  Hypertensive cerebrovascular injury
  Neuronal ceroid lipofuscinoses

Autoimmune diseases
Rheumatoid arthritis

Frostbite
Dupuytren's contracture(?)

Thalassemia and other chronic anemias
  treated with multiple blood trans-
  fusions
Nutritional deficiencies (kwashiorkor)

Sickle cell anemia
Favism
Fanconi's anemia
Hemolytic anemia of prematurity

Mineral dust pneumoconiosis
Asbestos carcinogenicity
Bleomycin toxicity
SO_2 toxicity
Paraquat toxicity

Pancreatitis
Nonsteroidal anti-inflammatory drug-
  induced gastrointestinal tract lesions
Oral iron poisoning

Allergic encephalomyelitis and other
  demyelinating diseases
Aluminum overload
  [Alzheimer's disease (?)]
Potentiation of traumatic injury
Muscular dystrophy
Multiple sclerosis

Alcoholism
　Including alcohol-induced iron overload
　　and alcoholic myopathy
Radiation
　Nuclear explosions　　　　　　　　　　Radiotherapy
　Accidental exposure　　　　　　　　　　Hypoxic cell sensitizers
Aging
　Disorders of premature aging
Heart and cardiovascular system
　Alcohol cardiomyopathy　　　　　　　　Atherosclerosis
　Keshan disease (selenium deficiency)　　Adriamycin cardiotoxicity
Kidney
　Autoimmune nephrotic syndromes　　　　Heavy metal nephrotoxicity
　Aminoglycoside nephrotoxicity　　　　　　(Pb, Cd, Hg)
Eye
　Cataractogenesis　　　　　　　　　　　Retinopathy of prematurity
　Ocular hemorrhage　　　　　　　　　　　(retrolental fibroplasia)
　Degenerative retinal damage　　　　　　Photic retinopathy
Skin
　Solar radiation　　　　　　　　　　　　Hypericin, other photosensitizers
　Thermal injury　　　　　　　　　　　　Contact dermititis
　Porphyria
Drug and toxin-induced reactions
　Bipyridyl herbicides　　　　　　　　　　Paracetamol and phenacetin
　Alloxan and streptozotocin　　　　　　　Halogenated hydrocarbons
　Substituted dihydroxyphenylalanines　　Antibiotics
　　and other phenolic compounds　　　　Sporidesmin
　Cigarette smoke　　　　　　　　　　　　"Spanish cooking-oil" syndrome
　Air pollutants
　Hemolytic and antimalarial drugs

Source: Ref. 11.

after the stimuli that initiated it have ceased (11). Most fatal tumors are malignant or cancerous.

The transformation of a normal cell into a neoplastic cell is considered to proceed through three phases: initiation, promotion, and progression (Fig. 6.21). The process is called *carcinogenesis*, and agents that induce it are called *carcinogens*. Initiation of a normal cell into a neoplastic or cancerous cell involves permanent alteration of the genetic information (mutation) in the cell. Although mutation per se does not necessarily lead to cancer, it is indicative of increased risk of neoplasia (228).

Initiation is followed by promotion, which can be a very slow process in humans.

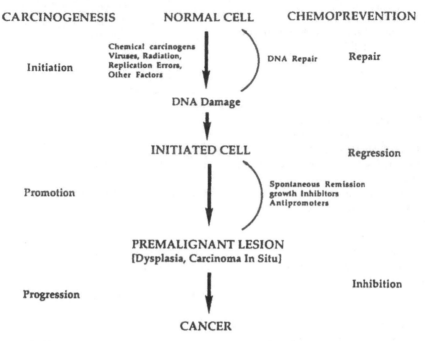

Fig. 6.21 A multistep carcinogenesis model. (From Ref. 228.)

For many common cancers, the latency period could be as long as 10–30 years (229). Because of the long time frame involved and because promotion is generally reversible (Fig. 6.21), this step is often a very attractive target for intervention with cancer chemoprotective agents.

Many carcinogens, when given in large doses, are both initiators and promoters. For example, ionizing radiation is a "complete carcinogen," being both an initiator and a promoter (230). It is therefore a reasonable possibility that some endogenous free-radical reactions, like those initiated by ionizing radiation, will result in tumor formation by serving as a continuous source of tumor initiators and promoters. Indeed, the dietary and environmental components mentioned above are all capable of generating active oxygen radicals or superoxides (231). Additional support for this possibility includes studies of the effect on cancer induction of dietary fat, antioxidants, and increasing age (232–235). The parallelism between cancer incidence and age is probably due, at least in part, to the increasing level of endogenous free-radical reactions with age, coupled with the apparently progressively diminishing capacity of the immune system to eliminate cancerous cells.

Because of the strong involvement of free-radical reactions in carcinogenesis, it is not surprising that a growing number of epidemiological surveys aim at evaluating the protective effect of food antioxidants. In this section, the possible role of vitamins E and C and carotenoids in the prevention of cancer is briefly described. Because of the large number of reported studies in this field and space limitations, only the salient features are described below.

Vitamin E

The anticarcinogenic properties of vitamin E appear to be primarily due to its roles as a lipid antioxidant and free-radical scavenger and as a blocker of nitrosation. It plays an important role in immunocompetence, inhibition of mutagen formation, and repair of membranes and DNA (236). Because of the possible association between these functions and inhibition of carcinogenesis, vitamin E may prove beneficial in cancer prevention.

Data on epidemiological studies dealing with vitamin E and cancer are summarized in Table 6.23. Eight prospective studies thus far have shown that higher serum levels of vitamin E are associated with a lower risk of cancer. In these studies, blood samples were collected from very large groups of people and stored. During subsequent years, cancer cases were identified and the blood levels of vitamin E from the stored samples were compared with those of matched controls without cancer. In all of these studies, low serum levels of vitamin E were associated with increased risks of cancers.

In contrast, several other studies failed to detect significant association between vitamin E levels in the serum and the risk of cancer. In this regard, the results from one such study conducted in Finland by the Alpha-Tocopherol Beta-Carotene Cancer Prevention Study Group (251) are particularly worrisome. In this study, the researchers did not find any reduction in the incidence of lung cancer among male smokers after 5–8 years of dietary supplementation with α-tocopherol or β-carotene. The results came from a large randomized, double-blind, placebo-controlled primary prevention clinical trial. Such trials are often considered the gold standard test of medical intervention. The study group suggested that public health recommendations about supplementation with vitamin E for cancer prevention would be premature at this time.

In a similar study involving 89,494 women who were 34–59 years old in 1980 and who did not have diagnosed cancer, Hunter et al. (250) studied the efficacy of vitamins C, E, and A on the risk of breast cancer. The study involved an 8-year follow-up period. Although the researchers did find an apparent weak inverse association between vitamin E intake and the incidence of breast cancer, it essentially vanished after vitamin A intake was corrected for.

Thus, although there is some evidence for an association between a low vitamin E status and subsequent risk of developing some form of cancer, recommendations regarding long-term, large-dose supplements of the vitamin for cancer prevention must await additional clinical intervention studies.

Table 6.23 Epidemiological Studies of Vitamin E and Cancer

Year	Findings related to serum vitamin E	Ref.
	Beneficial effects found	
1984	Lowest levels associated with a fivefold increase in breast cancer risk compared with highest levels.	237
1985	Low levels in subjects with precancerous gastric dysplasia.	238
1985	Low vitamin and selenium levels, but not low vitamin E alone, were associated with increased risk of fatal cancer.	239
1986	Lowest quintile associated with a 2.5 relative risk of developing lung cancer compared with subjects having highest serum level.	240
1987	Lowest quintile level associated with a fourfold increased risk of developing some type of cancer compared with highest quintile.	241
1987	Levels significantly lower in lung cancer patients and their offspring.	242
1988	Low levels associated with increased risk of developing cancer.	243
	No or adverse effects found	
1984	Low levels in men who later developed colon cancer, but the difference was not significant. Differences in blood lipid levels may account for most of the differences between groups.	244
1984	Levels 8% lower in subjects who later developed cancer, but the difference was not significant. Differences in blood lipid levels may account for most of the differences between groups.	245
1985	No association with later incidence of cancers of the lung, stomach, colon, rectum, or urinary bladder.	246
1985	No difference in level between ovarian tumor cases and controls.	247
1987	No association between level and later development of cancer.	248
1988	No association between level and later development of breast cancer.	249
1993	Large intakes of vitamin E did not protect women from breast cancer.	250
1994	No reduction in the incidence of lung cancer among male smokers after 5–8 years of dietary supplementation with vitamin E. The supplement may have both harmful and beneficial effects.	

Source: Updated from a table in Ref. 236.

Vitamin C

Vitamin C may also be useful in the prevention of cancer. Possible mechanisms of action in reducing cancer risk involve its antioxidant properties, enhancement of immune system functions, blocking formation of nitrosamines and fecal mutagens, and enhancement of hepatic clearance of toxins via the cytochrome P-450 enzyme system (252).

Representative epidemiological studies dealing with vitamin C and cancer risk are summarized in Table 6.24. In 1991, Block (7) reviewed the epidemiological evidence of a protective effect of vitamin C for non-hormone-dependent cancers. Of the 46 studies she reviewed in which a dietary vitamin C index was calculated, 33 found statistically significant protection, with high intake of the vitamin conferring approximately twofold protective effect compared with low intake. Of 29 additional studies that assessed fruit intake, 21 found significant protection.

At present, scientific evidence in support of the efficacy of vitamin C in the treatment of advanced cancer appears to be weak (252). Nevertheless, Block's (7) survey clearly suggests that a high dietary intake of vitamin C and/or vitamin C-rich foods may significantly reduce the risk of developing oral, esophageal, gastric, and colorectal cancers. Similarly, vitamin C and foods that are better sources of vitamin C than of β-carotene appear to have significant protective effect against lung cancer. Block (7) does suggest, however, that vitamin C, carotenoids, and other factors may act jointly in reducing the cancer risk.

β-Carotene (Carotenoids)

β-Carotene and carotenoids in general may act as anticarcinogens through their antioxidant and photoprotective properties against UV-induced carcinogenesis. β-Carotene is also responsible for immunoenhancement. For example, tumor immunity is stimulated in a dose-dependent manner in mice fed β-carotene (274). Similarly, experiments with animals have shown that high-dose supplemental β-carotene protects against radiation- and chemically induced cancers, blocks cancer progression, stimulates immune responses, and acts as an anticarcinogen by altering liver metabolism of carcinogens (196,275).

Among the antioxidant nutrients, β-carotene has attracted the most attention as a preventive agent for cancers. Over 50 epidemiological studies have been conducted during the last decade alone in different parts of the world. These studies have consistently demonstrated that a high intake of foods rich in β-carotene is associated with reduced risk of certain cancers, especially lung cancer. Increased β-carotene consumption has also been found to be associated with reduced risk of cancers of the cervix, esophagus, and stomach. These findings, however, are not as overwhelming as in the case of lung cancer. Representative examples from these epidemiological studies are summarized in Table 6.25.

Although in most epidemiological studies, β-carotene seems to be most protective against lung cancer, a recent Finnish study from the Alpha-Tocopherol, Beta Carotene Cancer Prevention Study Group (251), mentioned earlier in the context of vitamin E, casts serious doubts on the validity of its role. This study suggested that supplements of β-carotene markedly increased the incidence of lung cancer among heavy smokers in Finland (Fig. 6.22). The incidence of lung cancer was 18% higher among the 14,500 smokers who took β-carotene for over 6 years or

Table 6.24 Representative Epidemiological Studies of Vitamin C and Cancer Risk

Site and number of subjects	Findings	Ref.
Lung cancer		
1253 cases, 1274 controls	Reduced risk with vitamin C intake $\geq$140 mg/day	253
332 cases, 865 controls	Reduced risk with increased dietary vitamin C in males only	254
427 cases, 1094 controls	No effect of dietary vitamin C	255
364 cases, 627 controls	No significant effect of dietary vitamin C	256
Oral/pharyngeal cancer		
871 cases, 979 controls	Reduced risk with increased fruit and vitamin C intake	257
166 cases, 547 controls	Reduced risk with increased vitamin C intake	258
374 cases, 381 controls	Reduced risk with increased vitamin C intake	259
Breast cancer		
89,494 women	Large intakes of vitamin C did not protect women from breast cancer	250
Esophageal cancer		
743 cases, 1975 controls	Reduced risk with increased vitamin C and citrus fruit intake	260
120 cases, 250 controls	Reduced risk with increased intake of vitamin C-rich foods	261
Stomach cancer		
391 cases, 391 controls	Reduced risk with increased fruit and dietary vitamin C	262
564 cases, 1131 controls	Reduced risk with greater dietary intake of vitamin C	263
267 men from low- and 246 from high-risk towns	Plasma ascorbate and fruit intake lower in the high-risk area, but no direct relationship between ascorbate and atrophic gastritis	264
Bladder cancer		
164 cases, 314 controls	Reduced risk with increased vitamin C intake in most groups	265
Prostate cancer		
418 cases, 771 controls	No effect of dietary vitamin C	265
311 cases, 294 controls	Increased risk associated with higher vitamin C intake in men over 70 years of age	266
99 cases, 301 controls	Reduced risk with frequent citrus fruit consumption	267
Cervical cancer		
189 cases, 227 controls	Reduced risk with increased vitamin C intake	268
513 cases, 490 controls	No effect of dietary vitamin C	269
Colorectal cancers		
231 cases, 391 controls	No effect of dietary vitamin C	270
575 cases, 778 controls	No significant relationship to citrus fruit consumption	271
102 cases, 361 controls	Reduced risk with increased dietary vitamin C	272
11,888 older adults, 126 cases	Reduced risk in women with increased dietary vitamin C but not supplemented or total C	273

Table 6.25 Representative Epidemiological Studies of
β-Carotene and Various Cancers

Parameter measured[a]	Outcome[b]	Ref.
Lung cancer		
Serum β-carotene (P)	+	276
Dietary β-carotene (R)	+	254
Carrots and green leafy vegetables (R)	+	277
Dark yellow/orange vegetables (R)	+	278
Serum carotenes (P)	0	245
Dietary β-carotene (P)	0	279
Dietary β-carotene (R)	0	280
β-Carotene supplements (P)	0/–	251
Breast cancer		
Diet and serum β-carotene (R)	0	281
Serum carotenes (P)	0	245
Serum β-carotene (P)	+	237
Dietary β-carotene (R)	+	282
Cervical cancer		
Dietary β-carotene (R)	+	283
Dietary β-carotene (R)	+	284
Dietary β-carotene (R)	0	268
Esophageal cancer		
Dietary β-carotene (R)	+	285
Dietary carotene (R)	+	260
Fruits/carotene (R)	+/0	286
Stomach cancer		
Fruit and raw vegetables (R)	+	287
Fruits (R)	+	288
Serum β-carotene (P)	+	289
Fresh fruit and vegetables (R)	+	263
Colon cancer		
Dietary β-carotene (R)	+	270
Fruit and vegetables (R)	+	290
Fruit and vegetables (R)	0	291

[a]Letter in parentheses indicates study type. R = retrospective, P = prospective.
[b](+) Association found between high intake/status and reduced cancer risk; (0) no association found between intake/status and cancer risk; (–) increased cancer risk found with intake/status.

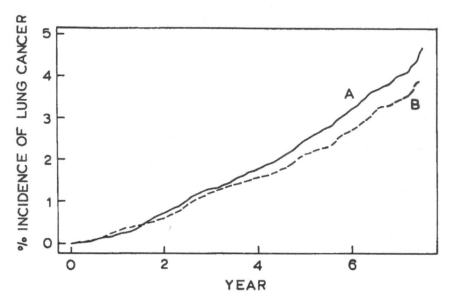

Fig. 6.22 Kaplan–Meier curves for the cumulative incidence of lung cancer among participants who received β-carotene supplements (A) and those who did not (B). Data are shown only through 7.5 years of follow-up because of the small numbers of participants beyond that time and are significant at the 0.01% probability level by the log-rank test. (Data from Ref. 311.)

more than among the 14,500 who did not. The probability that the increase was due to chance was less than 0.01% in this study. In clinical trials, a difference is taken seriously when there is less than a 1 in 20 probability that it happened by chance (292).

The researchers involved in this study warn that the benefits that have been seen in the epidemiological studies may have been overestimated and the dangers may have been underestimated or even unsuspected. This is further supported by the fact that in most prospective clinical trials with β-carotene, no strong association is seen between its intake and/or status and cancer risk. The only way to find out whether β-carotene is beneficial or harmful is to wait for the results from several other ongoing clinical trials of this antioxidant. Meanwhile, long-term intake of very large doses of β-carotene and other natural antioxidants must still be viewed with caution.

Since β-carotene also acts as a provitamin A in vivo, several studies have also investigated the effects of vitamin A and probable cancer risk. Multivariate analyses can also distinguish between the effects of carotenes (plant sources of vitamin A) and retinol (animal sources of vitamin A). Unlike β-carotene, vitamin A does not have any antioxidant properties. In 17 such studies reviewed by Gaby and Singh

(175), a consistent pattern of a lack of association between intake of preformed vitamin A and cancer incidence and a significant association between high intake of carotene-rich foods and lower incidence of certain cancers is seen. In addition, studies have shown a lack of association between serum retinol level and cancer risk and an association between serum β-carotene level and cancer risk, but a significant association between high serum β-carotene and reduced cancer risk.

6.4.2 Cardiovascular Diseases

Cardiovascular diseases are a leading cause of death globally. Many important forms of human cardiac disease, together with many cases of localized cerebral ischemia (stroke) are secondary to the condition of atherosclerosis, a disease of the arteries that is characterized by a local thickening of the intima, the innermost part of the blood vessel (11).

Atherosclerotic plaques can be initiated when the vascular endothelium is damaged, such as through oxidative injury or mechanical stress. Subsequent cellular reactions involving activated monocytes and macrophages that secrete $O_2^{\cdot-}$, H_2O_2, and hydrolytic enzymes could injure neighboring cells and lead to more endothelial damage and damage to smooth muscle cells.

The risk of coronary heart diseases can be predicted by classical risk factors such as hypercholesterolemia, hypertension, and smoking only to about 50% extent (293). Dietary factors also play a complementary role, with various diets being associated with a lower coronary mortality. Potential benefits are generally expected from essential PUFAs as well as from essential antioxidants such as vitamins C and E and selenium. Dietary PUFAs have long been recommended for lowering plasma cholesterol levels and consequently lowering the risk of atherosclerosis. They are, however, actually extremely subject to peroxidation reactions if a sufficient antioxidant defense, primarily in the form of vitamin E, is not available.

Evidence for the fact that free radicals may be involved in the etiology of coronary heart disease comes from several studies that suggest a correlation between antioxidant intake and function and various cardiovascular diseases (reviewed in Refs. 293 and 294). These include the following:

1. Hypoxia with subsequent reoxygenation produces a sudden burst of oxygen radicals that can overcharge protective enzymes such as SOD. Oxygen radicals can seriously damage arteries in animal models as soon as the physiological radical scavengers such as the vitamins E and C and GSH are exhausted.
2. Lipid peroxides have been found in atherosclerotic plaques, which suggests at least an increased susceptibility toward peroxidation.
3. Exogenous peroxidized PUFAs damage the endothelium and heart muscle cells and provoke proliferation of smooth muscle cells. This cytotoxicity can be prevented by antioxidants such as vitamins E and C.

4. Peroxides of low-density lipoproteins (LDL) and/or PUFA hydroperoxides (e.g., 15-hydroperoxytetraenoic acid) inhibit the arterial formation of the "antiatherosclerotic" prostaglandin.

5. Patients with coronary diseases generally tend to have increased plasma levels of thiobarbituric acid–reactive material, which indicates an increased susceptibility to lipid peroxidation. Lipid peroxides have also been demonstrated in human plasma (11,295). Peroxide- or alkenal-modified LDL, in contrast to normal LDL, is taken up by separate receptors known as "acetyl-LDL" or "scavenger" receptors. LDL bound to these receptors is taken up with enhanced efficiency, so that cholesterol rapidly accumulates within the monocytes or macrophages. These are later transformed into atherosclerotic, lipid-laden foam cells. This transformation is believed to be an initial step in the formation of the atherosclerotic plaque. Vitamin E, because of its antioxidant properties, prevents the oxidation of LDL.

6. Atherosclerotic lesions have been found in rodents, piglets, and primates with chronic marginal deficiency of vitamin C or E (296).

7. Vitamin E reduces platelet adhesion and aggregation.

8. Arrythmias are reduced by vitamin E in animal models (297).

9. Arteriographically established coronary diseases are generally inversely related to the blood status of vitamin C.

10. In several geographical areas, lower mortality rates due to coronary diseases are associated with vegetarian types of diets rich in natural antioxidants such as vitamin C and carotenoids. Conversely, in high-risk areas, it appears to be inversely correlated to the calculated vitamin C and carotenoid intake from fresh fruits and green vegetables.

If free radicals do indeed initiate atherosclerosis or contribute to its pathology, then an increased intake of antioxidants, especially lipid-soluble, chain-breaking antioxidants that accumulate in lipoproteins, might be expected to have a beneficial effect. Indeed, this fact was exploited in the development of probucol, a drug used clinically to lower blood cholesterol levels (11). Probucol is a powerful antioxidant whose structure is similar to that of several phenolic antioxidants. The roles of vitamins E and C in this regard are briefly described below.

Vitamin E

Of all the dietary antioxidants, vitamin E appears to play a major beneficial role, primarily because of its solubility in lipids. Vitamin E has been shown to inhibit platelet aggregation, intermittent claudication, and ischemic reperfusion injuries (236).

Recently, two large-scale studies in men and women suggested that the use of large doses of vitamin E supplements is associated with a significantly decreased risk of coronary heart disease (298,299). The study populations included almost

40,000 men and more than 80,000 women, and the participants were followed for 4 and 8 years, respectively. These studies were very carefully conducted. There was more than 95% follow-up, independent validation of the dietary question-naires, and careful documentation of endpoints. In both studies, the benefit of vitamin E was largely, if not entirely, confined to the subgroup of the populations taking large amounts of supplements.

In both these studies, the data did not provide a cause-and-effect relationship to coronary diseases. Thus, within the range of vitamin E intakes afforded by natural foodstuffs, even when supplemented by multivitamins at usual doses, there was little or no protective effect. These studies, however, did provide evidence of an association between a high intake of vitamin E and a lower risk of coronary heart disease in men and among middle-aged women taking vitamin E supplements.

In both these studies (298,299), the investigators addressed the important issue of possibly confounding variables. Thus, the probability of self-selection among the subjects taking vitamin supplements was considered. Indeed, the men in the upper fifth of the group in terms of vitamin E intake did exercise more and were more likely to take aspirin. Controlling for these variables by multivariate logistic regression analysis did not greatly affect the conclusions, however. These two large epidemiological studies, nevertheless, did support the hypothesis that oxidation of lipoproteins plays a part in atherogenesis.

Vitamin C

Vitamin C has also been studied as a factor in cardiovascular health. Because vitamin C is involved in collagen formation, antioxidant protection, and certain cellular reactions that facilitate lipid metabolism, a possible role for it in amelio-rating atherosclerosis has been hypothesized.

Vitamin C affects cholesterol metabolism. Its deficiency in guinea pigs results in decreased conversion of cholesterol to bile salts and in the formation of atherosclerotic lesions (300,301). Epidemiological data from several studies also indicate that vitamin C status is inversely associated with cholesterol levels and directly related to HDL levels. Leukocyte ascorbic acid was reported to be signif-icantly lower in patients with coronary artery disease than in normal individuals (302). A 12-year follow-up study of Swedish women found a negative correlation between vitamin C intake and serum triglycerides, body mass index, and waist-to-hip circumference ratio, all cardiovascular disease risk factors (303).

In a cross-cultural epidemiological study, Gey et al. (289) also found an association between lower vitamin C status and medium to high rates of coronary mortality. In a similar study on a comparatively large population (more than 11,000 men), Enstrom et al. (304) showed that both mortality from coronary heart disease and overall mortality were inversely correlated with vitamin C intake.

In contrast, in the two large recent epidemiological studies described earlier

(298,299), the data on vitamin C intake were essentially negative for both men and women. Further studies are required to ascertain its definitive role in the prevention of cardiovascular diseases. Nonetheless, available evidence suggests that vitamin C may improve cardiovascular health and reduce the risk of cardiovascular mortality, possibly through its effects on HDL cholesterol, blood pressure, platelets, maintenance of the vascular wall integrity due to its requirement for collagen synthesis, its ability to protect against free-radical damage to lipids and other molecules due to its antioxidant function, and its role in the regeneration of the active form of vitamin E.

Carotenoids

Compared to vitamins E and C, carotenoids appear to influence cardiovascular health only to a marginal degree. In the two large epidemiological studies described earlier, carotene intake was associated with a lower risk of coronary disease in men who smoked (298), whereas it had no apparent effect on risk in women (299).

The free-radical-dependent oxidative modification hypothesis of atherogenesis appears to have matured to a level strongly supported by experimental and epidemiologic data. However, it does not appear to have attained the status of a clinically validated hypothesis. Because we do have available a number of natural antioxidants, including vitamins E and C and carotenoids, that are generally considered not to be toxic, there is a growing trend that supports increased intakes of these antioxidants.

Recommendations for such increased intakes of these antioxidants also need to be viewed with caution. For example, except for one small treatment study in primates involving vitamin E, we do not even have data demonstrating the efficacy of the natural antioxidant vitamins in animals; all the other studies in animals were done with antioxidant drugs. We need to have at least some proof that includes valid clinical intervention trials demonstrating the magnitude of the benefit of such treatment to be obtained in humans.

At present, except for a preliminary report of an interim analysis of results in a subgroup of men in the Physician's Health Survey (305), no such data are available. Even in the above study, the primary hypothesis was that β-carotene would be anticarcinogenic. Because of the growing evidence that lipid oxidation may play a part in atherogenesis, it was decided to do a preliminary analysis looking for effects on coronary heart disease in a subgroup of subjects. That subgroup of 333 men already had clinical evidence of coronary heart disease when the study began. Those taking β-carotene had about 40% fewer cardiovascular events, and the difference was statistically significant. The complete data from this study are expected to be reported in the next 2–3 years.

According to Steinberg (306), yet another reason for caution is the lack of data showing that the long-term (and presumably lifetime) intake of very large doses of natural antioxidants will not be toxic. Most such studies in humans are generally limited to 6 months and rarely involve thousands of subjects. Moreover, such unproven preventive measures as large-dose supplements of natural antioxidants also need to be taken in conjunction with others such as cholesterol-lowering diets, regular exercise, and smoking cessation. It thus appears that reaching a consensus on the pathogenetic importance of oxidized LDL and thus on the value of antioxidant therapy may not be any easier than it was to reach a consensus on the importance of an elevated plasma cholesterol level as a risk factor for coronary heart disease. In this regard, long-term, double-bind clinical trials involving intakes of very large doses of natural antioxidants will certainly be helpful.

6.4.3 Aging

Aging in organisms is a result of normal developmental and metabolic processes. Humans have the longest lifespan among mammalian species. Although numerous hypotheses have been advanced, the nature of the causal mechanisms underlying the aging process is poorly understood and is a subject of intense debate. An overall view of the biological interactions involved in aging processes is schematically shown in Fig. 6.23. The overall process involves several feedback loops that can amplify a modest amount of initial damage into a disaster.

One of the more widely accepted theories of aging involves free radicals. First proposed by Harman (308), the free-radical theory suggests that normal aging results from random deleterious damage to tissues by free radicals produced during normal aerobic metabolism. The main assumption of the free-radical hypothesis of aging is that the normal level of antioxidant defense is not fully efficient, so that a fraction of the free radicals escape elimination. These free radicals then are capable of inflicting molecular damage. Some damage is irreparable and accumulates with age, thereby causing functional attrition associated with aging. It must be noted, however, that although this hypothesis is intuitively appealing as a result of the ubiquitous generation of the potentially deleterious free radicals, a direct causal link between free radicals and aging has not yet been thoroughly established. Nevertheless, the free-radical theory of aging is supported by several plausible explanations it provides for aging phenomena. Harman (235) cites the following reasons:

1. Inverse relationship between the average lifespans of mammalian species and their basal metabolic rates
2. Antioxidants, which increase the average lifespan of mice, depress body weight and fail to increase maximum lifespan
3. Clustering of degenerative diseases in the terminal part of the lifespan
4. Exponential nature of the mortality curve
5. Beneficial effects of caloric restriction on lifespan and degenerative diseases
6. Increase in autoimmune manifestations with age

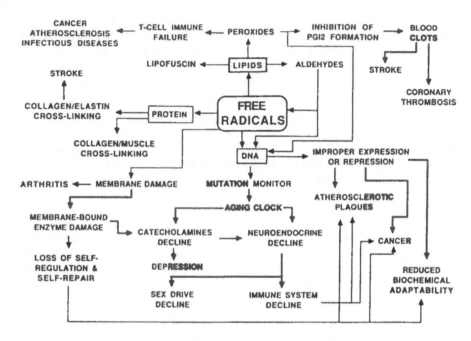

Fig. 6.23 A unified view of the aging process and the cellular and biochemical interactions involved. (After Ref. 307.)

In spite of the vast complexity of aging processes, relatively less complex processes such as longevity assurance and longevity determinant genes (LDGs) may exist that govern the aging rate. This concept was explored in detail by Cutler (309). Reviewing the available experimental data on the causative role of free radicals in aging processes and the hypothesis that antioxidants represent LDGs, he concluded that longer lived species have better antioxidant protective mechanisms in relation to rates of radical generation than do shorter lived species. Thus, a positive correlation in the tissue concentration of specific antioxidants such as SOD activity, carotenoids, α-tocopherol, and uric acid with lifespan of mammals is generally observed. These findings are summarized in Table 6.26.

Among the various antioxidant defenses, SOD activity, in conjunction with catalase, appears to be the most important enzymatic defense against the toxic effects of oxygen metabolism. Indeed, any organism that uses oxygen cannot live without SOD or an equivalent type of protective mechanism. Thus SOD activity of the human liver, expressed per unit metabolic rate, is higher than that of other primate species and much higher than that of other mammals (309). Since SOD appears to have a similar structure and enzymatic activity in all animals, Cutler (309,310) has suggested that a change in gene regulation has allowed synthesis of higher cellular SOD concentrations in humans, contributing to longevity.

Table 6.26 Summary of Major Experimental Results Searching for Unique Human Biological Characteristics as Related to Lifespan and Oxyradicals

---

I. Antioxidants showing significant positive correlation in tissue concentration per specific metabolic rate (SMR) vs lifespan, where human clearly has highest value
    1. SOD (Cu/Zn and Mn)
    2. Carotenoids (including β-carotene)
    3. α-Tocopherol
    4. Uric acid
    5. Ascorbate (specific tissues only)

II. Antioxidants showing significant negative correlations in liver tissue concentration per SMR vs lifespan, where human clearly has least value
    1. Catalase
    2. Glutathione
    3. Glutathione peroxidase

III. Tissue concentration of liver cytochrome P-450 and glutathione transferase generally decrease with increasing lifespan of mammalian species, where human has lowest amount

IV. Rate of spontaneous autoxidation of tissues decreases with increasing lifespan of mammalian species, where human tissues are most resistant to autoxidation.

V. Amount of steady-state oxidative damage in DNA (8-OHdG/dG) in liver tissue decreases with increasing lifespan of mammalian species

---

Source: Ref. 309.

If free radicals are indeed a causal factor in aging, then the enhancement of the primary defenses against them should reduce oxidative stress, decrease the rate of aging, and extend the lifespan. This concept recently was very elegantly elucidated experimentally by Orr and Sohal (311). These researchers investigated the effects of the overexpression of Cu/Zn/ SOD and catalase genes in *Drosophila melanogaster*. These two enzymes, acting in tandem, provide the primary enzymatic antioxidant defenses. SOD converts the $O_2^{\bullet -}$ radical to H_2O_2, and catalase breaks down H_2O_2 into water and oxygen, thus eliminating the possibility of the production of the highly reactive hydroxyl radical. Because GSH-Px, another enzyme involved in H_2O_2 removal, is absent in insects, SOD and catalase together contribute the first coordinated unit of defense against free radicals.

Working with transgenic lines of flies overexpressing both Cu/Zn SOD and catalase, Orr and Sohal (311) found that transgenic flies carrying three copies each of SOD and catalase genes exhibited as much as a one-third extension of lifespan, a longer mortality rate doubling time (MRDT), less protein oxidative damage, and a delayed loss in physical performance compared to diploid controls (Table 6.27). Furthermore, their results underscore the importance of an optimum balance

Table 6.27 Comparison of SOD and Catalase Activities, and Mortality Pattern Between Transgenic and Control Flies[a]

Group	Strain	SOD activity (units)	Catalase activity (units)	Mortality (days) Median	90%	100%	MRDT (days)
Control	Control	13.1 ± 0.6	19.2 ± 0.8	54.5	64	71	8.4
A	cat1.5 sod1.4	16.5 ± 0.3	33.2 ± 0.9	63	78	95	10.1
B	cat1.1 sod1.4	17.3 ± 0.1	27.5 ± 0.4	58	74	81	9.9
C	cat1.1 sod1.2	16.8 ± 0.5	31.6 ± 0.7	72.5	83	93	11.5

[a]Mortality data are based on initial samples of 100 flies in each group. Enzymatic activities were measured in whole-body homogenates of the flies and are mean ± SEM of three to six determinations. Mortality rate doubling time (MRDT) was measured as $\ln(2/G)$, where G is the slope of the Gompertz plot. All experiments were repeated one or more times with essentially similar results.

between SOD and catalase. The overexpression of Cu/Zn SOD alone or catalase alone had only a minor incremental effect (up to 10%) on the average lifespan and no effect on the maximum lifespan of *Drosophila*. Their results clearly suggest that a co-overexpression of the SOD and catalase genes does provide an efficient strategy for lowering the level of oxidative stress and extending lifespan.

Among the antioxidants, only uric acid, β-carotene, and vitamin E appear to play some role in slowing the aging process. Uric acid, a by-product of purine metabolism, shows an excellent positive correlation to maximum lifespan in humans. It is not, however, a normal constituent of the human diet.

In contrast, carotenoids and vitamin E show only a weak to moderate positive correlation (309,310). Serum levels of carotene do not appear to change with age in humans. Thus, aging does not appear to be a result of a loss of carotenoids. The protective action of carotenoids may, however, be attributed to their ability to provide resistance against a number of different types of cancers described earlier.

In Cutler's (309,310) studies, such important antioxidants as GSH-Px, GSH, and ascorbic acid failed to show any positive correlation with lifespan. In fact, some correlations of their tissue levels with species' lifespan potential tended to be negative. Cutler (308) warns, however, that it is important to measure the turnover rate of such antioxidants, as tissue levels alone are a measure only of potential protection and not of actual utilization of this protection.

The available experimental and epidemiological data do lend considerable support to the free-radical theory of aging. On this basis alone, it is reasonable to expect that the healthy human lifespan can be increased by at least 5–10 or more years through judicious selection of diets rich in natural antioxidants as well as synthetic antioxidant supplements.

6.4.4 Immune System

The immune system is responsible for protection against infection by invading pathogens such as bacteria, viruses, fungi, and protozoan parasites. In healthy individuals, these pathogens are destroyed by immune cells and their secretions. The immune system also responds similarly to cells of the body that have undergone changes that can lead to cancer. The altered cells are considered to be invaders and are destroyed. Conditions that depress immune functions consequently increase the risks of infection and the development of certain cancers. Conversely, factors that can enhance immunity may lower these risks (312,313).

The protective responses of the immune system require collaborations among at least four principal cell types. B lymphocytes make and secrete immunoglobulin antibodies. Cytotoxic or killer T lymphocytes and natural killer cells bind to and destroy the antigen-bearing cells found in tumors and virally infected target cells. Helper T cells release lymphokines that promote the proliferation and functional maturation of both T and B lymphocytes. Antigen-presenting cells, including those of the macrophage and dendritic series, are not themselves specific for any particular antigen but can ingest and process antigenic particles and then display the resulting fragments in a form that is stimulatory for T cells (313).

Certain cells of the immune system produce and use free radicals and reactive oxygen molecules, especially during the early stage of infection. However, if these reactive species are overproduced during this process, they may injure the immune cells themselves as well as neighboring cells and tissues. Proper functioning of the human immune system, therefore, depends on the intake of micronutrients that can act as antioxidants.

Certain environmental factors can decrease the immune functions. These include exposure to UV light, cigarette smoking, environmental pollutants, metabolism of certain drugs or xenobiotics, and infection with viruses such as HIV (312). The normal aging process is also associated with a loss of cell-mediated immune responses and a concomitant increase in infections and cancer incidence.

Cells of the immune system are highly dependent on a functioning cell membrane for secretion of lymphokines and antibodies, antigen reception, lymphocyte transformation, and contact cell lysis (314). Cell membrane fluidity is dependent on the concentration of PUFA, which in turn also increases the potential for membrane lipid peroxidation mediated by free radicals. Lipid peroxidation decreases membrane fluidity, which adversely affects immune responses (312).

In this regard, because of its lipid solubility and antioxidant properties, vitamin E is a potentially critical nutrient in the regulation of immune function. Lymphocytes and mononuclear cells, in fact, have the highest concentration of vitamin E of any cells in the body (109). Several parameters of the immune system are depressed in vitamin E-deficient animals. This includes T and B lymphocyte stimulation, mixed lymphocyte response, plaque-forming cell activities, interleukin-2 levels, and neutrophil and macrophage chemotaxis and adhesion (315).

Vitamin E supplementation has been shown to increase resistance to infection in numerous farm and laboratory animals (109,316–318). The requirement for optimum immune response in the rat is higher than for the prevention of classical vitamin E deficiency symptoms. This may partially explain why addition of vitamin E to diets nutritionally adequate in the vitamin not only enhances antibody formation but also reduces morbidity and mortality in response to infection with *E. coli* in chickens, *Mycoplasma pulmonaris* in rats, histomoniasis in turkeys, and *Chlamydia* in lambs (319). Similarly, neutrophil-induced damage to lung tissue following burn injury is reduced when rats are preinjected with vitamin E.

Several studies with human volunteers also indicate the possible beneficial effects of vitamin E on the functioning of the immune system. In a double-blind study examining the effects of vitamin E on immune responses, Meydani et al. (320) found significant improvement in several clinical indicators of immune function in healthy subjects 60 years of age and older. In an epidemiological study of 100 healthy elderly people, Chavance et al. (321) also found a statistically significant association between high vitamin E status (plasma levels > 1.35 mg/dL) and low incidence of infections over a 3-year period.

The beneficial effects of vitamin E on the immune function may be attributed to the inhibition of the production of certain immunosuppressive prostaglandins (322,323) or moderating granulocyte activation (324). Its antioxidative properties also explain the reduction in self-destruction of neutrophils during the oxidative burst. Its protective effect, therefore, may be associated with a reduction in the H_2O_2 generated by phagocytic cells of the immune system.

Vitamin C also plays a critical role in the maintenance of the immune system. It can regenerate the reduced antioxidant form of vitamin E (325). Since vitamin E can enhance both nonspecific and specific immune responses, its protection by vitamin C can indirectly affect overall immune function.

Vitamin C-deficient guinea pigs show depressed cell-mediated immune responses that include delayed hypersensitivity, cytotoxicity, inflammatory and autoimmune responses, and increased rejection of skin grafts (109). Vitamin C is also involved in the formation of collagen, secretory processes required by immune cells to release immunomodulatory substances, interferon production, and the enhancement of neutrophil function.

Several studies have also shown that carotenoids can enhance immune functions independently of any provitamin A activity (203,326,327). The mechanisms of immunoenhancement may involve the capacities of a number of carotenoids to quench singlet oxygen as well as their antioxidant properties.

Both in vitro and in vivo laboratory animal studies reviewed by Bendich (203) have shown that carotenoids, especially β-carotene, can protect phagocytic cells from autooxidative damage, enhance T and B lymphocyte proliferative responses, stimulate effector T cell functions, and enhance macrophage, cytotoxic T cell, and natural killer cell tumoricidal capacities, as well as increase the production of certain interleukins. The association of immunoenhancement with decreased tumor burden in animals given carotenoids may suggest a potential explanation for the epidemiological data linking lower carotenoid status with higher incidences of certain cancers. According to Bendich (203), β-carotene may be more important as a nutrient than simply serving as a precursor of vitamin A, since the latter is a relatively poor antioxidant and cannot quench singlet oxygen.

Thus available literature data suggest that supplementation of diets with vitamin C or E or β-carotene provides a safe and effective means to enhance clinically relevant immune functions. These antioxidant vitamins enhance immune responses that are involved in protection from infection and malignancies. An increased dietary intake over that commonly recommended for these nutrients in human nutrition may actually be beneficial in enhancing the activities of the immune system.

6.4.5 Inflammatory and Autoimmune Diseases

Pain and inflammation associated with several diseases as well as nutrient deficiencies have been known since ancient times. When human tissue is injured, an acute inflammatory response develops. It is generally characterized by swelling, warmth, pain, reddening, and partial immobilization and is induced by phagocytes and their products to ward off unwanted and potentially dangerous foreign particles such as bacteria (11).

Inflammatory cells and polymorphs contain at least three major components that are responsible for their bactericidal properties. These include the release of enzymes and other reactive components from lysosomes and specific granules, the generation of $O_2^{\cdot-}$ and the resultant family of reactive species, and the formation of oxidized arachidonic acid products (328). Prostaglandins, leukotrienes, and other lipid peroxides such as those of arachidonic, linolenic, and eicosapentaenoic acids therefore play an important role in the establishment of inflammatory reactions. Platelets, which are rich in unsaturated fatty acids, and polymorphs, which, under the influence of $O_2^{\cdot-}$, adhere to the vessel wall, also contribute strongly to the mechanism of vascular inflammation (329).

According to Del Maestro (328), there are four prerequisites for the possible role of oxygen-derived free-radical species during inflammation:

1. The presence of inflammatory cells such as polymorphs, macrophages, and monocytes
2. Appropriate stimuli that result in the induction of a respiratory burst with the generation of $O_2^{\cdot-}$ and H_2O_2
3. Low concentrations of scavenging enzymes in the extracellular space
4. The presence of metal complexes chelated such that hydroxyl radical and possibly other oxidizing species can be generated

In any given tissue, if these requirements are fulfilled, then free-radical-induced injury may be hypothesized as a cause of tissue injury.

Inflammation is normally a self-limiting event. However, anything causing abnormal activation of phagocytes has the potential to provoke a devastating response. It is now widely recognized that in the course of the inflammatory reactions, the respiratory burst in the phagocytic cells generates highly reactive oxygen species that are thought to play an important role in conditioning the defense capability of the host against microbial aggression through oxidative killing mechanisms (330–332). The primary oxidative products generated, $O_2^{\cdot-}$ and H_2O_2, then cooperate in forming additional and more reactive oxidants, including hydroxyl radical and halide-derived oxidant products, which in turn amplify the oxygen-dependent microbicidal system and are also effective as tumoricidal agents. Furthermore, since the extreme reactivity of the oxygen species is not surprising, they also can be implicated in the modulation of cell function, cellular injury, and tissue destruction.

One of the most commonly observed consequences of abnormal phagocyte actions is the development of autoimmune diseases. The human body has mechanisms to prevent the formation of antibodies against its own components. Any failure of these mechanisms allows the formation of "autoantibodies" that can bind to normal body components and provoke attack by phagocytic cells. Some examples of autoimmune diseases involve only a single tissue. Hashimoto's thyroiditis, which affects the thyroid gland; myasthenia gravis, which affects the neuromuscular system; and chronic autoimmune gastritis, which affects gastric parietal cells, belong to this class of autoimmune diseases. In contrast, in systemic lupus erythematosus, dermatomyositis, and autoimmune vasculitis, lesions are widespread. Rheumatoid arthritis, a disease characterized by chronic joint inflammation, especially in the hands and legs, has many features of an autoimmune disease, although its exact cause is unknown.

Halliwell (333) first proposed that oxidants might attack normal biomolecules to create new antigens and that this could be an origin of, or contribute to, autoimmunity. The oxidative metabolic pathways involved in inflammatory and autoimmune diseases can be retarded by the use of anti-inflammatory agents. The therapeutic effects of corticoids are thus attributed to their inhibition of the "arachidonic acid cascade" and those of nonsteroidal anti-inflammatory drugs (NSAIDS), to their inhibition of cyclooxygenase enzyme systems and thus the blocking of prostaglandin production.

Because of their antioxidant properties, anti-inflammatory drugs are capable of preventing oxidant damage at sites of inflammation (11). They might directly scavenge such reactive oxidants as $HO^{\cdot}$ and hypochlorous acid (HOCl). Some drugs can mediate oxidant reactions by altering their production by neutrophils, monocytes, and macrophages. Inhibition of $O_2^{\cdot-}$ production should then lead to the diminished formation of H_2O_2 and hence to decreased production of HOCl and possibly $HO^{\cdot}$.

The fact that gold compounds, which are potent singlet oxygen quenchers, relieve several types of rheumatism supported the hypothesis that antioxidant therapy might be a logical intervention to control the cumulative prooxidant pathological mechanisms involved in inflammatory and autoimmune diseases (226,334). In this regard, carotenoids, because of their strong singlet oxygen quenching properties, may have a potential therapeutic role in the treatment of rheumatism. An evaluation of the clinical efficacy of carotenoid supplements in the prevention, treatment, or recurrence of arthritis therefore warrants serious attention.

The production of collagen, an important constituent of connective tissue and a target for rheumatic damage, is known to be mediated and controlled by vitamins E and C (109). Anti-inflammatory effects of vitamin C are also known. It inhibits alkaline phosphatase and increases the net synthesis of sulfated proteoglycans, whereas vitamin E has been shown to inhibit activities of lysosomal enzymes, especially arylsulfatase A and acid phosphatase (335,336). In the latter case, the inhibition results from a direct interaction between the enzymes and vitamin E. Both vitamin C and vitamin E appear to stabilize sulfated proteoglycans, which are essential structures of the cartilage, and thus inhibit inflammation. They also seem to have analgesic properties similar to those of aspirin.

Thus antioxidant vitamins, when administered in amounts far above the normal doses, seem to show analgesic and anti-inflammatory properties that appear to be unrelated to a vitamin deficiency or their intrinsic vitamin nature. These activities derive from their classical nature as oxidants or antioxidants, for example, their ability to change the activity of certain enzymes involved in pain perception. Unlike the usual analgesics and anti-inflammatory drugs, natural antioxidants such as vitamin E, vitamin C, and carotenoids are well tolerated and, in particular, pose no difficulties in long-term use. The application of natural or synthetic antioxidants, alone or in combination, therefore appears to have potential in the treatment of inflammatory and autoimmune diseases.

6.4.6 Photoprotection

Antioxidant nutrients are involved in protecting both plants and animals from photooxidation. In green plants, the photosynthetic process produces large amounts of free radicals generated from the use of singlet oxygen (175,337). Antioxidant pigments such as β-carotene protect plants from the lethal oxidation caused by photosynthesis. Photosynthetic bacteria also synthesize carotenoids to prevent free-radical damage (338). Mutated bacteria that cannot synthesize carotenoids are destroyed by photooxidation upon exposure to sunlight. Carotenoids with a minimum of nine double bonds are essential for the protection of plant tissues from destruction by highly reactive oxygen species, including free radicals generated by singlet oxygen.

Carotenoids are also helpful in the treatment of patients with inherited light-sensitive skin disorders such as erythropoietic protoporphyria (EPP), congenital porphyria, polymorphous light eruption, actinic reticuloid, solar urticaria, hydroa aestivale, and prophyria cutanea tarda (339). EPP, the most serious of these disorders, is characterized by burning, redness, and swelling of skin when exposed to sunlight.

In a 7-year study, the effectiveness of β-carotene supplementation was examined in 133 patients with EPP. The results indicated that in 84% of the patients the ability to tolerate sunlight exposure was increased without development of symptoms (340,341). Since these original studies, several individuals suffering from photodamaging skin conditions have been treated successfully with β-carotene. The treatment often requires the use of β-carotene in high doses, up to 180 mg/day. Thus far, no serious side effects from such treatments have been reported in the literature.

Little information is available on the efficacy of vitamins C and E in the treatment of photosensitive skin disorders in humans.

6.4.7 Cataract

Senile cataract, a major cause of disability in the elderly, is one of the leading causes of blindness in the United States and the leading cause of preventable blindness throughout the world (342). It has been suggested that a 10-year delay in cataract formation could cut the need for lens extraction in half (343). Identifying the determinants of senile cataracts, therefore, can have a significant impact both on the well-being of the elderly and on the associated health costs.

Any opacity that occurs in the lens of the eye is referred to as a cataract. Its formation is primarily due to the changes in lens proteins over a period of decades. Exposure to light (especially ultraviolet, infrared, ionizing, and microwave radiation), oxygen, products of normal aging, diarrhea and dehydration, gender, blood pressure, blood sugar, education, occupation, vital capacity, and other demographic and biological parameters have been correlated with the risk of senile cataracts (344). These factors subject the proteins of the lens to extensive postsynthetic modifications including oxidation, racemization, and enzymatic and nonenzymatic glycosylation, which ultimately lead to their aggregation and thus the formation of a cataractous lens. Other potential targets for oxidative damage include lipids and DNA in the eye lens.

Several lines of evidence indicate that free radicals and oxidizing agents play an important role in the development of cataract. These include the following.

1. Agents such as X-rays, UV radiation, hyperbaric oxygen, and chemicals that either directly or indirectly induce free radicals can also induce cataract formation (345).
2. Free radicals have been detected in the lens (346).

3. Oxidative changes that occur concomitant with the development of browning of the lens have been produced experimentally through the use of photosensitizers. These chemicals act via free-radical mechanisms or through the generation of H_2O_2 or singlet oxygen (347).
4. Epidemiological studies have demonstrated a correlation between exposure to sunlight and cataract development. Cataracts are also more common and progress more rapidly in countries with comparatively strong sunlight exposure (348).
5. Agents that protect against free-radical or other oxidative damage can prevent or reduce the severity of some experimental cataracts; conversely, agents that inhibit enzymatic defenses against oxidative damage enhance cataract formation (349).
6. The lens is continually exposed to solar radiation, which can promote free-radical formation and alteration in lens function (350).

The human lens has a number of defense mechanisms, including a high concentration of antioxidant enzymes (SOD, GSH-Px, and catalase) and vitamin C, to protect it from the effects of oxidation. A reduction in the activity and/or concentration of each of these components is believed to play some role in cataract formation. The vitamin C content of the lens declines as cataract develops, falling to as little as 50% of that found in normal lenses (351,352). In vitro experimental studies have also shown that vitamin C protects the lens by inhibiting lipid photoperoxidation (348).

Epidemiological studies have also reported an inverse association between vitamin C status and cataract risk. In one such study comprising 175 cases with advanced cataract and 175 controls, in which the risk of senile cataracts among subjects taking regular daily supplements of ascorbic acid was compared to that of subjects who did not, the risk of senile cataract was found to be only 30% as high among the supplement users as in non-supplement users (352). In a similar but smaller study, Jacques et al. (353) also observed a 30% lower risk of cataracts for subjects with high plasma ascorbate levels (>90 μmol/liter) compared to those with levels below 40 μmol/liter. This difference in risk, however, was statistically insignificant.

The relationship between vitamin E and senile cataract has also received considerable attention. In vitro studies have shown that vitamin E prevents photoperoxidation of lens lipids (354). Its protective effects may be related to its ability to stabilize cell membranes by maintaining reduced glutathione levels in the lens and aqueous humor.

In epidemiological studies, Robertson et al. (352) reported that subjects taking an average of 400 IU of vitamin E daily in the form of supplements had 40% as high a risk of senile cataract as persons not consuming vitamin supplement. The combination of high vitamin E status with high blood levels of β-carotene and/or vitamin C also appears to be associated with a significant decrease in the risk of developing cataracts (353). However, unlike vitamin C or vitamin E, the human lens contains very little β-carotene. Nevertheless, β-carotene may protect against cataract formation through its reduction in the overall oxidative levels in the body.

6.4.8 Toxicants

Humans are constantly exposed to a multitude of toxic substances such as those associated with smoking, environmental pollution, alcohol consumption, side effects of pharmaceuticals, pesticides, and food poisons. The adverse effects of these substances on human health have been proved to be mediated by the intervention of free radicals at various levels. Antioxidant therapy has often proven beneficial in counteracting some of these adverse effects on human health (226, 355–357).

6.4.9 Metabolic Disorders

Free radicals have been implicated in the etiology of several genetic as well as acquired metabolic disorders. Examples include erythropoietic protoporphyria (EPP), described earlier, as a result of direct attack on double bonds by singlet oxygen, β-thalassemia and other hemoglobinopathies due to increased generation of oxygen radicals, sickle cell anemia, favism-related hemolysis, inborn errors of sulfur metabolism resulting in glutathione synthase deficiency and cystinosis with cystine accumulation, and hyperlipoproteinemia, which favors the formation of free radicals, thereby leading to arterial damage and platelet aggregation. Treatment with vitamin E, vitamin C, and carotenoids has been helpful for subjects suffering from such disorders (226,358–362).

6.4.10 Hyperoxygenation

Hyperbaric oxygen, if accompanied by poor reserves of antioxidant nutrients, can cause several serious conditions during the early days of infant life. These disorders include hyaline membrane disease, pneumothorax, apnea, pulmonary dysplasia, retrolental fibroplasia, and manifestations of intraventricular hemorrhagia and cerebral ischemia (226).

In adults, hyperbaric oxygen is capable of causing the decompression syndrome (363). It is associated with excessive production of arachidonic acid metabolites, prostaglandin endoperoxides, PGG_2, and thromboxane A, which favor platelet aggregation.

Disorders resulting from exposure to hyperbaric oxygen have been treated successfully with the administration of vitamin E and various carotenoids capable of acting as antioxidants (364–366).

6.4.11 Exercise and Stress

Oxygen consumption is greatly increased during exercise and stress-related situations, with the consequent generation of more $O_2^{\cdot-}$ and H_2O_2, primarily due to electron leakage from mitochondrial electron chains. Severe forced physical exercise can result in muscle damage, a decrease in mitochondrial respiratory control, loss of structural integrity of the sarcoplasmic reticulum, and increased levels of some markers of lipid peroxidation (11,367).

An increased antioxidant demand, therefore, may be seen under these situations. Exercise-training experiments in rats have shown vitamin E depletion at the level of skeletal muscle mitochondria (367). Although vitamin E supplementation does not improve athletic performance, it may play an important role in preventing exercise-induced muscle injury. An adequate vitamin C intake may also be beneficial for patients whose cardiac function is under stress (368). A sustained oxidation of glutathione during exercise and a compensatory reduction during recovery in humans also suggest an increase in the consumption of and requirement for antioxidants (226,368).

Stress situations such as psychopathological instability, surgery, strain, overwork, and trauma are also capable of generating excess free radicals. In such situations, vitamin C and vitamin E supplementation may be beneficial (226).

REFERENCES

1. J. H. Weisburger, in *Nutritional and Toxicological Consequences of Food Processing* (M. Friedman, Ed.), Plenum Press, New York, 1991, p. 137.
2. J. H. Weisburger, *Amer. J. Clin. Nutr.*, *53*: 226S (1991).
3. USDHHS, *The Health Consequences of Smoking: Cardiovascular Disease. A Report to the Surgeon General*, DHHS (PHS) Publ. 84-50204, U.S. Dept. of Health and Human Services, Office of Smoking and Health, Rockville, MD, 1983.
4. USDHHS, *The Health Consequences of Smoking: Cancer. A Report to the Surgeon General*, DHHS (PHS) Publ. 82-50179, U.S. Dept. of Health and Human Services, Office of Smoking and Health, Rockville, MD, 1983.
5. USDHHS, *Reducing the Health Consequences of Smoking: 25 Years of Progress. A Report to the Surgeon General*, DHHS (CDC) Publ. 89-8411, U.S. Dept. of Health and Human Services, Office of Smoking and Health, Rockville, MD, 1989.
6. D. Zaridze and R. Peto, *Tobacco: A Major International Health Hazard*, IARC Science Publ. 74, 1986, p. 3.
7. G. Block, *Amer. J. Clin. Nutr.*, *53*: 270S (1991).
8. T. F. Slater, in *Recent Advances in Biochemical Pathology. Toxic Liver Injury*, (M. V. Dianzani, G. Ugazio, and L. M. Sena, Eds.), Panminerva Medica, Torino, Italy, 1976, p. 381.
9. T. F. Slater, *Amer. J. Clin. Nutr.*, *53*: 394S (1991).
10. J. Bland, in *1986—A Year in Nutritional Medicine* (J. Bland, Ed.), Keats, New Canaan, CT, 1986, p. 293.

11. B. Halliwell and J. M. C. Gutteridge, *Free Radicals in Biology and Medicine*, Clarendon Press, Oxford, England, 1989.
12. T. F. Slater, *Biochem. J.*, *222*: 1 (1984).
13. B. Halliwell, in *Free Radicals and Food Additives* (O. I. Aruoma and B. Halliwell, Eds.), Taylor and Francis, London, 1991, p. 11.
14. I. A. Clark, W. B. Cowden, and N. H. Hunt, *Med. Res. Rev.*, *5*: 297 (1985).
15. W. A. Pryor, in *Free Radicals in Molecular Biology, Aging, and Disease* (D. Armstrong, R. S. Sohal, R. G. Cutler, and T. F. Slater, Eds.), Raven Press, New York, 1984, p. 13.
16. W. A. Pryor, *N.Y. Acad. Sci.*, *393*: 1 (1982).
17. W. A. Pryor, J. W. Lightsey, and D. F. Church, *J. Amer. Chem. Soc.*, *104*: 6685 (1982).
18. T. F. Slater, *Free Radical Mechanisms in Tissue Injury*, Pion, London, 1972.
19. S. W. Benson, *Thermochemical Kinetics*, John Wiley and Sons, New York, 1976.
20. S. W. Benson, *J. Chem. Phys.*, *40*: 1007 (1964).
21. A. Bendich, in *Micronutrients and Immune Functions* (A. Bendich and R. K. Chandra, Eds.), New York Acad. Sci., New York, 1990, p. 587.
22. A. T. Diplock, *Amer. J. Clin. Nutr.*, *53*: 189S (1991).
23. B. A. Freeman, in *Free Radicals in Molecular Biology, Aging, and Disease* (D. Armstrong, R. S. Sohal, R. G. Cutler, and T. F. Slater, Eds.), Raven, New York, 1984, p. 43.
24. B. Chance, H. Sies, and A. Boveris, *Physiol. Rev.*, *59*: 527 (1979).
25. O. Dionisi, T. Galeotti, T. Terranova, and A. Azzi, *Biochim. Biophys. Acta*, *403*: 292 (1975).
26. J. F. Turrens and A. Boveris, *Biochem. J.*, *191*: 421 (1980).
27. J. F. Turrens, B. A. Freeman, and J. D. Crapo, *Arch. Biochem. Biophys.*, *217*: 411 (1982).
28. A. Boveris and B. Chance, *Biochem. J.*, *134*: 707 (1973).
29. J. Capdevila, L. Parkhill, N. Chacos, R. Okita, and B. S. S. Masters, *Biochem. Biophys. Res. Commun.*, *101*: 1357 (1981).
30. C. G. Chignell, in *Spin Labeling*, Vol. 2 (L. J. Berliner, Ed.), Academic Press, New York, 1979, p. 223.
31. S. D. Aust, D. L. Roerig, and T. C. Pederson, *Biochem. Biophys. Res. Commun.*, *47*: 1133 (1972).
32. A. I. Archakov, G. I. Bachmanova, M. V. Isotov, and G. P. Kuznetsova, in *Microsomes, Drug Oxidations and Chemical Carcinogenesis* (R. W. Estabrook, Ed.), Academic Press, New York, 1980, p. 289.
33. C. Masters and R. Holmes, *Physiol. Rev.*, *57*: 816 (1977).
34. D. P. Jones, L. Eklow, H. Thor, and S. Orrenius, *Arch. Biochem. Biophys.*, *210*: 505 (1981).
35. D. Lonsdale, in *1986—A Year in Nutritional Medicine* (J. Bland, Ed.), Keats, New Canaan, CT, 1986, p. 85.
36. T. F. Slater, K. H. Cheeseman, and K. Proudfoot, in *Free Radicals in Molecular Biology, Aging, and Disease* (D. Armstrong, R. S. Sohal, R. G. Cutler, and T. F. Slater, Eds.), Raven Press, New York, 1984, p. 293.
37. C. Benedetto, M. V. Dianzani, M. Ahmed, K. H. Cheeseman, C. Connelly, and T. F. Slater, *Biochim. Biophys. Acta*, *677*: 363 (1981).

38. R. V. Bensasson, E. J. Land, and T. F. Truscott, *Flash Photolysis and Pulse Radiolysis*, Pergamon Press, Oxford, England, 1983.
39. D. C. Borg and K. Schaich, *Israel J. Chem.*, *24*: 38 (1984).
40. H. B. Demopoulos, *J. Environ. Pathol. Toxicol. 3*: 273 (1980).
41. H. B. Demopoulos, D. D. Pietronigro, and M. L. Seligman, *J. Amer. Coll. Toxicol.*, *2*: 173 (1983).
42. A. Schaefer, M. Komlos, and A. Seregi, *Biochem. Pharmacol.*, *24*: 1781 (1975).
43. B. M. Babior, *N. Engl. J. Med.*, *289*: 659 (1978).
44. H. Rosen and S. J. Klebanoff, *J. Exp. Med.*, *149*: 27 (1979).
45. M. J. Coon, *Nutr. Rev.*, *36*: 319 (1978).
46. N. J. Coon, *Methods Enzymol.*, *52*: 109 (1978).
47. M. J. Coon, *Methods Enzymol.*, *52*: 200 (1978).
48. E. M. Cranton and J. P. Frackelton, *J. Holistic Med.*, *6*: 6 (1984).
49. B. Halliwell and J. M. C. Gutteridge, *Arch. Biochem. Biophys.*, *246*: 501 (1986).
50. J. M. C. Gutteridge, D. A. Rowley, and B. Halliwell, *Biochem. J.*, *199*: 263 (1981).
51. J. M. C. Gutteridge and J. Stocks, *CRC Crit. Rev. Clin. Lab. Sci.*, *14*: 257 (1981).
52. J. E. Packer, T. F. Slater, and R. L. Wilson, *Nature (Lond.)*, *278*: 737 (1979).
53. T. Mann and D. Keilin, *Proc. Roy. Soc. (Lond.)*, *B126*: 303 (1938).
54. J. M. McCord and I. Fridovich, *J. Biol. Chem.*, *251*: 6049 (1969).
55. H. M. Hassan, in *Free Radicals in Molecular Biology, Aging, and Disease* (D. Armstrong, R. S. Sohal, R. G. Cutler, and T. F. Slater, Eds.), Raven Press, New York, 1984, p. 77.
56. J. M. McCord, B. B. Keele, and I. Fridovich, *Proc. Natl. Acad. Sci. (USA)*, *68*: 1024 (1971).
57. F. P. Tally, H. R. Godin, N. V. Jacobus, and S. L. Gorbach, *Infect. Immun.*, *16*: 20 (1977).
58. J. Hewitt and J. G. Morris, *FEBS Lett.*, *50*: 315 (1975).
59. J. I. Harris and H. M. Steinman, in *Superoxide Dismutases* (A. M. Michelson, J. M. McCord, and I. Fridovich, Eds.), Academic Press, New York, 1977, p. 225.
60. H. M. Steinman and R. L. Hill, *Proc. Natl. Acad. Sci. (USA)*, *70*: 3725 (1973).
61. H. M. Steinman, *J. Biol. Chem.*, *253*: 8708 (1978).
62. J. A. Tainer, E. D. Getzoff, K. M. Beem, J. S. Richardson, and D. C. Richardson, *J. Mol. Biol.*, *160*: 181 (1982).
63. J. K. Donnelly and D. S. Robinson, in *Oxidative Enzymes in Foods* (D. S. Robinson and N. A. M. Eskin, Eds.), Elsevier, London, 1991, p. 49.
64. H. J. Forman and I. Fridovich, *J. Biol Chem.*, *248*: 2645 (1973).
65. D. P. Malinowski and I. Fridovich, *Biochemistry*, *18*: 5909 (1979).
66. W. C. Stallings, K. A. Pattridge, R. K. Strong, and M. L. Ludwig, *J. Biol. Chem.*, *259*: 10695 (1984).
67. A. Gartner and U. Weser, in *Biomimetic and Bioorganic Chemistry*, Vol. 2 (F. Vogtle and E. Weber, Eds.), Springer, Berlin, 1986, p. 1.
68. I. Fridovich, *Science*, *201*: 875 (1978).
69. H. M. Hassan and I. Fridovich, *J. Bacteriol.*, *130*: 805 (1977).
70. H. M. Hassan and I. Fridovich, *Arch. Biochem. Biophys.*, *196*: 385 (1979).
71. W. Bors, *Program and Abstracts of Fifth Conference on Superoxide Dismutase*, Hebrew Univ. Jerusalem, Jerusalem, Israel, 1989, p. 45.

72. J. A. Fee, in *Oxidases and Related Redox Systems* (T. E. King, H. S. Mason, and M. Morrison, Eds.), Pergamon Press, Oxford, England, 1982, p. 101.
73. D. Touati, *Free Radical Biol. Med.*, 5: 393 (1988).
74. H. M. Hassan and I. Fridovich, *Rev. Infect. Dis.*, 1: 357 (1979).
75. J. D. Crapo and D. L. Tinerney, *Amer. J. Physiol.*, 226: 1401 (1974).
76. A. M. Michelson and M. E. Buckingham, *Biochem. Biophys. Res. Commun.*, 58: 1079 (1974).
77. A. Petkau, W. S. Chelack, and S. D. Plaskash, *Int. J. Radiat. Biol. Relat. Stud. Phys. Chem. Med.*, 29: 297 (1976).
78. J. J. Van Hemmon and W. J. A. Menling, *Biochim. Biophys. Acta*, 402: 133 (1975).
79. G. Czapski and S. Goldstein, *Free Radical Res. Commun.*, 4: 225 (1988).
80. J. K. Donnelly, K. M. McLellan, J. L. Walker, and D. S. Robinson, *Food Chem.*, 33: 243 (1989).
81. K. Grankvist, S. L. Marklund, and L. B. Taljedal, *Biochem. J.*, 199: 393 (1981).
82. C. Peeters-Joris, A. M. Vandevoorde, and P. Baudhuin, *Biochem. J.*, 150: 31 (1975).
83. W. H. Bannister and J. V. Bannister, *FEBS Lett.*, 142: 42.
84. S. L. Marklund, N. G. Westman, E. Lundgren, and G. Roos, *Cancer Res.*, 42: 1955 (1982).
85. D. S. Robinson, in *Oxidative Enzymes in Foods* (D. S. Robinson and N. A. M. Eskin, Eds.), Elsevier Applied Science, London, 1991, p. 1.
86. G. C. Mills, *J. Biol. Chem.*, 229: 189 (1957).
87. W. Nakamura, S. Hosoda, and K. Hayashi, *Biochim. Biophys. Acta*, 358: 251 (1974).
88. L. Flohe, B. Eisele, and A. Wendel, *Hoppe Seyler's Z. Physiol. Chem.*, 352: 151 (1971).
89. Y. C. Awasthi, E. Beutler, and S. K. Srivastava, *J. Biol. Chem.*, 250: 5144 (1975).
90. R. A. Sunde, H. E. Ganther, and W. G. Hoekstra, *Fed. Proc.*, 37: 757 (1978).
91. Y. C. Awasthi, D. D. Dao, A. K. Lal, and S. K. Srivastava, *Biochem. J.*, 177: 471 (1979).
92. J. T. Rotruck, A. L. Pope, H. E. Ganther, A. B. Swanson, D. G. Hafeman, and W. G. Hoekstra, *Science*, 179: 588 (1973).
93. S. H. Oh, H. E. Ganther, and W. G. Hoekstra, *Biochemistry*, 13: 1825 (1974).
94. L. Flohe, W. A. Gunzler, and H. M. Shock, *FEBS Lett.*, 32: 132 (1973).
95. J. W. Forstrom, J. J. Zakowski, and A. L. Tappel, *Biochemistry*, 17: 2639 (1978).
96. A. Wendel, B. Kerner, and K. Graupe, in *Functions of Glutathione in Liver and Kidney* (H. Sies and A. Wendel, Eds.), Springer-Verlag, Berlin, 1978, p. 107.
97. L. Flohe, G. Loschen, W. A. Gunzler, and E. Eicheie, *Hoppe Seyler's Z. Physiol. Chem.*, 353: 987 (1972).
98. L. Flohe and W. A. Gunzler, in *Glutathione* (L. Flohe, H. C. Benohr, H. Sies, H. D. Waller, and A. Wendel, Eds.), Academic Press, New York, 1974, p. 133.
99. H. E. Ganther, D. G. Hafeman, R. A. Lawrence, R. E. Serfass, and W. G. Hoekstra, in *Trace Elements in Human Health and Disease*, Vol. 2 (A. S. Prasad, Ed.), Academic Press, New York, 1976, p. 165.
100. C. Little and P. J. O'Brien, *Biochem. Biophys. Res. Commun.*, 31: 145 (1968).
101. R. A. Sunde and W. G. Hoekstra, *Nutr. Rev.*, 38: 265 (1980).
102. J. R. Stabel and J. W. Spears, in *Nutrition and Immunology* (D. M. Klurfeld, Ed.), Plenum Press, New York, 1993, p. 333.

103. P. F. Jacques and A. Taylor, in *Micronutrients in Health and in Disease Prevention* (A. Bendich and C. E. Butterworth, Eds.), Marcel Dekker, New York, 1991, p. 359.

104. H. Kappus, in *Free Radicals and Food Additives* (O. I. Aruoma and B. Halliwell, Eds.), Taylor and Francis, London, 1991, p. 59.

105. H. E. Evans, *Vitam. Horm.*, *20*: 379 (1963).

106. H. Dam, *Pharmacol. Rev.*, *9*: 1 (1957).

107. H. Dam and H. Granados, *Acta Physiol. Scand.*, *10*: 162 (1945).

108. H. Dam, *Vitam. Horm.*, *20*: 527 (1962).

109. L. J. Machlin, in *Handbook of Vitamins* (L. J. Machlin, Ed.), Marcel Dekker, New York, 1991, p. 99.

110. J. C. Bauernfeind, in *Vitamin E—A Comprehensive Treatise* (L. J. Machlin, Ed.), Marcel Dekker, New York, 1980, p. 99.

111. J. Moustgaard and J. Hyldgaard-Jensen, *Acta Agric. Scand.*, *19*: 11 (1971).

112. M. K. Horwitt, *Amer. J. Clin. Nutr.*, *27*: 939 (1974).

113. E. deDuve and O. Hayaishi, *Tocopherol, Oxygen and Biomembranes*, Elsevier, New York, 1978.

114. B. Lubin and L. J. Machlin, *Ann. N.Y. Acad. Sci.*, *581*: 393 (1982).

115. O. Hayaishi and M. Mino, *Clinical and Nutritional Aspects of Vitamin E*, Elsevier, New York, 1987.

116. A. T. Diplock, L. J. Machlin, L. Packer, and W. A. Pryor, *Ann. N.Y. Acad. Sci.*, *588*: 570 (1989).

117. A. T. Diplock, in *Fat Soluble Vitamins: Their Biochemistry and Applications* (A. T. Diplock, Ed.), Technomic, Lancaster, PA, 1985, p. 154.

118. P. Knekt, A. Aromaa, J. Maatela, R. Aaran, T. Nikkari, M. Hakama, T. Hakulinen, R. Peto, and L. Teppo, *Am. J. Clin. Nutr.*, *53*: 283S (1991).

119. W. A. Pryor, *Amer. J. Clin. Nutr.*, *53*: 391S (1991).

120. Anonymous, *Eur. J. Biochem.*, *46*: 217 (1974).

121. S. Kasparek, in *Vitamin E: A Comprehensive Treatise* (L. J. Machlin, Ed.), Marcel Dekker, New York, 1980, p. 7.

122. K. C. Ingold, G. W. Burton, D. O. Foster, M. Zuler, L. Hughes, S. Lacolle, E. Lusztyk, and M. Slaby, *FEBS Lett.*, *205*: 117 (1986).

123. J. C. Bauernfeind, in *Vitamin E: A Comprehensive Treatise* (L. J. Machlin, Ed.), Marcel Dekker, New York, 1980, p. 99.

124. M. Rechcigl, *Handbook of Nutritive Value of Processed Foods*, Vol. 1, CRC Press, Boca Raton, FL, 1984.

125. H. S. Olcott and H. A. Matill, *Chem. Rev.*, *29*: 257 (1941).

126. L. J. Machlin, *J. Am. Oil Chem. Soc.*, *40*: 368 (1963).

127. H. H. Draper, J. G. Bergan, M. Chin, A. Csallany, and A. V. Boara, *J. Nutr.*, *84*: 395 (1964).

128. A. L. Tappel, *Vitam. Horm.*, *20*: 493 (1962).

129. J. Green and J. Bunyan, *Nutr. Abstr. Rev.*, *39*: 321 (1969).

130. G. L. Catigani, in *Vitamin E: A Comprehensive Treatise* (L. J. Machlin, Ed.), Marcel Dekker, New York, 1980, p. 318.

131. H. J. Weiss, *N. Engl. J. Med.*, *293*: 531 (1975).

132. D. R. Phillips and M. A. S. Shuman, *Biochemistry of the Platelets*, Academic Press, Orlando, FL, 1986.

133. T. J. Rink, S. W. Smith, and R. Y. Tsien, *FEBS Lett.*, *148*: 21 (1982).
134. M. Stuart, *Ann. N.Y. Acad. Sci.*, *393*: 277 (1982).
135. R. V. Panganamala and D. G. Cornwell, *Ann. N.Y. Acad. Sci.*, *393*: 376 (1982).
136. J. S. C. Fong, *Experientia*, *32*: 639 (1976).
137. W. C. Hope, C. Dalton, L. J. Machlin, R. J. Filipski, and F. M. Vane, *Prostaglandins*, *10*: 557 (1975).
138. J. Lehmann, D. D. Rao, J. J. Canary, and J. T. Judd, *Amer. J. Clin. Nutr.*, *47*: 470 (1988).
139. R. O. L. Koff, D. R. Guptill, L. M. Lawrence, C. C. McKan, M. M. Mathias, C. F. Nockels, and R. P. Tengerdy, *Am. J. Clin. Nutr.*, *34*: 245 (1981).
140. A. T. Diplock and J. A. Lucy, *FEBS Lett.*, *29*: 205 (1973).
141. A. Bendich, L. J. Machlin, O. Scandurra, G. W. Burton, and D. D. M. Waynes, *Adv. Free Radical Biol. Med.*, *2*: 419 (1986).
142. E. Niki, *Chem. Phys. Lipids*, *44*: 227 (1987).
143. P. B. McCay, *Ann. N.Y. Acad. Sci.*, *570*: 32 (1989).
144. A. Szent-Gyorgyi, *Biochemi. J.*, *22*: 1387 (1928).
145. J. L. Svirbely and A. Szent-Gyorgyi, *Biochemi. J.*, *26*: 865 (1932).
146. C. G. King and W. A. Waugh, *Science*, *75*: 357 (1932).
147. W. N. Haworth and E. L. Hirst, *J. Soc. Chem. Ind.*, *52*: 645 (1933).
148. T. Reichstein, A. Grussner, and R. Oppenhauer, *Helv. Chim. Acta*, *16*: 1019 (1933).
149. C. G. King, *World Rev. Nutr. Diet.*, *18*: 47 (1973).
150. J. E. Halver, R. R. Smith, B. M. Tolbert, and E. M. Baker, *Ann. N. Y. Acad. Sci. 258*: 81 (1975).
151. U. Moser and A. Bendich, in *Handbook of Vitamins* (L. J. Machlin, Ed.), Marcel Dekker, New York, 1991, p. 195.
152. T. Reichstein and A. Grussner, *Helv. Chim. Acta*, *17*: 311 (1934).
153. IUPAC-IUB Commission on Biochemical Nomenclature, *Biochim. Biophys. Acta*, *107*: 1 (1965).
154. G. M. Jaffe, in *Handbook of Vitamins: Nutritional, Biochemical and Clinical Aspects* (L. J. Machlin, Ed.), Marcel Dekker, New York, 1984, p. 199.
155. K. Mikova and J. Davidek, *Chem. Listy*, *68*: 715 (1974).
156. K. Niemela, *J. Chromatogr.*, *399*: 235 (1987).
157. J. C. Brand, V. Cherikoff, A. Lee, and A. S. Truswell, *Lancet*, *2*: 873 (1982).
158. M. Oliver, in *The Vitamins*, Vol. I (W. H. Sebrell and R. S. Harris, Eds.), Academic Press, New York, 1967, p. 359.
159. B. K. Watt and A. L. Merrill, *Composition of Foods*, *Agric. Handbook* No. 8, U.S. Dept. Agric., Washington, D.C., 1975.
160. A. A. Paul and D. A. T. Southgate, *McCance and Widdowson's The Composition of Foods*, Spec. Rep. 297, Medical Research Council, London, 1978.
161. M. Levine and K. Morita, *Vitam. Horm.*, *42*: 1 (1985).
162. J. W. Erdman, Jr. and B. P. Klein, in *Ascorbic Acid: Chemistry, Metabolism and Uses* (P. A. Seib and P. M. Tolbert, Eds.), Am. Chem. Soc., Washington, D.C., 1982, p. 499.
163. M. J. Barnes and E. Kodicek, *Vitam. Horm.*, *30*: 1 (1972).
164. C. J. Bates, in *Vitamin C (Ascorbic Acid)* (J. N. Counsell and D. H. Hornig, Eds.), Applied Science, London, 1981, p. 1.

165. R. E. Hughes, in *Vitamin C (Ascorbic Acid)* (J. N. Counsell and D. H. Hornig, Eds.), Applied Science, London, 1981, p. 75.
166. M. B. Davies, J. Austin, and D. A. Partridge, *Vitamin C: Its Chemistry and Biochemistry*, Royal Society of Chemistry, Cambridge, England, 1991.
167. R. W. Fessenden and N. C. Verma, *Biophys. J.*, 24: 93 (1978).
168. B. E. Leibovitz and B. V. Siegel, *J. Gerontol.*, 35: 45 (1980).
169. R. S. Bodannes and P. C. Chan, *FEBS Lett.*, 105: 195 (1979).
170. C. W. M. Wilson, *Ann. N.Y. Acad. Sci.*, 258: 355 (1975).
171. R. Anderson, in *Vitamin C (Ascorbic Acid)* (J. N. Counsell and D. H. Hornig, Eds.), Applied Science, London, 1981, p. 249.
172. W. R. Thomas and P. G. Holt, *Clin. Exp. Immunol.*, 32: 370 (1978).
173. B. Kennes, *Gerontology*, 29: 305 (1983).
174. D. J. Sutor and C. S. Johnson, *FASEB J.*, 2: A851 (1988).
175. S. K. Gaby and V. N. Singh, in *Vitamin Intake and Health* (S. K. Gaby, A. Bendich, V. N. Singh, and L. J. Machlin, Eds.), Marcel Dekker, New York, 1991, p. 29.
176. L. W. Wattenburg, *Cancer Res.*, 45: 1 (1985).
177. J. A. Olson, in *Handbook of Vitamins* (L. J. Machlin, Ed.), Marcel Dekker, New York, 1991, p. 1.
178. Y. A. Ovchinnikov, *Retinal Proteins*, VSP Press, Utrecht, The Netherlands, 1987.
179. S. Moriguchi, L. Werner, and R. R. Watson, *Immunology*, 56: 169 (1985).
180. M. B. Sporn and A. B. Roberts, *Cancer Res.*, 43: 3034 (1983).
181. A. Bendich and L. Langseth, *Amer. J. Clin. Nutr.*, 49: 358 (1989).
182. J. A. Olson, *J. Nutr.*, 119: 105 (1989).
183. R. Peto, R. Doll, J. D. Buckley, and M. B. Sporn, *Nature (Lond.)*, 290: 201 (1981).
184. Anonymous, *Diet, Nutrition and Cancer*, Committee on Diet, Nutrition and Cancer, National Academy of Sciences, National Academy Press, Washington, DC, 1982.
185. O. Straub, in *Key to Carotenoids*, 2nd ed. (H. Pfander, Ed.), Birkhauser Verlag, Basel, Switzerland, 1987, p. 296.
186. S. L. Spurgeon and J. W. Porter, in *Biosynthesis of Isoprenoid Compounds*, Vol. 2 (J. W. Porter and S. L. Spurgeon, Eds.), Wiley, New York, 1983, p. 1.
187. J. Glover, *Vitam. Horm.*, 18: 371 (1960).
188. D. S. Goodman and H. S. Huang, *Science*, 149: 879 (1965).
189. J. A. Olson and O. Hayaishi, *Proc. Natl. Acad. Sci. (USA)*, 54: 1364 (1965).
190. Anonymous, *Recommended Dietary Allowances*, 9th rev. ed., Committee on Dietary Allowances, National Academy of Sciences, Washington, DC, 1980, p. 55.
191. J. A. Olson, *J. Nutr.*, 119: 94 (1989).
192. A. Bendich and J. A. Olson, *FASEB J.*, 3: 1927 (1989).
193. R. S. Parker, *J. Nutr.*, 119: 101 (1989).
194. N. I. Krinski and S. M. Deneke, *J. Natl. Cancer Inst.*, 69: 205 (1982).
195. G. W. Burton and K. U. Ingold, *Science*, 224: 569 (1984).
196. W. A. Pryor, T. Strickland, and D. F. Church, *J. Amer. Chem. Soc.*, 110: 2224 (1988).
197. C. S. Foote, R. W. Denny, L. Weaver, Y. Chang, and J. Peters, *Ann. N.Y. Acad. Sci.*, 171: 139 (1970).
198. C. S. Foote, Y. C. Chang, and R. W. Denny, *J. Amer. Chem. Soc.*, 92: 5116 (1970).
199. M. M. Matthew-Roth, *Photochem. Photobiol.*, 40: 63 (1984).

200. M. M. Matthew-Roth, M. A. Pathak, T. B. Fitzpatrick, L. C. Harber, and E. H. Kass, *N. Engl. J. Med.*, *282*: 1231 (1970).
201. H. N. Green and E. Mellanby, *Brit. J. Exp. Pathol.*, *11*: 81 (1930).
202. S. W. Clausen, *Trans. Am. Pediatr. Soc.*, *43*: 27 (1931).
203. A. Bendich, *J. Nutr.*, *119*: 112 (1989).
204. E. Seifter, G. Rettura, and S. M. Levinson, in *The Quality of Foods and Beverages* (G. Charalambors and G. Inglett, Eds.), Academic Press, New York, 1981, p. 335.
205. E. Seifter, G. Rettura, J. Padawer, and S. M. Levinson, *J. Natl. Cancer Inst.*, *68*: 835 (1982).
206. J. Schwartz, D. Suda, and G. Light, *Biochem. Biophys. Res. Commun.*, *136*: 1130 (1986).
207. N. I. Krinsky, *Am. J. Clin. Nutr.*, *53*: 238S (1991).
208. A. Bendich, *Nutr. Cancer*, *11*: 207 (1988).
209. S. Cunningham-Rundles, *Ann. N.Y. Acad. Sci.*, *587*: 114 (1990).
210. M. Bunk, *Nutr. Cancer*, *10*: 79 (1987).
211. M. A. Johnson and J. G. Fischer, *Food Technol.*, *48*(5): 112 (1994).
212. M. A. Johnson, J. G. Fischer, and S. E. Kays, *Crit. Rev. Food Sci. Nutr.*, *32*: 1 (1992).
213. NRC, *Diet and Health: Implications for Reducing Chronic Disease Risk*, Natl. Res. Council, National Academy Press, Washington, DC, 1989.
214. M. A. Johnson, in *Copper Bioavailability and Metabolism* (C. Kies, Ed.), Plenum Press, New York, 1989, p. 217.
215. R. F. Burk, *Pharmacol. Ther.*, *45*: 383.
216. R. F. Burk and K. E. Hill, *Biol. Trace Elem. Res.*, *33*: 151 (1992).
217. A. T. Diplock, *Med. Oncol. Tumor Pharmacother.*, 7: 193 (1990).
218. J. E. Seegmiller, in *Contemporary Metabolism*, Vol. 1 (E. Freinkel, Ed.), Plenum Press, New York, 1979, p. 1.
219. D. R. Farr and J. Loliger, Eur Patent Appl. 226 211, Feb. 19, 1986 (1987).
220. D. R. Farr, J. Loliger, and M. C. Savoy, *J. Sci. Food Agric.*, *37*: 804 (1986).
221. R. C. Smith and L. Lowing, *Arch. Biochem. Biophys.*, *223*: 166 (1983).
222. B. Halliwell, *Free Radical Res. Commun.*, *9*: 1 (1990).
223. B. N. Ames, *Proc. Natl. Acad. Sci.* (*USA*), *78*: 6858 (1981).
224. S. Al-Malaika, in *Free Radicals and Food Additives* (O. I. Aruoma and B. Halliwell, Eds.), Taylor and Francis, London, 1991, p. 151.
225. J. Pospisil, in *Developments in Polymer Stabilization*, Vol. 7 (G. Scott, Ed.), Applied Science, London, 1984, p. 1.
226. P. Bermond in *Food Antioxidants* (B. J. F. Hudson, ed.), Elsevier Science, Barkings, England, 1990, p. 193.
227. R. Doll and R. Peto, *J. Natl. Cancer Inst.*, *66*: 1192 (1981).
228. V. N. Singh and S. K. Gaby, *Amer. J. Clin. Nutr.*, *53*: 386S (1991).
229. I. D. Rotkin, *Cancer Treat. Rep.*, *61*: 173 (1977).
230. H. C. Pitot, *Cancer*, *49*: 1206 (1982).
231. B. N. Ames, *Science*, *221*: 1256 (1983).
232. Anonymous, *Lancet*, *1*: 1223 (1982).
233. D. Harman, *J. Gerontol.*, *26*: 451 (1971).
234. D. Harman, *J. Gerontol.*, *16*: 247 (1961).
235. D. Harman, in *Free Radicals in Molecular Biology, Aging, and Disease* (D. Arm-

strong, R. S. Sohal, R. G. Cutler, and T. F. Slater, Eds.), Raven Press, New York, 1984, p. 13.

236. S. K. Gaby and L. J. Machlin, in *Vitamin Intake and Health* (S. K. Gaby, A. Bendich, V. N. Singh, and L. J. Machlin, Eds.), Marcel Dekker, New York, 1991, p. 71.

237. N. J. Wald, J. Boreham, J. L. Hayward, and R. D. Bulbrook, *Brit. J. Cancer, 49*: 321 (1984).

238. W. Haenszel, P. Correa, A. Lopez, C. Cuello, G. Zarama, D. Zavala, and E. Fontham, *Int. J. Cancer, 36*: 43 (1985).

239. J. T. Salonen, R. Salonen, R. Lappetelainen, P. H. Maenpaa, G. Alftham, and P. Puska, *Brit. Med. J., 290*: 417 (1985).

240. M. S. Menkes, G. W. Comstock, J. P. Vuilleumier, K. J. Helsing, A. A. Rider, and R. Brookmeyer, *N. Engl. J. Med., 315*: 1250 (1986).

241. F. J. Kok, C. M. van Duijn, A. Hofman, R. Vermeeren, A. M. de Bruijn, and H. A. Valkenburg, *N. Engl. J. Med., 316*: 1416 (1987).

242. H. Miyamoto, Y. Araya, M. Ito, I. Hiroshi, H. Dosaka, T. Shimizu, F. Kishi, I. Yamamoto, H. Honma, and Y. Kawakami, *Cancer, 60*: 1159 (1987).

243. P. Knekt, A. Aromaa, J. Maatela, R. K. Aaran, T. Nikkari, M. Hakama, T. Hakulinen, R. Peto, E. Saxen, and L. Teppo, *Amer. J. Epidemiol., 127*: 28 (1988).

244. H. B. Stahelin, F. Rosel, E, Buess, and G. Brubacher, *J. Natl. Cancer Inst., 73*: 1463 (1984).

245. W. C. Willett, B. F. Polk, B. A. Underwood, M. J. Stampfer, S. Pressel, B. Rosner, J. O. Taylor, K. Schneider, and C. G. Hames, *N. Engl. J. Med., 310*: 430 (1984).

246. A. M. Y. Nomura, G. N. Stemmermann, L. K. Heilbrun, R. M. Salkeld, and J. P. Vuilleumier, *Cancer Res., 45*: 2369 (1985).

247. P. K. Heinonen, T. Koskinen, and R. Tuimala, *Arch. Gynecol., 237*: 37 (1985).

248. N. J. Wald, S. G. Thompson, J. W. Densem, J. Boreham, and A. Bailey, *Brit. J. Cancer, 56*: 69 (1987).

249. M. J. Russell, B. S. Thomas, and R. D. Bulbrook, *Brit. J. Cancer, 57*: 213 (1988).

250. D. J. Hunter, J. E. Manson, G. A. Colditz, M. J. Stampfer, B. Rosner, C. H. Hennekens, F. E. Speizer, and W. C. Willett, *N. Engl. J. Med., 329*: 234 (1993).

251. The Alpha-Tocopherol, Beta Carotene Cancer Prevention Study Group, *N. Engl. J. Med., 330*: 1029 (1994).

252. S. K. Gaby and V. N. Singh, in *Vitamin Intake and Health* (S. K. Gaby, A. Bendich, V. N. Singh, and L. J. Machlin, Eds.), Marcel Dekker, New York, 1991, p. 103.

253. E. T. H. Fontham, L. W. Pickle, W. Haenszel, P. Correa, Y. Lin, and R. T. Falk, *Cancer, 62*: 2267 (1988).

254. L. LeMarchand, C. N. Yoshizawa, L. N. Kolonel, J. H. Hankin, and M. T. Goodman, *J. Natl. Cancer Inst., 81*: 1158 (1989).

255. T. Byers, J. Vena, C. Mettlin, M. Swanson, and S. Graham, *Amer. J. Epidemiol., 120*: 769 (1984).

256. M. W. Hinds, L. N. Kolonel, J. H. Hankin, and J. Lee, *Amer. J. Epidemiol., 119*: 227 (1984).

257. J. K. McLaughlin, G. Gridley, G. Block, D. M. Winn, S. Preston-Martin, J. B. Schoenberg, R. S. Greenberg, A. Stemhagen, D. F. Austin, A. G. Ershow, W. J. Blot, and J. F. Fraumeni, *J. Natl. Cancer Inst., 80*: 1237 (1988).

258. M. A. Rossing, T. L. Vaughan, and B. McKnight, *Int. J. Cancer, 44*: 593 (1989).

259. S. Graham, C. Mettlin, J. Marshall, R. Priore, T. Rzepka, and D. Shedd, *Amer. J. Epidemiol.*, *113*: 675 (1981).
260. A. J. Tuyns, E. Riboli, G. Doornbos, and G. Pequignot, *Nutr. Cancer*, *9*: 81 (1987).
261. R. G. Ziegler, L. E. Morris, W. J. Blot, L. M. Pottern, R. Hoover, and J. F. Fraumeni, *J. Natl. Cancer Inst.*, *67*: 1199 (1981).
262. P. Correa, E. Fontham, L. W. Pickle, V. Chen, Y. Lin, and W. Haenszel, *J. Natl. Cancer Inst.*, *75*: 645 (1985).
263. W. C. You, W. J. Blot, Y. S. Chang, A. G. Ershow, Z. T. Yang, Q. An, B. Henderson, G. W. Xu, J. F. Fraumeni, and T. G. Wang, *Cancer Res.*, *48*: 3518 (1988).
264. M. L. Burr, C. J. Bates, G. Goldberg, and B. K. Butland, *Hum. Nutr. Clin. Nutr.*, *39C*: 387 (1985).
265. L. N. Kolonel, M. W. Hinds, A. M. Y. Nomura, J. H. Hankin, and J. Lee, *Natl. Cancer Inst. Monogr.*, *69*: 137 (1985).
266. S. Graham, H. Dayal, M. Swanson, A. Mittelman, and G. Wilkinson, *J. Natl. Cancer Inst.*, *61*: 709 (1978).
267. S. E. Norell, A. Ahlbom, R. Erwald, G. Jacobson, I. Lindberg-Navier, R. Olin, B. Tornberg, and K. L. Wiechel, *Amer. J. Epidemiol.*, *124*: 894 (1986).
268. R. Verreault, J. Chu, M. Mandelson, and K. Shy, *Int. J. Cancer*, *43*: 1050 (1989).
269. J. R. Marshall, S. Graham, T. Byers, M. Swanson, and J. Brasure, *J. Natl. Inst. Cancer*, *70*: 847 (1983).
270. D. W. West, M. L. Slattery, L. M. Robison, K. L. Schuman, M. H. Ford, A. W. Mahoney, J. L. Lyon, and A. W. Sorensen, *Amer. J. Epidemiol.*, *130*: 883 (1989).
271. C. La Vecchia, E. Negri, A. Decarli, B. D'Avanzo, L. Gallotti, A. Gentile, and S. Franceschi, *Int. J. Cancer*, *41*: 492 (1988).
272. L. K. Heilbrun, A. Nomura, J. H. Hankin, and G. N. Stemmermann, *Int. J. Cancer*, *44*: 1 (1989).
273. A. H. Wu, A. Paganini-Hill, R. K. Ross, and B. E. Henderson, *Brit. J. Cancer*, *55*: 687 (1987).
274. Y. Tomita, K. Hlmeno, K. Nomoto, H. Endo, and T. Hirohata, *J. Natl. Cancer Inst.*, *78*: 679 (1987).
275. T. K. Basu, N. J. Temple, and J. Ng, *J. Clin. Biochem. Nutr.*, *3*: 95 (1987).
276. J. E. Connett, L. H. Kuller, M. O. Kjelsberg, B. F. Polk, G. Collins, A. Rider, and S. B. Hulley, *Cancer*, *64*: 126 (1989).
277. L. C. Koo, *Nutr. Cancer*, *11*: 155 (1988).
278. R. G. Ziegler, T. J. Mason, A. Stemhagen, R. Hoover, J. B. Schoenberg, G. Gridley, P. Virgo, and J. Fraumeni, *Amer. J. Epidemiol.*, *123*: 1080 (1986).
279. D. Kromhout, *Amer. J. Clin. Nutr.*, *45*: 1361 (1987).
280. P. A. Holst, D. Kromhout, and R. Brand, *Brit. Med. J.*, *297*: 1319 (1988).
281. E. Marubini, A. Decarli, A. Costa, C. Mazzoleni, C. Andreoli, A. Barbieri, E. Capitelli, M. Carlucci, F. Cavallo, N. Monferroni, U. Pastorino, and S. Salvini, *Cancer*, *61*: 173 (1988).
282. T. E. Rohan, A. J. McMichael, and P. A. Baghurst, *Amer. J. Epidemiol.*, *128*: 478 (1988).
283. K. Brock, G. Berry, P. A. Mock, R. MacLennan, A. S. Truswell, and L. A. Brinton, *J. Natl. Cancer Inst.*, *80*: 580 (1988).

284. C. La Vecchia, S. Franceschi, A. Decarli, A. Gentile, M. Fasoli, S. Pampallona, and G. Tognoni, *Int. J. Cancer, 34*: 319 (1984).
285. A. Decarli, P. Liati, E. Negri, S. Franceschi, and C. LaVecchia, *Nutr. Cancer, 10*: 29 (1987).
286. L. M. Brown, W. J. Blot, S. H. Schuman, V. M. Smith, A. G. Ershow, R. D. Marks, and J. F. Fraumeni, *J. Natl. Cancer Inst., 80*: 1620 (1988).
287. E. Buiatti, D. Palli, A. Decarli, D. Amadori, C. Avellini, S. Bianchi, R. Biserni, F. Ciprani, P. Cocco, A. Giacosa, E. Marubini, R. Puntoni, C. Vindigni, J. Fraumeni, and W. Blot, *Int. J. Cancer, 44*: 611 (1989).
288. S. Kono, M. Ikeda, S. Tokudome, and M. Kuratsune, *Jpn. J. Cancer Res., 79*: 1067 (1988).
289. K. F. Gey, G. B. Brubacher, and H. B. Stahelin, *Amer. J. Clin. Nutr., 45*: 1368 (1987).
290. M. L. Slattery, A. W. Sorenson, A. W. Mahoney, T. K. French, D. Kritchevsky, and J. C. Street, *Natl. Cancer Inst., 80*: 1474 (1988).
291. G. A. Kune and S. Kune, *Nutr. Cancer, 9*: 1 (1987).
292. R. Nowak, *Science, 264*: 500 (1994).
293. K. F. Gey, *Bibl. Nutr. Dieta, 37*: 53 (1986).
294. K. F. Gey, in *Elevated Dosages of Vitamins* (P. Walter, H. Stahelin, and G. Brubacher, Eds.), Hans Huber, Toronto, 1989, p. 224.
295. Y. Yamamoto, M. H. Brodsky, J. C. Baker, and B. N. Ames, *Anal. Biochem., 160*: 7 (1987).
296. S. K. Liu, E. P. Dolensek, J. P. Tappe, J. Stover, and C. R. Adams, *J. Amer. Vet. Med. Assoc., 185*: 1347 (1984).
297. P. K. Singhal, N. Kapur, and R. E. Beamish, *Dev. Cardiovasc. Med., 60*: 190 (1985).
298. E. B. Rimm, M. J. Stampfer, A. Ascherio, E. Giovannucci, G. A. Colditz, and W. C. Willett, *N. Engl. J. Med., 328*: 1450 (1993).
299. M. J. Stampfer, C. H. Hennekens, J. A. Manson, G. A. Colditz, B. Rosner, and W. C. Willett, *N. Engl. J. Med., 328*: 1444 (1993).
300. G. C. Willis, *Can. Med. Assoc. J., 69*: 17 (1953).
301. G. C. Willis, *Can. Med. Assoc. J., 77*: 106 (1957).
302. J. Ramirez and N. C. Flowers, *Amer. J. Clin. Nutr., 33*: 2079 (1980).
303. L. Lapidus, H. Anderson, C. Bengtsson, and I. Bosaeus, *Amer. J. Clin. Nutr., 44*: 444 (1986).
304. J. E. Enstrom, L. E. Kanim, and M. A. Klein, *Epidemiology, 3*: 194 (1992).
305. J. M. Gaziano, J. E. Manson, P. M. Ridker, J. E. Buring, and C. H. Hennekens, *Circulation, 82*: Suppl. III-201 (1990) (Abstract).
306. D. Steinberg, *N. Engl. J. Med., 328*: 1487 (1993).
307. D. Pearson and S. Shaw, *Life Extension. A Practical Scientific Approach*, Warner Books, New York, 1982.
308. D. Harman, *J. Gerontol., 11*: 298 (1956).
309. R. G. Cutler, *Amer. J. Clin. Nutr., 53*: 373S (1991).
310. R. G. Cutler, in *Free Radicals in Molecular Biology, Aging, and Disease* (D. Armstrong, R. S. Sohal, R. G. Cutler, and T. F. Slater, Eds.), Raven Press, New York, 1984, p. 235.
311. W. C. Orr and R. S. Sohal, *Science, 263*: 1128 (1994).

312. A. Bendich, *J. Dairy Sci.*, *76*: 2789 (1993).
313. I. M. Roitt, J. Brostoff, and D. K. Male, *Immunology*, Grower Medical Publ., New York, 1989, p. 109.
314. S. N. Meydani and J. M. Blumberg, in *Micronutrients in Health and in Disease Prevention* (A. Bendich and C. E. Butterworth, Eds.), Marcel Dekker, New York, 1991, p. 289.
315. A. Bendich, in *Nutrition and Immunology* (R. K. Chandra, Ed.), Alan Liss, New York, 1988, p. 18.
316. A. Bendich, E. Gabriel, and L. J. Machlin, *J. Nutr.*, *116*: 675 (1986).
317. L. M. Corwin and R. K. Gordon, *Ann. N.Y. Acad. Sci.*, *393*: 437 (1982).
318. H. C. Meeker, M. L. Eskew, W. Scheuchenzuber, R. W. Scholz, and A. Zarkower, *J. Leukocyte Biol.*, *38*: 451 (1985).
319. G. O. Till, J. R. Hatherhill, W. W. Tourtellotte, M. J. Lutz, and P. A. Warel, *Amer. J. Pathol.*, *119*: 376 (1985).
320. S. N. Meydani, M. P. Barklund, S. Liu, M. Meydani, R. Miller, J. Cannon, F. Morrow, R. Rocklin, and J. Blumberg, *FASEB J.*, *3*: A1057 (1989).
321. M. Chavance, C. Brubacher, B. Herbeth, G. Vernhes, T. Mikstacki, F. Dete, C. Fournier, and C. Janot, in *Lymphoid Cell Function in Aging* (A. L. deWeek, Ed.), Eurage, Rajswik, The Neterlands, 1984, p. 123.
322. R. O. Likoff, D. R. Guptill, L. M. Lawrence, C. C. McKay, M. M. Mathias, C. F. Nockels, and R. P. Tengerdy, *Amer. J. Clin. Nutr.*, *34*: 245 (1981).
323. L. M. Lawrence, M. M. Mathias, C. F. Nockels and R. P. Tengerdy, *Nutr. Res.*, *5*: 497 (1985).
324. J. E. Lafuze, S. J. Weisman, L. A. Alperty, and R. L. Baehner, *Pediatr. Res.*, *18*: 536 (1984).
325. L. J. Machlin and A. Bendich, *FASEB J.*, *1*: 441 (1987).
326. A. Bendich, *Clin. Appl. Nutr.*, *1*: 45 (1991).
327. C. A. Leslie and D. P. Dubey, *Fed. Proc.*, *41*: 331 (1982).
328. R. F. Del Maestro, in *Free Radicals in Molecular Biology, Aging, and Disease* (D. Armstrong, R. S. Sohal, R. G. Cutler, and T. F. Slater, Eds.), Raven Press, New York, 1984, p. 87.
329. J. Emerit, *Cah. Nutr. Diet. (Paris)*, 22(1): 35 (1987).
330. J. A. Badwey and M. L. Karnovsky, *Ann. Rev. Biochem.*, *49*: 687 (1980).
331. D. Roos, in *The Cell Biology of Inflammation* (G. Weissmann, Ed.), Elsevier, Amsterdam, 1980, p. 119.
332. S. J. Weiss and A. Slivka, *J. Clin. Invest.*, *69*: 255 (1982).
333. B. Halliwell, *Cell Biol. Int. Rep.*, *6*: 529 (1982).
334. E. J. Corey, M. M. Mehrotra, and A. V. Khan, *Science*, *38*: 559 (1987).
335. F. A. Dolbeare and K. A. Martlage, *Proc. Soc. Exp. Biol. (NY)*, *139*: 540 (1972).
336. A. Hanck and H. Weiser, in *Vitamins: Nutrients and Therapeutic Agents* (D. Hanck and D. Hornig, Eds.), Hans Huber, Bern, The Netherlands, 1985, p. 189.
337. N. K. Krinsky, *Vit. Nutr. Inform. Sci.*, *1*: 1 (1992).
338. W. R. Sistrom, M. Griffith, and R. Y. Stainer, *J. Cell Comp. Physiol.*, *48*: 473 (1956).
339. M. M. Mathews-Roth, *Biochimie*, *68*: 875 (1986).
340. M. M. Mathews-Roth, M. A. Pathak, T. B. Fitzpatrick, L. C. Harber, and E. H. Kass, *N. Engl. J. Med.*, *282*: 1231 (1970).

341. M. M. Mathews-Roth, M. A. Pathak, T. B. Fitzpatrick, L. C. Harber, and E. H. Kass, *J. Am. Med. Assoc.*, *228*: 1004 (1974).
342. C. R. Dawson and I. R. Schwab, *Bull. WHO*, *59*: 493 (1981).
343. Anon., *Vision Research: A National Plan. The Report of the Cataract Panel*, Vol. 2, Part 3, U.S. Dept. HHS, NIH Publ. No. 83-2473, 1983.
344. P. F. Jacques and A. Taylor, in *Micronutrients in Health and in Disease Prevention* (A. Bendich and C. E. Butterworth, Eds.), Marcel Dekker, New York, 1991, p. 359.
345. R. D. Wiegand, J. G. Jose, L. M. Rapp, and R. E. Anderson, in *Free Radicals in Molecular Biology, Aging, and Disease* (D. Armstrong, R. S. Sohal, R. G. Cutler, and T. F. Slater, Eds.), Raven Press, New York, 1984, p. 317.
346. J. J. Weiter and E. D. Finch, *Nature (Lond.)*, *254*: 536 (1975).
347. J. D. Goosey, J. S. Zigler, and J. H. Kinoshita, *Science*, *208*: 1278 (1980).
348. S. D. Varma, D. Chand, Y. R. Sharma, J. F. Kuck, and R. D. Richards, *Current Eye Res.*, *3*: 35 (1984).
349. K. C. Bhuyan and D. K. Bhuyan, *Current Eye Res.*, *3*: 67 (1984).
350. J. F. R. Kuck, *Invest. Ophthalmol.*, *15*: 405 (1976).
351. D. B. Chandra, R. Varma, S. Ahmad, and S. D. Varma, *Int. J. Vit. Nutr. Res.*, *56*: 165 (1986).
352. J. M. Robertson, A. P. Donner, and J. R. Trevithick, *Ann. N.Y. Acad Sci.*, *570*: 372 (1989).
353. P. F. Jacques, S. C. Hartz, L. T. Chylack, R. B. McGrandy, and J. A. Sadowski, *Am. J. Clin. Nutr.*, *48*: 152 (1988).
354. S. D. Varma, N. A. Beachy, and R. D. Richards, *Photochem. Photobiol.*, *36*: 623 (1982).
355. L. Santamaria, A. Bianchi, I. Arnaboldi, L. Andreoni, and P. Bermond, *Experientia*, *39*: 1043 (1983).
356. J. J. Dougherty and W. G. Hoekstra, *Proc. Soc. Exp. Biol. Med.*, *169*: 201 (1982).
357. U. Korallus, C. Harzdorl, and J. Lewalter, *Int. Arch. Occup. Environ. Health*, *53*: 247 (1984).
358. E. I. Rachmilewitz, A. Kornberg, and M. Acker, *Ann. N. Y. Acad. Sci.*, *393*: 338 (1982).
359. N. A. Lachant and K. R. Tanaka, *Am. J. Med. Sci.*, *292*: 3 (1986).
360. I. Mavelli, *Eur. J. Biochem.*, *139*: 13 (1984).
361. J. D. Schulman, *Ann. Intern. Med.*, *93*: 330 (1980).
362. A. Fidanza, M. Andisio, and P. Mastroiacovo, *Int. J. Vit. Nutr. Res. (Suppl.)*, *23*: 153 (1982).
363. E. Reggiani and G. Odaglia, *Minerva Med.*, *72*: 1833 (1981).
364. H. M. Hitner and F. L. Kretzer, *Pediatrics*, *73*: 238 (1984).
365. M. Dehan, A. Lindenbaum, and F. Niessen, *Cah. Nutr. Diet.*, *22*(1): 15 (1987).
366. W. A. Colburn and R. A. Ehrenkranz, *Pediatr. Pharmacol.*, *3*: 7 (1983).
367. K. Gohil, L. Rothfuss, J. Lang, and L. Parker, *J. Appl. Physiol.*, *63*: 1638 (1987).
368. R. E. Hugues, in *Vitamin C (Ascorbic Acid)* (J. N. Counsell and D. H. Hornig, Eds.), Applied Science, London, 1982, p. 232.

7

Summary, Conclusions, and Future Research Needs

D. L. Madhavi
University of Illinois, Urbana, Illinois

S. S. Deshpande
Idetek, Inc., Sunnyvale, California

D. K. Salunkhe
Utah State University, Logan, Utah

Formation of various free radicals is a common process in biological and food systems. At the cellular level, normal metabolism of oxygen leads to the production of oxygen-derived free radicals which are eliminated efficiently by various cellular defense mechanisms. However, the biological system is also frequently subjected to prooxidant situations wherein the cellular concentration of an activated oxygen species is increased either by overproduction, because of some causative agents, or by a deficiency in the controlling mechanisms. The activated forms of oxygen produced are superoxide, hydroperoxy radical, singlet oxygen, hydroxyl radical, and hydrogen peroxide. Prooxidant states are induced in biological systems by various agents such as hyperbaric oxygen tension, radiation, transition metals, iron and copper, and xenobiotic metabolites. One of the major reactions of these active oxygen species is initiation of lipid oxidation at membranal and cytosolic levels which ultimately leads to cellular degeneration, chromosomal aberrations, mutations, and cytotoxicity. In food systems also, lipids are highly susceptible to oxidation resulting in the formation of cholesterol oxidation products, lipid hydroperoxides, and malonaldehyde. Some of the causative agents are heat, photosensitization, transition metals, metalloproteins, and radiation. Dietary lipid oxidation leads to the development of rancidity, protein damage, and oxidation of pigments resulting in a loss of sensory properties, nutritive value, and shelf life of food products.

The health implications of lipid oxidation products have long attracted the attention of food scientists, biochemists, and health professionals. Biological and dietary lipid oxidation products are implicated in the etiology of numerous diseases afflicting humankind including cancer, aging, atherosclerosis, ischemia, and rheumatoid arthritis. However, even after more than twenty years of research, the health effects of lipid oxidation products remain highly controversial. They have been linked to initiation of various diseases mainly by inference, and further investigations are necessary to establish a direct structure-activity relationship. It is also interesting to note that even after numerous investigations on the toxicity of dietary lipid oxidation products, there is a scarcity of specific regulations in the United States with regard to the control of rancidity in food products. In European countries, such regulations exist mainly for frying oils.

The identification and quantitative measurement of lipid oxidation products becomes very crucial in both of the systems not only to correlate the adverse biological effects, but also to identify rancid food products that are potentially toxic. An early diagnosis of impairments in cellular lipid oxidation could very well be a cutting-edge method in preventing certain diseases in humans. Lipid oxidation is generally measured by chemical or spectrophotometric methods which are often nonspecific and insensitive. There is a need for the development of methods which give an accurate estimation of the extent of oxidation. In recent years, measurement of by-products of lipid oxidation, particularly aldehydes has become more popular. In addition, specific assays have been developed to measure levels of the antioxidant enzymes: catalase, superoxide dismutase, glutathione peroxidase, glutathione reductase, and glutathione transferase in humans. These assays have immense potential in the prevention and therapy of various free-radical induced diseases.

Lipid oxidation in food and biological systems is prevented or delayed by various mechanisms. In food products, lipid oxidation is effectively delayed by various antioxidant compounds. With the advent of modern food processing methods, product formulations, and requirements for enhanced shelf life, use of antioxidants has become a technological necessity in the food industry. Antioxidants are employed in a wide variety of food products including fats, oils, fat-containing foods, and food packaging materials. Synthetic antioxidants such as butylated hydroxyanisole (BHA), butylated hydroxytoluene (BHT), and the gallates were introduced in the 1940s. In recent years, there has been an enormous demand for natural antioxidants mainly because of adverse toxicological reports on many widely used synthetic compounds. Thus, most of the recent investigations have been targeted toward identification of novel antioxidants from natural sources. Plant phenolic compounds such as flavonoids, sterols, lignanphenols, and various terpene related compounds are potent antioxidants. Many of these compounds have therapeutic properties and are known for their anticarcinogenic, antimutagenic, and cardioprotective activities. Numerous reports have appeared on the antioxidant activity and identification of specific active compounds from plant extracts. Since the antioxidant activities of natural extracts and compounds have been determined by a wide range of methods and varying endpoints, it has also become increasingly difficult to make a realistic assessment of the efficacy of various natural antioxidants. The antioxidant activity of natural antioxidants depends to a large extent on the mode of extraction, presence of inhibitors, nature, and concentrations of active components in the crude extracts. There is an urgent need for the standardization of evaluation methods in order to obtain meaningful information.

The synthesis, mode of action, and food applications of various synthetic and natural antioxidants approved for use in food products are well documented. Considerable efforts have also been made toward the development of novel compounds with superior antioxidant properties. *tert*-Butyl hydroquinone (TBHQ) was introduced in the 1970s and is being widely used in many food applications today. Some attempts were also made in the 1970s to introduce new synthetic polymeric compounds which are nonabsorbable and nontoxic. However, they are not being used on a large scale. Synthetic analogues or derivatives of α-tocopherol which have better antioxidant properties have also been introduced. Trolox-C has proven to be very effective in both animal fats and vegetable oils in contrast to α-tocopherol, which has poor activity in vegetable oils. The synthesis and properties of Troloxyl amino acids has been reported. Recently, Japanese investigators have reported the synthesis of a novel phosphatidyl derivative of α-tocopherol which is more efficient than a mixture of α-tocopherol and phosphatidylcholine in lard. In the area of natural antioxidants from plants, development of commercial processes are necessary to facilitate large-scale applications. Many of the natural antioxidants such as flavonols, flavones, tea leaf catechins, rosemary antioxidants and spice extracts have been reported to be more active than BHA, BHT, or the tocopherols in model systems. The food applications of these compounds need to be explored further.

The toxicological effects of food antioxidants have been the focus of controversy in recent years. Toxicological studies are mainly carried out to establish the no-effect level for an acceptable daily intake (ADI) for humans. However, safety assessments based on the various toxicological effects in animal testing need to be made with caution. Toxicological tests are generally performed with 100-times the level proposed for human consumption, which means the levels are very high compared to the concentration of the test substances in food products. The results are extrapolated based on several assumptions. One of the assumptions is, the metabolism and toxicity observed in animals in long-term or short-term studies at high doses is comparable to low levels of lifetime exposure in humans, which has not been established so far. Extrapolation of toxicological studies is further limited by complete lack of information on how some of these compounds will affect various segments of the population. Assessments are also made on the assumption that the compounds remain intact in the food product after processing and storage. However, degradation studies indicate that many of these compounds may be converted partly or wholly to other chemical forms which may have a totally different toxicological profile compared to the parent compound. Also, very little information is available on the possible interactions and reactions triggered by these degradation products in a complex food system.

The toxicological effects of BHA and BHT are ideal examples to illustrate the complexities of evaluating toxicological tests. In the 1980s, BHA was reported to induce forestomach papillomas and carcinomas in rats and hamsters which immediately raised several questions. Humans do not have a forestomach, but types of cells similar to rodent forestomach are found in other organs such as the esophagus. In monkeys, dogs, and pigs, who do not have a forestomach like humans, papillomas were not found in the stomach, esophagus or duodenum, even at higher doses of BHA. Thus, the relevance of toxic effects of BHA observed in rats to humans was questioned by many investigators, the Scientific Committee for Food (SCF), and the Joint FAO/WHO Expert Committee on Food Additives (JECFA). However, the temporary ADIs for BHA were based on no-effect levels for hyperplasia and endpoints in the forestomach. The major toxic effects reported for BHT are its adverse effects on blood clotting mechanisms, liver hyperplasia, and carcinogenicity. The hemorrhagic toxicity has been attributed mainly to BHT-quinone methide, one of the reactive metabolites of BHT. Studies have also indicated that addition of BHA can enhance the formation of quinone methide from BHT. Several natural food constituents such as eugenol, vanillin, and ferulic acid also enhance the metabolic activation of BHT. However, very little information is available on the metabolism of BHT or the formation of quinone methide in humans. The liver hyperplasia and carcinogenic effects observed in rats and mice are also highly species specific, or in some cases strain specific, and seem to depend on the metabolism of BHT.

In general, various antioxidants are readily absorbed and metabolized, with the exception of polymeric compounds such as Anoxomer or Ionox-330. The major metabolic pathways involve either conjugation reactions (with glucuronic acid or sulfate) or oxidation reactions (e.g. demethylation). In some instances the compounds are excreted unchanged as in the case of thiodipropionates. The metabolism of various antioxidants has been linked to their toxicological properties. Among the major synthetic phenolic antioxidants, BHA is metabolized mainly by conjugation reactions and to a lesser extent by oxidation reactions. The gallates and TBHQ are metabolized by conjugation pathways. They are also rapidly eliminated without any long-term tissue storage. On the other hand, BHT is excreted less rapidly from most of the species, mainly because of enterohepatic circulation. It is also mainly metabolized by oxidation reactions mediated by the microsomal monooxygenase system. This may explain the strong induction of the microsomal monooxygenase system by BHT, whereas other compounds are only weak inducers. The tocopherols are also metabolized by oxidation reactions. Interestingly, the hemorrhagic toxicity of BHT and tocopherols seems to be by comparative mechanisms and has been attributed to the formation of quinones in both the cases.

Of the various oxygen scavengers, it has been suggested that sulfites may function as cocarcinogens, however, this needs to be investigated further. Among the chelating agents, the adverse effects of ethylenediaminetetraacetic acid (EDTA), which is poorly absorbed, could be mainly due to the chelation of various essential metal ions. Studies have indicated that EDTA may not be genotoxic. However, it may enhance the genotoxic effects of certain environmental mutagens. In addition, the differences observed in the toxic effects of injected vs. dietary $CaNa_2EDTA$ remains to be explained. One of the research needs in the toxicology of food antioxidants is an evaluation of structure-activity relationships of the major metabolites and degradation products. Another important research need in this area is a complete toxicological evaluation of some of the important natural antioxidants, especially flavonoids and active compounds from spices, which undoubtedly have potential for large-scale food applications. At present, many of these compounds are not being used in food applications, mainly because of insufficient toxicological information.

The biological systems are endowed with certain powerful enzymes and small molecular weight compounds with antioxidant properties to prevent initiation of free radicals and lipid oxidation. The defense mechanisms operate at different levels. The first line of defense is naturally to prevent the generation of the reactive forms of oxygen. The cytochrome oxidase system of the electron transport chain is the foremost line of defense. It catalyzes the tetravalent reduction of oxygen without a significant release of any of these reactive intermediates. The second line of defense is provided by enzymes that catalytically scavenge the intermediates of oxygen reduction. The enzymes: superoxide dismutase, catalase, and glutathione peroxidase are three of such free radical control biomolecules that are ubiquitously distributed in all aerobic organisms on this planet. In addition to these antioxidant enzymes, low molecular weight nonprotein sulfhydrils such as glutathione, cysteine, and cysteinylglycine also play a significant role in the prevention of lipid oxidation. Many dietary constituents such as vitamins E, C, and probably A, β-carotene, and essential minerals such as selenium, copper, zinc, and manganese may directly function as antioxidants or are essential entities for the proper functioning of the various cellular defense mechanisms. Other diet derived nonnutritional antioxidant phytochemicals and synthetic antioxidants may also play a role in this regard. In this context, the report on the 'French Paradox' could be cited as an example. The marked decrease in the morbidity and mortality from coronary artery disease in a segment of the French population has been attributed to the consumption of red wines rich in flavonoids. However, the beneficial effects of phytochemicals and dietary antioxidants need to be elucidated further.

Recent advances in the medical and nutritional sciences have clearly shown the beneficial effects of consumption of certain types of foods, particularly leafy-green vegetables and certain fruits rich in the above-mentioned compounds against a number of chronic diseases. Consequently, vitamins E and C, β-carotene, and the antioxidant minerals have received renewed attention in recent years. The recent trend has been toward increasing the supplementation of these compounds. In this regard, the antioxidant minerals are likely to play a minor role, since the gap between the threshold level of their optimal intake and the one at which they are likely to exert toxic effects is minimal. Also, the concept of increasing intakes of antioxidant vitamins and β-carotene as opposed to eating a healthy diet in the prevention of certain types of cancer and cardiovascular disease will probably be a controversial issue through the rest of this century. There is still a scarcity of information on how these nutrients may act in combination, and whether they still provide the same beneficial effects as they seem to offer when studied in isolation in a number of dietary surveys and clinical trials.

Recommendations for increased intakes of antioxidant vitamins need to be viewed with caution as these compounds become prooxidants at high doses. Another major reason for such a caution is the lack of data showing that long-term (and presumably lifetime) intake of megadoses of these vitamins is not likely to induce any adverse effects in humans. At present, except for one small study in primates involving vitamin E, we do not have data demonstrating the efficacy of these antioxidant vitamins even in animals. In short-term clinical trials however, these compounds seem to have very little toxicity even in megadose quantities. However, such studies need to be interpreted in conjunction with lifestyles and habits. The recent Finnish study suggested that supplements of β-carotene markedly increase the incidence of lung cancer among heavy smokers in Finland indicating that we may be overestimating the benefits observed in epidemiological studies.

Although consumers are being increasingly exposed to media blitzes and commercial advertisements claiming the health benefits of dietary antioxidant supplements, the U.S. Food and Drug Administration (FDA) thus far has steadfastly refused to allow antioxidant claims on labels for foods or dietary supplements. To help resolve this issue, the FDA held a conference in November 1993 at the National Academy of Sciences in Washington D.C., to summarize the available scientific evidence on antioxidant vitamins' link with cancer and cardiovascular diseases and to determine if and how "significant scientific agreement" might be achieved to support such claims.

The FDA conference examined four topics: antioxidant vitamins and cardiovascular disease, vitamin E and cancer, β-carotene and cancer, and vitamin C and cancer. Although some evidence links these antioxidants to health effects, scientists pointed out that it is difficult to determine whether the effects are in fact due to a specific antioxidant or to other components of foods. The only conclusion reached was that the best public health policy is to encourage the consumers to eat plenty of fruits and vegetables rich in antioxidants to obtain any disease preventive effects, rather than to advocate taking antioxidant vitamin supplements.

Thus, researchers continue to debate whether antioxidant supplements are needed or if consumers can get sufficient amounts through consuming more fruits and vegetables. At least a sizeable segment of the U.S. population, however, is not waiting for what the medical community would classify as "definitive" proof before turning to supplements. Since 1988, the U.S. market for β-carotene supplements alone has grown from $7 to $82 million a year, while vitamin E supplements sales have risen from $260 to $338 million a year, according to an article in the June 7, 1993 issue of *Newsweek* magazine.

Index

479

About the Editors

D. L. MADHAVI is a Research Associate in the Department of Horticulture at the University of Illinois, Urbana. The author or coauthor of numerous professional papers, her research interests pertain to the biological activity of flavonoids and their expression in cell cultures. She is a professional member of the Institute of Food Technologists and a member of the American Society of Plant Physiologists and the Society for In Vitro Biology. Dr. Madhavi received the B.Sc. (1977) and M.Sc. (1979) degrees in botany from the University of Mysore, India, and the Ph.D. degree (1987) in biochemistry from the Central Food Technological Research Institute, University of Mysore, India.

S. S. DESHPANDE is Senior Scientist and Group Leader at Idetek, Inc., Sunnyvale, California. The author or coauthor of two books and over 70 professional papers, his research focuses on the development of immunodiagnostic kits based on polyclonal and monoclonal antibodies for the detection of environmental pollutants as well as drug residues, toxins, and pathogens in dairy and meat products. Dr. Deshpande received the B.Sc. degree (1977) in agriculture from Andhra Pradesh Agricultural University, Hyderabad, India, the M.Sc. degree (1979) in food technology from the Central Food Technological Research Institute, University of Mysore, India, and the Ph.D. degree (1985) in food science from the University of Illinois, Urbana.

D. K. SALUNKHE is Professor Emeritus in the Department of Nutrition and Food Science at Utah State University, Logan. The author or coauthor of 23 books and over 400 professional papers, book chapters, and reviews, and coeditor, with S. S. Kadam, of the *Handbook of Fruit Science and Technology* (Marcel Dekker, Inc.), he is a Fellow of the Institute of Food Technologists, among other organizations, and a former member of the editorial boards of the *Journal of Food Biochemistry*, the *International Journal of Plant Foods for Human Nutrition*, the *Journal of Food Science*, and the *Journal of Food Quality*. Dr. Salunkhe received the B.Sc. degree (1949) in agriculture from Poona University, Ganeshkhind, India, and the M.S. (1951) and Ph.D. (1953) degrees in crop science from Michigan State University, East Lansing.